Introduction to Physical Chemistry
SI edition

G I Brown BA, BSc
Assistant Master, Eton College

Longman

LONGMAN GROUP LIMITED
London
*Associated companies, branches and representatives
throughout the world*

*First published 1964
Second Edition 1972
Sixth impression 1977*

ISBN 0 582 32131 X

*Printed in Hong Kong by
Sing Cheong Printing Co Ltd*

Preface

THIS book is intended to provide a thorough introduction to physical chemistry and to arouse a student's interest in the subject. Its contents are based mainly on the requirements of such examinations as the G.C.E. (A and S levels), 1st M.B. and university scholarship examinations.

The minimum treatment of a topic, such as is required for the A level examination, has, generally, been extended to try to make the topic as relevant as possible. At points of special interest or significance, the discussion is taken beyond intermediate examination requirements. The aim has been to try to give, at a reasonably elementary level, as complete a picture of modern physical chemistry as possible.

The presentation is in numbered sections so that particular topics can be omitted or re-arranged to suit individual requirements, and it is hoped that the book, by selective use, may be of value to the rather weak A-level candidate, to the open scholarship winner, to first-year undergraduates, and to students in technical colleges. A very wide range of questions, totalling more than 750, is provided.

What might be called routine physical chemistry, based essentially on nineteenth-century discoveries, is adequately dealt with, but a more modern outlook is super-imposed by a simple treatment of such topics as atomic structure, atomic energy, chemical bonding, activation energy, reaction mechanisms and simple thermodynamics.

An attempt has been made to blend the old with the new, and the elementary with the advanced, to give a picture of simple physical chemistry both in its growth and development and in its present-day form.

Note to second (SI) edition

THE units in this edition have been changed into SI units (see Summary on p. 552), and the naming of chemicals has been brought up-to-date. Some mild compromise has been necessary to try to meet the different recommendations of various authoritative bodies.

The opportunity has been taken to make some other relatively minor alterations and improvements. The pagination remains essentially the same.

Acknowledgements

WE are grateful to the following for permission to reproduce copyright material:

Cambridge University Press for questions from University of Cambridge First M.B. Examinations; The Clarendon Press for questions from Oxford Preliminary Examinations; The Controller of Her Majesty's Stationery Office for six questions from Army, Navy and Air Force examination papers; the Joint Matriculation Board of the Universities of Manchester, Liverpool, Leeds, Sheffield and Birmingham for questions set by the Board in Chemistry examinations; the Oxford and Cambridge Schools Examination Board for nineteen questions set by the Board in examination papers; the Southern Universities' joint Board for School Examinations for questions set by the Board in G.C.E. examinations, and the Welsh Joint Education Committee for questions set in examination papers.

These authorities have also authorised the conversion of units to the new system, together with minor alterations to avoid unnecessary arithmetic. These changes have been undertaken by the author, and in no way imply the approval of the examining bodies concerned.

Contents

Chapter 1
Fundamental laws

1. Laws. Hypotheses. Theories. The first aim of a scientist, when investigating any problem, is to discover the facts, and this is usually done by carrying out experiments. If a large number of experiments all give similar results it is possible to summarise the results into a single statement, which is then known as a law. *A law is a concise statement summarising the results of a large number of separate experiments all leading to the same conclusion.*

Once a law has been enunciated, and checked by further experimental work, efforts are made to account for the law, or explain the facts summarised in the law. This is done by putting forward a *hypothesis*, which is an idea, or a collection of ideas, able to account for the facts. These first ideas are generally somewhat tentative, but if they become widely accepted as true, after consideration, discussion and modification, they are then restated in what is called a *theory*.

Finally, if a theory can be built up which effectively accounts for a variety of facts, it is often possible to use it to predict some new experimental results or facts. The theory can then be used and developed.

The early history of chemistry depended almost entirely on the process of discovering experimental laws, and then devising hypotheses and theories to account for them. Chemical theory rests very heavily on the atomic theory, the molecular theory, the kinetic theory and the ionic theory, and all these important theories had an experimental background as is indicated in the following very broad summary.

1774	Law of Conservation of Mass (Lavoisier)	
1799	Law of Constant Composition (Proust)	→ Dalton's Atomic
1803	Law of Multiple Proportions (Dalton)	Theory (1807)
1792	Law of Reciprocal Proportions (Richter)	

1808	Gay-Lussac's Law of Combining Volumes	→ Avogadro's hypothesis, i.e. Molecular Theory

1662	Boyle's Law	
1787	Charles's Law	→ Kinetic Theory
1846	Graham's Law of Diffusion	

1834	Faraday's Laws (and the results of other electrical measurements)	→ Ionic Theory

The laws in the first group of four are referred to as the Laws of Chemical Combination, whilst those in the second group of four are known as the Gas Laws. Many other chemical laws and theories are known, but those listed above are all of quite fundamental importance.

2. Law of conservation of mass. This law, usually attributed to Lavoisier in 1774, is commonly expressed in the statement that *matter is neither created nor destroyed in the course of a chemical reaction.*

The general idea behind the law had been expressed by Greek philosophers such as Anaxagoras (c. 450 B.C.), Lucretius (98–54 B.C.) and Plato (427–347 B.C.), and by Bacon (1561–1626), Boyle (1627–1691) and Jean Rey (1630), the last-named choosing the descriptive passage—'The mass with which each portion of matter was endowed at the cradle, will be carried by it to the grave'—to state the law. Lavoisier, however, was the first person to make any serious attempt to test the truth of the law experimentally.

The law means that the chemicals taking part in a chemical reaction have just the same mass both before and after the reaction has taken place.

FIG. 1. Landolt tube.

Proving the law involves the accurate measurement of the masses of chemicals before and after reaction, and such experimental work has been carried out mainly by Landolt (1831–1901), using a Landolt tube (Fig. 1). He sealed solutions which, when mixed, would react together, in the limbs of the tube, and the mass of the tube was then measured. It was then upturned so that the two solutions could mix, and, when the reaction was complete and the tube had retained its original temperature, the new mass was measured. The first and second masses were found to be identical within the severe limits of experimental error. Landolt carried out this experiment with fifteen pairs of solutions, choosing pairs which reacted with small heat changes. Typical reactions used were:

a. Hydriodic acid and iodic(v) acid solutions, which react to give iodine, a reaction which is made use of in volumetric analysis,

$$5I^-(aq) + IO_3^-(aq) + 6H^+(aq) \longrightarrow 3I_2(s) + 3H_2O(l)$$

b. Sodium sulphate(IV)* and iodine solutions, which react to give hydrogen iodide and sodium sulphate(VI), the iodine reacting as an oxidising agent,

$$SO_3^{2-}(aq) + I_2 + H_2O(l) \longrightarrow SO_4^{2-}(aq) + 2H^+(aq) + 2I^-(aq)$$

Landolt determined the maximum experimental error by carrying out a series of blank experiments in which tubes containing liquids or solutions which did not interact were used. The maximum change in mass recorded on mixing the solutions was 30 μg. This change was detected in an original total mass of about 350 g, giving an experimental error of less than 1 part in 10000000.

In forty-eight experiments, based on the fifteen pairs of solutions which did interact, Landolt found an average change in mass of 12 μg. There was an increase in mass in twenty-three experiments and a decrease in twenty-five. In only two experiments did the change in mass exceed 50 μg.

Landolt concluded that 'the final result of the whole investigation is that in all the fifteen decompositions involved it has not been possible to establish a change of mass. The observed deviations from absolute equality before and after reaction are due to external physical causes and are not the result of chemical reactions.'

* Sulphate(IV) is used to replace sulphite. The older sulphate name becomes sulphate(VI). Similarly nitrite becomes nitrate(III) and nitrate becomes nitrate(V).

Manley, in 1912, achieved even greater accuracy than Landolt and showed that any change in mass in the reaction between barium chloride and sodium sulphate(VI) solutions must be less than 1 part in 100000000.

The validity of the law of conservation of mass is, of course, assumed whenever a qualitative chemical experiment is carried out.

3. Mass and energy. Neither Landolt nor Manley, however, were able to investigate any reaction which involved any great evolution of heat or light. It is now thought that if a reaction involving a large evolution of heat could be studied, and if a sensitive enough balance were available (which it isn't), a small decrease in mass would be discernible as a result of a chemical reaction. This is because it is now known that matter can, in fact, be converted into energy.

That this might be possible was first suggested by Einstein as part of his theory of relativity. He predicted, theoretically, that matter and energy were related by the expression

$$
\begin{array}{cccc}
\text{Energy} & = & \text{Mass} & \times & \text{(Velocity of light)}^2 \\
E & = & m & \times & c^2 \\
\text{(in joules)} & & \text{(in kilogrammes)} & & (2.997925 \times 10^8\,\text{m s}^{-1})^2
\end{array}
$$

and since Einstein's theoretical prediction this relationship has been shown to be true experimentally (p. 148).

The large value of c, and the correspondingly larger value of c^2, mean that it is possible to obtain a large amount of energy from very little mass. Thus 1 g of matter would yield about 30.55×10^6 kilowatts of electricity, or would produce as much energy as 4.2×10^6 kg of fuel oil.

The foregoing statements presuppose a complete conversion of matter into energy, and it is not generally possible to bring about such a complete conversion. A partial conversion of some matter into energy is brought about, however, in an atomic pile (p. 151) or an atomic bomb (p. 150), or, on a smaller scale, in any chemical reaction which gives out energy in the form of heat or light or electricity.

Because the quantitative relationship between mass and energy is known, it is possible to calculate how much energy is obtainable from a given amount of matter, and vice versa. Such calculations show that the energy liberated in any ordinary chemical reaction requires a conversion of only about 100×10^{-12} kg of matter into energy. Such a small change of mass cannot, of course, be detected on any chemical balance, and for ordinary chemical reactions the slight conversion of mass into energy, which may take place, results in an unimportant change in mass which cannot be detected on a balance.

Whilst, therefore, the change in mass which might take place during a chemical reaction is of no significance when the ordinary gravimetric (mass) aspects of a reaction are being studied, or when the ordinary methods of chemical analysis are being applied, it is of great importance in atomic energy considerations (p. 149).

The idea of matter being converted into energy under certain conditions can be replaced by regarding matter as a form of energy. Jeans, for example, described matter as 'bottled energy'. In some changes, a little of the energy may be released, but the sum total of matter plus energy in any isolated system is always constant. The law of conservation of mass can, therefore, be expressed with absolute precision if it is regarded as a law of conservation of mass plus energy, and this can be expressed in the statement that *the sum of the quantity of matter and energy in an isolated system is always constant.*

4. Law of constant composition or definite proportions. This law states that *all pure samples of the same chemical compound contain the same elements combined together in the same proportions by mass.* It means that pure specimens of a compound, no matter how, or where, or when, or by whom they are made will always have identical compositions.

The law was first stated by Proust in 1799, and subsequently verified, mainly by Stas, in 1865. Stas obtained silver in a variety of ways, e.g. by electrolysis of silver cyanate (AgCNO) or by reducing silver nitrate(v) using milk-sugar (maltose, $C_{12}H_{22}O_{11}$), and then converted the various samples of silver into silver chloride. He did this either by direct reaction at red heat with chlorine, or by treating the silver with nitric(v) acid and then precipitating silver chloride by adding hydrogen chloride gas, hydrochloric acid or ammonium chloride solution. In all cases he obtained the same mass of pure silver chloride from equal masses of silver, the accuracy being of the order of 1 part in 100000.

In a school laboratory, the law of constant composition is most conveniently illustrated by making various samples of copper(II) oxide and by determining the percentage of copper they each contain by reducing the oxides to copper in a stream of hydrogen. The copper(II) oxide samples can be made, for example, by (a) heating powdered copper in oxygen; (b) treating copper with concentrated nitric(v) acid, and heating the resulting copper(II) nitrate(v); (c) heating copper(II) carbonate; or (d) precipitating copper(II) hydroxide and heating it.

Before Stas's work, Berthollet, who followed Lavoisier as the leading French chemist, severely criticised Proust's enunciation of the law of constant composition, but these criticisms are now known to have been based on inaccurate analyses of impure substances and on imprecise distinctions between mixtures, compounds and solutions. In particular, Berthollet suggested that lead and oxygen could combine in almost any proportions, but such suggestions were due to the fact that he was making and analysing mixtures of the various definite compounds which are now known to be formed between lead and oxygen.

The controversy which ensued between Proust and Berthollet ended, mainly, in Proust's favour, and did, at least, establish clearer ideas as to the differences between mixtures and compounds. As so often happens, however, Berthollet's views were not entirely wrong, even though they were, at the time, based on false evidence. For, nowadays, it is recognised that the composition of some compounds is variable. Such compounds are known as *Berthollide compounds* and are said to be non-stoichiometric, *stoichiometry* being a rather old-fashioned word for that part of chemistry which deals with the gravimetric or volumetric composition of substances. Various examples of such non-stoichiometric compounds are given on p. 212.

The discovery of isotopes (p. 85) also has led to the realisation that two samples of the same compound, which appear to be chemically alike, may contain different isotopes of the same element and, in consequence, have different percentage compositions.

5. Law of multiple proportions. This law states that *when two elements A and B combine together to form more than one compound the several masses of A which combine with a fixed mass of B are in a simple ratio.*

The law was first put forward by Dalton in 1803, during the time in which he was developing his atomic theory. At first, it was, essentially, a hypothesis and not a summarised statement of experimental results, but, to test the hypothesis, Dalton investigated existing experimental results and obtained new ones. All the available evidence supported the statement made.

The law means that if, in one compound between A and B, 10 g of A are combined with 25 g of B, then, in any other compound between A and B, the mass of A combining with 25 g of B will always be some simple multiple of 10, i.e. 5, 10, 20 . . . g.

The early proof of the law depended largely on the analysis of such compounds as the two oxides of carbon (analysed by Lavoisier), dinitrogen and nitrogen oxides (analysed by Davy) and methane and ethane (analysed by Dalton). But the law became more firmly established as a result of Berzelius's analysis of iron(II) sulphide and iron pyrites, of the oxides of iron, copper, lead and sulphur, and of the chlorides of copper.

In a school laboratory it is most convenient to use the two oxides of copper, and to analyse them by reduction with hydrogen. It will be found that in copper(II) oxide (black copper oxide) 31.8 g of copper are combined with 8 g of oxygen, whilst in copper(I) oxide (red copper oxide) 63.6 g of copper are combined with 8 g of oxygen. The ratio 31.8:63.6, i.e. 1:2, is simple.

Because carbon forms such a very wide range of compounds the law of multiple proportions does not apply to such compounds, unless the meaning of the term simple ratio is severely strained.

6. Law of reciprocal proportions. The law of constant composition deals only with chemical combination between two elements to form one compound. The law of multiple proportions extends the picture by dealing with chemical combination between two elements to form more than one compound. The law of reciprocal proportions deals with the situation existing when more than two elements combine between themselves.

Hydrogen will combine with oxygen, for example, to form water, and with chlorine to form hydrogen chloride. Oxygen will also combine with chlorine to form an explosive gas called chlorine dioxide, ClO_2. In a more complicated example, copper, sulphur and oxygen will combine to form copper(I) and copper(II) oxides, copper(I) and copper(II) sulphides, sulphur dioxide and trioxide and copper(II) and copper(I) sulphates(VI).

Combination between elements is, in fact, very extensive and varied, and accounts for the very large number of compounds which can be formed. Analysis of such compounds shows that there is a connection between the masses of the elements combining together, and the nature of this connection was first stated by Richter, in 1792, as the law of reciprocal proportions. This states that *the masses of A, B, C, etc., which combine separately with some fixed mass of another element, are the masses in which A, B, C, etc., will combine with each other, or simple multiples of them.*

7. Law of equivalents. The laws of constant composition and multiple proportions, and particularly the law of reciprocal proportions led, historically, to the important conception of combining or equivalent masses. If each element is allotted a number, or numbers, known as its combining or equivalent mass, or masses, the three laws just mentioned can all be brought into the general statement that *elements always combine together in the ratio of their combining or equivalent masses.* This statement is known as the law of equivalents.

The combining or equivalent masses of an element are, clearly, very important numerical values, for they control, numerically, the way in which elements will combine. The equivalent of one element will combine with the equivalent of another, if they combine at all.

In deciding to allot numerical values for the combining or equivalent masses of each element it is simply a matter of choosing, arbitrarily, a fixed mass of some element or other against which to measure, experimentally, the combining or equivalent masses of all other elements. Any mass of any element could be chosen; modern practice, decided by the International Committee on Relative Atomic Masses, is to take 8 parts by mass of oxygen.

On this basis, *the equivalent mass of an element is defined as the number of parts by mass of the element which will combine with (or replace) 8 parts by mass of oxygen.*

It is important to realise that the choice of 8 parts by mass of oxygen is arbitrary. At one time, Berzelius used 100 parts by mass of oxygen as the standard, but this gave clumsily high equivalent masses for many elements. Thomson used 1 part by mass of oxygen, but, on this scale, some elements had fractional equivalent masses.

Hydrogen has the smallest of all equivalent masses, and, for that reason, it was, for some time, accepted as the standard, 1 part by mass of hydrogen being chosen. For some time, too, it was thought that 1 part by mass of hydrogen combined with 8 parts by mass of oxygen, so that accepting 1 part by mass of hydrogen as the standard was the same as accepting 8 parts by mass of oxygen. On a financial analogy it is the same thing as saying that an article is valued at £1 or 100 new pence.

More accurate experiments showed, however, that it is 1.0080 g of hydrogen (and not 1 g) which combines with 8.0000 g of oxygen, so that the definition of equivalent mass in terms of 1 part by mass of hydrogen is not exactly the same as the definition in terms of 8 parts by mass of oxygen. For approximate work, there is no significant difference, but, for accurate work, a choice had to be made between the hydrogen scale and the oxgyen scale. 1.0080 parts by mass of hydrogen could have been accepted as the standard, but 8.0000 parts by mass of oxygen was, in the end, universally accepted, mainly because many more elements will form more stable compounds with oxygen than with hydrogen, thereby facilitating *direct* measurements of equivalent masses on the oxygen scale.

The equivalent mass of hydrogen is 1.0080 because this is found, experimentally, to be the mass (in grammes) of hydrogen which combines with 8.0000 of oxygen. Similarly the equivalent mass of chlorine is 35.457 because this is the mass of chlorine which will combine with 8.0000 g of oxygen. By the law of equivalents it follows that hydrogen and chlorine will combine in the ratio of 1.0080 g to 35.457 g, if they combine at all. Thus

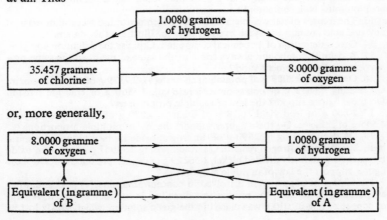

or, more generally,

The experimental measurement of the equivalents of elements is dealt with on pages 10–15.

QUESTIONS ON CHAPTER 1

1. 'It is, I think amongst our own countrymen that we discover the fathers of chemical philosophy; for Bacon, Boyle, Hooke and Newton present unequivocal claims to that distinctive title.' What are the claims of the four scientists mentioned?

2. 'Chemistry is a French science. Its founder was Lavoisier of immortal memory.' What was Lavoisier's contribution to chemical development?

3. The word 'law' is used in many different ways. Illustrate the different meanings it may have.

4. 'In no walk of life is it possible to get something for nothing.' Comment.

5. Explain the importance of (a) the law of conservation of mass, (b) the law of constant composition, to someone who has no knowledge of science.

6. Suggest ten pairs of solutions which could be used in Landolt's experiment. Choose pairs a s widely different as possible, and write equations for the reactions they undergo.

7. If 57.060 kJ are evolved when 1 dm³ of M hydrochloric acid is neutralised by 1 dm³ of sodium hydroxide solution, what loss in mass takes place?

8. If a complete conversion of matter into energy could be achieved what mass of matter would be required to raise the temperature of 1000 kg of water by 50 K?

9. If the relative atomic mass of helium is 4.0039 and that of deuterium 2.0147, how many joules would be evolved if 1 mol of helium could be formed by fusing deuterium atoms into helium atoms?

10. Approximately how many kilogrammes of water could be converted into steam if the total heat energy available from 1 kg of matter could be used?

11. Calculate the energy equivalent of 1 g of matter in (a) joules, (b) calories, (c) MeV, (d) kilowatt-hours.

12. How is it possible to distinguish between a mixture and a compound? To what extent is it true to say that a substance which has a definite, fixed composition must be a compound?

13. The masses of various metallic oxides which neutralise a constant mass of any one acid contain the same mass of oxygen. Illustrate this statement.

14. Give a definition of a chemical compound. Criticise the definition you give.

15. A definite compound always contains the same elements in the same proportions. Is this statement valid?

16. On heating 4.5087 g of potassium chlorate(v), 2.7379 g of chloride remain. On heating 6.6672 g of potassium chlorate(vii), 3.5860 g of chloride remain. Do these figures support the law of multiple proportions?

17. 10.00 g of lead yield 10.78 g of litharge. 4.870 g of lead(iv) oxide yield 4.545 g of litharge. Do these figures support the law of multiple proportions?

18. On decomposing 5.62690 g of dinitrogen oxide by heating an iron spiral in it, 3.58150 g of nitrogen remain. On decomposing 2.93057 g of nitrogen oxide by reaction with finely divided nickel, 1.56229 g of oxygen are removed. Do these figures support the law of multiple proportions?

19. 4.231 g of copper were dissolved in concentrated nitric(v) acid and, after evaporation to dryness, the residue was dissolved in hydrochloric acid. A strip of copper of mass 8.010 g was placed in the green solution, which slowly lost its

colour. When the colour had completely disappeared the mass of the strip of copper was found to be 3.824 g. Explain the reactions. Do the figures support the law of multiple proportions?

20. Two hydrates of the same salt contain 10.17 per cent and 36.14 per cent of water. Do these figures illustrate the law of multiple proportions?

21. State the law of multiple proportions. Three oxides of a metal, X, are found to contain 24.96, 29.95 and 49.95 per cent of oxygen respectively. Show by calculation that these results are in agreement with the law of multiple proportions. Given that the relative atomic mass of the element X is 47.9, deduce the simplest formula of these oxides. (O. Prelims.)

22. The percentage compositions by mass of four compounds are (a) nitrogen, 11.6, chlorine, 88.4, (b) nitrogen, 46.7, oxygen, 53.3, (c) oxygen, 22.2, iron, 77.8, (d) iron, 44.1, chlorine, 55.9. Show that these compositions are in accordance with the law of reciprocal proportions. (O. Schol.)

Chapter 2
Measurement of equivalent masses of elements

ALL compounds contain elements chemically combined together, and values of the equivalent masses of the elements were of paramount importance when quantitative aspects of chemical combination were considered. Moreover, as will be seen in Chapter 3, it was from experimental values of the equivalent masses of the elements that the relative atomic masses of most elements were originally derived. A great many methods of measuring equivalent masses have, therefore, been devised. The topic is, nowadays, of historical interest, for accurate values of the equivalent masses of all the common elements are well established, but much of the quantitative work of the last century was concerned with equivalent mass measurement, and some classical experiments were carried out in this field.

Typical experimental procedures are outlined below.

1. Equivalent mass of an element by synthesis of its oxide. In this method, a known mass of an element is completely converted into its oxide and the mass of the oxide is measured. The following (over-simplified) result may be obtained,

$$10 \text{ g of element A} \longrightarrow 14 \text{ g of oxide of A}$$

It follows that 10 g of A will combine with 4 g of oxygen, or that 20 g of A will combine with 8 g of oxygen. The equivalent of A is, therefore, 20.

The experimental method used depends on the nature of A, particularly whether its oxide is a solid or a gas, and the main difficulty is in ensuring that a complete conversion to oxide is obtained. Typical methods are outlined below.

a. *For calcium and magnesium.* Both these reactive metals form solid oxides simply on strong heating in air in a crucible. Some nitride is also formed, but both this and any unchanged metal can be converted into the oxide by treatment with dilute nitric (v) acid and further heating.

b. *For magnesium, zinc, tin, lead and copper.* A known mass of the metal is first converted into its nitrate(v) by treatment with concentrated nitric(v) acid. Subsequent heating decomposes the nitrate(v) into the oxide and the mass of this oxide can be measured. The reactions can be brought about in a crucible.

c. *For carbon and sulphur.* The oxides formed by carbon and sulphur are gaseous. A known mass of the element is placed in a porcelain boat in a combustion tube, and the element is heated whilst excess oxygen is passed over. The carbon dioxide or sulphur dioxide formed is passed through weighed tubes containing potassium hydroxide. This absorbs any carbon or sulphur dioxide and the increase in the mass of the absorption tubes gives the mass of oxide formed. The loss in mass of the carbon or sulphur in the porcelain boat must be measured to find how much of the element has been converted into the oxide.

2. Equivalent mass of hydrogen. Because of the early difficulties and differences caused by defining equivalent mass in relation to either hydrogen or oxygen, the combining ratio between hydrogen and oxygen, i.e. the gravimetric composition of water, was, historically, of great importance. It was not easy to measure accurately, mainly because both hydrogen and oxygen are gases. In principle, hydrogen and oxygen were combined to form water, but the measurement of the masses involved was difficult. The most notable experiments were those of Dumas (1842), Morley (1895), Noyes (1908) and Burt and Edgar (1915).

a. *Dumas's method*. Dumas passed a stream of pure dry hydrogen over a known mass of hot copper(II) oxide contained in a hard glass bulb. The hydrogen was passed for 10–12 hours, and the water formed was collected in a bulb connected to a variety of U-tubes containing drying agents such as anhydrous calcium chloride, solid potassium hydroxide, concentrated sulphuric(VI) acid soaked in pumice, or phosphorus(V) oxide. The increase in mass of the water-collecting 'unit' gave the mass of water formed, and the decrease in mass of the copper(II) oxide gave the mass of oxygen contained in this amount of water. By subtraction, the mass of hydrogen in the water was obtained, but the mass of the hydrogen and the oxygen in the water formed were not measured directly.

The result obtained by Dumas, as a mean of nineteen determinations, was that hydrogen and oxygen combined together in the ratio 1.002:8. This ratio was accepted for almost fifty years, though 1.002 is now known to be too low a figure.

b. *Morley's method*. Morley measured the mass of pure, dry hydrogen and oxygen in two separate glass globes and then led the two gases into an evacuated combustion tube, of known mass, where they were burnt at platinum jets by passing electric sparks between two adjacent electrodes. During the burning the tube was immersed in cold water and, afterwards, it was placed in a freezing mixture to convert the water formed inside the tube into ice. The mixture of gases remaining uncombined within the tube was pumped out and analysed, and the masses of oxygen and hydrogen found were subtracted from the original masses of gases which had been led into the tube. The mass of ice formed in the tube was also measured.

In this way, Morley found the masses of both the oxygen and hydrogen used and the water formed; the masses served to check each other. The mean of twelve results, according to Morley, was that hydrogen and oxygen combined in the ratio of 1.0076:8.

c. *Noyes's method*. Noyes passed pure, dry hydrogen into an evacuated tube containing palladium, which is able to absorb hydrogen. The mass of the palladium and the absorbed hydrogen was taken. The tube was now heated, and pure, dry oxygen was passed in, whereupon the oxygen and hydrogen reacted to form water, which was collected in a cooled tube. The mass of water formed was measured and the loss in mass of the palladium gave the mass of hydrogen contained in this water. The mass of oxygen in the water was obtained by subtraction. Noyes's ratio was 1.0078:8.

d. *Burt and Edgar's method*. Burt and Edgar adopted a different principle, for they measured the volumes of hydrogen and oxygen used, and converted these volumes into masses by using density values obtained by Morley in 1896.

Carefully measured volumes of hydrogen and oxygen were led into an explosion vessel, where they were exploded by passing an electric spark. Excess hydrogen was used, and the volume of hydrogen remaining after the explosion was measured by separating the gas from the water formed, by freezing. The result obtained was 1.0077:8.

3. Equivalent mass of an element by analysis of its oxide. This method is applicable if the oxide of an element can be reduced, fairly easily, to the element. For many metallic oxides this can be done by heating the oxide in a stream of hydrogen.

. If 1 g of the oxide yields 0.8 g of metal it follows that 0.8 g of metal has been combined with 0.2 g of oxygen in the oxide. This means that 32 g of metal combine with 8 g of oxygen, i.e. the equivalent of the metal is 32.

Dinitrogen and nitrogen oxides can be reduced by reaction with hot metals, and the mass of nitrogen obtained from a known mass of oxide gives the equivalent mass of nitrogen.

A very accurate determination of the equivalent mass of nitrogen was carried out in this way by Gray in 1905. He obtained nitrogen oxide, containing nitrogen and other oxides of nitrogen, by reaction of ethanoic acid on a mixture of potassium nitrate(III) and potassium hexacyanoferrate(II)

$$Fe(CN)_6{}^{4-}(aq) + NO_2{}^-(aq) + 2H^+(aq) \longrightarrow Fe(CN)_6{}^{3-} + NO(g) + H_2O(l)$$

The nitrogen oxide was purified by cooling the mixture with liquid air and fractionating.

Known masses of nitrogen oxide were then passed into a combustion bulb in which they were decomposed by hot, finely divided nickel in a porcelain boat. The increase in mass of the nickel gave the mass of oxygen in the original mass of nitrogen oxide, and the mass of nitrogen gas produced was measured by absorbing it on cold charcoal.

In a typical series of experiments, 2.93057 g of nitrogen oxide was found to contain 1.56229 g of oxygen. The mass of nitrogen in the oxide was 1.36819 g by direct measurement, or 1.36828 g by difference, giving a mean of 1.36824 g. Taking this mean figure gives an equivalent mass for nitrogen of 7.00063.

4. Equivalent mass of an element by reaction with hydrogen. The main reason why hydrogen was eventually abandoned as the basis for defining equivalent mass was that very few elements combine directly with hydrogen to form stable compounds. Measurements of equivalent masses by direct combination with hydrogen are, therefore, rare, but the equivalent of chlorine was measured, in this way, by Edgar in 1908, and by Noyes in the same year.

Edgar used a known mass of pure chlorine (measured as a liquid) and a known mass of pure hydrogen (measured by absorption on palladium) and exploded the two gases in an evacuated vessel by sparking. The hydrogen chloride formed was condensed by cooling with liquid air and its mass was measured. The mass of chlorine combining with 1.0080 g of hydrogen gives the equivalent mass of chlorine, the accepted modern value being 35.457.

Noyes's method was very similar except that he used the chlorine in a known mass of potassium chloroplatinate. On passing hydrogen over the heated chloroplatinate, hydrogen chloride was formed and its mass was measured after condensation.

5. Equivalent mass of an element by displacement of hydrogen. In this method, a known mass of a metal is made to react with a suitable chemical and the mass of the hydrogen produces is obtained by measuring its volume and using the fact that 1 litre (1 dm³) * of hydrogen at s.t.p. weighs 0.09 g.

If 1 g of a metal yields 500 cm³ * (500 × 10⁻⁶ m³) of hydrogen at s.t.p. it follows that 1 g of metal displaces 0.045 g of hydrogen. 1 g of hydrogen will therefore be displaced by 22.2 g of metal; the equivalent of the metal must be 22.2.

The method is available for metals which will react with acids or alkalis to produce hydrogen. Magnesium, aluminium, zinc, iron and tin can be used reacting with acids, and aluminium and zinc reacting with alkalis. The method can also be applied to potassium, sodium and calcium, using their reaction with ethanol.

6. Equivalent mass by replacing one metal by another. In this method, one metal, A, is made to replace another, B, e.g.

$$Zn(s) + Cu^{2+}(aq) \longrightarrow Zn^{2+}(aq) + Cu(s)$$

If the equivalent mass of either A or B is known, then the equivalent mass of the other metal can be obtained since the equivalent (in grammes) of A will replace the equivalent (in grammes) of B.

The equivalent of copper, for example, is 31.5. If it is found, experimentally, that 1.031 g of zinc will replace 1 g of copper, it follows that 32.5 g of zinc will replace 31.5 of copper. The equivalent of zinc must, therefore, be 32.5.

This is not a good method as there are many difficulties in obtaining an accurate result.

7. Equivalent mass of an element by conversion of one compound into another. In this method, which is the one most frequently used to obtain exact values, two equivalents must, in general, be known, before a third can be measured.

To obtain the equivalent mass of sodium, for example, a known mass of sodium chloride is dissolved in water and excess silver nitrate(v) solution is added. The precipitated silver chloride

$$Cl^-(aq) + Ag^+(aq) \longrightarrow AgCl(s)$$

is washed and dried and its mass is measured.

---

* The litre was originally defined as 1.000028 dm³ but this definition was abolished in 1964 and the term litre was reintroduced as a special name for the cubic decimetre. The term litre is so well entrenched both in everyday and chemical use that it will probably be continued to be used. In this book, however, cubic decimetre is preferred.

The millilitre (ml) is a sub-multiple of a sub-multiple of the basic SI unit of volume (m³). On these grounds, cm³ is preferred to ml in this book. Both will probably continue to be used for some time.

In sodium chloride, the equivalent of sodium is combined with that of chlorine; in silver chloride, the equivalent of silver is combined with that of chlorine. If the equivalent of sodium is taken as x, and if the equivalents of silver and chlorine are 107.880 and 35.457 respectively, then $(x + 35.457)$ g of sodium chloride would yield $(107.880 + 35.457)$ g of silver chloride. If, by experiment, it is found that a g of sodium chloride yield b g of silver chloride, then

$$\frac{a}{b} = \frac{(x + 35.457)}{(107.880 + 35.457)}$$

from which, a and b being unknown, the value of x can be calculated.

Substances other than silver chloride, e.g. barium sulphate(VI), can be precipitated, so that the method is widely applicable.

8. Equivalent mass of silver. In the method described in the preceding section, precipitation of silver chloride is most commonly used, and calculation of the results required demands a knowledge of the accurate equivalent mass of silver and chlorine. These values were obtained by Stas in 1860, using a method first suggested by Berzelius. The method involves three stages:

a. Potassium chlorate(V) is heated to give potassium chloride and oxygen,

$$2KClO_3(s) \longrightarrow 2KCl(s) + 3O_2(g)$$

122.592 g of potassium chlorate(V) were found to yield 74.592 g of potassium chloride and 48 g of oxygen. As potassium chlorate(V) contains six equivalents of oxygen combined with one of potassium and one of chlorine, the equivalent mass of potassium chloride must be 74.592.

b. Addition of excess silver nitrate(V) solution to a known mass of potassium chloride dissolved in water gives a precipitate of silver chloride. The results showed that 74.592 g of potassium chloride produced 143.397 g of silver chloride, so that the equivalent of silver chloride must be 143.397.

c. On heating pure silver in chlorine, silver chloride is formed, and Stas found that 143.397 g of silver chloride are formed from 107.943 g of silver and 35.454 g of chlorine. Stas took these to be the values of the equivalent masses of silver and chlorine. The equivalent mass of potassium must be 74.592 − 35.454. i.e. 39.138.

By methods similar to this, Stas determined the equivalents of ten elements in the years between 1857 and 1872, and his values were accepted as being the most accurate available at that date.

A discrepancy arose later, however, between Stas's value for the equivalent mass of chlorine and the value determined by direct combination between chlorine and hydrogen (p. 12). Thereupon, it was realised that Stas's work had been inaccurate because (a) the potassium chlorate(V) used had not been entirely free from potassium chloride, and (b) precipitated silver chloride always adsorbs (p. 498) other salts from solution.

Attempts to obtain still greater accuracy were, therefore, made by Richards (1868–1928) and by Honigschmidt and Sachtleben (1929).

Richards converted pure silver into pure silver nitrate(v) by treatment with nitric(v) acid, and he found that 1 g of silver gave 1.574 79 g of silver nitrate(v). On accepting the value of 14.008 for the relative atomic mass of nitrogen (this is the value obtained by the gas-density method, p. 56), Richards calculated the value of the equivalent mass of silver, x, from the following relationship,

$$\frac{x}{x + (14.008 + 48)} = \frac{1}{1.574\,79}.$$

This gives a value of 107.880 for the equivalent mass of silver, and this is the accepted modern value.

Its accuracy has been confirmed by Honigschmidt and Sachtleben. They heated barium chlorate(vii), $Ba(ClO_4)_2$, to form barium chloride and oxygen. Silver chloride was then precipitated from the barium chloride. All the masses were measured *in vacuo* to avoid errors arising from absorption of gases.

9. Equivalent mass of an element by volumetric analysis. A normal solution is defined as one containing 1 gramme-equivalent (p. 49) of solute per cubic decimetre of solution. It follows that the mass, in grammes, of an element which will just react with 1 dm³ of a normal solution must be the equivalent mass of the element, and this is the basis on which equivalent masses can be measured by volumetric analysis.

A known mass of a metal (0.24 g of magnesium, for example) is treated with a known excess of a standard solution of an acid (50 cm³ N hydrochloric acid, for example). When the reaction ceases, the excess of acid remaining is estimated by titration with a standard alkali. If, in the example taken, 30 cm³ of N hydrochloric acid remained after reaction with the magnesium it follows that 0.24 g of magnesium must have reacted with 20 cm³ of N hydrochloric acid. 12 g of magnesium will, therefore, react with 1 dm³ of N hydrochloric acid, so that the equivalent of magnesium must be 12.

10. Equivalent mass of an element by electrolysis. Faraday's laws of electrolysis (p. 89) show that 96487 coulomb will liberate 1 gramme-equivalent of any element. If, therefore, the mass, in grammes, of any element liberated by a definite number of coulombs be measured the mass which would have been liberated by 96487 coulomb is readily calculated to give a value for the equivalent mass of the element.

QUESTIONS ON CHAPTER 2

1. (*a*) Equal masses of different metals displace equal volumes of hydrogen from the same acid. (*b*) Equal masses of the same metal displace equal volumes of hydrogen from different acids. Comment on these statements.

2. Three elements, A, B and C, have equivalent masses in the ratio of 2:3:5.

If the equivalent mass of C is 40, what are the equivalents of A and B? What are the percentages by mass of (i) A in AC, (ii) C in BC and (iii) B in AB?

3. Show that the equivalent mass of an element with an atomicity of a, a valency of v, and a vapour density of d, is given by $2d/(a \times v)$.

4. A base contains 60 per cent by mass of metal. What is the equivalent mass of the metal and of the base? 2 g of the base are added to 125 cm³ of N acid. After the reaction has ended, what volume of N/2 alkali will be needed to neutralise the excess acid?

5. The equivalent mass of a metal is 21. What volume of hydrogen at s.t.p. will be required to reduce 8.7 g of the oxide of the metal?

6. 0.569 g of aluminium foil were dissolved in 100 cm³ of 0.96M hydrochloric acid. The excess acid required 35.1 ml of 0.94M sodium hydroxide solution for neutralisation. Calculate the equivalent mass of aluminium.

7. 0.2 g of iron liberate 82.4 cm³ of hydrogen on reaction with excess acid. The hydrogen volume was measured over water at 20°C and 10^5N m^{-2} (750 mm) pressure, the saturated vapour pressure of water at 20°C being 2.33 kN m⁻² (17.5 mm). Calculate the equivalent mass of iron.

8. 0.375 g of zinc liberated 135.3 cm³ of dry hydrogen at 15°C and 104 kN m⁻² (780 mm) pressure by reaction with excess acid. Calculate the equivalent mass of zinc.

9. What are the main reasons why it is much more difficult to measure the mass of a gas than a solid or a liquid?

10. Explain in detail how you would measure the equivalent mass of (a) phosphorus, (b) iodine, (c) sodium in a school laboratory.

11. Outline four methods which could be used in a school laboratory for measuring the equivalent mass of copper. Which method would give the most accurate result? Which of the methods could not be used for calcium? Why?

12. If 6.35 g of copper are dissolved in concentrated nitric(v) acid and the solution is heated until there is no further change 7.95 g of oxide remain. What is the equivalent mass of copper? Write equations for the reactions involved.

13. 0.65 g of a mixture of iron and cadmium gave 150 cm³ of dry hydrogen at s.t.p. on reaction with excess hydrochloric acid. What mass of iron was there in the mixture?

14. 3.78 g of a mixture of copper and copper(II) oxide was heated in hydrogen and a loss in mass of 0.6 g was observed. What percentage of copper was there in the original mixture?

15. 0.1827 g of the chloride of an element could be converted into 0.1057 g of the oxide of the element. Calculate the equivalent mass of the element.

16. If 1.158 g of silver chloride can be precipitated from a solution containing 1.201 g of radium chloride, what is the equivalent mass of radium?

17. Calculate the average equivalent mass of tellurium from the following data: (a) 2.997 g of tellurium dioxide yield 2.396 g of tellurium on reduction, (b) 0.803 g of tellurium form 2.817 g of tellurium tetrabromide, (c) during electrolysis, 0.307 g of tellurium were deposited by the same current and in the same time as 1.041 g of silver.

18. Iron can exert valencies of 2 and 3 and has, therefore, two equivalent masses. How can they both be measured experimentally?

19. 1.788 g of hydrogen chloride will precipitate 7.048 g of silver chloride when passed into excess silver nitrate(v) solution. Assuming the equivalent masses of silver and hydrogen, calculate the equivalent mass of chlorine.

20. If b g of silver, dissolved in nitric(v) acid, require a g of sodium chloride for complete precipitation as silver chloride, and if a' g of sodium chloride yield

b' g of silver chloride, show that the equivalent mass of chlorine is equal to

$$107.9 \cdot \left\{ \frac{a \cdot b'}{b \cdot a'} - 1 \right\}$$

where 107.9 is the equivalent mass of silver.

21. 10 cm³ of dry chlorine, measured at 15°C and 100 kN m⁻² (750 mm) pressure, were absorbed by a solution of potassium iodide. The liberated iodine was sufficient to produce 0.1982 g of silver iodide. If silver iodide contains 54.04 per cent of iodine and the equivalent of chlorine is 35.5, calculate the equivalent of iodine.

22. 15.00 g of silver when heated in sulphur vapour gave 17.228 g of silver sulphide. 9.002 g of silver sulphate(VI) on reduction in hydrogen gave 6.230 g of silver. Assuming that silver sulphate(VI) contains four equivalents of oxygen (equivalent, 8.000) calculate the equivalents of silver and sulphur with the accuracy justified by the data. (C. Schol.)

Chapter 3
Relative atomic masses

1. Introduction. The hypothesis that matter is made up, in some way, of atoms was first put forward in a speculative way by many Greek philosophers such as Democritus and Lucretius; and, later, by Boyle and Newton. Dalton, however, was the first person to revive the older notions, to restate and extend them, and to apply them to the interpretation of the laws of chemical combination described in Chapter 1.

What began a very long time ago as speculation was developed, by Dalton, into a fertile theory, once described as 'the greatest generalisation in chemistry', and certainly a very solid cornerstone of chemical foundations. Berzelius said of the atomic theory that 'it represented the greatest advance that chemistry had yet made in the course of its development into a science'.

Like all theories, Dalton's atomic theory has grown and changed with the times and has, in fact, had to be modified as new facts, quite unknown to Dalton, have come to light. In this chapter, some of the simpler aspects of the theory are described. Later chapters introduce necessary extensions and modifications.

2. Statement of the theory. Like many of the notable contributions to the advancement of science, the atomic theory is essentially simple, and the ideas of it may be summarised in four parts as follows:

a *All matter is composed of atoms.*
b *Atoms are indivisible, indestructible and cannot be created.*
c *All the atoms of any one element are the same, but they are different from those of any other element.*
d *Compounds are formed by the chemical combination of atoms in small, whole numbers.* The result is a small group of atoms chemically combined together. Dalton called such a group a 'compound atom'; we now call it a molecule.

On this basis, an atom is the smallest particle of an element that can possibly exist; a molecule is a small group of atoms chemically combined together. As will be seen (p. 46) an element does not necessarily exist as single atoms under normal conditions, for two, three or more atoms of an element may link together and the particle so formed will be a group of atoms, i.e., a molecule.

3. The atomic theory and the laws of chemical combination. One of the very early advantages of the atomic theory was that it was able to account for the laws of chemical combination, by arguments summarised as follows:

a. *The law of conservation of mass.* Any chemicals will, on Dalton's atomic theory, contain a certain fixed number of atoms combined together in a certain way, and the atoms concerned will have a certain fixed mass.

When chemicals react together, the atoms rearrange themselves but, as atoms can neither be created nor destroyed, there will be the same total number of atoms both before and after reaction, and, therefore, there will be no change in mass as a result of the reaction. *A chemical reaction is*, in fact, *simply a rearrangement of atoms*. Why atoms should want to rearrange themselves is another matter and involves a study of reaction mechanisms (p. 336).

b. *The law of constant composition.* Dalton accounted for the law of constant composition in the statements that all the atoms of any one element are alike and that they combine in small, whole numbers with the different atoms of other elements.

In copper(II) oxide, for example, there are, let us say, x atoms of copper and y atoms of oxygen. The percentage by mass of copper in copper(II) oxide is, therefore,

$$\frac{\text{Mass of } x \text{ atoms of copper}}{\text{Mass of } x \text{ atoms of copper} + \text{Mass of } y \text{ atoms of oxygen}}$$

Because all copper atoms are alike and all oxygen atoms are alike, it follows that the percentage by mass of copper in copper(II) oxide must be constant, i.e. that copper(II) oxide has a constant composition.

A similar argument can be applied to demonstrate the truth of the law of constant composition in all other cases.

c. *The law of multiple proportions.* If atoms, A, of one element combine with atoms, B, of another in small whole numbers to form more than one compound, then some of the various compounds that might possibly be formed can be represented as

AB	A_2B	AB_2
Cpd. 1	Cpd. 2	Cpd. 3

It is clear, from these formulae, that, for a fixed mass (1 atom) of A, there is half as much B in compound 2 as in compound 1, and twice as much B in compound 3 as in compound 1. Thus the masses of B combining with a fixed mass of A are in the proportion,

Cpd. 1	Cpd. 2	Cpd. 3
1	$\frac{1}{2}$	2

i.e. the masses of B are in a simple ratio, $2:1:4$, as in accordance with the law of multiple proportions.

d. *The law of equivalents.* When three different elements are considered, the application of the ideas of the atomic theory is a little more complicated.

The simplest case concerns three atoms, represented as A, B and C, combining to form compounds AB, AC and BC.

If the masses of the atoms A, B and C are taken as a, b and c units, the state of affairs in the compounds AB, AC and BC is

1 atom of	1 atom of
A	B
a units	b units

Compound AB

1 atom of	1 atom of
A	C
a units	c units

Compound AC

1 atom of	1 atom of
B	C
b units	c units

Compound BC

The masses of B and C combining with a units of A are b and c respectively, and the masses of B and C combining together are in the proportion $b:c$. This is in accordance with the law of equivalents.

4. Relative atomic masses. Dalton's statement that all the atoms of any one element are alike but different from those of any other element, enabled him to represent any atom by a symbol. He used geometrical shapes, but these were replaced by letters, first used by Berzelius in 1818.

Because the atomic theory had followed the formulation of gravimetric laws, and because values of combining or equivalent masses of the elements were known, the masses of the atoms of the elements was of particular significance. Dalton recognised that the direct measurement of the absolute mass of such an infinitely small particle as a single atom was not possible, and he was more concerned with the *relative* masses of different atoms. It is this relative mass which is now known as the relative atomic mass (A_r); the older term was atomic weight.

But, as with equivalent masses, which are relative combining masses, it is necessary, for relative atomic masses, to decide some standard of reference. Historically, various choices were made, the most obvious, perhaps, being to take the hydrogen atom, i.e. the lightest known atom, as the standard. On this basis, now known as the *hydrogen scale*, the relative atomic mass of an element was defined by the statement

$$\text{Relative atomic mass of an element} = \frac{\text{Mass of 1 atom of the element}}{\text{Mass of 1 atom of hydrogen}}$$

As will be seen (p. 31), this definition was first replaced by one based on the *oxygen scale* in which $\frac{1}{16}$ of the mass of an oxygen atom was taken as the standard, i.e.

$$\text{Relative atomic mass of an element} = \frac{\text{Mass of 1 atom of the element}}{\frac{1}{16} \times \text{Mass of 1 atom of oxygen}}$$

Relative atomic mass was then redefined, first in terms of $\frac{1}{16}$ of the mass of one atom of the ^{16}O isotope (p. 131)

$$\text{Relative atomic mass of an element} = \frac{\text{Mass of 1 atom of the element}}{\frac{1}{16} \times \text{Mass of 1 atom of } ^{16}O}$$

and then in terms of $\frac{1}{12}$ of the mass of one atom of the ^{12}C isotope (p. 131)

$$\text{Relative atomic mass of an element} = \frac{\text{Mass of 1 atom of the element}}{\frac{1}{12} \times \text{Mass of 1 atom of } ^{12}C}$$

The problem confronting chemists at the start of the nineteenth century was not, however, the precise way in which to define relative atomic mass, but the way in which relative atomic masses and equivalent masses were related. Equivalent masses, which could be measured fairly easily, gave the relative masses in which elements combined together. Dalton said that chemical combination took place between atoms; what, then, were the relative masses of the atoms concerned?

5. Relationship between the relative atomic mass and the equivalent mass of an element. The gravimetric analysis of water shows that 1 g of hydrogen combines with 7.94 g of oxygen, i.e. 1.0080 g of hydrogen with 8 g of of oxygen (p. 7). Now, if water is made up of equal numbers of hydrogen and oxygen atoms, i.e. if its formula is HO, then the mass of the oxygen atom relative to the hydrogen atom must be 7.94. Alternatively, if, in water, there are twice as many hydrogen atoms as oxygen atoms, i.e. if the formula is H_2O, then each atom of oxygen must have a mass of 15.88 (2 × 7.94) relative to the hydrogen atom to give the measured gravimetric composition. If water had a formula H_3O, then the relative mass of the oxygen atoms would be 3 × 7.94, i.e. 23.82.

We all know, today, that the formula of water is, in fact, H_2O, but chemists in the earlier part of the last century did not know this, and before they could deduce any correct values of relative atomic masses from equivalent masses they had to know something about the numbers of atoms which combined together in different compounds. In modern languge, they had to know something about the valencies of the elements.

In the example of water, just quoted, the equivalent of oxygen (on the hydrogen scale) is 7.94. But the relative atomic mass of oxygen may be any multiple of 7.94 depending on the number of oxygen and hydrogen atoms which combine together. If water is HO, then the relative atomic mass of oxygen is 7.94; if water is H_2O, the relative atomic mass is 15.88; if water is H_3O, it is 23.82; and so on.

This can be expressed, for oxygen, in the statement

$$\text{Relative atomic mass of oxygen} = \text{Equivalent mass of oxygen} \times \begin{array}{l}\text{No. of hydrogen atoms} \\ \text{with which 1 atom of} \\ \text{oxygen will combine}\end{array}$$

or, quite generally, as

$$\text{Relative atomic mass of an element} = \text{Equivalent mass of the element} \times \begin{array}{l}\text{No. of hydrogen atoms} \\ \text{with which 1 atom of} \\ \text{element will combine}\end{array}$$

The number of hydrogen atoms with which 1 atom of an element will combine is now known as the valency of the element, and it will be seen that this is the very important connecting link between relative atomic masses and equivalent masses, i.e.

$$\text{Relative atomic mass} = \text{Equivalent mass} \times \text{Valency}$$

6. Early decisions on relative atomic mass values. Dalton, Berzelius and contemporary chemists did not fully understand the significance of the valency of an element, for the term was not accurately used until introduced by Frankland in 1853. But they did realise that it was necessary to know the formula of a substance, i.e. the number of atoms combined together, before they could obtain relative atomic masses from equivalent masses.

Because they had no means of allotting valency values or correct chemical formula, i.e. because they simply did not know whether the formula of water was HO or H_2O or H_3O or HO_2 or HO_3, they resorted, very largely, to guesswork.

Dalton assumed, correctly in some cases but wrongly in others, that if two elements A and B formed only one compound then that compound would be made up of one atom of A for every one atom of B, i.e. its formula would be AB. On this simple basis, he took the formula for water as HO. The table of atomic weights which he drew up, based on this over-simplified principle, contained some values which were correct but many which were wrong.

Berzelius attempted to allot formulae to compounds which would indicate their chemical behaviour in the simplest way, and which would ensure that similar substances had similar formulae. Because he knew that the masses of oxygen which would combine with a fixed mass of iron in iron(II) and iron(III) oxides were in the ratio 2:3 he gave these oxides formulae of FeO_2 and FeO_3. And because zinc oxide was similar to iron(II) oxide he said its formula was ZnO_2.

In Berzelius's own words 'to hit upon what is true is a matter of luck' but, as luck would have it, Berzelius was not always wrong, and in a table of relative atomic masses drawn up by him in 1818, after 10 years of laborious work, he got at least some of the values right.

The period was, however, one of chemical chaos. This was relieved, to some extent, by the application of Dulong and Petit's Rule and Mitscherlich's Law of Isomorphism to the problem of fixing relative atomic masses, but the matter was more fully clarified by Cannizzaro's work of 1858 (p. 49).

7. Dulong and Petit's rule. By using Berzelius's values for relative atomic masses, and adjusting them where necessary (usually by dividing them by two) Dulong and Petit, in 1819, discovered that, for solid elements, particularly for metals,

$$\text{Relative atomic mass} \times \text{Specific heat capacity} \simeq 26.8 \times 10^3$$

The product of the relative atomic mass and specific heat capacity is known as the *atomic heat capacity*; its value is 26.8×10^3 when specific heat capacity is expressed in $J\,kg^{-1}\,K^{-1}$ (or 6.4 if $cal\,g^{-1}\,K^{-1}$ are used). Dulong and Petit expressed their rule in words by stating that 'the atoms of all simple substances have the same capacity for heat'.

By applying the rule, Dulong and Petit were able to fix many relative atomic masses. Essentially, the rule enables the valency of an element to be discovered. Consider, for example, an element, A, with a specific heat capacity of 1.13×10^3 J kg^{-1} K^{-1}, and an accurate equivalent mass of 12.1.

By Dulong and Petit's rule, the approximate relative atomic mass must be 26.8×10^3 divided by 1.13×10^3, i.e. 23.7.

The approximate relative atomic mass divided by the accurate equivalent mass gives a value of 1.96 for the approximate valency. The valency of an element must, however, by definition, be a whole number. The real valency of A is, therefore, 2.

Since the accurate equivalent mass is 12.1, the accurate relative atomic mass must be 12.1×2, i.e. 24.2.

The stages may be summarised as follows:

a. $\dfrac{26.8 \times 10^3}{\text{Specific heat capacity}} \simeq$ Approximate relative atomic mass.

b. $\dfrac{\text{Approximate relative atomic mass}}{\text{Accurate equivalent mass}} \simeq$ Approximate valency.

c. The nearest whole number to the approximate valency value gives the accurate valency.

d. Accurate equivalent mass × Accurate valency = Accurate relative atomic mass.

8. Limitations of Dulong and Petit's rule. The atomic heat capacity of an element is only approximately constant at 26.8×10^3, and some elements, at ordinary temperatures, deviate very considerably from this rule. These elements are of low relative atomic mass and high melting point, and include beryllium, boron, silicon and carbon.

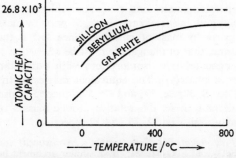

FIG 2. Variation of atomic heat capacity with temperature for some elements which do not obey Dulong and Petit's rule at room temperature.

The specific heat capacity of an element varies, however, with temperature, so that the atomic heat capacity of an element depends, also, on temperature.

The value of the atomic heat capacity of any element rises, in fact, as the temperature is increased, from zero to a maximum value of approximately 26.8×10^3. Elements which 'obey' Dulong and Petit's rule at normal temperatures are those whose atomic heat capacity values have reached the maximum value at ordinary room temperatures. For elements such as beryllium, boron, silicon and carbon, the maximum value of the atomic heat capacity is not reached except at high temperatures. It is only at such temperatures that Dulong and Petit's rule holds for these elements, and, even then, the maximum value of the atomic heat capacity is rather less than 25×10^3 (Fig. 2)

The way in which atomic heat capacity varies with temperature for different elements depends on the way in which the particular crystal structures concerned can absorb energy.

9. Mitscherlich's law of isomorphism. Berzelius had, in attempting to allot formulae to different compounds, always been aware of the desirability of representing compounds which clearly resembled each other by similar formulae. In 1819, Mitscherlich, a pupil of Berzelius, observed that some substances with similar chemical formulae formed very similar crystals; he used the word isomorphous (same shape) to describe such substances, and called the phenomenon isomorphism. Moreover, he stated, in what is now known as Mitscherlich's law of isomorphism, that *substances which have similar chemical compositions crystallise in similar forms, i.e. are isomorphous.*

Typical examples of isomorphous substances are listed, and the phenomenon is discussed in more detail on p. 210.

. There are many exceptions to Mitscherlich's Law, and it is doubtful whether it warrants the name of a law at all, but it did play a small part in establishing some relative atomic mass values and in checking others. Typical examples are described below:

a. *The relative atomic masses of manganese, chromium, selenium and sulphur.* Potassium manganate(VII) (permanganate) and potassium chlorate (VII) (perchlorate) are isomorphous, and, in these compounds, 55 parts by mass of manganese replace 35.5 parts by mass of chlorine. Since 35.5 is the relative atomic mass of chlorine, that of manganese must be 55.

Moreover, potassium manganate(VI) is isomorphous with potassium chromate(VI), selenate(VI) or sulphate(VI). The equivalent masses of the elements concerned are Mn, 55; S, 32; Se, 79; and Cr, 52. Since 55 is also the relative atomic mass of manganese, the relative atomic masses of sulphur, selenium and chromium must be 32, 79 and 52 repectively.

b. *The relative atomic mass of chromium.* Berzelius originally gave iron(II) and iron(III) oxides the formulae FeO_2 and FeO_3 (p. 22). By a similar argument he gave chromium(III) oxide and chromium(VI) oxide (chromium trioxide) the formulae CrO_3 and CrO_6. When he realised, however, that chromates(VI) and sulphates(VI) were isomorphous, he accepted CrO_3 as the formula for chromium(VI) oxide because of its similarity to sulphur trioxide best represented as SO_3. Chromium(III) oxide must, therefore, be Cr_2O_3, which suggested to Berzelius that iron(III) oxide ought to be Fe_2O_3 and not FeO_3. Similarly, iron(II) oxide must be FeO, and not FeO_2, and zinc oxide, ZnO and not ZnO_2.

Such arguments made Berzelius publish a revised table of relative atomic masses, in 1827, in which the values of some metals were halved as compared with his 1818 list.

On accepting the formula of chromium(VI) oxide as CrO_3, the relative atomic mass of chromium was easily allotted from its equivalent mass, for the valency of chromium, here, is 6.

c. *The relative atomic mass of vanadium.* The relative atomic mass of vanadium was fixed by Roscoe when the similarity between lead phosphate(v), lead arsenate(v) and lead vanadate(v) became evident as a result of the discovery of the isomorphism of the three minerals pyromorphite, mimetite and vanadinite, with formulae $Pb_5(XO_4)_3Cl$ where X is P, As and V respectively. On this evidence, vanadium must be pentavalent, so that its relative atomic mass could be obtained from its equivalent mass.

10. The periodic table. Once the relative atomic masses of some elements become known, interesting numerical relationships were soon noticed, and these eventually led to the periodic table in which the elements are arranged in order of their relative atomic masses.

The periodic table has always played a very large part in the study of chemistry. In this chapter, some of the simpler points connected with the table are considered. A discussion of the deeper significance of the table, depending on atomic structure, is given on p. 114.

Even before many relative atomic masses were known with great precision, Döbereiner noticed, in 1829, that certain groups of three chemically similar elements had relative atomic masses which were approximately in arithmetic progression. Such groups became known as *Döbereiner's triads* and are exemplified, using modern values, by

Chlorine	Bromine	Iodine	Lithium	Sodium	Potassium
35.5	79.9	126.9	6.9	23	39.1

Sulphur	Selenium	Tellurium	Calcium	Strontium	Barium
32	79	127.6	40.1	87.6	137.4

Other similar, but, at the time, mysterious numerical relationships using both relative atomic and equivalent masses also came to light, and led to Newland's *Law of Octaves* in 1864.

Newlands arranged all the elements he knew in ascending order of relative atomic mass and assigned to the elements a series of ordinal numbers which he called atomic numbers. He then noticed that elements with similar chemical properties had atomic numbers which differed by seven or some multiple of seven. In other words Newlands discovered that the chemical properties of elements were often found to be similar for every eighth or sixteenth element, like the notes in octaves of music.

Newlands's ideas were not, however, widely accepted and were subjected

	Period	Group 0	Group 1 A	Group 1 B	Group 2 A	Group 2 B	Group 3 A	Group 3 B	Group 4 A	Group 4 B
	1		1 H 1.00797							
Typical elements	2 1st short period	2 He 4.0026	3 Li 6.939		4 Be 9.0122		5 B 10.811		6 C 12.01115	
	3 2nd short period	10 Ne 20.179	11 Na 22.9898		12 Mg 24.312		13 Al 26.9815		14 Si 28.086	
	4 1st long period	18 A 39.948	19 K 39.102		20 Ca 40.08		21 Sc 44.956		22 Ti 47.90	
				29 Cu 63.546		30 Zn 65.37		31 Ga 69.72		32 Ge 72.59
	5 2nd long period	36 Kr 83.80	37 Rb 85.47		38 Sr 87.62		39 Y 88.905		40 Zr 91.22	
				47 Ag 107.868		48 Cd 112.40		48 In 114.82		50 Sn 118.69
	6 3rd long period	54 Xe 131.30	55 Cs 132.905		56 Ba 137.34		57 La 138.91 58–71 The Rare Earths *		72 Hf 178.49	
				79 Au 196.967		80 Hg 200.59		81 Tl 204.37		82 Pb 207.19
	7	86 Rn (222)	87 Fr (223)		88 Ra (226)		89 Ac (227) 90–103 The Acti-nons †			

*	The Rare Earths or Lanthanons	58 Ce 140.12	59 Pr 140.907	60 Nd 144.24	61 Pm (147)	62 Sm 150.35	63 Eu 151.96

†	The Actinons	90 Th 232.038	91 Pa (231)	92 U 238.03	93 Np (237)	94 Pu (242)	95 Am (243)

THE PERIODIC TABLE (after Mendeléef)

Group 5 A	Group 5 B	Group 6 A	Group 6 B	Group 7 A	Group 7 B	Group 8		
7 N 14.0067		8 O 15.9994		9 F 18.9984				
15 P 30.9738		16 S 32.064		17 Cl 35.453				
23 V 50.942		24 Cr 51.996		25 Mn 54.9380		26 Fe 55.847	27 Co 58.9332	28 Ni 58.71
	33 As 74.9216		34 Se 78.96		35 Br 79.904			
41 Nb 92.906		42 Mo 95.94		43 Tc (99)		44 Ru 101.07	45 Rh 102.905	46 Pd 106.4
	51 Sb 121.75		52 Te 127.60		53 I 126.9044			
73 Ta 180.948		74 W 183.85		75 Re 186.2		76 Os 190.2	77 Ir 192.2	78 Pt 195.09
	83 Bi 208.98		84 Po (210)		85 At (210)			

64 Gd 157.25	65 Tb 158.924	66 Dy 162.50	67 Ho 164.930	68 Er 167.26	69 Tm 168.934	70 Yb 173.04	71 Lu 174.97
96 Cm (247)	97 Bk (247)	98 Cf (249)	99 Es (254)	100 Fm (253)	101 Md (256)	102 No (254)	103 Lw (258)

International relative atomic masses (1967)

to some ridicule, until, in 1869, they were essentially restated by Mendeléef and by Lothar Meyer. The periodic recurrence of similar chemical properties when the elements are arranged in ascending order of relative atomic mass, which was the essential point of what Mendeléef called the law of periodicity, was supported by the periodic relationships noticed by Lothar Meyer in plotting the atomic volumes (relative atomic masses/densities) of the elements against their relative atomic masses.

11. Mendeléef's form of periodic table. Mendeléef's original periodic table arrangements were, naturally, incomplete because he only knew a limited number of elements. He had, of necessity, to leave many gaps in his arrangements for elements not, at that time, discovered. Mendeléef predicted both the eventual discovery of such elements and their probable properties, and it was a real triumph that the many gaps he had originally left were, in his lifetime, filled by newly discovered elements having the properties he had predicted.

A modernised version of Mendeléef's table is given on pages 26–7, and the following points are worthy of attention.

a. *Groups and periods.* The elements fall into nine vertical groups, and elements in the same group have certain chemical and physical similarities. In some groups the similarities are very marked; in others, less so.

The horizontal groups are known as periods, those containing eight elements being called short periods, and the others long periods.

b. *Typical elements.* The elements in the short periods at the head of each vertical group are known as typical elements; their chemistry is, on the whole, typical of the group. Below the typical elements, the groups are usually subdivided into sub-groups, A and B. The similarity between elements in different sub-groups, but in the same group, is not very marked, but it is noticeable that it is the sub-group A elements which resemble the typical elements in the left-hand groups, whereas sub-group B elements are more like the typical elements in the right-hand groups.

c. *Transitional elements.* The three groups of elements placed in group 8 were originally called transition, or transitional, elements. They show some resemblances both to the elements preceding them, and to those following them, and in that way they link up the first and second halves of the long periods. The term transition element is, however, now used in a wider sense, as explained on p. 116.

d. *Group valencies.* The numerical valency of an element is often equal either to the group number in which the element occurs, or to 8 minus the group number. Such a relationship was expressed, in 1904, in Abegg's rule of eight. It is by no means universally true, though, as will be seen in

Chapter 15, the number 8 plays an important part in valency considerations.

e. *Anomalous placings.* In order to maintain the chemical similarities in the Mendeléef form of periodic table, a few elements have to be placed in their wrong relative atomic mass order. In the following pairs of elements, for example, the first element, with the lower relative atomic mass, is placed after the second in the periodic table.

Potassium	Argon	Nickel	Cobalt
39.10	39.95	58.71	58.93

Iodine	Tellurium	Protoactinium	Thorium
126.90	127.60	231	232.94

f. *Position of hydrogen.* Hydrogen has some likeness both to the alkali metals (lithium, sodium, potassium, rubidium and caesium) and the halogens (fluorine, chlorine, bromine and iodine), and it can be placed at the head of Group 1 or Group 7, though neither place is entirely satisfactory for it on purely chemical grounds.

g. *Rare earths, lanthanons or lanthanides, actinons or actinides.* The rare earths, lanthanons or lanthanides, with marked chemical similarity, all occupy one single position in the periodic table in Group 3. The actinons or actinides provide a similar series.

h. *Noble gases.* The discovery of a whole new series of elements could not have been predicted from the early forms of the periodic table. The noble gases, discovered at the turn of the nineteenth and twentieth centuries, provided such a series which had to be fitted into the periodic table. This was conveniently done by placing them in Group 0, a position which suggested a valency of zero in keeping with their chemical inactivity.

12. Thomsen–Bohr periodic table. Very many variations on the periodic table theme have been put forward, each emphasising some particular features. Perhaps the main drawbacks to the Mendeléef table are the division of groups into sub-groups, and the isolation of a limited number of transition elements in Group 8.

To overcome such difficulties Thomsen proposed a form of periodic table, in 1885, which was developed, by Bohr and others, into the arrangement shown on p. 30.

On this scheme, some rather artificial chemical similarities suggested by Mendeléef's arrangement are avoided, and the arrangement is more in keeping with the known atomic structures of the elements, as explained on p. 116.

THE PERIODIC TABLE
(after Thomsen and Bohr)

1A	2A		3A	4A	5A	6A	7A	8			1B	2B	3B	4B	5B	6B	7B	O
1 H																		2 He
3 Li	4 Be												5 B	6 C	7 N	8 O	9 F	10 Ne
11 Na	12 Mg												13 Al	14 Si	15 P	16 S	17 Cl	18 Ar
19 K	20 Ca	21 Sc	22 Ti	23 V	24 Cr	25 Mn	26 Fe	27 Co	28 Ni		29 Cu	30 Zn	31 Ga	32 Ge	33 As	34 Se	35 Br	36 Kr
37 Rb	38 Sr	39 Y	40 Zr	41 Nb	42 Mo	43 Tc	44 Ru	45 Rh	46 Pd		47 Ag	48 Cd	49 In	50 Sn	51 Sb	52 Te	53 I	54 Xe
55 Cs	56 Ba	57 * La	72 Hf	73 Ta	74 W	75 Re	76 Os	77 Ir	78 Pt		79 Au	80 Hg	81 Tl	82 Pb	83 Bi	84 Po	85 At	86 Rn
87 Fr	88 Ra	89 † Ac																

← Transitional Elements →

* The Rare Earths or Lanthanons

58 Ce	59 Pr	60 Nd	61 Pm	62 Sm	63 Eu	64 Gd	65 Tb	66 Dy	67 Ho	68 Er	69 Tm	70 Yb	71 Lu

† The Actinons

90 Th	91 Pa	92 U	93 Np	94 Pu	95 Am	96 Cm	97 Bk	98 Cf	99 Es	100 Fm	101 Md	102 No	103 Lw

13. Relative atomic masses from periodic table. Once an element can be placed in the periodic table its probable valency and its approximate relative atomic mass can be predicted, and such predictions were useful, historically, in fixing the relative atomic masses of some elements.

Beryllium, for instance, is, in many ways, similar to aluminium. It was therefore, originally thought to be trivalent and to have a relative atomic mass of 14.1 because its measured equivalent mass was 4.7. Mendeléef suggested, however, that beryllium ought to be placed in Group 2 of the periodic table and that its relative atomic mass should be 9.4. This was confirmed by the discovery that the oxides of zinc and beryllium are isomorphous. Similarly the only possible position for indium in the periodic table showed that it must be trivalent with a relative atomic mass of 114.5.

14. Chemical and physical relative atomic masses. When the relative atomic mass of an element is measured on either the hydrogen or oxygen scale definitions, given on page 20, the result is known as a *chemical relative atomic mass.*

The hydrogen scale was used originally, but, as equivalent mass came to be defined in terms of oxygen and because relative atomic masses were derived from equivalent masses, it was agreed to define relative atomic mass in terms of the oxygen rather than the hydrogen scale. If the mass of one atom of hydrogen was exactly equal to $\frac{1}{16}$th of an oxygen atom there would be no numerical difference between chemical relative atomic masses on the hydrogen or oxygen scale. But it is 1.0080 g of hydrogen which combines with 8.0000 g of oxygen, and not exactly 1 g of hydrogen. On the hydrogen scale, therefore, the relative atomic mass of hydrogen is 1.0000, whilst that of oxygen is 15.880. On the oxygen scale, the relative atomic mass of hydrogen is 1.0080 and that of oxygen is 16.000. When great accuracy is not required, it is convenient to take the values as 1 and 16 respectively.

For real accuracy, however, a further extension is necessary. Not all atoms of oxygen have the same mass, for isotopes exist (p. 131). When this was realised it became necessary to stipulate which isotope of oxygen was to be used in defining relative atomic masses on the oxygen scale. The ^{16}O isotope was chosen and relative atomic masses measured against this standard are known as *physical relative atomic masses.* Their values differ slightly from the corresponding chemical relative atomic masses.

The most modern values of relative atomic masses are measured against the standard of a ^{12}C atom (the carbon scale), as explained on page 131.

15. Summary of methods of measuring relative atomic masses. Chemical relative atomic masses are obtained from measured values of equivalent or relative molecular masses.

a. *If the equivalent mass is known*, it is really a matter of deciding the valency of the element in order to fix the relative atomic mass, for

Relative atomic mass = Equivalent mass × Valency

This might be possible,

  i by using Dulong and Petit's rule (p. 22),
  ii by using Mitscherlich's law of isomorphism (p. 24),
iii by using the position of the element in the periodic table (p. 31).

b. *If the relative molecular mass is known*, the fixing of the relative atomic mass is a matter of deciding on the atomicity of the element for

Relative molecular mass = Relative atomic mass × Atomicity

This might be possible,

  i by using Cannizzaro's method (p. 49). This method also requires a knowledge of the equivalent mass of the element concerned to give a really accurate result;
  ii by measuring the ratio of the specific heat capacities of a gaseous element (p. 77). This ratio is 1.67 for monatomic gases, 1.40 for diatomic gases and 1.33 for triatomic gases (p. 78). This method is particularly useful for the noble gases as it does not involve any chemical reaction.

c. *The measurement of physical relative atomic masses* is described on page 130.

QUESTIONS ON CHAPTER 3

1. Dalton wrote that 'we might as well attempt to introduce a new planet into the solar system, or to annihilate one already in existence, as to create or destroy a particle (atom) of hydrogen'. To what extent is such a statement true today?

2. Explain what is meant by the terms hydrogen scale, oxygen scale and carbon scale.

3. A metallic oxide contains 52.94 per cent by mass of the metal, and 0.18 g of an impure sample of the metal gave 200 cm^3 of hydrogen at s.t.p. when treated with excess acid. The specific heat capacity of the metal is 0.992 J g^{-1} K^{-1}. Calculate (*a*) the purity of the sample of metal, (*b*) the accurate relative atomic mass of the metal.

4. 1 g of a metallic bromide was dissolved in water and excess silver nitrate(v) solution was added. The resulting precipitate of silver bromide weighed 1.88 g. If the specific heat capacity of the metal concerned was 0.67 J g^{-1} K^{-1}, calculate its relative atomic mass.

5. When 0.560 g of a metal was treated with dilute acid it liberated 575 cm^3 of hydrogen at 16°C and 10^5 N m^{-2} (750 mm) pressure. The specific heat capacity of the metal was 1.042 J g^{-1} K^{-1}. Calculate its relative atomic weight and the percentage of oxygen in its oxide. (The vapour pressure of water vapour at 16°C is 1.866 × 10^3 N m^{-2} or 14 mm.) (Army and Navy.)

6. Give an account of three important methods which have been used to determine the relative atomic mass of the elements.

0.1 g of a metal M of specific heat capacity $0.88 \text{ J g}^{-1}\text{K}^{-1}$ dissolved in acid to give 173 cm^3 of hydrogen measured at 27°C and 8×10^4 N m^{-2} (600 mm) pressure. 0.1 g of its anhydrous chloride volatilised completely to give 12.05 cm^3 of vapour measured under the same conditions. Calculate the relative atomic mass of the metal M and the formula of the chloride. (O. & C.)

7. The equivalent of a volatile metal is 100.3 and its specific heat capacity is $0.138 \text{ J g}^{-1}\text{K}^{-1}$; 0.25 g of the metal occupies 79.5 cm^3 at 500°C and 101.3×10^3 kNm^{-2} (760 mm). Calculate the relative atomic mass of the metal and the relative molecular mass of the metal in the vapour phase. (Oxf. Prelims.)

8. The specific heat capacity of a metal at 15°C is $1.799 \text{ J g}^{-1}\text{K}^{-1}$. The metal forms a chloride which contains 88.65 per cent of chlorine, and which has a vapour density of 40. Discuss the evidence for the relative atomic mass of the metal afforded by these data. (C. Schol.)

9. 11.72 g of a non-volatile oxide were heated with excess of sodium carbonate and the resulting loss of mass, due to escape of carbon dioxide, was 2.228 g. In another experiment 0.1037 g of the element when treated with ammoniacal silver nitrate(v) yielded 0.3657 g of silver. Derivatives of the element were isomorphous with chromates(vi). What conclusions do you draw from these data regarding the equivalent and relative atomic masses of the element and the formula of the oxide? (C. Schol.)

10. 0.7994 g of anhydrous beryllium chloride were dissolved in water and all the chlorine was quantitatively precipitated as silver chloride (2.867 g). Calculate the equivalent of beryllium.

The specific heat capacity of beryllium is $1.758 \text{ J g}^{-1}\text{K}^{-1}$ at 20°C. The element forms a volatile compound with the monobasic acid $C_5H_7O_2H$ (acetyl acetone) whose vapour density at 180°C is 103.5. In its general chemical behaviour beryllium shows similarities to both aluminium and to magnesium but does not form salts isomorphous with those of either aluminium or magnesium. Calculate the relative atomic mass of beryllium and explain why there was for a long time doubt whether the element belongs to group II or group III. (B.)

11. Two chlorides of a metallic element with a specific heat capacity of $0.138 \text{ J g}^{-1}\text{K}^{-1}$ contain 35.52 per cent and 42.35 per cent of chlorine respectively. What is the relative atomic mass of the element?

12. Plot the boiling points of (i) the noble gases, (ii) the alkali metals, (iii) the halogens, against their atomic numbers. What conclusions can you draw? Can you give any explanation for your conclusions?

13. Draw graphs to show the way in which (a) the melting point, (b) the atomic volume, i.e. relative atomic mass/density, of elements changes with the atomic mass of the element concerned.

14. State and explain the principles which have been used to determine the relative atomic masses of (a) noble gas, (b) a solid non-metallic element, e.g. carbon or sulphur and (c) a heavy metal. (O. & C.)

15. What are the essentials of a good relative atomic mass determination? Illustrate your answer by reference to some one important piece of work in this field. Why is it important to know the relative atomic mass of the elements with a high degree of accuracy? (O. Schol.)

16. Define relative atomic mass and describe briefly one physical and one chemical method for determining this quantity. What connection has the relative atomic mass of an element with the masses of the atoms of that element? Why do isotopes of an element have the same chemical properties? (O. & C. S.)

17. What is meant by the statement that zinc is a bivalent element? Show

whether the statement is consistent with the following data: (a) addition of excess silver nitrate(v) to a solution containing 0.4089 g of anhydrous zinc chloride precipitated 0.8600 g of silver chloride; (b) the specific heat capacity of zinc is $0.398 \text{ J g}^{-1} \text{ K}^{-1}$; (c) when heated with iodoethane, zinc forms a volatile compound, $Zn(C_2H_5)_x$, the vapour density of which is 62. How does zinc stand in relation to magnesium, calcium and cadmium in the electrochemical series? (W.)

18. Define the terms chemical relative atomic mass, atomic number and equivalent mass. Give (a) the electrical charges on the ions formed in solution by elements of atomic number 16 and 30 respectively; (b) the electronic formula of the hydride of an element X of atomic number 19.

A solution containing 0.250 g of an alum, $K_2SO_4 . M_2(SO_4)_3 . 24H_2O$, gave 0.234 g of barium sulphate(vi) on treatment with excess barium chloride solution. Calculate the relative atomic mass of the element M. (W.)

19. The vapour density of an anhydrous metallic chloride is 40. The sulphate(vi) of the metal crystallises with four molecules of water of crystallisation. 1.771 g of the hydrated sulphate(vi) on ignition yield 0.251 g of metallic oxide. Calculate the relative atomic mass of the metal. (C. Schol.)

20. 0.256 g of the sulphate(vi) of a certain element yielded 0.525 g of barium sulphate(vi), when it was treated in aqueous solution with excess of barium chloride. The vapour density of the chloride of the element was about 67. Suggest possible values for the relative atomic mass of the element. What further experiments could be made in order to confirm the value assigned? (O. & C.)

21. Give two methods by which the accurate relative atomic mass of an element may be determined. 5.21 g of the chloride of a divalent metal were dissolved in distilled water and to the solution an excess of silver nitrate(v) was added. The weight of the silver chloride obtained was 7.17 g. Calculate the relative atomic mass of the metal. (O. & C.)

22. What is the chemical evidence for the existence of atoms and molecules? Explain, in outline, how you would determine the relative atomic masses of (a) hydrogen, (b) silver, (c) nitrogen.

23. When 1.148 g of the crystals of a compound were heated, ammonia and water were evolved and 0.936 g of the oxide of an element was left. The oxide was unchanged by further heating in oxygen. It was soluble in an aqueous solution of sodium hydroxide, when it was found by the method of back titration that 0.520 g of sodium hydroxide was equivalent to 0.936 g of the oxide. The vapour density of the fluoride of the element was 105. Suggest a relative atomic mass for the element. (C. Schol.)

Chapter 4
Gases

An experimental study of the gravimetric aspects of chemical combination between elements led to Dalton's atomic theory. The molecular theory and the kinetic theory arose from a study of the behaviour of gases. In this chapter, some of the important experimental aspects of a study of gases are described. The full theoretical significance of the work is explained in Chapters 5 and 7.

1. Boyle's law, 1662. This law states that *the volume of a given mass of a gas is inversely proportional to the pressure, if the temperature remains constant.* Expressed mathematically,

$$p \propto \frac{1}{V} \qquad \text{or} \qquad pV = \text{a constant value}$$

Because pV has a constant numerical value, at a given temperature, a straight, horizontal line should result if pV values are plotted against p.

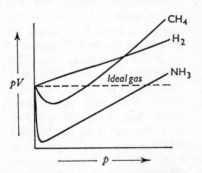

FIG. 3. pV—p curves at 0°C, showing deviation from Boyle's law.

Over a wide range of pressures, particularly at high pressures, Regnault and Amagat showed, however, that the pV—p curves are not, in fact, straight, horizontal lines. This means that gases do not strictly obey Boyle's law. The original data on which Boyle based his law was rather scanty.

The way in which gases deviate from Boyle's law is shown by the shape of the pV—p curves as shown in Fig. 3. Each curve is plotted for figures obtained at a definite temperature; such curves are known as *isothermals*. Hydrogen and the inert gases give curves which rise steadily; other gases give decreasing pV values at first. Any gas which would give a straight, horizontal line is regarded as a *perfect* or an *ideal gas*, but there are no known examples, and the concept of a perfect or an ideal gas is a theoretical one.

The deviation from the behaviour expected of a perfect gas is most marked at high pressures, at low temperatures, and for gases which can be liquefied easily. A further treatment is given on page 72.

2. Charles's law, 1787. This law, which is sometimes known as Gay-Lussac's law, states that *the volume of a given mass of a gas increases, or decreases, by approximately $\frac{1}{273}$ of its volume at 0°C for each degree Celsius rise, or fall, in temperature, if the pressure remains constant.* Expressed mathematically,

$$V_t = V_0\left(1 + \frac{t}{273}\right) \quad \text{or} \quad V_t = \frac{V_0(t + 273)}{273}$$

where V_t is the gas volume at $t\,^{\circ}\text{C}$ and V_0 the volume at $0\,^{\circ}\text{C}$.

This expression shows that if a gas could be obtained at approximately $-273\,^{\circ}\text{C}$ it would have a zero volume; the gas would have vanished! In practice, however, all gases liquefy and/or solidify at temperatures above $-273\,^{\circ}\text{C}$, and Charles's law does not apply to liquids or solids. Gas volumes cannot, therefore, be measured at temperatures right down to $-273\,^{\circ}\text{C}$, but, if the volume of a given mass of gas is measured at different temperatures, extrapolation shows that the gas would have no volume if it could be obtained at a temperature of $-273\,^{\circ}\text{C}$ (Fig. 4).

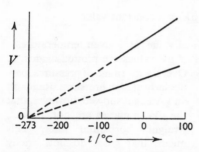

FIG. 4. Variation in the volume of a given mass of gas, at a fixed pressure, with temperature.

3. Temperature scales. On the Celsius scale of temperature the upper and lower fixed points are determined by the boiling point of water and the melting point of ice under a pressure of $101\,325\ \text{N m}^{-2}$. The melting point of ice is called $0\,^{\circ}\text{C}$ and the boiling point of water $100\,^{\circ}\text{C}$. One hundredth of the interval between these two points is called a degree Celsius ($1\,^{\circ}\text{C}$). The use of this scale was originated by Celsius in 1742, and it has been very widely used. The main inaccuracy, which is very small, is due to the difficulty involved in an accurate measurement of the melting point of ice.

The temperature at which a gas would, theoretically, have no volume (see section 2) was taken as the zero for an alternative scale known as the absolute absolute scale. On this scale $0\,^{\circ}\text{C}$ was approximately $273\,^{\circ}\text{A}$.

A similar scale has been adopted as the SI temperature scale; it is referred to as the absolute, the Kelvin or the thermodynamic scale. The unit of temperature difference, known as the kelvin ($1\ \text{K}$) is defined as $\frac{1}{273.16}$ of the thermodynamic temperature of the triple point of water ($273.16\ \text{K}$, see p. 525). The triple point of water has been chosen because of its reproducibility.

As the triple point of water is $0.01\,^{\circ}\text{C}$ it follows that

$$\text{Temperature in K} = \text{Temperature in }^{\circ}\text{C} + 273.15$$

so that the melting point of ice at $101\,325\ \text{N m}^{-2}$ ($0\,^{\circ}\text{C}$) is $273.15\ \text{K}$.

Notice that the $^{\circ}$ sign is not used on the Kelvin scale. 1 K means a temperature of 1 Kelvin above absolute zero, or the temperature interval between n K and $(n + 1)$ K. This temperature interval of 1 K is equal to that of 1°C so that the latter need no longer be used.

If the mathematical expression of Charles's law is written using K instead of °C, then

$$V_T = \frac{V_0 T}{273} \quad \text{or} \quad V_T = kT \quad \text{or} \quad \frac{V_T}{T} = k$$

where k is a constant, and where T represents temperature in K.

This means that *the volume of a given mass of gas is proportional to the absolute temperature, if the pressure remains constant*, which is another way of stating Charles's law.

Because real gases are not perfect they deviate from Charles's law as they do from Boyle's law.

4. The gas equation. Charles's and Boyle's laws can be combined together into one general expression which is known as the gas equation.

If a given mass of gas has a volume of V_1 at a pressure of p_1, it will have a volume, V_x, at a pressure of p_2, and, if the temperature remains constant,

$$V_1 \times p_1 = V_x \times p_2 \text{ (by Boyle's law)}$$

If, now, the temperature at which V_x is measured is changed from T_1 K to T_2 K then the new volume of the gas, V_2, will be given by the relationship,

$$\frac{V_x}{T_1} = \frac{V_2}{T_2} \text{ (by Charles's law)}$$

Substituting the value of V_x from one of these equations into the other gives the expression

$$\frac{p_1 \times V_1}{T_1} = \frac{p_2 \times V_2}{T_2}$$

and, in more general form, this can be expressed as

$$\frac{pV}{T} = \text{a constant} \quad \text{or} \quad pV = kT$$

which is known as the gas equation. k is a numerical constant, and it must be remembered that temperature, in this equation, is expressed in K.

5. The molar gas constant. The value of k in the gas equation depends on the mass of gas considered and on the units in which p and V are measured. If 1 mol (p. 48) of gas is considered, k is generally written as R, and the gas equation then becomes

$$p \times V = R \times T \text{ (for 1 mol of gas)}$$

R is known as the *molar gas constant*.

In SI units, pressure is measured in newtons per square metre, i.e. force per unit area, and volume in cubic metres. The corresponding value of the molar gas constant is 8.3143 J K^{-1} mol^{-1}.

Knowing this value * the generalised relationship for n mol,

$$p \times V = n \times R \times T \text{ (for } n \text{ mol)}$$

enables any of the four quantities p, V, n or T to be calculated if three of them are known.

6. Gaseous diffusion. Diffusion means, literally, spreading out, and one of the most striking characteristics of gases is their ability to diffuse. One of the essential differences between a gas and a solid or a liquid, for instance, is that a gas will always spread out or diffuse so as to occupy fully any container. This diffusion takes place in all directions, even against gravity, and enables gases to pass through porous substances, or through very small apertures, though the escape of a gas through a single hole is often referred to as *effusion*. Gases, in simple language, leak well.

Gaseous diffusion may be demonstrated very readily by placing a gas jar full of air over one full of nitrogen dioxide. Though nitrogen dioxide has a greater density than air, it will diffuse upwards and its spreading out can be observed because of its colour.

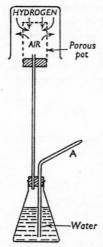

The fact that all gases are completely miscible, provided there is no chemical reaction between them, also indicates the readiness with which they diffuse, as does the short time-lag between an escape of, say, ammonia or hydrogen sulphide and its detection, by smell, at some distance.

7. Graham's law of gaseous diffusion. That gases diffuse at different rates can be demonstrated by using the apparatus shown in Fig. 5. If a beaker full of hydrogen is inverted over the porous pot, a jet of water is pushed out of the tube, A. This is because the hydrogen diffuses into the porous pot more quickly than the air inside it diffuses out of the pot. If carbon dioxide is used instead of hydrogen, the pressure inside the pot is reduced, and air is sucked in through A. Carbon dioxide must, therefore, diffuse less rapidly than air.

FIG. 5.

* The value of R is expressed in units of work per degree per mole. This is because

$$R = \frac{\text{Pressure} \times \text{Volume}}{T} = \frac{\text{Force} \times \text{Volume}}{\text{Area} \times T} = \frac{\text{Force} \times \text{Length}}{T} = \frac{\text{Work}}{T}$$

1 joule is the work done when a force of 1 newton acts through a distance of 1 metre.

If pressure is measured in atmospheres and volume in cubic decimetres, the value of V will be 22.413 when p is 1 and T is 273. R will, therefore, have a value of $1 \times 22.413/273$, i.e. 0.082 dm³-atmosphere K^{-1} mol^{-1}.

Since 1 calorie is equal to 4.18 joule, the value for R of 8.314 3 J K^{-1} mol^{-1} is approximately equal to 2 calorie K^{-1} mol^{-1} and this simple numerical value is sometimes useful.

Graham, in 1846, compared the rates at which various gases diffused through porous pots, and also the rates of effusion through a small aperture. His results were summarised in his law of diffusion which states that *the rate of diffusion (or effusion) of a gas, at a constant temperature and pressure, is inversely proportional to the square root of its density.* In mathematical form,

$$\text{Rate of diffusion or effusion} \propto \sqrt{\frac{1}{\text{Density}}}$$

By comparing the rate of diffusion of two gases, A and B, it is possible to measure the density of one if that of the other is known, for

$$\frac{\text{Rate of diffusion of A}}{\text{Rate of diffusion of B}} = \sqrt{\frac{\text{Density of B}}{\text{Density of A}}}$$

As hydrogen has the lowest density of any gas it diffuses most rapidly.

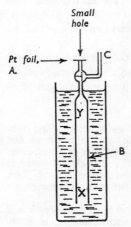

Small hole

Pt foil, A.

FIG. 6. Apparatus for comparing rates of effusion.

8. Comparison of rates of effusion. Fairly accurate results can be obtained for the relative rates of effusion of two gases using the apparatus shown in Fig. 6.

A is a piece of platinum foil sealed over the end of tube B and containing a small hole through which a gas from B can effuse into the atmosphere. The tube B is immersed in water or some other suitable liquid. By first allowing the tube to fill, up to the three-way tap, with liquid, it can be filled with gas through C and the three-way tap. The gas is then allowed to effuse through the hole in the platinum foil, the time being taken for the liquid level in B to rise from X to Y as the gas escapes from the tube into the atmosphere. The experiment is repeated with a second gas in the tube B.

The measurements give the times taken for a definite volume of each gas to effuse under the same conditions. If these times are t_1 for gas 1 and t_2 for gas 2, then, by Graham's law,

$$\frac{t_2}{t_1} = \sqrt{\frac{\text{Density of gas 2}}{\text{Density of gas 1}}}$$

If the density of one of the gases is known, that of the other can be easily calculated.

The errors in such a method arise, mainly, from the difficulty in getting the tube B absolutely full of the gas in question, and from the fact that the pressure of the gas in the tube varies slightly throughout the experiment. More complicated apparatus is necessary to overcome these difficulties.

9. The density of a gas. There are two ways of expressing the density of a gas or vapour. It can be expressed as a *density*, in g dm^{-3} or other similar units e.g. kg m^{-3}, or, as a *relative density*, by comparing the mass of any volume of the gas or vapour with that of an equal volume of hydrogen. The relative density of a gas or vapour is also known as the *vapour density* and, by definition,

$$\text{Relative (vapour) density of a gas or vapour} = \frac{\text{Mass of any volume of gas or vapour}}{\text{Mass of an equal volume of hydrogen}}$$

Measurement of the density of a gas is made by measuring the mass of an evacuated globe, filling it with gas at a known temperature and pressure, and measuring the new mass. This gives the mass of the gas required to fill the globe. The volume of the globe is then measured by filling it with water, and finding the mass. The density is obtained by dividing the mass of the gas by its volume.

Measuring the mass of gases, however, is always difficult for a reasonable volume of gas is not very heavy. Moreover, evacuation of vessels is necessary for, normally, a vessel is full of air though it may be spoken of as being empty.

Regnault tried, in 1845, to eliminate or reduce the likely errors by using compensating globes. Two similar globes are hung from the arms of a balance and enclosed in a draught-proof chamber. Globe A, being fitted with a stop-cock, is the one which is evacuated or filled with gas; globe B, full of air and sealed, is the compensating globe. Regnault used globes of about 50 dm^3 capacity, but, with modern balances, 2 dm^3 globes suffice. Any variation in atmospheric conditions affects both globes equally.

In making a measurement, globe A is first of all dried by passing dry air and then the dry gas under investigation through it. This is to replace any moisture film inside the globe with a film of gas. The globe is then evacuated and is counterpoised against the globe B together with a few weights on the balance pan above B. Globe A is then filled with gas, under known conditions of temperature and pressure. The additional weights which have to be added to the pan above globe B enables the mass of the gas in globe A to be obtained.

The volume of globe A is determined by filling it with water and measuring the mass, again making corrections for temperature and for the mass of air displaced by the globe.

Lord Rayleigh repeated many of Regnault's density determinations, between 1885 and 1892, with even greater accuracy. It was in this work that he discovered the discrepancy between the density of atmospheric nitrogen (1.2572 g dm^{-3}) and nitrogen prepared from chemical sources (1.2560 g dm^{-3}), an observation which led to the discovery of the noble gases.

When relative densities of gases, and not densities, are required to be measured by this direct method, the density of hydrogen must also be obtained. The relative density of any gas, X, is then given by

$$\text{Relative density of gas, X} = \frac{\text{Density of X}}{\text{Density of hydrogen}}$$

10. Measurement of gas density by buoyancy method. Gas densities can be measured by comparing the buoyancy of a small, evacuated quartz bulb in gases at different pressures.

The apparatus used is known as a *microbalance*. A quartz bulb is attached to one end of a quartz balance beam, the beam being supported on a knife-edge, or, better, by a torsion fibre. At the other end of the beam there is a counterweight and a pointer, the whole being enclosed in a glass case (Fig. 7).

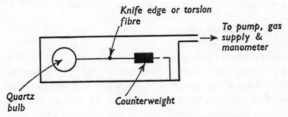

FIG. 7. Diagrammatic arrangement in a microbalance.

The glass container is first evacuated and then filled with the gas under test at such a pressure that the beam balances with the pointer at the zero mark. The pressure, p_g, is measured. After a second evacuation, oxygen, or some other reference gas, is introduced and the pressure, p_0, necessary to maintain the pointer on the zero mark, is recorded.

If the density of the gas under test is ρ_g at s.t.p. it will be $\rho_g p_g$ at a pressure of p_g. The upthrust on the quartz bulb will, therefore, be $\rho_g p_g$ multiplied by the volume of the bulb. When supported in oxygen (with density ρ_0 at s.t.p.), at a pressure of p_0 the upthrust will be $\rho_0 p_0$ multiplied by the volume of the bulb.

As the balance beam rests in the same position in both cases, the upthrust on the bulb must be the same whichever gas is being used, so that

$$\rho_g p_g = \rho_0 p_0$$

By measuring p_g and p_0, and knowing the density of oxygen, ρ_g can be calculated.

The measurement is a relative one against a standard gas of known density. Only pressure measurements have to be made, and very small quantities of gas can be used.

11. Dalton's law of partial pressures. Dalton found, in 1801 that *the total pressure exerted by a mixture of gases, which do not interact, is the sum of the pressures which each gas would exert if it were present alone in the entire volume occupied by the mixture.* The pressures exerted by each gas separately are known as their partial pressures, and the above statement is called Dalton's law of partial pressure.

Dalton arrived at the law experimentally, but it can be deduced from the gas equation. If a mixture of gases consisting of n_1 mol of A and n_2 mol of B occupy a volume, V, then, if p_1 and p_2 are the partial pressures of A and B,

$$p_1 \times V = n_1 \times R \times T$$
$$\text{and } p_2 \times V = n_2 \times R \times T$$

If the total pressure exerted by the mixture of gases is p, then,

$$p \times V = (n_1 + n_2) \times R \times T$$

and, therefore, p must be equal to $p_1 + p_2$.

The law of partial pressure is not widely used, but it is often necessary to find the pressure exerted by a dry gas from a knowledge of the pressure exerted by a mixture of the dry gas with water vapour, i.e. by the wet gas. In this case,

$$\frac{\text{Total pressure of dry}}{\text{gas plus water vapour}} = \frac{\text{Pressure}}{\text{of dry gas}} + \frac{\text{Pressure of}}{\text{water vapour}}$$

The pressure of the water vapour, at any given temperature, can be obtained from tables, so that the pressure of the dry gas is easily obtained.

The partial pressure of a gas in a gaseous mixture is also used as a convenient measure of its concentration (p. 307).

12. Gay-Lussac's law of combining volumes. Cavendish had found, in 1781, that water was formed by sparking a mixture of hydrogen and air, and he had estimated that 2 volumes of hydrogen combined with 1 volume of oxygen. In 1805, Gay-Lussac and Humboldt repeated the determination, and again obtained the ratio 2:1. Struck by the apparent simplicity of this ratio, Gay-Lussac examined other data relating to the volumes in which gases reacted together, and made further investigations.

His results led, in 1808, to the law of combining volumes which states that *when gases react they do so in volumes which bear a simple ratio to each other and to the volume of any gaseous product, all volumes being measured under the same conditions of temperature and pressure.*

Typical results, used or obtained by Gay-Lussac (but with modified figures in some cases), are:

a	Oxygen	+	Hydrogen	$\longrightarrow$	Steam
	1 vol		2 vols		2 vols
b	Nitrogen	+	Hydrogen	$\longrightarrow$	Ammonia
	1 vol		3 vols		2 vols

As with other laws relating to gases Gay-Lussac's law has been shown, by modern work, to be not strictly accurate, simply because real gases are not perfect. But the impact of the law, coming, as it did, very shortly after Dalton had put forward his atomic theory, was of immense importance, as will be seen in the following chapter.

QUESTIONS ON CHAPTER 4

1. Use the gas equation to calculate (a) the volume occupied by 200 g of chlorine at 15°C and 54.71 kN m^{-2}, (b) the volume occupied by 200 g of carbon dioxide at 98.65 kN m^{-2} and 30°C.

2. If the gas constant is expressed per molecule instead of per mole it is sometimes called the Boltzmann constant. What is the value of this constant.

3. Derive the relationship between the density of a gas in grammes per dm^3 (d), the gas pressure in atmospheres (p), the relative molecular mass of the gas (M$_r$),the absolute temperature (T) and the gas constant (R). If dry air is regarded as 21 per cent by volume of oxygen, 1 per cent of argon and 78 per cent of nitrogen, what will its density in g dm^{-3} be at 20°C and 98.65 kN m^{-2} pressure?

4. The density of solid carbon dioxide is 1.53 g cm^{-3}. What volume of gas is obtained, at 15°C and 101.325 kN m^{-2} pressure, from 1 cm^3 of solid carbon dioxide?

5. State Graham's law of diffusion of gases, and give a qualitative explanation of it in terms of the kinetic theory. Describe an experiment to demonstrate gaseous diffusion.

A certain volume of ethanoic acid vapour was found to diffuse in 580.8 seconds, whilst the same volume of oxygen, with the same experimental conditions, took 300 seconds to diffuse. What deductions can be made as a result of this experiment? (N.)

6. Discuss the diffusion of gases in terms of the kinetic theory. Illustrate the use of this phenomenon for (a) the determination of relative molecular mass, (b) the separation of gaseous mixtures. (O. Schol.)

7. Describe some simple experiments which illustrate gaseous diffusion. A given volume of a certain gas was found to diffuse in two-thirds of the time taken by an equal volume of hydrogen chloride under the same physical conditions. What is the relative molecular mass of this gas? (Army and Navy.)

8. A compound gave, on analysis, X = 90.35 per cent and H = 9.65 per cent, and was found to diffuse through a porous plug with two-thirds of the velocity of nitrogen. What is the molecular formula of the compound? (X = 28.1; N = 14.0.)

9. Calculate the ratio of the rate of diffusion of a gas at 91°C and 0°C at a constant pressure.

10. If the density of hydrogen at s.t.p. is x g dm^{-3} what will it be at 10°C and 100 kN m^{-2} pressure?

11. Compare the rates of diffusion of (a) ammonia and hydrogen, (b) hydrogen and deuterium, (c) oxygen and ozone and (d) ^{238}UF$_6$ and ^{235}UF$_6$. Why is the result obtained in (d) of practical importance?

12. A mixture of gases at s.t.p. contains 65 per cent nitrogen, 15 per cent carbon dioxide and 20 per cent oxygen by volume. What is the partial pressure of each gas in $kN\,m^{-2}$?

13. The partial pressures of the components of a mixture of gases are 26.64 $kN\,m^{-2}$ (oxygen), 34 $kN\,m^{-2}$ (nitrogen) and 42.66 $kN\,m^{-2}$ (hydrogen). What is the percentage by volume of oxygen in the mixture?

14. Prove that the fraction of the total gass pressure exerted by one component of a gaseous mixture is equal to the fraction of the total number of molecules provided by that component.

15. The mole is a fundamental concept of chemistry. Review some of the chemical quantities which are generally defined in terms of moles.

Chapter 5
Molecular theory

1. The importance of Gay-Lussac's law. When Gay-Lussac's law was first put forward (1808), Dalton's atomic theory held the stage. It is important to realise, too, that this stage was not well adorned. Chemical symbols had only just been introduced, and then only in geometrical shapes; no chemical formulae were known with any degree of certainty; relative atomic masses could only be guessed; chemists of the day had to make use of very rudimentary equipment; and the number of chemists actively engaged in research was not very great.

Dalton's atomic theory had been so successful in accounting for the gravimetric laws of chemical combination, that it seemed hopeful that it ought to account, too, for volumetric laws. This was particularly so because of the close numerical similarity which seemed to exist between the laws of multiple proportions, as interpreted by Dalton's atomic theory, and Gay-Lussac's law of combining volumes.

Atoms, according to Dalton, combined together in small, whole-number ratios; gases, according to Gay-Lussac, combined, by volume, in small whole-number ratios.

It seemed more than likely that there must be some relation between atoms and volumes of gases, and Dalton suggested that equal volumes of all gases, under the same conditions, contained the same number of atoms. Such a suggestion was not tenable, however, for it was incompatible with the view that atoms were indivisible.

In the reaction between hydrogen and chlorine, for example,

$$\text{1 volume of hydrogen} + \text{1 volume of chlorine} \longrightarrow \text{2 volumes of hydrogen chloride}$$

and, on Dalton's assumption that equal volumes of all gases contain equal numbers of atoms,

$$n \text{ atoms of hydrogen} + n \text{ atoms of chlorine} \longrightarrow 2n \text{ compound atoms (molecules) of hydrogen chloride}$$

or $$\text{1 atom of hydrogen} + \text{1 atom of chlorine} \longrightarrow \text{2 compound atoms (molecules) of hydrogen chloride}$$

But 1 compound atom (molecule) of hydrogen chloride must contain at least 1 atom of hydrogen (it would not be hydrogen chloride, otherwise), and, therefore, 2 compounds atoms (molecules) of hydrogen chloride must contain at least 2 atoms of hydrogen.

In the above reaction, however, on Dalton's assumption, 2 compound atoms (molecules) of hydrogen chloride must be formed from 1 atom of hydrogen. 1 atom of hydrogen must, therefore, be capable of providing two atoms of hydrogen, but because atoms are indivisible this is absurd.

The same predicament was reached in all reactions in which the volume of a gaseous product was greater than the volume of any one of the gaseous elements from which it had been formed.

Dalton reacted to this impasse unfavourably, by suggesting that Gay-Lussac's law was not true because it was based on inaccurate measurements, but Berzelius adopted a more moderate attitude and considered that the difficulty could be overcome by modifying the atomic theory in some way.

2. Avogadro's hypothesis. The problem was solved by a bold and imaginative suggestion made by Avogadro in 1811, and now known as Avogadro's hypothesis. This states that *equal volumes of all gases, under the same conditions of temperature and pressure, contain equal numbers of molecules* (not atoms, as supposed by Dalton).

The significance of this is that Avogadro conceived the idea that *elements* might, sometimes, exist as groups of atoms (molecules) rather than as single atoms. In modern terms, the idea of the *atomicity of an element, i.e. the number of atoms in one molecule of an element,* was born, and the indivisible atom which had controlled Dalton's thinking too rigidly was joined by another equally important particle, the molecule, which could be divided.

Having rejected Gay-Lussac's law, Dalton also rejected Avogadro's hypothesis, and yet this was the hypothesis which was able to account for Gay-Lussac's law, and, also, to help in settling Dalton's other main difficulty, that of deciding how many atoms were present in the molecule of a compound.

The full potentiality of Avogadro's hypothesis was not, however, exploited until 1858, when Cannizzaro, a pupil of Avogadro, developed the ideas and, in so doing, placed theoretical chemistry on firmer foundations than those on which it had rested in the early part of the nineteenth century.

3. Interpretation of Gay-Lussac's law. The atomicity of hydrogen. Dalton was not able to account for the experimental result

$$\begin{array}{ccc} \text{1 volume of} \\ \text{hydrogen} \end{array} + \begin{array}{c} \text{1 volume of} \\ \text{chlorine} \end{array} \longrightarrow \begin{array}{c} \text{2 volumes of} \\ \text{hydrogen chloride} \end{array}$$

but Avogadro's hypothesis can do so. For, by applying the hypothesis,

$$\begin{array}{c} n \text{ molecules} \\ \text{of hydrogen} \end{array} + \begin{array}{c} n \text{ molecules} \\ \text{of chlorine} \end{array} \longrightarrow \begin{array}{c} 2n \text{ molecules of} \\ \text{hydrogen chloride} \end{array}$$

$$\text{or} \quad \begin{array}{c} \text{1 molecule} \\ \text{of hydrogen} \end{array} + \begin{array}{c} \text{1 molecule} \\ \text{of chlorine} \end{array} \longrightarrow \begin{array}{c} \text{2 molecules of} \\ \text{hydrogen chloride} \end{array}$$

The number of atoms in the molecules of hydrogen, chlorine and hydrogen chloride cannot be finally decided from this result, but the reaction can be expressed as

$$H_{2x} + Cl_{2y} \longrightarrow 2H_x Cl_y \qquad \text{(i)}$$

where x and y must have whole number values. The molecule of hydrogen must, therefore, be H_2 or H_4 or H_6 etc. That it is H_2 is supported by four lines of evidence:

a. 1 volume of hydrogen is never known to produce more than 2 volumes of any gaseous product in any reaction. In other words, 1 molecule of hydrogen is never known to split up into more than two atoms. The atom of hydrogen is, therefore, half the molecule, so that the molecule is probably H_2, i.e. hydrogen is diatomic.

b. Hydrogen chloride dissolves in water to form hydrochloric acid, and this acid forms no acid salts. This means that the acid is monobasic, and that its correct formula is probably HCl_y. If this is so, the value of x in equation (i) is 1 and hydrogen is diatomic.

c. Cannizzaro's method, described on page 49.

d. The value of the ratio of the specific heat capacities of hydrogen indicates that it is diatomic (p. 78).

4. $M_r = 2 \times$ Relative density (d). This important relationship was derived by Cannizzaro by a straightforward application of Avogadro's hypothesis.

The relative molecular mass, M_r, previously known as the molecular weight, of an element or compound is defined in much the same way as the relative atomic mass (A_r) of an element is. Either the hydrogen or, more accurately, the ^{12}C scale may be used (p. 31 and 130). Thus,

$$\frac{M_r \text{ of an element}}{\text{or compound}} = \frac{\text{Mass of 1 molecule of element or compound}}{\text{Mass of 1 atom of hydrogen}}$$

or $$\frac{M_r \text{ of an element}}{\text{or compound}} = \frac{\text{Mass of 1 molecule of element or compound}}{\frac{1}{12} \times \text{Mass of 1 atom of } ^{12}C}$$

The relative density of a gas or vapour is defined (p. 40) as

$$\frac{\text{Relative density of}}{\text{gas or vapour } (d)} = \frac{\text{Mass of a volume of gas or vapour}}{\text{Mass of an equal volume of hydrogen}}$$

all volumes being measured at the same temperature and pressure. This definition can be rewritten, by applying Avogadro's hypothesis, as,

$$d = \frac{\text{Mass of } n \text{ molecules of gas or vapour}}{\text{Mass of } n \text{ molecules of hydrogen}}$$

or $$d = \frac{\text{Mass of 1 molecule of gas or vapour}}{\text{Mass of 1 molecule of hydrogen}}$$

And, because hydrogen is diatomic, the mass of one molecule of hydrogen must be twice the mass of one atom of hydrogen. Therefore,

$$d = \frac{\text{Mass of 1 molecule of gas or vapour}}{2 \times \text{Mass of 1 atom of hydrogen}}$$

$$= \frac{M_r \text{ of gas or vapour}}{2}$$

i.e. $M_r = 2 \times d$

This relationship is best remembered in this simple form, but if the definition of relative molecular mass on the ^{12}C scale had been used to derive the relationship it would have been

$$M_r = 2.016 \times d$$

The importance of the relationship is that it enables values of relative molecular masses to be obtained from values of relative density, and, as will be seen (p. 57), the experimental mesurement of relative density is not difficult.

5. The mole. The number of atoms in 1 mol of ^{12}C is obtained by dividing 12 g by the mass of an individual atom of ^{12}C. The answer, known as the *Avogadro constant*, has a value of 6.02252×10^{23} mol^{-1} and this is used in defining an amount of substance known as a mole.

A mole is the amount of a substance containing as many elementary particles as there are in 12 g *of* ^{12}C. Or, alternatively, *a mole is the amount of substance which contains* 6.02252×10^{23} *elementary particles.*

The particles concerned may be molecules, atoms, ions, radicals or electrons. As a mole is an amount it can be expressed in a variety of different units of mass or volume. So far as mass is concerned, the recommended units, on the SI system, are kilogrammes. Thus the mass of:

1 mol of hydrogen, H_2, is 2.0160×10^{-3} kg
2 mol of hydrogen, H_2, is 4.0320×10^{-3} kg
1 mol of hydrogen ion, H^+, is 1.0080×10^{-3} kg
1 mol of sulphate ion, SO_4^{2-}, is 96×10^{-3} kg
1 mol of electrons, e^-, is 0.54860×10^{-6} kg

There are two main advantages in expressing quantities in moles, i.e. of using molar quantities. In the first place, the number of moles of a substance is directly proportional to the number of elementary particles. 64 kg of oxygen does not, at first sight, appear to be twice as much as 2 kg of hydrogen. On a mass basis it isn't twice as much; but on a molar basis it is, for there are twice as many molecules in 64 kg of oxygen as there are in 2 kg of hydrogen.

Secondly, as is explained in the following section, 1 mol of any gas or vapour occupies the same volume under the same conditions of temperature and pressure.

6. Molar volume. Since 1 mol of any gas contains, by definition, the same number of molecules it must also, by Avogadro's hypothesis, occupy the same volume at s.t.p. It is, therefore, only necessary to find the volume occupied by 1 mol of any gas to know the volume occupied by 1 mol of any other.

The density of hydrogen at s.t.p. has been measured experimentally (p. 40) and found to be 0.089 g dm^{-3} or 0.089 kg m^{-3}. Therefore, taking 1 mol of hydrogen as 2.016 g, it follows that it must occupy 2.016/0.089, i.e. 22.4 dm^3 or 2.016 × 10^{-3}/0.089 i.e. 22.4 × 10^{-3} m^3, at s.t.p. 1 mol of any other gas must also occupy 22.4 dm^3 at s.t.p. and this value is known as the molar volume.

22.4136 dm^3 mol^{-1} or 22.4136 × 10^{-3} m^3 mol^{-1} are more accurate figures.

7. Gramme-equivalent. Gramme-atom. Gramme-molecule. Neither equivalent mass nor relative atomic or molecular masses have any units; they are simply numbers. The numbers can, however, be expressed in grammes (or any other units) and can be used, in this way, to denote definite amounts of substance. The following numerical examples will illustrate the usage,

	Hydrogen, H$_2$	*Oxygen*, O$_2$
Equivalent mass	1.0080	8 g
1 gramme-equivalent	1.0080 g	8 g
Relative atomic mass	1.0080	16
1 gramme-atom	1.0080 g	16 g
Relative molecular mass	2.0160	32
1 gramme-molecule	2.0160 g	32 g

Since the introduction of the mole as the unit for the amount of a substance the use of the gramme-equivalent, gramme-atom and gramme-molecule is no longer necessary. The terms will, however, be found in older writings and it is possible that they may linger on despite all recommendations that their use should be discontinued.

8. Cannizzaro's method for fixing relative atomic masses. One of the most important developments of Cannizzaro's use of Avogadro's hypothesis was his method of fixing relative atomic masses. The values for such important elements as carbon and nitrogen could not be fixed by either Dulong and Petit's rule or by Mitscherlich's law of isomorphism, whereas Cannizzaro's method is directly applicable to elements, such as these, which form a large number of volatile or gaseous compounds.

Cannizzaro took as his definition of relative atomic mass *the smallest mass of an element found in* 1 *mol of any of its compounds*. In this way he approached the idea of an atom and of relative atomic mass in a different way from that adopted by Dalton.

His method may be summarised as follows, as applied, for example, to finding the relative atomic mass of carbon:

a. Measure the relative densities (p. 57) of a large number of volatile or gaseous compounds of carbon.

b. Double the values obtained in *a* to obtain values for the mole of the substances concerned.

c. Analyse the substances to find the percentage, by mass, of carbon which they contain.

d. Use the carbon percentage figure to calculate the mass of carbon in 1 mol of each of the substances.

e. Take the smallest mass of carbon ever found in 1 mol of any carbon compound as the relative atomic mass of carbon, on the basis that at least some of the compounds of carbon will contain only 1 atom of carbon per molecule, and, therefore, only 1 mol of carbon per mol of compound.

Some typical figures (below) for carbon compounds will illustrate the method. The lowest mass of carbon in 1 mol of any carbon compound is found to be 12 g, and Cannizzaro, therefore, took the relative atomic mass of carbon to be 12.

It is now known that values obtained in this way are not absolutely accurate, for the calculation makes use of the relationship $M_r = 2 \times d$ which is only strictly accurate for perfect gases.

Accurate relative atomic masses can be obtained, however, by multiplying the equivalent mass of the element concerned by the small whole number which gives a value nearest to that of the approximate relative atomic mass. If the Cannizzaro value for the relative atomic mass of an element is, for example, 14, and if the equivalent mass of the element is 4.669, then the accurate relative atomic mass will be 4.669×3, i.e. 14.007.

Compound	Relative density	M_r	1 mol	Percentage by mass of carbon in compound	Mass of carbon in 1 mol
Methane .	8	16	16 g	75.0%	12 g
Ethane . .	15	30	30 g	80.0%	24 g
Ethyne . .	13	26	26 g	92.3%	24 g
Benzene . .	39	78	78 g	92.3%	72 g
Propane . .	22	44	44 g	81.8%	36 g
Carbon dioxide	22	44	44 g	27.3%	12 g

By his method Cannizzaro was able to find the probable relative atomic masses of such elements as hydrogen, carbon, silicon, nitrogen, phosphorus, arsenic, oxygen, sulphur, chlorine, bromine and iodine. What is more, his method indicated the number of atoms to be found in one molecule of various compounds, which is what Dalton and earlier workers had not been able to do.

In this way, for instance, Cannizzaro produced convincing evidence that hydrogen was diatomic. For he was able to show that the mass of hydrogen in 1 mol of hydrogen was twice the smallest mass of hydrogen found in 1 mol of any hydrogen compound.

Similarly he showed that steam contained 2 atoms of hydrogen and 1 atom of oxygen per molecule, so that its formula must be H_2O.

Cannizzaro devised, then, not only a method of obtaining relative atomic masses, but also a method of assigning firm chemical formulae to a large number of volatile compounds. In 1858 this was, indeed, a great step forward.

9. The Avogadro constant. The number of molecules in 1 mol of a substance or the number of atoms in 12 g of ^{12}C, is known as the Avogadro constant. It is equal to 6.02252×10^{23} mol^{-1} and can be measured in a variety of ways. Two of the simplest and most accurate methods will be considered.

a. *From measurement of the charge on the electron.* 96487 n coulomb (C) are required to deposit 1 mol of a metal from its ions, M^{n+}, during electrolysis (p. 89). This involves the discharge of L ions each with a charge of $n \times e$ coulomb, where L is the Avogadro constant and e is the charge on the electron. The value of e is 1.60210×10^{-19} C, so that

$$L = \frac{96487}{1.602 \times 10^{-19}} = 6.022 \times 10^{23}$$

b. *From measurement of the lattice spacing in a crystal.* In this method, the measured density of a crystalline substance is equated to the density of the crystal calculated from the measured lattice spacing.

The interionic distance in the crystal of potassium chloride, for instance, is found to be 3.1454×10^{-8} cm by X-ray analysis (p. 215). The volume of the small cube formed by four K^+ and four Cl^- ions will, therefore, be $(3.1454 \times 10^{-8})^3$ cm^3.

Four ion-pairs of potassium chloride will be found within this small cube, but each ion will be shared equally with seven other similar, and surrounding, cubes. The mass of the material in the small cube can, therefore, be regarded as $4m/8$, i.e. $m/2$, where m is the actual mass of an ion-pair of potassium chloride. m will be equal to M/L, where M is the relative molecular mass of potassium chloride and L is the Avogadro constant, so that the mass of the small cube will be $M/2L$ g. Its calculated density will be $M/2L$. $(3.1454 \times 10^{-8})^3$ g cm^{-3}.

As the measured density of the crystal is 1.9893 g cm^{-3},

$$\frac{M}{2L\,(3.1454 \times 10^{-8})^3} = 1.9893$$

which leads to a value for L of 6.024×10^{23}.

10. Summary of uses of Avogadro's hypothesis. The main uses of Avogadro's hypothesis may be summarised as follows:

a. It introduces the idea of a molecule of an element or, in other words, the idea of the atomicity of an element.

b. It accounts for Gay-Lussac's law of combining volumes.

c. It leads to the important relationship, $M_r = 2 \times d$. This provides one of the main methods of obtaining relative molecular mass values from measurements of relative density (p. 57).

d. It leads to the ideas of the mole and molar volume, and the very important results that 1 mol of any substance contains the same number of molecules and 1 mol of a gas occupies 22.4 dm^3 at s.t.p.

e. It leads to Cannizzaro's method of determining relative atomic masses and molecular formulae.

f. It enables the numbers of molecules of gases participating in a reaction to be related directly to the gas volumes, and vice versa. If one molecule of nitrogen and three molecules of hydrogen react, for example, to form two molecules of ammonia, Avogadro's hypothesis leads to the result that one volume of nitrogen will react with three volumes of hydrogen to form two volumes of ammonia, so long as the volumes are measured under the same conditions of temperature and pressure. Being able to interchange mols for vols, so far as gases are concerned, is very useful.

g. The Avogadro constant enables the number of entities (molecules, ions, atoms, etc.) in any amount of a substance to be obtained. The number of molecules in x gramme of a substance of relative molecular mass, M_r, will be $6.02252 \times 10^{23} \times x/M_r$.

QUESTIONS ON CHAPTER 5

1. If you were provided with supplies of hydrogen and of chlorine explain, in detail, what you would do to illustrate the truth of Gay-Lussac's law.

2. Explain fully the meaning of the following terms: atom, molecule, atomicity. gramme-equivalent, molecular formula.

3. Why do you think that (a) hydrogen, (b) nitrogen and (c) oxygen are di-

atomic? What is the evidence for the monatomicity of neon? How would you attempt to discover the atomicity of mercury in the vapour state?

4. Write short biographical sketches on (a) Avogadro, (b) Dalton, (c) Gay-Lussac and (d) Cannizzaro.

5. Explain fully why Avogadro's hypothesis was first put forward and why it eventually became so important.

6. Show how Avogadro's hypothesis can be deduced from the basic assumptions of the kinetic theory.

7. What evidence is there to support the essential validity of Avogadro's hypothesis?

8. Chlorine is said to be a diatomic gas. Is this true at all temperatures? Is I_2 a reasonable formula for iodine over a wide range of temperature?

9. A mixture of 10 cm³ of a gaseous hydrocarbon and 100 cm³ of oxygen (excess) was exploded. The volume after explosion was 75 cm³, and this was reduced to 35 cm³ on treatment with potassium hydroxide solution. Deduce the molecular formula of the hydrocarbon and give its possible structural formulae. Explain the principles on which your calculation is based.

10. The densities of three gaseous compounds of an element Y are respectively 1.25, 1.875 and 2.50 g dm⁻³. Each compound contains six-sevenths of its weight of Y. Calculate the probable relative atomic mass of Y.

11. The measured vapour densities of water, dinitrogen oxide, nitrogen dioxide, ethanoic acid, carbon dioxide, carbon monoxide, and ozone are 9, 22, 15, 30, 22, 14 and 24 respectively, and the percentage by mass of oxygen in these substances are 88.1, 36.4, 53.33, 53.33, 72.75, 57.14 and 100 respectively. What is the probable approximate relative atomic mass of oxygen? What further measurements would be necessary to find the accurate relative atomic mass?

12. Deduce the probable relative atomic mass of carbon from the following data:

Compound	Relative density	% by mass of C
Carbon monoxide . .	14	42.9
Carbon dioxide . .	22	27.3
Ethane	15	80.0
Ethene	14	85.7
Benzene	39	92.3
Toluene	46	91.3
Propane	22	81.8

13. How many gas molecules will there be in an X-ray tube of volume 2 dm³ if the temperature is 17°C and the pressure inside the tube is $1.333\,22 \times 10^{-3}$ N m⁻² (10^{-5} mm of mercury)?

14. A vacuum tube at 27°C was evacuated until it contained only 4×10^{15} molecules. At this stage the pressure was 1.6×10^{-3} N m⁻² (1.2×10^{-5} mm of mercury). What was the volume of the tube?

15. How many atoms are there in one-tenth of a gramme-equivalent of (a) aluminium, (b) carbon, (c) oxygen, (d) hydrogen, (e) magnesium, (f) helium?

16. To what extent do you believe in the existence of molecules? Summarise the evidence on which your views are based.

17. What type of particle, and what number of particles, would be found in (a) 1 mol of methane, (b) half a mole of iodine at a temperature high enough for dissociation to be complete, (c) 1 mol of fully ionised barium chloride, (d) 1 mol of fully dissociated phosphorus pentachloride?

18. Explain why it is necessary to know the equivalent mass of the element concerned when using Cannizzaro's method of relative atomic mass determination to obtain really accurate values.

19. Two flasks of equal volume are connected by a narrow tube of negligible volume. Initially each flask is at 27°C and the system contains 0.10 mol of helium. The pressure is 202.65 kN m^{-2} One of the flasks is then put in a thermostat at 227°C whilst the other is kept at 27°C. Calculate the overall final pressure and the number of moles of helium in each flask.

If the system contained 0.10 mol of hydrogen iodide (gas) at 27°C and 202.65 kN m^{-2} pressure, and the same procedure was followed, would you expect the same result?

20. Explain what is meant by the terms density and relative density, and describe briefly how you would measure the relative density of nitrogen and of propan-2-one (acetone).

Discuss briefly why the measurement of relative density is important in chemistry.

What conclusions do you draw from the observation that the relative density of sulphuric(vi) acid is 24.5 at 450°C?

21. Discuss the various methods that have been adopted to find the numerical value of the Avogadro constant.

22. The Avogadro constant should not be referred to as Avogadro's number because it is not a number. Comment on this statement.

23. Write short biographical sketches of Dalton, Gay-Lussac and Avogadro.

24. What will be the mass of a mixture containing 0.67 mol of hydrogen and 0.33 mol of oxygen?

Chapter 6
Measurement of relative molecular mass

1. Summary of methods. The method adopted for measuring the relative molecular mass of either an element or a compound depends, mainly, on whether the substance is a gas, a volatile liquid or solid, or a non-volatile solid which is soluble in a suitable solvent.

a. *For gases.* The relative molecular mass of a gas is generally obtained by measuring its relative density and then multiplying the value by 2. The relative density may be obtained by direct measurement (p. 40), or by measuring the rate of diffusion of the gas as compared with that of a gas of known density (p. 40). For accurate results the method of limiting densities (p. 56) must be applied.

For gaseous hydrocarbons, the relative molecular mass can be found by measuring the volume changes when the hydrocarbon is sparked with excess of oxygen and the resultant carbon dioxide is absorbed in alkali (p. 57).

b. *For volatile liquids or solids.* Here the relative molecular mass is obtained by measuring the relative density of the vapour obtained from the volatile liquid or solid, and then using the relationship $M_r = 2 \times d$. There are three well known experimental methods; (i) Victor Meyer's, (ii) Dumas's, (iii) Hofmann's.

Steam distillation (p.252) can also be used as a method of relative molecular mass measurement.

c. *For non-volatile solids in solution.* The methods, here, depend on the measurement of vapour pressure, osmotic pressure, freezing point lowering or boiling point elevation. They are described on pp. 265, 268, 273 and 288.

RELATIVE MOLECULAR MASSES OF GASES

2. From density measurements. The relative density or the density of a gas can be measured either by Regnault's direct method (p. 40) or by comparing the rate of diffusion of the gas against that of a gas of known density under the same conditions and applying Graham's law (p. 38).

If the relative density of a gas is measured, then its relative molecular mass value is obtained by doubling the relative density figure. If the density of a gas is measured the relative molecular mass is obtained by calculating the mass of 22.4 dm³ of the gas.

Such relative molecular mass values are quite accurate enough for many purposes, but they are not absolutely accurate as the relationships used in their calculation are only really true for ideal or perfect gases. To obtain a more accurate relative molecular mass value for a gas, such as may be required in calculating accurate relative atomic masses, it is necessary to use the method of limiting densities.

3. Method of limiting densities. The gas equation, (p. 37),

$$pV = nRT$$

applies for n mol of a gas. The number of moles, n, is equal to W/M_r, where W is the weight of the gas and M is its relative molecular mass. Therefore,

$$pV = \frac{WRT}{M_r} \quad \text{or} \quad M_r = \frac{WRT}{pV}$$

Such equations hold accurately for an ideal or perfect gas, but pV varies with pressure for real gases so that the relative molecular mass M_r, of a real gas also varies with pressure.

The deviation from the gas laws gets less and less as the pressure is lowered, and, at zero pressure, all gases obey the gas laws (p. 72). To obtain really accurate relative molecular mass values for a gas it is, therefore, necessary to try to estimate the experimental results which would be obtained if it was practicable to make measurements on a gas at zero pressure. This can only be done by a graphical method involving extrapolation down to zero pressure.

By measuring values of pV for a gas at different pressures, all at 0°C, and plotting the pV values against p, the value at zero pressure, i.e. $p_0 V_0$, can be found by extrapolation. At a pressure of 101.325 kN m^{-2} (1 atmosphere), i.e. at s.t.p., the corresponding value is $p_1 V_1$, and the ratio $p_1 V_1 / p_0 V_0$ is used in finding accurate relative molecular masses as follows:

$$M_r \text{ of gas at } p_1 \, (M_1) = \frac{WRT}{p_1 V_1}$$

$$\text{Accurate } M_r \text{ of gas at zero pressure} \, (M_0) = \frac{WRT}{p_0 V_0}$$

$$\therefore \frac{M_1}{M_0} = \frac{p_0 V_0}{p_1 V_1} \quad \text{or} \quad M_0 = M_1 \times \frac{p_1 V_1}{p_0 V_0}$$

For ammonia, the measured density at s.t.p. is 0.771 69 g dm^{-3}, which, taking the molar volume as 22.414 dm^3, gives a molecular weight of 17.297.

The $p_1 V_1 / p_0 V_0$ ratio obtained from $pV - p$ graph is 759 90/771 69, so that the accurate relative molecular mass of ammonia is given by 17.297 × 759 90/771 69, i.e. 17.032. The difference between this value and that obtained at s.t.p. is about 2 per cent.

The density of a gas in g dm^{-3} at s.t.p. is given by $M_1/22.414$ and the value of $M_0/22.414$, which is known as the *limiting density* of the gas, is related to it as follows:

$$\text{Limiting density} = \text{Density at s.t.p.} \times \frac{p_1 V_1}{p_0 V_0}$$

The limiting density of ammonia is, therefore, 0.759 90 g dm^{-3}.

4. Relative molecular masses of gaseous hydrocarbons. In this method, a known volume of a gaseous hydrocarbon is mixed with excess oxygen in a eudiometer, and the mixture is exploded by sparking. The products are carbon dioxide, steam and excess oxygen, but, on cooling, the steam condenses and the volume of water formed may be neglected as compared with the gas volumes. The decrease in volume on sparking, after the condensation of the steam, is measured.

Addition of potassium hydroxide solution absorbs all the carbon dioxide formed so that its volume is also measured. All volumes must be measured at, or converted to, the same temperature and pressure.

The reaction between any hydrocarbon and oxygen can be represented by the equation

$$C_xH_y \quad + \quad (x + y/4)O_2 \quad = \quad xCO_2 \quad + \quad y/2H_2O$$

| 1 vol | $(x + y/4)$ vol | x vol | $y/2$ vol of steam |

(Negligible vol of water)

the volumes of gases taking part being as shown.

If the original volume of hydrocarbon is taken as 1 volume, then:

a. Decrease in volume on sparking, after condensation of steam
$$= (1 + x + y/4) - x$$
$$= (1 + y/4)$$

b. Volume of carbon dioxide, i.e. decrease in
volume on adding potassium hydroxide $\quad = x$

The value of y can be obtained from the measured values of a. and the value of x from b.

Once the values of x and y are found, the molecular weight of the hydrocarbon is equal to $(12x + y)$.

RELATIVE MOLECULAR MASSES OF VOLATILE LIQUIDS OR SOLIDS

5. Victor Meyer's method. In this method a known mass of a volatile liquid is quickly vaporised by raising its temperature well above its boiling point, and the volume occupied by the resulting vapour is obtained by measuring the volume of air it displaces.

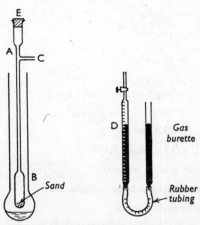

The apparatus used is shown in Fig. 8. For normal use the inner tube, A, is made of glass, and the outer jacket, B, of glass or copper. A liquid, whose boiling point is considerably higher than that of the substance whose relative density is to be measured, is placed in B, and gently boiled to provide a jacket of hot vapour round the lower half of tube A. Meanwhile the mass of a small

Fig. 8. Victor Meyer's apparatus for measuring vapour density.

bottle with a ground-glass stopper, known as a Victor Meyer bottle, is measured, first empty and then about three-quarters full of the substance whose relative density is to be measured.

When the jacket of hot vapour around tube A has built up, the delivery tube, C, is connected to a gas burette. If the liquid level, D, does not remain steady it means that thermal equilibrium has not been established, i.e. that air in tube A is still expanding or contracting. It is necessary to wait until the level D does remain steady and, when this is so, the two levels in the two limbs of the gas burette are equalised (to ensure that the pressure inside tube A is equal to atmospheric pressure) and the level of D is recorded.

The rubber stopper, E, is now removed, the Victor Meyer bottle, of known mass, is dropped into tube A, and the stopper is quickly replaced. Sand or asbestos at the bottom of tube A prevents breakage. At the higher temperature at the bottom of tube A, the volatile liquid rapidly vaporises, blows the stopper out of the Victor Meyer bottle, and pushes air from tube A into the gas burette. The level, D, goes down and its lowest level is recorded, again with the two levels in the limbs of the gas burette equalised.

The atmospheric pressure and room temperature must be recorded, and, if water is used in the gas burette, the pressure of saturated water vapour at room temperature must be obtained from tables.

After use, and whilst still hot, the vapour remaining in tube A must be blown out by inserting a long tube so that the apparatus will be ready for use again.

Typical results, using propan-2-one (acetone) (b.p. = 56°C) as the volatile liquid and water (b.p. = 100°C) in tube B, are as follows.

Mass of propan-2-one	= 0.124 g
Atmospheric pressure	= 102.2 kN m^{-2}
Room temperature	= 13°C
Volume of moist air displaced at 13°C and 766 mm.	= 50 cm^3
Saturated water vapour pressure at 13°C	= 1.5 kN m^{-2}
∴ Pressure of dry air displaced	= 102.2 − 1.5 = 100.7 kN m^{-2}
∴ Volume of dry air displaced at s.t.p.	= $50 \times \dfrac{100.7}{101.3} \times \dfrac{273}{286}$ cm^3
	= 47.4 cm^3

∴ 47.4 cm^3 of propan-2-one vapour at s.t.p. weigh 0.124 g

∴ 22400 cm^3 of propan-2-one vapour at s.t.p. weigh $0.124 \times \dfrac{22400}{47.4}$ g

∴ **relative molecular mass of propan-2-one = 58.58**

The following more detailed points are of interest:

i. A rapid vaporisation of the volatile liquid in tube A is essential. If vaporisation is slow, some vapour may diffuse upwards into the colder part of A and condense. This would cause the displacement of too small a volume of air. To ensure rapid vaporisation it is desirable to have a liquid in tube B whose boiling point is about 50 K above that of the volatile liquid in A.

ii. The air displaced from the bottom of tube A by the vapour is hot air, but its temperature need not be known for the air collected in the gas burette is measured at room temperature and atmospheric pressure. All that is required is that it should be measured at some definite temperature and pressure so that its volume at s.t.p., which is what is required, can be calculated. It is not necessary, therefore, to know the temperature of the vapour in tube B.

iii. Correction for water vapour pressure can be avoided if mercury is used in the gas burette.

Victor Meyer's method is the most important one for measuring relative densities as it is simple to carry out and very adaptable. Only small quantities of volatile liquid are needed, and the method can be used over a wide range of temperature. Measurements can be made at low temperatures by using liquefied gases in tube B, and high temperature readings can be taken, up to about 2000°C, using inner tubes of porcelain or iridium and electrical heating.

6. Dumas's method. In this method, a large glass bulb is filled with vapour and both the mass and volume of the vapour are measured.

A bulb of the shape shown in Fig. 9 is used and its mass when full of air is first measured. It is then warmed slightly and its mouth is placed under the volatile liquid whose relative density is to be measured. On cooling, a few cubic centimetres of the liquid are drawn into the bulb. The bulb is then clamped in a heating bath containing a liquid whose boiling point is well above that of the volatile liquid in the bulb. The heating liquid is boiled, and when all the volatile liquid has vaporised the bulb is quickly sealed at the tip using a blow-pipe flame.

FIG. 9. A Dumas bulb.

The sealed bulb is now full of vapour at the temperature of the boiling liquid in the heating bath and at atmospheric pressure. Both these values must be measured, along with the room temperature.

When the bulb is removed from the heating bath, the vapour inside it condenses, and the total mass is measured giving the mass of the bulb full of vapour. The tip of the bulb is now broken off under water, whereupon the bulb fills with water. Measuring the mass of the bulb full of water gives the mass of water which it contains and hence the internal volume can be found.

The mass of air in the supposedly empty bulb must be taken into account in relation to the mass of vapour, but is negligible as compared with the mass of water.

Typical results using ethoxyethane (ether), and a water bath, are as follows:

Mass of bulb full of air	$= 18.320$ g
Mass of bulb full of ethoxyethane vapour	$= 18.432$ g
Temperature of heating bath	$= 100°C$
Room temperature	$= 13°C$
Atmospheric pressure	$= 100.8$ kN m^{-2}
Mass of bulb full of water	$= 112.4$ g
$\therefore$ Mass of water	$= 94.08$ g
$\therefore$ Internal volume of bulb	$= 94.08$ cm^3
$\therefore$ Volume of air in bulb at s.t.p.	$= 94.08 \times \dfrac{273}{286} \times \dfrac{100.8}{101.3}$ cm$^3 = 89.32$ cm^3
Density of air at s.t.p.	$= 0.0013$ g cm^{-3}
$\therefore$ Mass of air in bulb	$= 89.32 \times 0.0013$ g $= 0{\cdot}1161$ g
$\therefore$ Mass of evacuated bulb	$= 18.320 - 0.1161$ g $= 18.2039$ g
$\therefore$ Mass of ethoxyethane vapour in bulb	$= 18.432 - 18.2039$ g $= 0.2281$ g
Volume of ethoxyethane vapour at s.t.p.	$= 94.08 \times \dfrac{273}{373} \times \dfrac{100.8}{101.3}$ cm$^3 = 68.50$ cm^3
Relative molecular mass of ethoxyethane	$= \dfrac{0.2281 \times 22400}{68.50}$
	$= \mathbf{74.6}$

Dumas's method is, essentially, a gravimetric method, and as such, is capable of good accuracy. It requires, however, more of the volatile liquid than Victor Meyer's method, and its use at high temperature, using porcelain bulbs, is not very convenient.

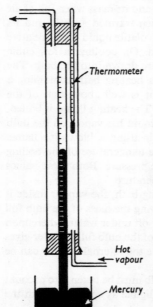

FIG. 10. Hofmann's apparatus for measuring vapour density.

7. Hofmann's method. This method is useful when the substance used decomposes at its normal boiling point because vaporisation is brought about under reduced pressure and, therefore, at a lower temperature.

A barometer tube, filled with mercury and graduated in cm^3, is surrounded by a jacket of hot vapour from a boiling liquid (Fig. 10). A known mass of volatile liquid in a Victor Meyer bottle is inserted under the bottom of the barometer tube so that it rises up into the space above the mercury. Here the liquid vaporises, and the volume of vapour produced is recorded from the level of the mercury. This is the volume of vapour at the temperature of the external jacket, which must be measured, and at a pressure given by subtracting the final level of the mercury column from its initial level. The corresponding volume at s.t.p. can readily be calculated, and, as the mass of the vapour is known, the relative density can easily be calculated.

The method can give accurate results if all the

factors, such as the expansion of the tube and the mercury at higher temperatures, are taken into account, but it is cumbersome, not easy to read, and not readily adaptable for use at high temperatures.

8. Effect of dissociation on relative density values. When a solid, liquid or gas splits up, reversibly, on heating into other molecules or atoms it is said to undergo *thermal dissociation*. Examples of substances which dissociate in this way are provided by phosphorus pentachloride, dinitrogen tetroxide and iodine, the nature of the dissociation being shown by the equations,

$$PCl_5(g) \rightleftharpoons PCl_3(g) + Cl_2(g) \qquad N_2O_4(g) \rightleftharpoons 2NO_2(g) \qquad I_2(g) \rightleftharpoons 2I(g)$$

The extent of the dissociation is measured by the *degree of dissociation*, usually represented by the symbol α, which *is the fraction, or percentage, of the original undissociated molecules which have dissociated*. The degree of dissociation is constant at any one temperature and pressure, for an equilibrium mixture is set up, but the degree of dissociation may change with temperature and pressure. When the dissociation is complete, α is equal to 1 or 100 per cent.

When dissociation leads to a change in the number of molecules present, it affects relative density measurements. The measured relative density of ammonium chloride, for example, is always found to be half the value which would be expected if ammonium chloride existed as NH_4Cl molecules in the vapour state. This is because ammonium chloride is completely dissociated, in the vapour state, into ammonia and hydrogen chloride,

$$NH_4Cl(g) \longrightarrow NH_3(g) + HCl(g)$$

This complete dissociation doubles the number of molecules as compared with the number which would be present if there was no dissociation. The vapour, therefore, occupies twice the volume (there are twice the number of molecules), but, as dissociation does not cause any change in mass, the resulting relative density is halved.

If the degree of dissociation is below 1 or 100 per cent, the measured relative density as compared with that calculated on the basis of no dissociation can readily be calculated. If the degree of dissociation of phosphorus pentachloride, for example, is α, at a certain temperature, and if 1 mol of phosphorus pentachloride be considered, the state of affairs as a result of dissociation can be summarised in the equation,

$$PCl_5(s) \rightleftharpoons PCl_3(g) + Cl_2(g)$$
$$(1 - \alpha) \text{ mol} \qquad \alpha \text{ mol} \qquad \alpha \text{ mol}$$

This means that the 1 mol which would have been present if no dissociation had taken place is replaced by $(1 + \alpha)$ mol. This causes an increase in volume, and it follows that

$$\frac{\text{Volume as a result of dissociation}}{\text{Volume if no dissociation}} = \frac{1 + \alpha}{1}$$

and, as increasing volume leads to lowering of relative density,

$$\frac{\text{Relative density as a result of dissociation}}{\text{Relative density if no dissociation}} = \frac{1}{1 + \alpha}$$

In numerical terms, the relative density of phosphorus pentachloride, if it did not dissociate, would be half its relative molecular mass, i.e. 208.5/2 or 104.25. The measured relative density of phosphorus pentachloride at 200°C is, in fact, found to be 70. It follows that

$$\frac{70}{104.25} = \frac{1}{1 + \alpha}$$

so that α is equal to 0.489 or 48.9 per cent. Different values would be found at different temperatures.

Notice that the effect of thermal dissociation on relative density measurements only becomes apparent when the dissociation produces a change in the number of molecules. Hydrogen iodide dissociates,

$$2HI(g) \rightleftharpoons H_2(g) + I_2(g)$$

but, as there is no change in the number of molecules, the measurement of relative density is not affected.

Notice, too, that the formula derived above applies only to dissociation in which 1 molecule splits up into 2. In the more general case of 1 molecule splitting up into n molecules, i.e.

$$A_n \rightleftharpoons nA$$
$$(1 - \alpha) \text{ mol} \qquad n\alpha \text{ mol}$$

it follows that,

$$\frac{\text{Relative density as a result of dissociation}}{\text{Relative density if no dissociation}} = \frac{1}{1 - \alpha + n\alpha}$$

9. Effect of association on relative density values. *Association is the joining together of two or more like molecules to form a single molecule of higher relative molecular mass.*

Association in the vapour state causes a decrease in the number of molecules and a corresponding increase in the measured relative density. The relative density of ethanoic (acetic) acid, $CH_3 . COOH$, for instance, at certain temperatures, has a measured value of 60, giving a relative molecular mass for ethanoic acid of 120. The expected relative molecular mass of ethanoic acid on the basis of the formula, $CH_3 . COOH$, would be 60. The measured relative molecular mass of 120 comes about because ethanoic acid is fully associated into double molecules,

$$2CH_3 . COOH(g) \longrightarrow (CH_3 . COOH)_2(g)$$

in the vapour state at certain temperatures.

Other examples of substances giving abnormal relative density results because of association are iron(III) chloride ($FeCl_3$), aluminium chloride ($AlCl_3$), and the oxides of phosphorus (P_2O_3 and P_2O_5). In each case, relative density measurements show that the molecules actually existing in the vapour state have twice the relative molecular mass of the simple molecules given in brackets. Iron(III) chloride, for instance, exists as Fe_2Cl_6, and so on.

QUESTIONS ON CHAPTER 6

1. For oxygen, the relative atomic mass is twice the equivalent mass, and the relative molecular mass is four times the equivalent mass. For hydrogen, the equivalent mass and the relative atomic mass are equal, but the relative molecular mass is twice this value. For neon, the relative atomic and molecular masses are alike, and the equivalent mass has no significance. Account for these facts.

2. State Avogadro's hypothesis. Outline the experimental method and theoretical deductions which led to the conclusion that the hydrogen molecule contains two, and only two, atoms. Explain how it is possible to use this conclusion to find the relative atomic mass of an element which forms volatile compounds.

196.5 cm^3 of the vapour of a volatile metal, measured at 600°C and 101.325 $kN\ m^{-2}$ pressure, weighed 0.55 g. Calculate the relative molecular mass of the vapour of the metal. (N.)

3. Describe how you would measure the relative molecular mass of ethoxyethane (ether) by Victor Meyer's method, paying due attention to the necessary experimental precautions. Using this method, a liquid whose normal boiling-point was 56°C was volatilised in a jacket at 100°C. 0.200 g of the liquid caused the displacement of 85.0 cm^3 of air, measured at 17°C over water at a pressure of 10^5 N m^{-2}. Calculate the relative molecular mass of the substance. (Vapour pressure of water at 17°C $= 2 \times 10^3$ N m^{-2}.) (N.)

4. Calculate the relative molecular mass of chloroform from the following data obtained by Victor Meyer's method. 0.220 g of trichloromethane (chloroform) displaced 45 cm^3 of air measured at 20°C and 100.7 $kN\ m^{-2}$ pressure. The saturated vapour pressure of water at 20°C $= 2.319 \times 10^3$ N m^{-2}.

5. The rate of diffusion of a volatile metallic fluoride containing 32.39 per cent of fluorine is 13.27 times as slow as that of hydrogen. Establish the formula of this fluoride and the relative atomic mass of the metal. (W.)

6. Explain the reasoning which leads to the conclusion that the relative molecular mass of a gas is twice its relative density. The chloride of a certain element has a relative density of 69 and contains 77.5 per cent of chlorine. Comment on these figures. (Army and Navy.)

7. State the laws governing the behaviour of a perfect gas and explain how the changes concerned are summed up in the expression $pv = RT$, where p is the external pressure, v is the volume of the gas, R is constant and T is the Absolute temperature.

A Victor Meyer determination showed that the volume of a vapour measured at s.t.p. corresponding to a weight of 0.3050 g of a substance was 56.9 cm^3. The percentage composition of the substance as determined by analysis was C = 10.04 per cent, H = 0.84 per cent and Cl = 89.12 per cent. What is the formula of the substance?

8. Define relative atomic mass and relative molecular mass. Quote two pieces of evidence that the atomicity of the hydrogen molecule is two and give one important application of this fact.

When 0.1963 g of an organic acid chloride is added to water the products require for complete neutralisation 50.0 cm³ 0.1M sodium hydroxide. The same mass of the compound is found to occupy approximately 74 cm³ when in the vapour state at 100°C and 100 kN m⁻² pressure. Calculate the exact relative molecular mass of the compound and suggest its structural formula. (N.)

9. Deduce the relation between the relative density of a gas and its relative molecular mass. Indicate the assumptions on which this relation is based. What is the significance of the ratio of relative molecular mass to the density of any gas?

100 cm³ of iodine vapour at 600°C took 125 s to diffuse through a porous plug whereas an equal volume of oxygen under similar conditions took 50 s. Calculate the degree of dissociation of the iodine vapour at 600°C. (W.)

10. Iodine is ten per cent dissociated at 900°C at atmospheric pressure. If 0.254 g of iodine were dropped into a bulb of a Victor Meyer apparatus at 900°C, what volume of air would be collected at s.t.p. and what would be the apparent relative molecular mass? (W.)

11. Explain what is meant by the molar volume of a substance. At 50°C and 98.65 kN m⁻² pressure 1 dm³ of partially dissociated dinitrogen tetroxide (N_2O_4) was found to have a mass of 2.5 g. What is the degree of dissociation of the compound under these conditions. (Army and Navy.)

12. The apparent relative molecular mass of iodine at 55°C was found to be 165; calculate the degree of dissociation of iodine into atoms at this temperature.

13. 2.25 g of phosphorus pentachloride vaporised completely at 200°C to occupy a volume of 618 cm³ at one atmosphere pressure. Calculate the relative density, and from this the degree of dissociation of phosphorus pentachloride under these conditions.

14. Describe one method of determining the relative molecular mass of a volatile substance, deriving any relationships on which the method depends. 0.1 g of ammonium carbamate ($NH_4 . COONH_2$) was completely vaporised at 100°C and 101.325 kN m⁻² pressure. The vapour occupied a volume of 117 cm³. What can you infer from this? (O. & C.)

15. The relative density of steam at 2000°C is 8.9. What is the percentage dissociation into hydrogen and oxygen?

16. What evidence, other than the abnormal vapour density results, is there for the thermal dissociation of (a) N_2O_4, (b) PCl_5 and (c) NH_4Cl?

17. Write equations representing as many examples of thermal dissociation as you can think of.

18. If a gas of density d_1 has a degree of dissociation, x, and gives, on dissociation, a gaseous mixture of density d_2, prove that $x = (d_1 - d_2)/(n - 1)d_2$, where n is the number of molecules obtained from the complete dissociation of 1 molecule of the original gas.

19. The specific heat capacity of a metal X is 0.226 J g⁻¹ K⁻¹ and it forms two oxides containing 11.88 per cent and 21.23 per cent of oxygen. It forms a volatile chloride the relative density of which is approximately 130. Calculate the relative atomic mass of X; assign formulae to the two oxides and deduce the molecular formula of the chloride, stating the principles on which your calculations are based.

Give one example (with supporting reasons) of an element for which the Cannizzaro method of relative atomic mass determination would be appropriate. (W.)

20. The relative density of nitrogen dioxide at 60°C and 101.325 kN m⁻² is 28.3. Find the percentage dissociation under these conditions, and calculate the dissociation constant, K_p, in terms of the partial pressures of N_2O_4 and NO_2. At what pressure would the relative density be 32.5 at this temperature? (B.S.)

21. A sample of pure ethanoic (acetic) acid weighing 0.410 g is heated in a

glass bulb, volume 448 cm³, provided with a manometer. At 127°C the pressure in the bulb was 25.33 kN m⁻² and the acid had all vaporised, no other gas being present. The temperature was then raised to 227°C when the pressure rose to 59.46 kN m⁻². Calculate the relative molecular mass of ethanoic acid under these two conditions. If you think that a chemical reaction occurred on heating from 127°C to 227°C write an equation for it. The ethanoic acid may be recovered unchanged at the end of the experiment. (B.)

22. When 0.1 g of a pink crystalline metallic chloride was vaporized at 1200°C, 77.5 cm³ of vapour was produced, this volume being measured at the experimental temperature and pressure.

The salt was only very sparingly soluble in water yielding, in the cold, a violet solution. A large volume of this solution which contained 0.0053 g of salt was treated with silver nitrate(v) solution, the precipitate obtained having a mass of 0.0143 after drying. Calculate the relative atomic mass of the metal. Assuming that the experimental data are accurate only to ±7 per cent, attempt to identify the metal with the aid of the information provided (Cl = 35.5 Ag = 108 Cr = 52 Mn = 55 Fe = 56 Co = 59 Ni = 59). (C. Schol.)

23. What do you consider to be the relative advantages and disadvantages of Victor Meyer's, Dumas's and Hofmann's method of measuring relative density?

Chapter 7
Kinetic theory

THE kinetic theory was developed, between 1860 and 1890, mainly by Clausius, Clerk Maxwell and Boltzmann. The theory is mainly useful in accounting for the known properties of gases, but it also clarifies many problems concerned with liquids and solids.

1. Outline of the theory. The ideas of the kinetic theory were not put forward by one individual, or at one time, but the ideas underlying the theory, which finally accumulated, may be summarised as follows:

a. Matter is made up of particles. These particles may be small groups of atoms (molecules), or, in monatomic gases, single atoms.

b. The particles in a gas are in continual, rapid, random motion in straight lines in every direction. They continually collide with each other and with the walls of any containing vessel. The pressure exerted by a gas on the walls of its containing vessel is due to bombardment by the moving particles.

Though the same idea of random motion applies to the particles in a liquid the motion is greatly decreased, and it is still further decreased in a solid.

c. The particles in a gas are separated from each other by distances which are large as compared with the size of the particles. In a liquid, the particles are closer together, and they are still closer in a solid. In simple treatments the particles are regarded as points.

d. The particles are regarded as being perfectly elastic, so that the collisions they undergo in a gas do not result in any change in the total amount of kinetic energy of the gas.

The *average* kinetic energy of the individual particles remains constant, at a constant temperature, though the kinetic energy of one particular particle may vary enormously depending on the nature of the collisions it undergoes.

e. Increase in temperature causes the motion of the particles to increase, the average kinetic energy of the particles in a gas being proportional to the absolute temperature of the gas.

2. The fundamental kinetic equation. The quantitative applications of the kinetic theory are based on the fundamental equation,

$$pV = \tfrac{1}{3}mnu^2$$

where p is the pressure exerted by the gas, V is its volume, m is the mass of one molecule, n is the total number of molecules of gas present, and u is the root mean square velocity.

The root mean square is a rather unusual way of expressing an average value. The average of such quantities as 1, 2, 3, 4 and 6 would, most commonly, be taken as the sum of the quantities divided by five, i.e. 16/5 or 3.2. This is correctly described as the *arithmetic mean* of the quantities.

The root mean square of the quantities is given by

$$\sqrt{\frac{1^2 + 2^2 + 3^2 + 4^2 + 6^2}{5}} = \sqrt{\frac{66}{5}} = \sqrt{13.2} = 3.633$$

The root mean square velocity of gas molecules, u, is, therefore, obtained from the expression,.

$$u = \sqrt{\frac{u_1^2 + u_2^2 + u_3^2 + u_4^2 \ldots + u_n^2}{n}}$$

where n is the total number of molecules concerned, and $u_1, u_2, u_3, u_4 \ldots u_n$ are their individual velocities.

The root mean square of the velocities of the gas molecules is used in the fundamental equation so that the expression $\frac{1}{2}mu^2$ will accurately give the average kinetic energy per molecule.

3. Derivation of $pV = \frac{1}{3}mnu^2$. The equation is derived by considering n molecules, each of mass m g contained in a cube of side l cm.

The velocity of any one single molecule, u_1, can be resolved into components, x, y and z acting in directions parallel to the three sides of the cube. If this is done, then,

$$u_1^2 = x^2 + y^2 + z^2$$

The pressure on one face of the cube is due to all the components in the same direction as x; that on the other faces is due to the components in the same directions as y and z.

In the direction of x, the molecule with total velocity u_1 travels x cm in 1 s. It can, however, only travel l cm before colliding with a wall, and for each l cm it travels it will undergo one collision with a wall. In one second, therefore, it will undergo x/l collisions.

Before collision, the momentum of the single particle, in the direction of x, will be mx. After collision, which is perfectly elastic, the momentum will have the same value, but opposite sign. The change of momentum for every single collision is, therefore, $mx - (-mx)$, i.e. $2mx$.

As there are x/l collisions per second, the change of momentum per second, in the direction of x, will be

$$2mx \times \frac{x}{l} = \frac{2mx^2}{l}$$

In the direction of y, the change of momentum will, similarly, be $\frac{2my^2}{l}$; and in the direction of z, it will be $\frac{2mz^2}{l}$.

The total change of momentum per second, caused by one single molecule, will, therefore, be

$$\frac{2mx^2}{l} + \frac{2my^2}{l} + \frac{2mz^2}{l}$$

i.e. $\frac{2m}{l}(x^2 + y^2 + z^2)$ or $\frac{2mu^2}{l}$

For n molecules the total change of momentum per second will be

$$\frac{2m}{l}(u_1^2 + u_2^2 + u_3^2 \ldots + u_n^2) \quad \text{or} \quad \frac{2mnu^2}{l}$$

Now, by Newton's law, the force exerted on a surface by bombarding particles is equal to the rate of change of momentum of the particles. The force exerted on the faces of the cube by molecular bombardment is, therefore,

$$\frac{2mnu^2}{l}$$

Because pressure is force per unit area, and because the area of the six faces of the cube is $6l^2$, the pressure exerted by the gas will be

$$\frac{2mnu^2}{6l^3} \quad \text{i.e.} \quad \frac{mnu^2}{3l^3}$$

But l^3 is the volume of the cube, i.e. the volume of gas under consideration, and, therefore,

$$p = \frac{mnu^2}{3V} \quad \text{or} \quad pV = \tfrac{1}{3}mnu^2$$

If u is in ms^{-1}, m in kg and V in m^3, then p will be in newton metre^{-2}

4. Kinetic theory and Boyle's law. According to the kinetic theory, the average kinetic energy of the molecules of a gas is proportional to the Kelvin or absolute temperature. This average kinetic energy is equal to $\tfrac{1}{2}mnu^2$, and, therefore,

$$\tfrac{1}{2}mnu^2 = kT$$

For a given mass of gas at a constant temperature, $\tfrac{1}{2}mnu^2$ will be constant, and $\tfrac{1}{3}mnu^2$, i.e. pV must also be constant.

Thus Boyle's law is derived from the kinetic theory.

5. Kinetic theory and the gas equation and Charles's law. As explained in section 4,

$$\tfrac{1}{2}mnu^2 = kT$$

and, therefore,

$$\tfrac{1}{3}mnu^2 = k'T$$

But, by the fundamental kinetic equation, $\tfrac{1}{3}mnu^2 = pV$ and, therefore,

$$pV = k'T$$

which is the statement of the gas equation (p. 37).

At constant pressure,

$$V = k''T$$

which is the statement of Charles's law.

Both the gas equation and Charles's law are, in this way, derived from the kinetic theory.

As

$$\tfrac{1}{2}mnu^2 = kT$$

it follows that, at absolute zero, i.e. $T = 0$, u would also be zero, i.e. *the molecules in a gas at absolute zero would be at rest*.

The fundamental kinetic equation can be rewritten in the form,

$$pV = \tfrac{2}{3} \times \tfrac{1}{2}mnu^2$$

or, if $\tfrac{1}{2}mnu^2$ is written as E (the kinetic energy of the molecules),

$$pV = \tfrac{2}{3}E$$

For one mol of gas, however, the gas equation gives

$$pV = RT$$

and, therefore, for one mol,

$$RT = \tfrac{2}{3}E$$

R is the molar gas constant (p. 37) and it has a value of 8.3143 J K^{-1} mol^{-1}. The kinetic energy of the molecules in 1 mol of a gas is, therefore, approximately $12.5T$ joule, where T is the Kelvin or absolute temperature.

(The value of R in cal K^{-1} mol^{-1} is approximately 2 (p. 37) so that the energy of molecules in 1 mol of a gas is approximately $3T$ cal.)

6. Kinetic theory and Dalton's law of partial pressure. For a gas, $pV = \tfrac{2}{3}E$, where E is the kinetic energy of the molecules. If a mixture of gases is considered,

$$E = E_A + E_B$$

where E_A and E_B are the kinetic energies of the components, A and B, of the mixture.

Therefore, for a mixture of gases,

$$pV = \tfrac{2}{3}(E_A + E_B)$$

If the pressure exerted by A is p_A, then $p_AV = \tfrac{2}{3}E_A$ and, similarly, for B, $p_BV = \tfrac{2}{3}E_B$.

It follows, then, that

$$p = p_A + p_B$$

which is the statement of Dalton's law of partial pressure.

7. Kinetic theory and Graham's law of diffusion. From the fundamental kinetic equation it follows that

$$u^2 = \frac{3pV}{mn}$$

and as the density of a gas, ρ, is equal to $\frac{mn}{V}$, that

$$u^2 = \frac{3p}{\rho} \quad \text{or} \quad u = \sqrt{\frac{3p}{\rho}}$$

At a constant pressure, then, $u \propto \sqrt{\frac{1}{\rho}}$, and if the rate of diffusion of a gas is taken as being proportional to u, it follows that

$$\text{Rate of diffusion} \propto \sqrt{\frac{1}{\rho}}$$

which is Graham's law.

8. Kinetic theory and Avogadro's hypothesis. If equal volumes of two gases, both at the same pressure, are considered, then, for the first gas,

$$pV = \tfrac{1}{3}m_1 n_1 u_1{}^2$$

and, for the second,

$$pV = \tfrac{1}{3}m_2 n_2 u_2{}^2$$

If the gases are also at the same temperature, then

$$\tfrac{1}{2}m_1 u_1{}^2 = \tfrac{1}{2}m_2 u_2{}^2$$

It follows that n_1 must equal n_2, and this is the statement of Avogadro's hypothesis.

9. Molecular velocities. The fundamental kinetic equation enables molecular velocities to be calculated. The value obtained is the root mean square velocity, and not necessarily the velocity of any one molecule.

For hydrogen, at s.t.p., the density is equal to 0.09 kg m^{-3} and from the equation derived in section 7.

$$u = \sqrt{\frac{3p}{\rho}}$$

the root mean square velocity is equal to

$$\sqrt{\frac{3 \times 101\,325}{0.09}} \text{ i.e. } 1838 \text{ m s}^{-1}$$

This speed, about that of a rifle bullet, is over a mile per second, but hydrogen, with the lowest density, has the highest root mean square velocity of any gas. At s.t.p., the root mean square velocities in m s^{-1}, of other typical gases are,

Oxygen	460	Water vapour	610
Nitrogen	493	Carbon dioxide	390
Argon	410	Mercury vapour	180

The root mean square velocity of a gas is proportional to the square root of the Kelvin temperature. This means that the velocities at $1000°C$ are rather more than twice the velocities at $0°C$.

10. Distribution of molecular velocities. The result of the numerous molecular collisions within a gas is that the velocities of the individual molecules vary enormously. Most molecules have a velocity close to the average, but some molecules may acquire considerably higher or lower velocities as a result of a series of favourable or unfavourable collisions.

The distribution of velocities amongst the molecules of a gas was calculated, by Maxwell and Boltzmann, from the laws of probability. The distribution of kinetic energy follows the same pattern as the distribution of velocities.

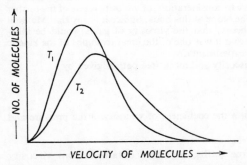

FIG. 11. The distribution of molecular velocities in a gas at a temperature, T_1, and a higher temperature, T_2.

Maxwell's distribution law may be expressed, in a simplified form, as

$$n = n_0 e^{-\frac{E}{RT}}$$

where n_0 is the total number of molecules, and n is the number having an energy greater than the value E.*

The result of such a distribution law† is conveniently shown graphically

* e, which is numerically equal to 2.718 28 . . . , is the base of Napierian logarithms, just as 10 is the base of ordinary logarithms. If $y = 10^x$, then $x = \log_{10}y$; if $y = e^x$, then $x = \log_e y$. $\log_e y$ is often written as $\ln_y$. To convert ordinary logs to Napierian logs, multiply by 2.302 6.

† That the law should take this form can be seen quite readily. If the chance of one favourable collision is $1/x$, the chance, P, of n favourable collisions in a row will be equal to $(1/x)^n$, i.e.

$$P = \left(\frac{1}{x}\right)^n$$

$$\therefore \ \ln P = n\ln 1/x \quad \text{or} \quad \ln P = yn \quad \text{or} \quad P = e^{-yn}$$

The chance of a molecule having an energy greater than a value, E, will, therefore, be proportional to e^{-zE} for E will be proportional to n.

as in Fig. 11. It will be seen that, at any temperature, most molecules have about an average energy but that some have small or high energies. At higher temperatures the number of molecules with high kinetic energy increases whilst the number with average energy decreases.

11. Mean free path. Gases diffuse at speeds very much less than those of their root mean square velocities. This is because there are so many intermolecular collisions, which cause such random motion. A swarm of bees can be in constant motion and not make much forward progress, and it is difficult to work one's way through a jostling crowd. The mean distance travelled by a molecule in a gas before it collides with another molecule is known as its mean free path.

Numerical values can be obtained from measurements of viscosity, and the application of the kinetic theory to considerations of viscosity is one of the further extensions which are beyond the scope of this book. Suffice it to say that Maxwell predicted, from the kinetic theory, that the viscosity of gases would be independent of their densities and that it was one of the great triumphs of the kinetic theory when this was verified experimentally.

The relationship between viscosity and mean free path is given by

$$l = \eta \sqrt{\frac{3}{p \times \rho}}$$

where l is the mean free path, η the coefficient of viscosity, p the pressure and ρ the density.

Some values of mean free paths, at s.t.p., obtained from viscosity measurements, are summarised below

Hydrogen	11.23	Ammonia	4.41
Oxygen	6.47	Carbon dioxide	3.97
Helium	17.98	Carbon monoxide	5.84

The values are given in micrometres. At low pressures the mean free paths rise to several centimetres.

The number of collisions undergone by one molecule in one second is obtained by dividing the mean velocity by the mean free path. Values of the order of 10^9 to 10^{10} are obtained at s.t.p. For hydrogen, at s.t.p., the value is 10.33×10^9, and even at low pressures there may be about 2×10^5 collisions per second.

12. Deviations from gas laws. A perfect gas may be defined as one which fully obeys Boyle's law, but real gases are not perfect and all the gas laws are only approximately true. All the gas laws can, however, be derived from the kinetic theory, and if the gas laws are not absolutely true then the tenets of the kinetic theory must only be approximate.

The simple kinetic theory is, in fact, an over-simplification, mainly on two counts. First, it regards gas molecules as having no size in comparison with the total gas volume, i.e. it regards molecules as pin-points. Secondly, simple kinetic theory considerations disregard cohesive forces between molecules.

Both these factors have been taken into account by many workers in efforts to modify the gas laws and produce relationships which will be absolutely accurate.

van der Waals, for example, in 1873, replaced the gas equation, $pV = RT$, by an equation,

$$\left(p + \frac{a}{V^2}\right)(V - b) = RT \text{ (for 1 mol of gas)}$$

which is known as van der Waals' equation; a and b are numerical constants for any given gas, and R is the gas constant.

Cohesive forces between molecules mean that a molecule near the surface of a gas experiences an inwards attractive force, and the force decreases the pressure compared with that which would be exerted if it did not exist. The total inward pull will be proportional to the number of surface molecules being pulled, and to the number of molecules in the interior that are doing the pulling. Both these factors are proportional to the density of the gas concerned, and, for a given mass of gas, at a fixed temperature and pressure, are inversely proportional to the volume of the gas. The inward pull may, therefore, be written as a/V^2, and this must be added to the pressure to make up for the inward forces. Hence the $\left(p + \frac{a}{V^2}\right)$ term in van der Waals' equation.

The fact that molecules have a finite size means that the volume in which they are free to move is less than the total volume which the gas occupies. Instead of V in the gas equation, then, van der Waals wrote $(V - b)$, where b represents the volume actually occupied by the molecules themselves; it is called the excluded volume.

van der Waals' equation fits the experimental facts better than the simple gas equation, but agreement with all experimental data is still not completely exact, and many other equations of state, as they are called, have been suggested. The best known, which still retain a reasonably simple form, are those of Berthelot and of Dieterici, both of which try to account for the fact that the cohesive force between molecules depends on temperature.

Berthelot's equation $\quad \left(p + \frac{A}{TV^2}\right)(V - B) = RT$

Dieterici's equation $\quad p(V - b')e^{a'/RTV} = RT$

More complicated equations give still closer agreement with experimental results, but are more difficult to handle.

The values of a and b in van der Waals' equation are constants for any one gas, but different gases have different values. The numerical values can be obtained by fitting the equation to experimental results of p, V and T, or, more commonly, from the critical constants of gases.

13. Critical constants. When a gas is cooled to a low enough temperature and then compressed it can be converted into a liquid. For each gas, however, this can only be done below a certain temperature known as the critical temperature. If a gas is above its critical temperature, no degree of compression will liquefy it.

The critical temperature is generally denoted by T_c. *The pressure which just suffices to liquefy a gas at its critical temperature is called the critical pressure, p_c. The volume occupied by* 1 *mol of gas at its critical temperature and pressure is known as the molar critical volume, V_c. The values of T_c, p_c*

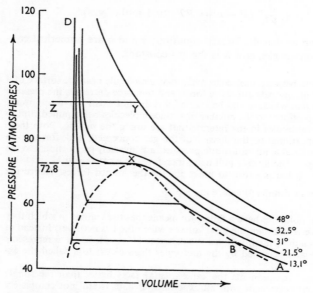

Fig. 12. pV isothermals for carbon dioxide.

and V_c, known as the critical constants, are constants for any one gas, but their values differ from gas to gas.

The behaviour of gases in the neighbourhood of their critical constants was first investigated by Andrews in 1869. He obtained p and V values for carbon dioxide at different temperatures and plotted p against V at each temperature to give a series of isothermals. The results are shown graphically in Fig. 12.

If pV was equal to a constant, i.e. if Boyle's law held, then p plotted against V should give a rectangular hyperbola. At high temperatures the experimental curve approaches a rectangular hyperbola, but, at low temperatures, the isothermals are discontinuous, splitting up into three parts.

If the curve ABCD is considered, the part AB represents the effect of increasing pressure in decreasing the volume of a gas. Between B and C, however, there is a very large volume change with no change in pressure. This represents the liquefaction of the gas. At C, liquefaction is complete, and the curve CD represents the effect of pressure on the volume of liquid produced. The steepness of CD as compared with AB shows that a liquid is much less compressible than a gas.

The ends of the horizontal portions of different isothermals form an approximate parabola when joined together, and the apex of this gives the point at which the horizontal portion of the isothermals just disappears. This represents the critical point, X.

For carbon dioxide the relevant isothermal occurs at 31°C, so that this is the critical temperature for carbon dioxide. The critical pressure is 7.372 MN m^{-2}. Carbon dioxide can, therefore, be liquefied at room temperature under a big enough pressure.

Other critical temperatures are

Helium −268°C	Sulphur dioxide 157°C	Ammonia 131°C
Hydrogen −239°C	Carbon monoxide −139°C	Oxygen −119°C
Nitrogen −147°C	Carbon dioxide 31°C	Chlorine 141°C

Gases whose critical temperatures lie below room temperature are sometimes referred to as permanent gases. This name originated a century ago when it was found impossible to liquefy these gases simply by increasing the pressure.

Above the critical temperature, isothermals do not show any discontinuity, and it is not possible to detect the precise point at which a liquid becomes a gas or vice versa. If a liquid, represented by point Z in Fig. 12, be heated at constant pressure until its temperature reaches 48°C its condition at different temperatures will be represented by the line ZY. At Z, the substance concerned is a liquid; at Y, it is a gas. The change has taken place smoothly and continuously, and this is expressed by referring to the *continuity of state*.

14. Critical constants and van der Waals' equation. If van der Waals' equation is written in the form

$$pV^3 - (RT + pb)V^2 + aV - ab = 0$$

it represents a cubic equation in V. For any given values of p and T, therefore, there will be three values of V, since R, a and b are constants for 1 mol of a given gas.

If p is related to V by an equation like van der Waals' then p plotted against V would give curves like that shown in the full lines in Fig. 13. Such a curve is very similar to the experimental p—V curves obtained by Andrews (Fig. 12) except that the horizontal portion (dotted line) is replaced by a curved portion. The full line curves in Fig. 13 can give three values of V for any one value of p, but at a temperature when the humps disappear, i.e. at the critical temperature where there is no horizontal portion corresponding to the humps, all the three values of V will be identical.

At the critical temperature, therefore, V is equal to V_c so that

$$(V - V_c)^3 = 0$$

or

$$V^3 - 3V_c V^2 + 3V_c^2 V - V_c^3 = 0$$

Van der Waals' equation, at the critical point, gives, however,

$$V^3 - \left(\frac{RT_c}{p_c} + b\right) V^2 + \frac{aV}{p_c} - \frac{ab}{p_c} = 0$$

It follows, then, that

$$3V_c = \frac{RT_c}{p_c} + b; \qquad 3V_c^2 = \frac{a}{p_c}; \qquad V_c^3 = \frac{ab}{p_c}$$

or

$$V_c = 3b; \qquad p_c = \frac{a}{27b^2}; \qquad T_c = \frac{8a}{27Rb}$$

Such relationships enable values of a and b for any gas to be obtained, or, if a and b values are known the critical constants can be calculated.

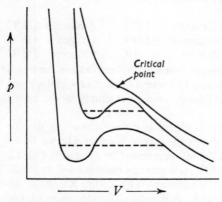

FIG. 13. The full lines represent the curves expected from a plot of p against V from van der Waals' equation.

15. Corresponding states. If the actual T, p and V values of any gas are expressed as fractions of the T_c, p_c and V_c values for the gas, the resulting fractions are known as the reduced T, p and V values respectively, and the symbols θ, π and ϕ are used. That is,

$$\frac{T}{T_c} = \theta \qquad \frac{p}{p_c} = \pi \qquad \frac{V}{V_c} = \phi$$

The use of these terms enables an equation of state to be obtained which contains no constants which only apply to one particular substance. Such an equation can be applied to any gas or liquid.

The equation is derived by substituting p, V and T in the van der Waals' equation by πp_c, ϕV_c and θT_c. This gives

$$\left(\pi p_c + \frac{a}{\phi^2 V_c^2}\right) (\phi V_c - b) = R\theta T_c$$

and, on substituting the values of V_c, p_c and T_c (obtained in section 14),

$$\left(\frac{\pi a}{27b^2} + \frac{a}{9b^2\phi^2}\right)(3b\phi - b) = \frac{8aR\theta}{27Rb}$$

or

$$\left(\pi a + \frac{3a}{\phi^2}\right)(3b\phi - b) = 8\theta ab$$

or

$$\left(\pi + \frac{3}{\phi^2}\right)(3\phi - 1) = 8\theta$$

This equation is known as the reduced equation of state. It applies to all gases and liquids, and agrees with experimental results with fair accuracy.

When substances have the same reduced p, V and T values they are said to be in corresponding states, and the fact that two substances will have the same reduced temperature, if they have the same reduced pressure and volume, is sometimes known as the law of corresponding states.

16. The specific heat capacity of gases. The specific heat capacity of a substance is the number of joules required to raise the temperature of 1 kg of the substance through 1 K; the units are, therefore, J kg^{-1} K^{-1}. Alternatively, the quantity may be expressed per gramme in J g^{-1} K^{-1} units. The molar heat capacity of a gas is the number of joules required to raise the temperature of 1 mol of the gas through 1 K.

For a gas, there is a marked difference between the value of the molar heat capacity measured at constant pressure (c_p) and that measured at constant volume (c_v). If measured at constant pressure, the gas is allowed to expand whilst being heated; if it did not expand the pressure would not remain constant. But, if measured at constant volume, no expansion is allowed and there is an increase in pressure.

When a gas expands it has to do work against the external pressure, and if expansion is allowed, i.e. if the measurement of specific heat capacity is at constant pressure, then the heat put into the gas is used, partly, to raise the temperature of the gas and, partly, to provide the energy for the external work which must be done in expansion. At constant volume, the heat put into the gas all goes into raising the temperature of the gas. For these reasons the value of c_p is greater than that of c_v.

a. *The value of c_v.* The kinetic energy associated with the molecules in 1 mol of a gas is approximately equal to 12.5 T joule, where T is the absolute temperature (p. 69). A rise in temperature of 1 K would, therefore, cause an increase in kinetic energy of 12.5$(T + 1) - 12.5T$, i.e. of 12.5 joule.

This assumes that all the heat provided goes into increasing the kinetic energy of the molecules, but this is only true for monatomic molecules and for measurement at constant volume. At constant pressure, it has already been pointed out that some of the heat supplied is used in doing external work. For polyatomic molecules, some of the heat may be used in causing, or increasing, either the rotational energy or the vibrational energy of the molecule.

The value of c_v for a monatomic gas will, therefore, be 12.5 joule, but for a polyatomic gas it will be $(12.5 + x)$ joule. The value of x will vary from gas to gas and will be larger for more complex molecules which have greater opportunities for internal rotation and vibration. For most diatomic gases, x is equal to R, or approximately 8.3, and for triatomic gases it is equal to $3R/2$, or approximately 12.5, but there is some doubt about the theoretical establishment of these values.

b. *The value of* c_p. c_p is greater than c_v, the difference being due to the work done in expanding. This work done is equal to the pressure multiplied by the increase in volume, i.e.

$$\text{Work done} = \text{Pressure} \times \text{Increase in volume}$$

At a pressure of p, the work done for a change in volume of V to V_1 is $p(V_1 - V)$ or $pV_1 - pV$.

But, for 1 mol of gas, $pV_1 = RT_1$ and $pV = RT$ so that the work done is equal to $(RT_1 - RT)$ or $R(T_1 - T)$. If the rise in temperature is 1 K the work done will, therefore, be R, or 8.314 joule.

For a monatomic gas, c_p is, therefore, approximately $12.5 + 8.3$, i.e. 20.8 J. For a polyatomic gas it is $(20.8 + x)$, with x approximately equal to 8.3 or 12.5 for diatomic and triatomic molecules respectively.

c. *The ratio* c_p/c_v *or* γ. Measurement of the ratio, c_p/c_v, can be made directly without measuring c_p and c_v individually, and its value gives very useful information about the atomicity of a gaseous element.

To sum up, the normal values, in $J\ K^{-1}\ mol^{-1}$, are

	c_p	c_v	c_p/c_v
Monatomic gas	20.8	12.5	1.67
Diatomic gas	29.1	20.8	1.40
Triatomic gas	33.3	25	1.33

c_p/c_v is most conveniently measured by Kundt's method, which depends on measuring the velocity of sound in a gas, or by Clement and Desormes method. A text book of physics must be consulted for details of these methods.

The chemical importance of c_p/c_v measurements is that they provide a method of obtaining the atomicity of a gaseous element. This enables the relative atomic mass of the element to be obtained from its relative molecular mass for

$$\text{Relative molecular mass} = \text{Relative atomic mass} \times \text{Atomicity}$$

The method is particularly useful for finding the relative atomic masses of the noble gases.

17. Kinetic theory and liquids. Most liquids, like all gases, are miscible, and liquids can, too, be compressed, though to a much smaller extent than gases and not in a regular way covered by any laws. Solutes, too, will diffuse completely throughout a solvent when a solution is made.

These facts, together with the continuous transition which can take place between a gas and a liquid (p. 75) suggest that the molecules in a liquid are in a state of random motion, as in a gas, but the motion is very much less than it is in a gas and the molecules in a liquid are very much closer together. Liquids, in fact, lie midway between the disorderly, scattered array found in a gas and the orderly, compact design found in a crystalline solid, (p. 209). On the one hand, the structure of a liquid may be regarded as being something like that of a very imperfect gas; on the other, it may be viewed as that of a very disordered crystal.

The random motion of molecules in a liquid is shown, very directly, in the phenomenon of *Brownian motion*, first observed by a botanist, Robert Brown, in 1827. He found that very small pollen grains immersed in water underwent a curiously irregular motion which could be observed under a microscope. It can be seen very conveniently by viewing a drop of Indian ink on a microscope slide. Such motion is due to molecular bombardment of the small grains by molecules of the liquid in which the grains are suspended, and this simple observation provides one of the most direct lines of evidence that the general ideas of the kinetic theory are correct.

18. Surface tension. Cohesive forces between molecules in a gas have to be taken into account in modifying the simple gas laws, but they are not very strong forces. In a liquid, however, the molecules are much closer together and the cohesive forces are much stronger.

Molecules on the surface of a liquid are subjected to an inward force, which accounts for the surface tension of liquids. This causes liquids to behave as though their surfaces consisted of an elastic skin, and one sign of this is that liquid surfaces usually assume the smallest possible area. This means that freely suspended liquids form into spherical drops, for a sphere has the minimum surface-to-volume ratio. Further details of surface tension are given on p. 490.

19. Vapour pressure. Although the inward forces on the molecules at the surface of a liquid may be quite strong, some molecules with kinetic energy higher than the average will have sufficient energy to escape from the surface. This is the phenomenon of evaporation. The escaping molecules constitute the vapour above the liquid.

If evaporation takes place into a closed space, the concentration of vapour will build up, and the molecules in the vapour will exert a vapour pressure. Molecules will escape from the liquid surface to form vapour, but some vapour molecules will also be attracted by the liquid and will condense, i.e. pass back into the liquid. Eventually, as many molecules will be leaving the liquid as will be passing back into it, and the rate of evaporation will be equal to the rate of condensation; a dynamic or kinetic equilibrium will be set up. At this stage, the vapour reaches its maximum concentration and exerts its maximum pressure. The vapour is said to be saturated, and the pressure it exerts is known as the saturated vapour pressure.

20. Vapour pressure and temperature. The saturated vapour pressure exerted by the vapour above a liquid is independent of the amount of liquid present, so long as there is enough to form a saturated vapour, but it increases with temperature. Clearly, at a higher temperature, more molecules will have sufficient energy to escape from the liquid surface. When the temperature reaches a high enough value and the saturated vapour pressure becomes equal to the external pressure, the liquid changes rapidly and completely into vapour and is said to boil.

The boiling point of a liquid is defined as the temperature at which its saturated vapour pressure becomes equal to the external pressure on the liquid. The boiling point of a liquid depends, then, on the external pressure. Water, for instance, boils at 100°C under a pressure of 101.325 kN m^{-2} i.e. at 100°C the saturated vapour pressure of water is 101.325 kN m^{-2}, but, under an external pressure of 1.6 kN m^{-2}, water will boil at 14°C. The change of vapour pressure with temperature for a typical liquid is shown graphically on p. 272.

21. Latent heat of evaporation. Escape of molecules of higher than average kinetic energy from the surface of a liquid reduces the mean kinetic energy of the liquid which, therefore, becomes cooler. Evaporation, in fact, causes cooling, and this is the basis of refrigeration.

If the temperature of a liquid is to be maintained, when it evaporates, heat must be supplied, and this is the latent heat of evaporation. This is a measure of the energy which has to be supplied to cause molecules to escape from a liquid surface, i.e. to evaporate.

22. Kinetic theory and solids. The molecular motion in a liquid is much less than it is in a gas, but there is no really orderly array of molecules in either a gas or a liquid.

When a liquid is cooled, its molecules lose energy and molecular motion decreases until a point is reached when the cohesive forces between molecules draw the molecules together into a solid structure with a definite, orderly array. This is the kinetic picture of crystallisation from a liquid.

In a solid crystal, molecular motion is limited to vibration or oscillation about a fixed position, and the molecules, atoms or ions are close together. This close-packing of the particles is shown by the very great difficulties encountered in compressing a solid.

The slight motion in a solid, at ordinary temperatures, is shown by the fact that two solids when placed in close contact may diffuse very slightly into each other, and also by the fact that solids can exert a vapour pressure. The vapour pressure of a solid, at normal temperatures, is generally so small as to be negligible, but some solids, e.g. iodine and naphthalene, exert considerable vapour pressures at temperatures below their melting points. On heating, such solids are converted directly into vapours, and, on cooling the vapour formed it condenses directly into a solid. Such a change from solid to vapour or from vapour· to solid, without an intermediate liquid stage, is known as sublimation (p. 528).

The majority of solids melt into a liquid on heating. In the process of melting, it is thought that the vibrational energy of a solid increases as it is heated and that the particles of the solid vibrate more and more until the cohesive forces can no longer hold them together; the particles break loose, and the solid melts.

In non-crystalline, or amorphous, solids, e.g. glass and pitch, there is no regular pattern of molecules, and they have no fixed melting point. They are, in fact, super-cooled liquids (p. 526). Many so-called amorphous solids, e.g. amorphous sulphur and copper(II) oxide, appear to have no crystalline structure but are, in fact, made up of random mixtures of micro-crystalline aggregates.

23. Summary of uses of the kinetic theory. Some of the more complicated applications of the kinetic theory have not been treated in this chapter, but it will be obvious that the kinetic theory can join the atomic, molecular and ionic theories as one of the foundation stones of physical chemistry.

a. *As applied to gases.* The kinetic theory provides both a qualitative picture of the state of molecular chaos in a gas, and enables some of the matters concerned with a study of gases to be treated quantitatively.

i It provides a fundamental equation for gases,
$$pV = \tfrac{1}{3}mnu^2$$
ii It accounts for the rapid diffusion and ready compressibility of gases.
iii It accounts for Boyle's law, Charles's law, the gas equation, Dalton's law of partial pressure, Graham's law of diffusion and Avogadro's hypothesis.
iv It enables molecular velocities to be calculated, and the laws of probability, as applied by Maxwell and Boltzmann, show how the molecular velocities are distributed throughout a large number of molecules. The distribution of kinetic energy follows the same pattern.
v It provides reasonable suggestions as to why real gases are not perfect.
vi It suggests why c_p/c_v values for elementary gases have constant values depending on the atomicity of the gas.

b. *As applied to liquids.* The kinetic theory as applied to liquids accounts for—

i The diffusion of liquids and solutes.
ii The small compressibility of liquids as compared with gases.
iii Brownian motion.
iv Surface tension and surface energy.
v Evaporation and vapour pressure.
vi The effect of temperature on vapour pressure; boiling and the latent heat of evaporation.

c. *As applied to solids*. The kinetic theory as applied to solids accounts for—

i The very, very slow diffusion of solids.

ii The incompressibility of solids.

iii The sublimation of solids.

iv The melting of solids

QUESTIONS ON CHAPTER 7

1. What assumptions are made in the kinetic theory of gases? How does the theory explain the following facts: (a) that equal volumes of all gases at the same temperature and pressure contain the same number of molecules, (b) that the vapour pressure of a liquid depends only on its temperature, (c) that gases diffuse at rates inversely proportional to the square roots of their densities. (O. Prelims.)

2. Derive the fundamental kinetic theory equation by considering a gas in (i) a cylinder, (ii) a sphere rather than a gas in a cube.

3. Both temperature and pressure are manifestations of molecular motion. Explain what this means in non-technical language.

4. Elemental iodine, at room temperature, is a black crystalline solid, consisting of an array of iodine molecules in a regular pattern. Show by the kinetic theory what changes occur when iodine is heated from room temperature to about 2000°C, always at atmospheric pressure. You are recommended to refer, in your description, to the essential difference between a solid and a liquid, a liquid and a gas, and between a perfect and an imperfect gas. Indicate as far as you can what forces of cohesion are involved in each change the iodine undergoes. (B.)

5. Calculate the root mean square velocities of (a) chlorine, (b) deuterium and (c) methane, at s.t.p.

6. Explain what is meant by the gas constant. If the volume of a mole of gas at s.t.p. is 22.4 dm³, calculate the gas constant in joule per kelvin. Calculate the root mean square velocity of carbon dioxide at 23°C.

7. The equation $pV = RT$ represents the behaviour of a perfect gas. Give some account of the deviations from this equation which are observed experimentally. How are these deviations accounted for by van der Waals' equation?

8. Write short notes on the following terms: (a) root mean square, (b) critical temperature, (c) the reduced equation of state, (d) Brownian motion.

9. If 1 cm³ of oxygen at s.t.p. weighs 143 mg, calculate the root mean square velocity of oxygen at s.t.p.

10. Give an account of the methods which have been used for liquefying gases.

11. The molar volume of hydrogen, at s.t.p., is 22430 cm³, but the comparative figure for ammonia is 22094. Comment.

12. If the densities of nitrogen in the solid, liquid and gaseous states be taken as 1.03, 0.80 and 0.00125 g cm⁻³ respectively, calculate the molar volumes of nitrogen in each state. Comment on your results.

13. Explain carefully why the specific heat capacity of a gas measured at constant pressure is greater than the specific heat capacity measured at constant volume.

14. Outline the assumptions underlying the kinetic theory of gases and comment on their shortcomings.

A study of the rate of effusion of a gas through a pinhole yielded the result that the root mean square velocity was 2.1×10^3 m s⁻¹. Given that the gas was originally at 27.3°C, what is its probable identity? What would be the effect of adding chlorine to the gas? (R = 8 J mol⁻¹ K⁻¹). (C. Schol.)

Chapter 8
Outlines of atomic structure

DALTON pictured an atom as a minute, indivisible particle, and such an idea led to many important developments. In more recent years the elucidation of the more detailed structure of atoms has played a major part in chemical progress.

The development of today's ideas from those of Dalton, 150 years ago, is a long story of great scientific achievement; it is, however, a complex story. A concise summary of the main features of the modern viewpoint will be given first, with a more detailed treatment in the following chapters. The chronological table on the back cover may help the reader to keep his bearings.

1. Fundamental particles. Atoms are made up, essentially, of three fundamental particles, which differ in mass and electrical charge as follows:

	Mass	*Charge*
Electron	1/1840 unit	−1 unit
Proton	1 unit	+1 unit
Neutron	1 unit	Not charged

The existence of electrons in atoms was first suggested, by J. J. Thomson, as a result of experimental work on the conduction of electricity through gases at low pressures, which produces cathode rays and X-rays, and a study of radioactivity. The term electron had, however, been introduced earlier by Johnstone Stoney (1881) in a rather different sense; it had resulted from considerations of Faraday's laws governing the passage of electricity through solutions. Accurate measurement of the charge and mass of an electron has been very important.

An atom is electrically neutral, and if it contains negatively charged electrons it must also contain some positively charged particles. Protons are positively charged, and the supposition that they existed within atoms came about as a result of Lord Rutherford's experiments in which he bombarded elements with the α- and β-rays given off by radioactive elements. The neutron was discovered in 1932 by Sir James Chadwick by bombarding beryllium with α-rays.

Other particles such as the positron, mesons and anti-proton must be considered in dealing with atomic structures and atomic processes but they are, for the most part, of less significance, particularly from a chemical point of view.

2. The nuclear atom. Rutherford's bombardment experiments indicated that an atom consisted of a heavy, positively charged central nucleus with electrons distributed around it. The nucleus contains protons and neutrons (except that there are no neutrons in a hydrogen atom), and its positive charge is neutralised by the negative charge of the electrons round the nucleus so that the atom as a whole is electrically neutral.

Loss of electrons gives rise to positively charged ions (cations); gain of electrons gives negatively charged ions (anions).

In certain conditions the nucleus of an atom can be split so that one element can be transmuted into another, and energy, called atomic energy, may be released in the process.

3. Atomic number and relative atomic mass. The value of the positive charge on the nucleus or of the number of electrons in the atom of any element is equal to the ordinal number of that element in the periodic table arrangement. This number is called the atomic number, Z.

In passing from one atom to the next in the periodic table there is a unit increase of positive charge on the nucleus of the atoms concerned and a consequent addition of one electron. Such an arrangement was supported by the results of Moseley's investigation of X-ray spectra.

The approximate relative atomic mass, A_r, of an element is obtained by adding up the number of protons and neutrons in the atom concerned for, by comparison, the mass of the electrons in an atom is very small.

Typical atomic structures may be represented as follows:

Hydrogen $\begin{array}{l} A_r = 1 \\ Z = 1 \end{array}$ Helium $\begin{array}{l} A_r = 4 \\ Z = 2 \end{array}$ Lithium $\begin{array}{l} A_r = 7 \\ Z = 3 \end{array}$

(1p) 1e   (2p 2n) 2e   (3p 4n) 3e

Beryllium $\begin{array}{l} A_r = 9 \\ Z = 4 \end{array}$ Carbon $\begin{array}{l} A_r = 12 \\ Z = 6 \end{array}$ Nitrogen $\begin{array}{l} A_r = 14 \\ Z = 7 \end{array}$

(4p 5n) 4e   (6p 6n) 6e   (7p 7n) 7e

4. Electrons in atoms. The chemical properties of an atom depend, very largely, on the arrangement of the electrons around the nucleus, for chemical combination depends on interaction between the outer electrons of the atoms combining together.

The working out of the detailed arrangements of electrons rested mainly on the ideas of the quantum theory, first put forward by Planck in 1900, and on the interpretation of a mass of spectroscopic data originally undertaken by Bohr.

Bohr regarded the electron as a negatively charged particle and was able to account for many spectroscopic results by allotting the various electrons in an atom to different orbits arranged around the nucleus. Electrons in different orbits had different energies, and transfer of electrons from orbit to orbit accounted for the absorption or emission of energy, which could be related, in simple cases, to spectroscopic results.

Such an orbital distribution of electrons also accounted for the chemical properties of elements in relation to their position in the periodic table.

Each element in Group 1 is found, for example, to have one electron in the outermost orbit. The electron arrangements in Group 1 are represented as:

Li 2.1

Na 2.8.1

K 2.8.8.1	Cu 2.8.18.1
Rb 2.8.18.8.1	Ag 2.8.18.18.1
Cs 2.8.18.18.8.1	Au 2.8.18.32.18.1
Fr 2.8.18.32.18.8.1	

The elements in Group 2 have two electrons in the outermost orbit, and so on.

Further details about the arrangement of electrons are known, and provide greater insight into the properties of different elements.

5. Isotopes. Because the chemical properties depend, in the main, on the arrangement of electrons it is possible to have atoms with the same chemical properties, i.e. the same arrangements of electrons, but with different relative atomic masses, i.e. different nuclear structures. Such atoms are known as isotopes, and typical examples of the atomic structures of some common isotopes are given below:

Hydrogen, or protium, ^1_1H	Heavy hydrogen, or deuterium, ^2_1H or D	Tritium, ^3_1H or T
(1p) 1e	(1p 1n) 1e	(1p 2n) 1e

Uranium-238, $^{238}_{92}U$	Uranium-235, $^{235}_{92}U$	Uranium-234, $^{234}_{92}U$
(92p 146n) 92e	(92p 143n) 92e	(92p 142n) 92e

In the symbols given to each isotope the superscript shows what is known as the mass number (A), whilst the subscript gives the atomic number (Z).

The existence of isotopes was first discovered when the way in which naturally occurring radioactive elements disintegrated was worked out

(p. 124) by Rutherford, Soddy and Russell, and when positive rays were analysed.

Most naturally occurring elements have been found to consist of isotopes, and Aston's mass spectrograph has been of particular importance in detecting these isotopes. Isotopes which do not occur naturally can be made by atomic transmutations, and some of them, especially the radioactive ones, are very useful.

6. The wave nature of the electron. Davisson and Germer observed, in 1927, that a stream of moving electrons could be diffracted by crystals acting as diffraction gratings. Since it is only possible to account for diffraction phenomena in terms of waves, it is necessary to assume that electrons, like light, have a dual nature. They may appear to be small, negatively charge particles, as envisaged by Thomson, Rutherford and Bohr; but they also have a wave-like nature, similar to that of light or X-rays.

The waves associated with moving electrons can be treated mathematically by a specialised technique known as wave-mechanics and developed, mainly, by Schrödinger. The idea of tiny, negatively charged particles existing in fixed orbits around the nucleus of an atom is replaced by the idea of 'clouds' of electrical charge, of varying charge density, around the nucleus. The electron in a specific, limited orbit is replaced by a more diffuse region of electric charge referred to as an atomic orbital. The shape of such orbitals, and the distribution of charge within them, is of major importance.

7. The atomic nucleus. When an atom is split, it is the nucleus of the atom which splits into two, or more, smaller nuclei. In naturally occurring radioactive elements, this nuclear splitting is going on all the time, and cannot be stopped. Rays, known as α-, β- and γ-rays are given off.

It is now possible to bring about a great number of nuclear reactions in which one nucleus is changed into another, so that the alchemist's dream of converting, say, lead into gold can now be achieved, though not commercially. This has made possible the preparation of many isotopes not occurring naturally, and radioactive isotopes of naturally occurring elements, which are not themselves radioactive, can be made. Such isotopes are said to be artificially radioactive. It has also been possible to make trans-uranium elements with heavier nuclei than that of uranium, which has the heaviest nucleus of all naturally occurring elements.

Moreover, by splitting some big nuclei, particularly those of uranium-235 and plutonium into smaller nuclei, atomic energy can be released by nuclear fission. Similarly, energy can be obtained from nuclear fusion by uniting the nuclei of two light isotopes. In such matters, consideration of nuclear structure and nuclear stability plays a big part.

8. Historical development. A chronological list of the major discoveries in the field of atomic structure is given on the back cover. It will be seen that a great many discoveries are involved. Whilst it is not necessarily best to treat the matter entirely historically, it may, nevertheless, help the reader to refer to this list from time to time.

Chapter 9
The electron

THE term electron was first introduced, by Johnstone Stoney, to describe a certain quantity of electricity, following a consideration of the application of Faraday's Laws to the passage of electricity through solutions.

Further experimental work on the passage of electricity through gases at reduced pressures led to the discovery of cathode rays (p. 90) consisting of very small, negatively charged particles, and it is these particles which are now known as electrons. Similar particles were also found to be emitted, as β-rays (p. 98), by radioactive substances, or by heated wires, as in thermionic valves, or when light of appropriate wavelengths falls on to metals, as in the photoelectric effect.

The possibility of obtaining electrons from such widely different sources led J. J. Thomson to suggest that electrons must be a component part of atoms, and the part played by electrons in atomic structure is one of the major themes of the following chapters. In developing this theme it will be seen that the electron has to be regarded as a minute, negatively charged particle, and, also, that the wave nature of the electron has to be taken into account (p. 117).

CONDUCTION OF ELECTRICITY THROUGH SOLUTIONS

1. Faraday's laws of electrolysis. That amber rubbed with fur, and glass rubbed with silk, will attract small pieces of dry paper has been known for a very long time, and the word electric, originating from the Greek for amber (ἤλεκτρον), was first used by Gilbert in 1600. It was also realised, very early, that amber rubbed with fur would attract glass rubbed with silk, and two varieties of electricity, known as resinous and vitreous, were supposed to exist. Ingenious frictional machines were developed to build up high electric charges.

But electric currents of any magnitude only became obtainable following the chance observation by Galvani, in 1791, of the twitching of a partially dissected frog's leg when the nerve and leg were connected through a metal dipping into the body fluids. This observation was developed by Volta, in 1800, into the Voltaic pile, and a new era in science began.

In 1800, Nicholson and Carlisle decomposed water into hydrogen and oxygen by passing an electric current through it, and in the first decade of the nineteenth century Davy isolated sodium and potassium, and other similar metals, by passing electric currents through their molten hydroxides. Such activities gave a great fillip to electrical experimentation, and to the view that chemical combination between atoms was electrical in nature.

The results of much measurement on the decomposition of solutions of salts, acids and alkalis by electricity were eventually summarised in 1834, by Faraday, who began his scientific work as Davy's laboratory assistant, in his Laws of Electrolysis. A modernised version of these laws states them in two parts.

a. *First law. The mass of a substance liberated at an electrode during electrolysis is proportional to the quantity of electricity passed.*

The quantity of electricity is measured in coulombs, 1 coulomb being the quantity passing when 1 ampere flows for 1 second, i.e.,

No. of coulombs = No. of amperes × No. of seconds

The law may, therefore, be summarised as

$$\text{Mass of substance liberated} \propto \text{Current in amperes} \times \text{Time in seconds}$$

or $$m \propto I \times t$$
or $$m = e \times I \times t$$

The value of e for any element is an important quantity known as the electrochemical equivalent. It is the mass of an element liberated by 1 coulomb. Typical values are 1.044×10^{-8} kg C^{-1} for hydrogen, 111.8×10^{-8} kg C^{-1} for silver and 32.9×10^{-8} kg C^{-1} for copper.

b. *Second law. When the same quantity of electricity is passed through solutions of different electrolytes the masses of the substances liberated at the electrodes are in the ratio of their equivalents.* That is

$$\frac{\text{Mass of A deposited}}{\text{Mass of B deposited}} = \frac{\text{Equivalent of A}}{\text{Equivalent of B}}$$

0.001 118 g of silver are liberated by 1 coulomb and, as the equivalent of silver is 107.880, it follows that 107.880/0.001 118, i.e. 96487 coulomb are required to liberate 107.880 g of silver. This is also the quantity of electricity required to liberate 1 gramme equivalent of any other element. It has been used as a distinct unit known as the Faraday, i.e. 1 Faraday = 96487 C. Modern usage it to call 96487 C mol^{-1} the *Faraday constant*.

The theory eventually put forward to account for Faraday's laws, and other electrical measurements, is now known as the Ionic Theory and is explained in Chapter 33.

2. The electron. One of the most significant conclusions to be drawn from Faraday's laws did not materialise until 1881. 96487 C of electricity is required to liberate 1.0080 g of hydrogen or 8 g of oxygen, i.e. $\frac{1}{2}$ mol of hydrogen or $\frac{1}{4}$ mol of oxygen. As 1 mol contains 6.02252×10^{23} molecules, and as both hydrogen and oxygen are diatomic, it follows that 96487 C is required to liberate 6.02252×10^{23} atoms of hydrogen or 3.01126×10^{23} atoms of oxygen. By taking other similar examples it can be seen that every atom of a monovalent element requires a definite small quantity of electricity to liberate it during electrolysis. A divalent atom requires twice the quantity of electricity; a trivalent atom, three times the quantity; and so on.

Such conclusions, published about 1881 by Helmholtz and Johnstone Stoney, strongly suggested that electricity, like matter, was not continuous but was made up of definite small units. Johnstone Stoney called the quantity of electricity required to liberate one atom of a monovalent element an electron. Its value was equal to $96487/6.02252 \times 10^{23}$ C, i.e. 1.60210×10^{-19} C.

This was the origin of the word electron, and little did Helmholtz or Johnstone Stoney realise what an incredibly large part this remarkably small quantity of electricity was destined to play in the development of science.

CONDUCTION OF ELECTRICITY THROUGH GASES

3. Cathode rays. Gases, at normal pressures, are good insulators, but, if a sufficiently high voltage is used, a current can be passed between two electrodes separated by a gas under reduced pressure. The observed effect depends on the pressure, and a remarkable series of changes takes place as the pressure is progressively lowered by pumping more and more gas out of the discharge tube. It is interesting to reflect on the part played by the development of pumps in scientific progress.

At a pressure of about 0.25 N m^{-2} the tube is occupied by a dark space (known as the Crookes' dark space after the man who first studied the effects), whilst the glass of the tube fluoresces, the colour of the fluorescence depending on the nature of the glass.

The nature of the dark space within the discharge tube was, originally, in much doubt, but it is now known to consist of cathode rays (a name first used by Goldstein in 1876) made up of a stream of electrons passing from the cathode to the anode. Such a conclusion is supported by the following experimental evidence:

a. When a thick metallic shape is placed in the discharge tube between the cathode and the anode, a well-defined shadow of the shape is formed (Hittorf, 1869). This shows that the cathode rays travel in straight lines.

b. The cathode rays will penetrate a thin sheet of metal.

c. A small, light paddle wheel pivoted between the cathode and anode is caused to rotate by the cathode rays, and it can be rotated in the opposite direction by reversing the current. This shows that the cathode rays can cause mechanical movement, and that they move in a direction away from the cathode.

d. Cathode rays cause fluorescence when they impinge on most substances. In particular, they cause very clear fluorescence on a screen coated with zinc sulphide. Such a screen is, therefore, useful in observing cathode rays.

e. A beam of cathode rays is deflected by an external magnetic field in a direction both at right angles to the direction of the beam and the direction of the lines of force of the magnetic field (Plücker, 1858). Application of Fleming's left-hand rule shows that the cathode rays behave like a flow of negative charge.

f. A beam of cathode rays is also deflected by an electrical field, again in a direction indicating negative charge.

g. If the cathode rays are led into a metal cylinder connected to an electroscope or electrometer outside the discharge tube, the cylinder is found to become negatively charged (Perrin, 1895). This again shows that the cathode rays are negatively charged.

h. Cathode rays passed through a supersaturated vapour cause the formation of a fog or vapour trail in their path by producing nuclei for condensation of the vapour.

It is on such evidence that cathode rays came to be regarded as a stream of negatively charged particles now known as electrons.

4. Measurement of charge, mass and velocity of electrons. J. J. Thomson, in 1897, measured the velocity of cathode rays, and the value of the charge/mass ratio for the electrons of which they are composed. He did this by

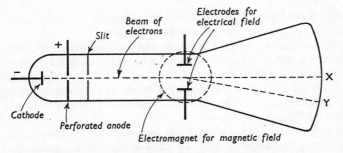

FIG. 14. Thomson's method of measuring e/m and v for electrons.

subjecting a beam of cathode rays, obtained by passing them through a perforated anode and a slit, to the effect of both magnetic and electric fields so arranged as to deflect them in opposite directions. The deflection caused to the beam of electrons could be observed on a zinc sulphide strip at the end of the tube remote from the cathode (Fig. 14).

Application of the magnetic field caused the beam to move in an arc of a circle whilst passing through the field so that the beam was deflected from X to Y. If the magnetic field is of strength, H, and the electrons in the beam are carrying a charge, e, and moving with velocity, v, the magnetic force exerted on each electron will be Hev. This will be balanced by a centrifugal force of mv^2/r, where m is the mass of an electron and r the radius of the circle in which the electrons move in the magnetic field. Thus,

Magnetic force on electrons = Centrifugal force on electrons

or
$$Hev = \frac{mv^2}{r}$$

or
$$\frac{e}{m} = \frac{v}{rH} \qquad \text{(i)}$$

The value of r can be obtained from the dimensions of the apparatus and the distance XY.

By applying an electric field of the required strength, at right angles to the magnetic field, the deflected beam can be brought back on to its original path. When this is done, the magnetic force (Hev) on the electrons must be balanced by the electrical force, which is given by Ee, where E is the strength of the electrical field. Thus

Magnetic force = Electrical force

or
$$Hev = Ee$$

or
$$v = \frac{E}{H}$$

Knowing the values of E and H, v can be calculated, and substitution into equation (i) gives the value of e/m for an electron.

With different gases, different pressures and different potential differences between cathode and anode, v varied, though it was of the order of $\frac{1}{10}$th of the velocity of light, i.e. about 3×10^7 m s^{-1}. But the value of e/m, 1.7588×10^{11} C kg^{-1}, was constant under all conditions, and this suggested that the electron was a quite definite, individual particle which was probably a component of all matter.

This experiment could only give a value for the ratio e/m. If the charge on the electron was assumed to be the same as that on a univalent ion (p. 90) it followed that the mass of an electron must be approximately $\frac{1}{1840}$th of that of a hydrogen atom. No smaller particle has ever had a bigger future.

5. Determination of the charge on an electron. Thomson's experiment, described in the preceding section, gives only a value of the ratio e/m. This determination was followed by a direct measurement of e by Townsend, and by Thomson himself, and, later (1909), by a more accurate method by Millikan, known as the oil-drop method.

Millikan observed the motion of a charged drop of non-evaporating oil

between two circular plates of metal set accurately parallel by glass, in-
sulating separators.

Oil drops, from a spray, were allowed to pass through a small hole in the
upper plate (Fig. 15). The drops became charged, by friction, in the spray-
ing process, and could be further charged, as required, by passing a beam
of X-rays (p. 96) between the plates. The drops were illuminated from one
side and viewed through a microscope at right angles to the direction of
illumination. Single drops, appearing as a bright spot, could be observed
in this way.

With the plates earthed, an oil drop falls under the influence of gravity.

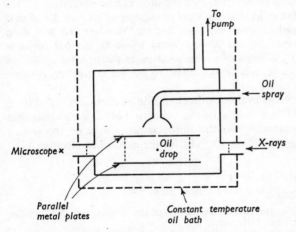

FIG. 15. Diagrammatic representation of Millikan's oil-drop apparatus.

Measurement of the terminal velocity of the drop, and application of
Stokes's Law, enables the mass of the drop to be calculated.*

The gravitational fall of a charged drop, between the plates, can be
stopped by applying a potential difference of several thousand volts be-

---

* Stokes's law states that the retarding force, F, on a spherical particle of radius, r,
falling through a gas of viscosity, η, with a velocity of v is given by,

$$F = 6\,\pi\eta r v$$

If v is the terminal velocity of an oil drop, the corresponding retarding force will be
equal to the mass of the drop minus the upthrust on it, i.e. $\frac{4}{3}\pi r^3 g(\rho_1 - \rho_2)$ where g
is the acceleration due to gravity and ρ_1 and ρ_2 are the densities of the oil and the gas
respectively.

Thus, for terminal velocity,

$$6\,_c\pi\,\eta r v = \tfrac{4}{3}\pi r^3 g(\rho_1 - \rho_2)$$

If the radius of an oil drop be calculated from experimental measurements, its mass
can readily be obtained from its volume and its density. For small drops, however,
such as those used in Millikan's experiment, a slightly modified form of Stokes's law
must be used for real accuracy.

tween the plates, the upper plate being made positive. By adjusting the potential difference a single drop can be made to move up or down, or be held stationary.

For a stationary drop, the upward force on the drop due to the electrical field must be equal to the mass of the drop minus the upthrust on it. The upward force on the drop due to the electrical field is equal to the electrical field multiplied by the charge on the drop. Thus, for a stationary drop,

$$\text{Electrical field} \times \text{Charge on drop} = \text{Mass of drop} - \text{Upthrust on drop}$$

from which the value of the charge on the drop can be obtained.

Millikan found that the charge on any single drop of oil varied, and that it could be changed by treatment with X-rays. Whatever charge a drop might have, however, it was always found to be an integral multiple of 1.60×10^{-19} C. Change of gas or pressure in the apparatus did not affect this basic value, which was taken to be the charge on a single electron.

The presently accepted value for the charge on an electron is $(1.60210 \pm 0.00007) \times 10^{-19}$ C, which is, of course, equal to the charge on a univalent ion (p. 90). In conjunction with the most accurate e/m value, this gives a value of $(9.1091 \pm 0.0004) \times 10^{-31}$ kg for the mass of an electron.

QUESTIONS ON CHAPTER 9
(Questions on Faraday's laws will be found on p. 448.)

1. Write a short biographical sketch on one of the following: J. J. Thomson, Davy, Millikan, Crookes, Helmholtz.

2. Give an account on either (a) thermionic valves, or (b) the photoelectric effect.

3. Write short notes on the following terms: electron volt, electron spin, electron alloy, electron diffraction, electron affinity.

4. Describe, with diagrams, what is observed when the pressure is slowly reduced inside a discharge tube connected to a source of high potential difference

5. Use Fleming's left-hand rule to show the direction of the force acting on an electron when it moves at right angles to a magnetic field. Draw a diagram, and give an explanation of it.

6. All electrons have the same charge and the same mass. In what other ways might they differ from each other?

7. Outline some modern developments which would probably not have been possible without the discovery of the electron.

8. If a stream of electrons each of mass m, charge e, and velocity 3×10^7 m s^{-1} is deflected 2 mm in passing for 100 mm through an electrostatic field of 1.8 V mm^{-1} perpendicular to their path, find the value of e/m in coulomb per kilogramme.

9. A charged particle is found to follow a circular track of radius 200 mm when it enters a perpendicular magnetic field of 7.5×10^{-4} Wb m^{-2} at a velocity of 26.4×10^6 m s^{-1}. What is the charge/mass ratio for the particle in C kg^{-1}?

10. An electron passing into an electric field perpendicular to its direction of

motion is deflected in a parabolic path, whereas a similar magnetic field deflects the electron in a circular path. Does this invalidate the argument on which Thomson's method for measuring e/m for an electron was based?

11. Describe, in your own words, Millikan's method of measuring the charge on an electron.

12. Why is it that Faraday's laws might be said to have led to the discovery of the electron?

Chapter 10
Radioactivity and the nuclear atom

1. X-rays. Röntgen discovered, in 1895, that wrapped photographic plates became fogged when left near a discharge tube, and this is now known to be due to the emission of a very penetrating radiation, known as X-rays, from solids bombarded by cathode rays.

Besides being very penetrating and affecting photographic plates, X-rays travel in straight lines and cause luminescence of, for example, zinc sulphide. The rays can, moreover, ionise a gas when they are passed through it, thus causing it to conduct electricity. X-rays are not, however, deflected by magnetic or electrical fields.

The real nature of X-rays was not discovered until 1912. In particular, it was not known whether they consisted of a stream of particles, like cathode rays, or whether they were wave-like, similar to light.

If like light, then X-rays ought to be capable of being diffracted if a suitable diffraction grating was available, but early attempts to diffract X-rays, using ruled gratings, were not successful. In 1912, von Laue suggested that the wavelength of X-rays might be too small to give diffraction patterns with a ruled grating, but that the regular, close-spaced array of planes of atoms within a crystal might serve as a diffraction grating for such short wavelengths. Friedrich tested this suggestion experimentally, and found that a copper sulphate crystal would, in fact, diffract X-rays. This result established the nature of X-rays as light-like radiation of very small wavelength.

This was the beginning of X-ray crystallography (p. 215), which enables crystal structures to be investigated by X-rays, and which, using crystals of known structure, also enables the wavelengths of X-rays and other similar radiation to be determined.

When a solid is bombarded by cathode rays to form X-rays, the wavelengths of the X-rays produced vary. Bombardment of every element gives a general, or white, X-radiation, but each element also gives X-rays of a wavelength characteristic of the particular element. This was discovered by Moseley, in 1913, when he investigated the spectra of X-rays emitted by different elements (p. 114), thereby elucidating the meaning of atomic numbers.

RADIOACTIVITY

2. Radioactive elements. In the year following the discovery of X-rays, Becquerel (1896) found, whilst investigating various fluorescent substances, that uranium and uranium compounds would also emit a penetrating radiation capable of affecting wrapped photographic plates, and he called this phenomenon radioactivity.

The discovery was followed, in 1898, by the isolation of two more strongly radioactive elements, polonium and radium, by M. and Mme Curie. The Curies examined the radioactivity of a uranium mineral called pitchblende and found that it was much more radioactive than would be expected from its uranium content. A prolonged and tedious extraction process led to the isolation of polonium and radium from pitchblende. 1000 kg of the mineral gave only about 0.2 g of radium, but this element was about 2 million times more radioactive than uranium. Mme Curie also found, in 1898, that thorium was radioactive, and, in 1900, actinium was found to be radioactive.

The naturally occurring elements now known to be radioactive are polonium, radon, radium, actinium, thorium, protoactinium and uranium. All the trans-uranium elements (p. 140) are also radioactive, and artificial radioactivity can be induced into elements not naturally radioactive (p. 135).

3. α-, β- and γ-rays. Radioactive substances emit three different types of ray, which can be separated and investigated because they are affected differently by magnetic and electric fields. If the radiation from a radio-active substance is passed through a magnetic field it is split up into what are now known as α-, β- and γ-rays.

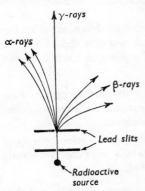

FIG. 16. Effect of a magnetic field, acting normally to the plane of the paper, on α-, β- and γ-rays.

The α-rays are deflected in a direction indicating that they are positively charged; the β-rays are deflected in the opposite direction, and must be negatively charged; the γ-rays are not deflected at all (Fig. 16).

Furthermore, the β-rays are deflected to a much greater extent than the α-rays, which indicates that the β-rays are much lighter than the α-rays, and the β-ray beam is dispersed more than the α-ray beam, which indicates that the β-rays consist of particles of more widely varying velocity than the α-rays.

The γ-rays are the most penetrating, and the α-rays the least.

a. *α-rays.* Rutherford measured the charge/mass ratio for α-rays by the method of applying magnetic and electric fields similar to that used by Thomson in measuring e/m for cathode rays (p. 91). The result agreed with the supposition that α-rays were made up of helium atoms bearing two positive charges, i.e. of helium nuclei, and this was proved in two ways.

First, some radon was sealed in a glass tube thin enough to allow the α-rays emitted to escape through the glass walls into a surrounding, evacuated tube. After some time, an electric discharge was passed through the outer tube and analysis of the resulting spectrum (p. 103) showed that helium was present.

Secondly, Rutherford made a direct measurement of the charge on an alpha particle. α-rays cause a glow when they fall on a zinc sulphide screen, and, if the intensity of radiation is not too great, each α-particle causes a definite flash which can be seen through an eye-piece. In this way, in what is known as a spinthariscope, α-particles can be counted by counting the number of flashes.

Rutherford measured the number of α-particles falling on a given area in a given time, and then measured the charge imparted to a metal screen of the same area in the same time by the same radioactive source. He was, in this way, able to calculate the charge carried by each α-particle and this came out to be twice the charge on a hydrogen ion. In conjunction with the measured charge/mass ratio for an α-particle this showed that the mass of an α-particle was four times that of a hydrogen atom. The α-particle is, therefore, a helium atom bearing two positive charges, i.e. a helium nucleus (p. 84).

The velocity of α-particles emitted by radioactive substances varies, but it is about $\frac{1}{20}$th of the velocity of light. α-rays can penetrate about 70 mm of air at atmospheric pressure, or a thin sheet of aluminium a small fraction of a millimetre thick.

b. **β-rays.** β-rays are found to be very similar to cathode rays in their general properties and this, coupled with measurement of their charge/mass ratio, shows that they consist of a stream of electrons.

The velocity of β-particles from radioactive sources varies from 3 per cent to 99 per cent the velocity of light, and β-rays will penetrate several millimetres of aluminium or about 3 mm of lead.

c. **γ-rays.** γ-rays are not affected by electric or magnetic fields and they are extremely penetrating electro-magnetic radiation similar to X-rays but of much shorter wavelength. They will penetrate about 150 mm of lead, and travel with the speed of light.

The nature of α-, β- and γ-rays is summarised on p. 99.

THE NUCLEAR ATOM

4. Bombardment of matter by α- and β-rays. The existence of cathode rays and β-rays, obtained from such different sources, and yet both consisting of electrons, provides evidence that electrons are a component part of atoms. This suggestion is also supported by the fact that electrons are given off from a hot metallic wire, as in a thermionic valve, and when light of appropriate wavelength falls on to metals, as in the photoelectric effect, first observed by Hertz in 1887.

	Nature	Electrical charge	Mass of particle	Velocity	Relative penetration
α-Rays.	Helium nuclei	+2 units	4 units	c. 1/20th velocity of light	1
β-Rays.	Electrons	−1 unit	1/1840th unit	3–99% velocity of light	100
γ-Rays.	Electro-magnetic radiation	No charge		Velocity of light	10000

But, if an atom contains electrons (negatively charged) it must also contain an equal number of some positively charged particles to give it electrical neutrality. The first atomic model, suggested by Thomson, envisaged an arrangement of electrons within a sphere of positive electricity.

Such an idea was replaced, however, in 1911, by a nuclear atom, a suggestion made by Rutherford as a result of experiments he directed on the scattering of α- and β-rays by metallic foil.

a. *Bombardment by β-rays.* When a parallel beam of β-rays is directed towards a thin sheet of metal, the β-rays pass through, but the beam is diverged. This observation is explained as being due to the repulsion of the electrons in the β-rays by the electrons in the atoms of the foil (Fig. 17).

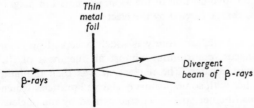

FIG. 17. The divergence of a beam of β-rays on passing through thin metal foil.

The measured divergence shows, moreover, that the number of electrons in an atom necessary to cause this divergence must be about half the relative atomic mass of the atom concerned. Magnesium, with relative atomic mass of 24, is shown, for example, to have about 12 electrons in its atom.

As the mass of the electron is very small, the electrons in an atom cannot contribute very much to the total mass of the atom, and the main mass of an atom must be situated in the positively charged part of the atom. This viewpoint was firmly supported by the results of experiments in which α-rays were used as bombarding particles instead of β-rays.

b. *Bombardment by α-rays.* In these experiments, carried out by Geiger and Marsden, a parallel beam of α-particles, from a radioactive source, was directed towards a platinum or gold foil. The effect was observed by picking up the α-particles on a zinc sulphide screen, and it was found that the α-particles passed right through the metallic foil but that they were deflected from their course in many directions. The average deflection was less than one degree, but some α-particles were deflected through a much greater angle, and much to everyone's surprise at that time, a very few (about 1 in 20000) were deflected right back through angles greater than 90° (Fig. 18). This back-deflection was not simply a reflection from the front surface of the foil, for the number of back-deflections increased as thicker pieces of foil were used.

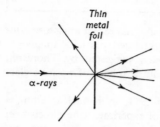

As the α-particle is heavy, it could only be deflected through large angles by a near collision with something of its own mass and charge. Thomson's sphere of positive electricity would be far too diffuse to cause the observed result.

Rutherford was able to show that the observations of the deflections of α-particles could be accounted for if it was assumed that an atom had a positively charged nucleus, of diameter about 10^{-12} mm, carrying a positive charge equal in number to about half the relative atomic mass of the atom concerned. Thus it was shown that the charge necessary on the positive portion of the atom was equal in magnitude to the total negative charge carried by the electrons.

Fig. 18. The scattering of α-particles by a sheet of thin metal.

5. The nucleus. The idea of a nuclear atom was widely accepted by 1914. A positively charged nucleus, contributing almost all the mass of the atom, was surrounded by electrons. The nucleus, it was then thought, contained electrons together with some positively charged particles which became known as protons, the net positive charge carried by the nucleus being equal to the total negative charge of the extra-nuclear electrons.

The following typical atomic structures illustrate the stage of development reached about 1914,

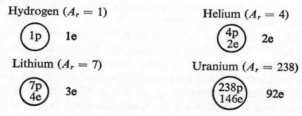

Hydrogen ($A_r = 1$)   Helium ($A_r = 4$)

Lithium ($A_r = 7$)   Uranium ($A_r = 238$)

It will be seen that the proton, represented as p, is the nucleus of a hydrogen atom.

Such structures provided two immediate problems. First, the packing together of positively charged protons and negatively charged electrons in unequal numbers in the nucleus of an atom was not satisfactory arrangement so far as electrical forces and energy considerations were concerned (p. 144). Secondly, little was known about the arrangement of the extra-nuclear electrons.

The arrangement of these electrons is of major importance from the point of view of chemical combination between atoms (p. 158), and the structure of the nucleus is important in considerations of atom splitting. For when an atom splits, either naturally in radioactive elements, or artificially, it is the nucleus of the atom which splits. Consideration of nuclear stability is, therefore, of great importance.

The immediate difficulty of having protons and electrons together in a nucleus was resolved after the discovery by Chadwick, in 1932, of a neutral particle known as the neutron. Chadwick bombarded the element, beryllium, with α-rays from a radioactive substance and found that particles were emitted which were not affected by magnetic or electric fields. These particles are neutrons, with zero charge and unit mass.

With the discovery of the neutron it was no longer necessary to postulate the existence of protons and electrons together in an atomic nucleus. Instead, the nucleus was thought to consist of protons and neutrons, the atomic structures given above being replaced by

Hydrogen ($A_r = 1$)

（1p） 1e

Helium ($A_r = 4$)

（2p 2n） 2e

Lithium ($A_r = 7$)

（3p 4n） 3e

Uranium ($A_r = 238$)

（92p 146n） 92e

Further details of nuclear stability are given on pages 141–5, and the problems associated with the arrangement of electrons around the nucleus are discussed in Chapter 11.

QUESTIONS ON CHAPTER 10

1. Write an account on the production and uses of X-rays.
2. How was it shown that (a) β-rays consisted of electrons, and (b) α-rays consisted of helium nuclei?
3. Draw a diagram showing how Fleming's left-hand rule can be used to predict the effect of a magnetic field on (i) an electron, (ii) an α-particle, (iii) a proton.

4. Explain the difference between (a) a β-particle and an α-particle, (b) γ-rays and X-rays, (c) a proton and a deuteron.

5. What were the major contributions to the elucidation of atomic structure of (i) Rutherford, (ii) Thomson, (iii) Moseley, (iv) Röntgen?

6. Explain how it is possible to get an approximate value for the Avogadro constant by measuring the rate of α-particle emission from a known mass of radium using a spinthariscope and, also, the rate of helium formation from a known mass of radium.

7. By placing a minute, but known, mass of radium close to a zinc sulphide screen and counting the flashes on the screen it was estimated that 1 g of radium emitted 34000 million alpha particles per second. The volume of helium obtained from decaying radium was found to be 0.156 cm³ at s.t.p. per gramme of radium per year. What value do these figures give for the Avogadro constant?

8. How many protons and how many neutrons are there in the nuclei of the following isotopes: ^{19}F, ^{75}As, ^{70}Ge, ^{16}O, ^{1}H, ^{2}H, ^{235}U, ^{238}U and ^{239}Np?

9. Write down the nuclear structures of the isotopes formed when (a) $^{239}_{92}U$ loses first one and then a second β-particle, and (b) $^{239}_{92}U$ loses first one and then a second α-particle.

10. Give the numbers of protons, neutrons and electrons which will be found in the following atoms: $^{27}_{13}Al$, $^{235}_{92}U$, $^{209}_{83}Bi$, $^{4}_{2}He$, $^{3}_{2}He$, $^{1}_{1}H$, $^{2}_{1}H$, $^{3}_{1}H$, $^{16}_{8}O$, $^{17}_{8}O$, $^{18}_{8}O$, $^{35}_{17}Cl$, $^{37}_{17}Cl$.

11. Write down the nuclear structures of all the naturally radioactive elements.

12. Why was the supposition that an atomic nucleus contained protons and electrons unsatisfactory, and how was this difficulty overcome?

Chapter 11
The arrangement of electrons in atoms

THE chemical properties of an atom are largely controlled by its electrons, for it is interaction between the electrons of two or more atoms that leads to chemical combination between the atoms (p. 158). The detailed arrangement of electrons within an atom is, therefore, of fundamental importance. The working out of these arrangements is an amazing feat similar to the fitting together of a most complicated jig-saw puzzle or the solution of an elaborate cipher.

The solution of the problems involved comes from interpretation of spectroscopic data and the application of the ideas of the quantum theory. It also involves the relative position of the various elements in the periodic table, and a consideration of the wave-like nature of electrons.

SPECTRA

1. Types of spectra. As opposed to particle-like rays, wave-like radiation can cause interference patterns, can be diffracted, if a suitable diffraction grating can be found (p. 215), and can be 'sorted out' into its component wavelengths by using a spectrometer.

In this way, visible light is readily shown to be made up of different

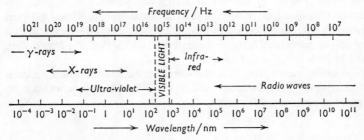

FIG. 19. The electro-magnetic spectrum. The boundaries between one type of radiation and another are not very sharply defined.

coloured lights, each colour corresponding to a group of waves of different wavelengths and frequencies. Blue light, at one end of the visible spectrum, has a shorter wavelength (0.45 μm or 450 nm or 4500 Å) than red light, at the other end (0.75 μm or 750 nm or 7500 Å).* Frequencies are expressed in vibrations per second, s^{-1}, or hertz (Hz).

Visible light represents only a small part of all radiation, and a more complete representation of the possible types of radiation is given in Fig. 19.

Characteristic spectra can be obtained from substances by causing them to emit radiation. This can be done in a variety of ways, e.g. by heating a

---

* Wavelengths in, or near to, the visible region are conveniently expressed in micrometres (μm) or nanometres (nm); 1 nm is 10^{-9} m, and 1 μm is 10^{-6} m. The older unit used for this purpose was the Ångstrom, 1 Å being equal to 10^{-10} m.

substance or by subjecting it to electrical stimulation or excitation by using an electric arc, spark or discharge; and a variety of so-called *emission spectra* can be obtained.

If the light from an electric filament lamp is examined by a spectroscope it is found to consist of all visible wavelengths, i.e. it gives a *continuous spectrum*. The light from a mercury vapour lamp, however, is made up of a limited number of wavelengths and its spectrum, known as a *line spectrum*, consists of a series of sharply defined lines each corresponding to a definite wavelength.

These are particular examples of the general fact that incandescent solids produce continuous spectra, whilst the excited vapour of an element gives rise to a line spectrum. These line spectra are characteristic of atoms. *Band spectra*, consisting of bands of overlapping lines of varying intensity, are characteristic of molecules.

Absorption spectra are formed when light of particular wavelengths is absorbed by passage through a substance. Black lines are found in the spectrum corresponding to the wavelengths absorbed.

2. Spectral series. So far as the historical development of atomic structure is concerned a study of the line spectrum of hydrogen is of great importance. This spectrum consists of lines corresponding to widely different frequencies, but, over a period of time, starting in 1885, it was found that many of the numerous lines could be fitted into series and hence were related to each other in some way.

These series, known after their discoverers as the Balmer (1885), Paschen (1896), Lyman (1915), Brackett (1922) and Pfund (1925) series, can be expressed in one overall formula:

$$\frac{1}{\lambda} = R_H(1/n^2 - 1/m^2)$$

where λ is the wavelength, R_H a constant, known as the Rydberg constant, and n and m have integral values as follows:

Series	n	m	Main spectral region
Lyman	1	2, 3, 4, etc.	Ultra-violet
Balmer	2	3, 4, 5, etc.	Visible
Paschen	3	4, 5, 6, etc.	Infra-red
Brackett	4	5, 6, 7, etc.	Infra-red
Pfund	5	6, 7, 8, etc.	Infra-red

It is a remarkable experimental fact that so many apparently, at first sight, unrelated lines in a spectrum can be expressed by a simple formula. Line spectra of the alkali metals are also made up of similar series of lines known as the sharp, principal, diffuse and fundamental series. The lines in these series can be related in a single formula as for the hydrogen spectrum.

Suggestions, by Bohr, as to how these series of lines originated from energy changes within the atom led to the elucidation of the arrangement of electrons in atoms (p. 112).

QUANTUM THEORY

3. Outline of theory. Progress in working out the arrangement of extra-nuclear electrons in an atom came, first, when Bohr, in 1914, applied the ideas of the quantum theory, first put forward by Planck in 1900,. to the interpretation of spectroscopic data.

The essential idea of the quantum theory is that the energy of a body can only change by some definite whole-number multiple of a unit of energy known as the quantum. This means that the energy of a body can increase or decrease by 1, 2, 3, 4 . . . n quanta, but never by $1\frac{1}{2}$, $2\frac{3}{4}$, 107.3, etc., quanta. It is rather like the fact that our currency can only change by 1, 2, 3, 4 . . . n halfpennies, but not by $1\frac{1}{2}$, $2\frac{3}{4}$, 107.3 halfpennies.

Unlike the halfpenny, however, the value of the quantum is not fixed, but is related to the frequency of radiation which, by its emission or absorption, causes the change in energy. This relationship is expressed as

$$E \quad\quad = \quad\quad h \quad\quad \times \quad\quad \nu$$

(Value of a   (Planck constant   (frequency of
quantum in joules) $= 6.6256 \times 10^{-34}$ J s)   radiation/s^{-1} or Hz)

or, in terms of the wavelength (λ) of the radiation and the velocity of light (c), as

$$E = \frac{hc}{\lambda}$$

so that it is a simple matter to calculate the value of the quantum corresponding to any known frequency.

This idea was originally developed by Planck in considering a vibrating body changing in energy by emitting or absorbing radiation of a frequency equal to the frequency of the vibrating body, but Einstein showed that the idea held more generally and that if the energy of a body changed from a value E_1 to a value E_2 by emission or absorption of radiation of frequency ν, then

$$E_1 - E_2 = nh\nu = \frac{nhc}{\lambda}$$

where n is an integer. It was this generalised statement of the quantum theory that was used by Bohr in his interpretation of spectra.

4. Bohr's interpretation of spectral series. Rutherford assumed that the electrons circulated round the nucleus of an atom in orbits, rather as planets circulate round the sun, and atoms were pictured as minute solar systems. The electrical forces of attraction between the negatively charged electrons and the positively charged nucleus were just counterbalanced by centrifugal forces.

Bohr pointed out, however, that charged particles could not circulate in an orbit without having an acceleration towards the centre of the orbit, and, according to the accepted electro-dynamic theory of the time, an electric charge must radiate energy when it is accelerated.

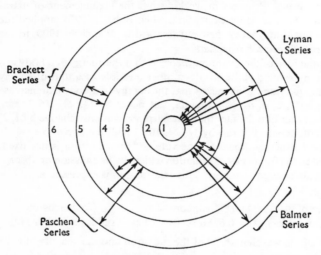

FIG. 20. Illustration of the energy changes which an electron can undergo in moving from one energy level to another in an atom. The diagram is not to scale, for the radii of the various stationary states shown are, in fact, proportional to the.squares of the principal quantum numbers allotted to them.

If Rutherford's idea is correct, then an atom would radiate energy continuously; in doing so it would undergo spontaneous destruction. Moreover, the continuous emission of radiation would not account for the formation of line spectra.

To deal with these difficulties, Bohr put forward suggestions which, in effect, deny the truth of older electrodynamic theories as applied to the motion of electrons. These are:

a. that the electrons in an atom could only rotate in certain selected orbits and that they did not, in these orbits, radiate energy. Such orbits were called *stationary states*, and

b. that each stationary state corresponds to a certain energy level, i.e. that an electron in a certain stationary state had a certain energy, and that emission of radiant energy was caused by the movement of an electron from one stationary state to another of lower energy. Conversely, absorption of energy took place by an electron moving into a stationary state of higher energy.

Bohr now applied the ideas of the generalised quantum theory (p. 105) to the change in energy when an electron moves from one stationary state to another. Thus, if the energy of one stationary state is given as E_1, and

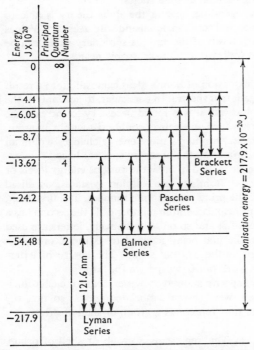

FIG. 21. Some of the energy levels in the hydrogen atom.

that of the next stationary state with lower energy as E_2, an electron passing from the first to the second would cause an energy change of $E_1 - E_2$, and an emission of radiation of frequency, ν, where

$$E_1 - E_2 = h\nu$$

Similarly, absorption of radiation of this frequency would cause an electron to pass from energy level E_2 to the higher energy level E_1.

On this view the series observed in the line spectrum of hydrogen are explained by the various limited energy changes which an electron can undergo in moving between the various stationary states. The general idea is made clear by a study of Figs 20 and 21 which show the various energy changes leading to the various lines in the spectrum.

The atom is in its normal or ground state when the electron is in the stationary state of least energy; when in any other state the atom is said to be excited. On excitation the electron moves into stationary states of higher energy content, and it is the return of the electron to stationary states of lower energy which results in the emission of radiant energy and the formation of spectral lines. It is like lifting a ball up and letting it fall again, both the lifting and falling being done in definite stages.

For hydrogen, and hydrogen-like, spectra the Bohr theory is able to account for the observed spectral series in detail and with accuracy. Values for the radii and the energy levels of the various stationary states can be calculated, and these fit in well with the spectroscopic results.

5. Quantum numbers. The extension of Bohr's ideas came about as a result of a more detailed investigation of spectra. In particular, to account for the greater number of lines observed, it was found necessary to increase the number of possible orbits in which an electron could exist within an atom. In other words, it was necessary to allow more energy changes within an atom to explain the formation of all the observed spectral lines.

The term quantum number is used to label the various energy levels or orbits. The number allotted to Bohr's original stationary states, visualised as circular orbits, is called the principal quantum number. The first orbit, nearest the nucleus, has a principal quantum number of 1; the second has a principal quantum number of 2, and so on. Alternatively, letters are used to characterise the orbits, the first being referred to as the K orbit, the second as the L, the third as the M, and so on. The choice of letters originates from Moseley's work on X-ray spectra (p. 114).

A more detailed study of spectra showed, however, that for each principal quantum number there were several associated orbits, so that the principal quantum number now represents a group or shell of orbits.

a. *Principal quantum number.* This represents a group or shell of orbits, but the total number of electrons that can occupy any shell, i.e. that can have the same principal quantum number, is limited and is given by $2n^2$ where n is the principal quantum number concerned. Thus

Shell	K	L	M	N
Principal quantum number (n) . .	1	2	3	4
Maximum number of electrons . .	2	8	18	32

b. *Subsidiary quantum number.* This represents the various subsidiary orbits within a shell; they may be visualised as elliptical orbits. Thus in any one shell there are various subsidiary orbits denoted as the 1, 2, 3, 4 . . . or the s, p, d, f . . . orbits. The number of orbits in any one shell is, however, limited, and in dealing with the structures of atoms in their normal states, it is only necessary to consider the orbits listed at the bottom of page 111. The 1s orbit may be referred to as the K.1 orbit, but the former is the commonest usage.

c. *Spin quantum number*. The Pauli or Exclusion principle states that all the electrons in any atom must be distinguishable, or that no two electrons in a single atom can have all their quantum numbers alike.

To agree with the principle it is assumed that electrons spin on their axes; this assumption also accounts for the splitting of spectral lines observed with a spectroscope of good resolving power. Because of electron spin, a spin quantum number is allotted to an electron to characterise it.

The shell of principal quantum number 1 can hold two electrons. As these cannot be exactly alike, because of the Pauli principle, it is assumed that they have different spins. In general, if two electrons occupy the same orbit they must have different spins, and, as electrons in the same orbit can only differ by having different spins, it follows that no orbit can contain more than two electrons.

d. *Magnetic quantum number*. For electrons in orbits of principal quantum number greater than 1, a further complication arises. The shell with principal quantum number of 2 can contain a maximum of eight electrons. Of these, two, with opposed spins, will be in the 2s level. The remaining six will be in the 2p level, but for all six to be different it is necessary to subdivide the 2p level still further, into three, so that each of the three subdivisions may contain two electrons. These three 2p orbits may be envisaged as in different planes and can be denoted as 2p_x, 2p_y and 2p_z orbits (Fig. 22).

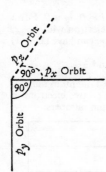

This subdivision of p orbits into three, and a similar subdivision of d orbits into five and of f orbits into seven, necessitates a fourth quantum number known as the magnetic quantum number. For simple purposes this is not of great importance since the energy of an electron in any of the three p orbits is the same unless the atom is placed in a strong magnetic field. When this happens the three p orbits take up different positions with respect to the lines of force of the field and attain slightly different energy levels. This accounts for the splitting of spectral lines when the source of emission is placed in a magnetic field (the Zeeman effect) or in an electric field (the Stark effect).

FIG. 22. Illustration of the directional arrangement of p orbits.

e. *Summary*. To sum up, four quantum numbers are required to characterise completely any particular electron in a particular orbit. In a simple way this corresponds to a normal post office address. To characterise a particular Mr. X it is necessary to allot a particular address to him, e.g. Mr. X, 114, High Street, Eton, Bucks. The county corresponds to the principal quantum number, the town to the subsidiary quantum number, the street to the spin quantum number, and the street number to the magnetic quantum number.

The idea of quantum numbers has been presented from a pictorial point of view. What quantum numbers are allowed arises from the solution of the Schrödinger equation (p. 118) and those quantum numbers which are allowed can be summarised as follows:

 i n = principal quantum number. The allowed values are 1, 2, 3, 4. . . .
 ii l = subsidiary quantum number. The allowed values of l depend on the value of n. When n is 1, l is 0, i.e. there are only s electrons in the shell with principal quantum number 1.
　　When n is 2, l can be 0 or 1, i.e. the shell with principal quantum number of 2 contains both s and p electrons.
　　When n is 3, l can be 0, 1 or 2, i.e. s, p and d electrons. When n is 4, l may be 0, 1, 2 or 3, i.e. s, p, d and f electrons.
　　It will be seen that $l = 0$ for an s electron, 1 for a p electron, 2 for a d electron and 3 for an f electron.
 iii m = magnetic quantum number. For an s electron with $l = 0$, m is 0. For a p electron, with $l = 1$, m can be -1, 0 or $+1$. For a d electron with $l = 2$, m can be -2, -1, 0, $+1$ and $+2$, and so on. In general, m can have $(2l + 1)$ different values.
 iv s = spin quantum number. The allowed values are $+\frac{1}{2}$ and $-\frac{1}{2}$.

The maximum number of electrons with the same principal quantum number can be summarised as follows, bearing in mind the fact that no two electrons in the same atom can have the same values for the four quantum numbers.

n	1	2				3								
l	0	0	1			0	1			2				
m	0	0	-1	0	$+1$	0	-1	0	$+1$	-2	-1	0	$+1$	$+2$
s	$+\frac{1}{2}\ -\frac{1}{2}$	$+\frac{1}{2}\ -\frac{1}{2}$	$+\frac{1}{2}\ -\frac{1}{2}$	$+\frac{1}{2}\ -\frac{1}{2}$	$+\frac{1}{2}\ -\frac{1}{2}$	$+\frac{1}{2}\ -\frac{1}{2}$	$+\frac{1}{2}\ -\frac{1}{2}$	$+\frac{1}{2}\ -\frac{1}{2}$	$+\frac{1}{2}\ -\frac{1}{2}$	$+\frac{1}{2}\ -\frac{1}{2}$	$+\frac{1}{2}\ -\frac{1}{2}$	$+\frac{1}{2}\ -\frac{1}{2}$	$+\frac{1}{2}\ -\frac{1}{2}$	$+\frac{1}{2}\ -\frac{1}{2}$

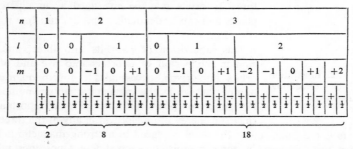

    2   8   18

6. The arrangement of electrons in orbits. The arrangement of electrons in an atom in its normal state is that which makes its energy a minimum, i.e. it is the most stable arrangement, within the limitations already mentioned.
　　The energy levels of the various orbits can be obtained from spectroscopic data, and the detailed arrangements of electrons can be built up from a knowledge of these energy levels, from the position of the

atom concerned in the periodic table, and from the allowable quantum numbers.

The relative energy levels of orbits in an atom, as obtained from spectroscopic measurements, are shown in Fig. 23. The precise positioning of one

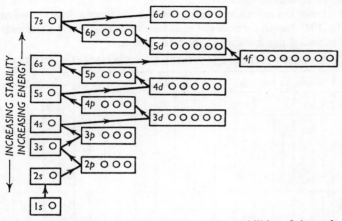

Fig. 23. Showing the approximate comparative stabilities of the various orbits in an atom.

orbit to another depends on the atomic number of the element concerned. The order given is that for elements of low atomic number, and it changes to some extent with elements of higher atomic number. This is because of the changing nuclear attraction for the electrons as the positive charge on the nucleus changes.

In Fig. 23 each circle represents an orbit which can be occupied either by a single electron or by two electrons with different spins. The circles enclosed within a rectangle represent orbits of equal energy. In passing along the periodic table the electrons occupy the orbits in energy order, as shown by the arrows in Fig. 23.

The order is conveniently remembered as follows:

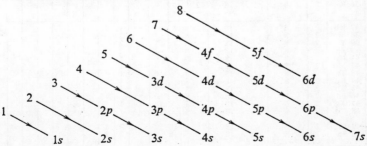

The one electron in a hydrogen atom will occupy the most stable orbit, i.e. the $1s$ orbit, and the second electron in the helium atom will occupy the same orbit but will have a different spin. The $1s$ orbit is now full; a third electron will occupy the next most stable orbit, i.e. the $2s$ orbit, and so on. The best 'seats' are occupied first.

On this basis, but not differentiating between the three p, five d or seven f levels (p. 109), the arrangement of electrons in the atoms of the elements in their normal states is shown in the table below.

THE ARRANGEMENT OF ELECTRONS IN THE ATOMS

	1s	2s	2p	3s	3p	3d	4s	4p	4d	4f	5s	5p	5d	5f	6s	6p	6d	7s	
1 H	1																		
2 He	2																		1s Full
3 Li	2	1																	
4 Be	2	2																	
5 B	2	2	1																2s Full
6 C	2	2	2																
7 N	2	2	3																
8 O	2	2	4																
9 F	2	2	5																
10 Ne	2	2	6																2p Full
11 Na	2	8		1															
12 Mg	2	8		2															
13 Al	2	8		2	1														3s Full
14 Si	2	8		2	2														
15 P	2	8		2	3														
16 S	2	8		2	4														
17 Cl	2	8		2	5														
18 A	2	8		2	6														3p Full
19 K	2	8		8			1												
20 Ca	2	8		8			2												4s Full
21 Sc	2	8		8		1	2												
22 Ti	2	8		8		2	2												
23 V	2	8		8		3	2												
24 Cr	2	8		8		5	1												
25 Mn	2	8		8		5	2												
26 Fe	2	8		8		6	2												
27 Co	2	8		8		7	2												
28 Ni	2	8		8		8	2												
29 Cu	2	8		8		10	1												
30 Zn	2	8		8		10	2												3d Full
31 Ga	2	8		18			2	1											
32 Ge	2	8		18			2	2											
33 As	2	8		18			2	3											
34 Se	2	8		18			2	4											
35 Br	2	8		18			2	5											
36 Kr	2	8		18			2	6											4p Full
37 Rb	2	8		18			8				1								
38 Sr	2	8		18			8				2								5s Full
39 Y	2	8		18			8		1		2								
40 Zr	2	8		18			8		2		1								
41 Nb	2	8		18			8		4		1								
42 Mo	2	8		18			8		5		1								
43 Tc	2	8		18			8		6		1								
44 Ru	2	8		18			8		7		1								
45 Rh	2	8		18			8		8		1								
46 Pd	2	8		18			8		10										
47 Ag	2	8		18			8		10		1								
48 Cd	2	8		18			8		10		1								4d Full
49 In	2	8		18			18				2	1							
50 Sn	2	8		18			18				2	2							

Transition elements (19–30); Transition elements (37–48)

These electronic arrangements are commonly written in the form $1s^2$, $2s^2$, $2p^6$, $3s^2$, $3p^6$, which is the argon arrangement.

A more detailed arrangement of electrons can be predicted by application of the rule of maximum multiplicity. This empirical rule, suggested by Hund and sometimes known as the Hund rule, states that the distribution of electrons in a free atom between the three *p*, five *d* and seven *f* orbits is such that as many orbits as possible are occupied by a single electron before any pairing of electrons with opposed spins takes place. Thus if three

OF THE ELEMENTS IN THEIR NORMAL STATES

Group	Element	1s	2s 2p	3s 3p 3d	4s 4p 4d	4f	5s	5p	5d	5f	6s	6p	6d	7s	
	51 Sb	2	8	18	18		2	3							
	52 Te	2	8	18	18		2	4							
	53 I	2	8	18	18		2	5							
	54 Xe	2	8	18	18		2	6							**5*p* Full**
	55 Cs	2	8	18	18		8				1				
	56 Ba	2	8	18	18		8				2				**6*s* Full**
Transition elements; Rare earths (or lanthanons)	57 La	2	8	18	18		8		1		2				
	58 Ce	2	8	18	18	2	8				2				
	59 Pr	2	8	18	18	3	8				2				
	60 Nd	2	8	18	18	4	8				2				
	61 Pm	2	8	18	18	5	8				2				
	62 Sm	2	8	18	18	6	8				2				
	63 Eu	2	8	18	18	7	8				2				
	64 Gd	2	8	18	18	7	8		1		2				
	65 Tb	2	8	18	18	9	8				2				
	66 Dy	2	8	18	18	10	8				2				
	67 Ho	2	8	18	18	11	8				2				
	68 Er	2	8	18	18	12	8				2				
	69 Tm	2	8	18	18	13	8				2				
	70 Yb	2	8	18	18	14	8				2				
	71 Lu	2	8	18	18	14	8		1		2				**4*f* Full**
	72 Hf	2	8	18	32		8		2		2				
	73 Ta	2	8	18	32		8		3		2				
	74 W	2	8	18	32		8		4		2				
	75 Re	2	8	18	32		8		5		2				
	76 Os	2	8	18	32		8		6		2				
	77 Ir	2	8	18	32		8		7		2				
	78 Pt	2	8	18	32		8		9		1				
	79 Au	2	8	18	32		8		10		1				
	80 Hg	2	8	18	32		8		10		2				**5*d* Full**
	81 Tl	2	8	18	32		18				2	1			
	82 Pb	2	8	18	32		18				2	2			
	83 Bi	2	8	18	32		18				2	3			
	84 Po	2	8	18	32		18				2	4			
	85 At	2	8	18	32		18				2	5			
	86 Rn	2	8	18	32		18				2	6			**6*p* Full**
	87 Fr	2	8	18	32		18				8			1	
	88 Ra	2	8	18	32		18				8			2	**7*s* Full**
	89 Ac	2	8	18	32		18				8		1	2	
Actinons	90 Th	2	8	18	32		18				8		2	2	
	91 Pa	2	8	18	32		18			2	8		1	2	
	92 U	2	8	18	32		18			3	8		1	2	
	93 Np	2	8	18	32		18			5	8			2	
	94 Pu	2	8	18	32		18			6	8			2	
	95 Am	2	8	18	32		18			7	8			2	
	96 Cm	2	8	18	32		18			7	8		1	2	
	97 Bk	2	8	18	32		18			9	8			2	
	98 Cf	2	8	18	32		18			10	8			2	
	99 Es	2	8	18	32		18			11	8			2	
	100 Fm	2	8	18	32		18			12	8			2	
	101 Md	2	8	18	32		18			13	8			2	
	102 No	2	8	18	32		18			14	8			2	

electrons are to occupy the three p orbits in any one shell one will go into each of the three available orbits.

A more detailed representation of atomic structures, showing the spins of the electrons, is made possible by the application of this rule, and the electronic arrangements in the atoms having atomic numbers of 1 to 10 are shown in Fig. 24.

		$1s$	$2s$	$2p_x$	$2p_y$	$2p_z$
1	H	↑				
2	He	↑↓				
3	Li	↑↓	↑			
4	Be	↑↓	↑↓			
5	B	↑↓	↑↓	↑		
6	C	↑↓	↑↓	↑	↑	
7	N	↑↓	↑↓	↑	↑	↑
8	O	↑↓	↑↓	↑↓	↑	↑
9	F	↑↓	↑↓	↑↓	↑↓	↑
10	Ne	↑↓	↑↓	↑↓	↑↓	↑↓

FIG. 24. Illustration of the spins of electrons in simple atoms.

THE PERIODIC TABLE

The periodic table of the elements was first drawn up by listing the elements in the order of their relative atomic masses, as described on pp. 25–31. At first, there was no understanding as to why such an arrangement should fit elements into groups with similar chemical properties. The deeper significance of the arrangement came, first, from a study of X-ray spectra and, more completely, from a knowledge of the arrangements of the electrons in different atoms.

7. X-ray spectra. When a solid anode is bombarded by cathode rays to produce X-rays, the X-rays given out are of varying wavelength. Bombardment of every element gives a general, or white, X-radiation which shows up in an X-ray spectrum as a continuous background. Superimposed on this background, however, are a number of lines characteristic of the particular element being used as the anode. These lines occur in groups, known as the K, L, M, N . . . groups, and there are various lines within each group.

By measuring the wavelengths of corresponding lines in the X-ray spectra of as many elements as possible, Moseley discovered, in 1913, that the square root of the frequency of corresponding lines for different elements gave almost a straight line when plotted against the atomic numbers of the elements concerned (Fig. 25).

This suggested that the atomic number of an element is really of more significance than the relative atomic mass in the periodic table arrangement, and, on this basis, Moseley was able to make some adjustments in the older periodic table orders. Cobalt, for example, has a higher relative

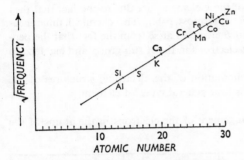

FIG. 25. The relationship between the atomic numbers of some simple elements and the square root of the frequency of the Kα lines in their X-ray spectra.

atomic mass than nickel and ought to follow it in the periodic table so far as relative atomic mass is concerned. Moseley found, however, that X-ray spectral lines of nickel had higher frequencies than those of cobalt, and he therefore placed nickel after cobalt.

8. Origin of X-rays. X-rays originate from atoms bombarded by fast-moving electrons. It is now accepted that the bombarding electrons knock electrons out of the inner shells of the bombarded atoms. The displaced electrons are then replaced by electrons passing from outer to inner levels, with corresponding emission of radiation. A displaced electron from the K shell may be replaced by one from the L, M, N . . . shells, and, possibly, by a free electron from outside the atom. Such electrons shifts give rise to the group of K lines in an X-ray spectrum, and other groups of lines are formed similarly.

The X-rays of continuously varying wavelength which provide the background to an X-ray spectrum are thought to originate from collisions of bombarding electrons with atomic nuclei.

9. Electronic structures and the periodic table. Once the arrangement of electrons in different atoms had been worked out, the significance of the periodic table became much clearer. In particular, all the atoms in any one group of the table were often found to have the same number of electrons in their outermost orbit. For example, the elements in Group 1 have the structures given on the next page.

Li 2.1

Na 2.8.1

K 2.8.8.1	Cu 2.8.18.1
Rb 2.8.18.8.1	Ag 2.8.18.18.1
Cs 2.8.18.18.8.1	Au 2.8.18.32.18.1
Fr 2.8.18.32.18.8.1	
(*A* sub-group)	(*B* sub-group)

The chemical similarities of these elements are due to the fact that they all have one electron in the outermost orbit. The chemical differences between the sub-group *A* and *B* elements arise from the fact that the penultimate orbit contains eight electrons in the *A* sub-group and eighteen in the *B* sub-group.

Moreover, a complete examination of the electronic structures of the atoms of the elements reveals four general types of element.

a. *The noble gases.* The arrangement of the electrons in the atoms of the noble gases is as follows:

	1	2		3			4				5			6	
	s	*s*	*p*	*s*	*p*	*d*	*s*	*p*	*d*	*f*	*s*	*p*	*d*	*s*	*p*
He	2														
Ne	2	2	6												
A	2	2	6	2	6										
Kr	2	2	6	2	6	10	2	6							
Xe	2	2	6	2	6	10	2	6	10		2	6			
Rn	2	2	6	2	6	10	2	6	10	14	2	6	10	2	6

In all these atoms the orbits contain the maximum number of electrons. The shell is not necessarily complete but all the sub-levels within a shell are complete. Such an arrangement accounts for the stability of the noble gases.

b. *Atoms with all except the outer shell complete.* These include all the elements, in Groups 1–7 B, 1 A and 2 A in the Thomsen–Bohr arrangement of the periodic table given on page 30.

Such elements may be regarded as normal elements. They have a more or less fixed valency.

c. *Atoms with the two outermost shells incomplete.* These include all the elements in Groups 3–7 A and 8 in the Thomsen–Bohr arrangement. The elements are known as *transition elements.* They form coloured ions, which are paramagnetic, exhibit variable valency, and often possess marked catalytic activity.

d. *Atoms with the three outermost shells incomplete*. These elements are known as the *rare earths* or *lanthanons* or lanthanides, and as the *actinons* or actinides. They are series of elements with very similar chemical properties, and are placed in Group 3 A in the Thomsen-Bohr arrangement.

The electronic structures given on page 112 shows that the lanthanon and actinon series have the same number of electrons in their two outermost orbits, and that the series are formed by a filling up of the $4f$ or $5f$ levels. Electrons in such inner levels play only a small part in determining the chemical properties of an element and that is why the lanthanons and actinons are so chemically similar within their series.

PARTICLES AND WAVES

10. Dual nature of light. The word 'ray' is used somewhat indiscriminately to describe both a stream of particles and wave-like radiation, but the usage is, perhaps, valid when it is realised that particles may have a wave-like aspect whilst waves may have a particle-like aspect.

Newton originally regarded light on a corpuscular theory, but Huyghens introduced the wave theory of light. The wave theory is essential in accounting for interference and diffraction phenomena, whereas the corpuscular theory is necessary to explain, for instance, the photoelectric effect.

In this effect, metals give off electrons when illuminated with light of appropriate wavelengths. Moreover, there is a simple relationship between the energy of the emitted electrons and the frequency of the incident radiation. This result was interpreted, by Einstein in 1905, as meaning that radiation could be regarded as made up of small 'packets' of energy, known as photons. The energy of a photon was dependent on the frequency, according to the basic equation of the quantum theory (p. 105).

$$E = h\upsilon$$

which explained why the energy of the emitted electrons in the photoelectric effect was related to the frequency of the radiation.

11. The wave nature of an electron. So far, the electron has been regarded as a tiny, negatively charged particle, but very important results come from a consideration of the wave nature of an electron.

de Broglie first suggested, in 1924, that moving electrons had waves of definite wavelength associated with them, and this theoretical prediction was demonstrated experimentally when Davisson and Germer showed, in 1927, that a stream of electrons could be diffracted by crystals acting as simple diffraction gratings, just as light- or X-rays can. As it is only possible to account for diffraction in terms of waves, it is necessary to assume that a stream of electrons behaves as a wave-like radiation such as light-or X-rays.

The de Broglie relationship between moving electrons and the waves associated with them can be expressed mathematically as

$$\lambda = \frac{h}{mv}$$

where λ is the wavelength of the waves, h is the Planck constant, m is the mass of the electron, and v is the velocity of the electron. If this expression is written as

$$\text{momentum} \times \lambda = h$$

it relates the particle-like aspect of an electron, i.e. momentum, to the wave-like aspect, i.e. wavelength.

The expression is true for all particles, but it is only with very small particles that the wave-like aspect is of any significance. The wavelength of a large particle will be so small that its wave-like properties will not be measurable or observable. An electron of energy 1.6×10^{-19} J or 1 eV (p. 142) has an associated wavelength of 1.2 nm; an α-particle from radium, a wavelength of 6.6×10^{-6} nm; and a golf ball, travelling at 30 m s^{-1}, a wavelength of 4.9×10^{-25} nm.

It is not easy to obtain a pictorial idea of this new conception of an electron, but it is possible to treat the wave nature of an electron mathematically. This is done by a specialised technique known as wave mechanics or quantum mechanics.

So far as atomic structure is concerned, the idea of tiny, negatively charged particles existing in fixed orbits around the nucleus of an atom is replaced by the idea of charge clouds of varying charge density existing in a wave pattern around the nucleus. The wave pattern for an electron, or a group of electrons, can be expressed in the form of a mathematical equation, which involves a wave function (ψ), the total energy (W) and the potential energy (V) of the system, the mass of the electron (m), the Planck constant (h) and the co-ordinates of the system. The equation, developed by Schrödinger, is written as follows for a single-electron system,

$$\frac{\delta^2\psi}{\delta x^2} + \frac{\delta^2\psi}{\delta y^2} + \frac{\delta^2\psi}{\delta z^2} + \frac{8\pi^2 m}{h^2}(W - V)\psi = 0$$

The equation can only be solved satisfactorily when the total energy of the system has certain definite values, and these values correspond to those of the more precise stationary states in the Bohr model of the atom. The allowable wave functions corresponding to the energy values which allow the Schrödinger equations to be solved are known as characteristic functions or eigenfunctions. From these eigenfunctions, the probability of finding an electron in the region around a nucleus can be calculated, and the region in which the electron might have any probability of existing is referred to as an *orbital*.

An electron can no longer be said to occupy a specific, limited orbit. Instead, it must be thought of as existing in a much more diffuse region known as an atomic orbital. The relation between this new idea and the older one of stationary states is that the position of greatest probability for finding the electron closely corresponds to that of the more definite orbits previously envisaged.

12. Representation of atomic orbitals. As there is a probability of finding an

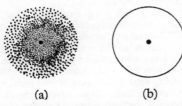

(a)   (b)

FIG. 26. Two methods of representing a $1s$ atomic orbital. (*a*) Showing the variation of charge density by shading. (*b*) The boundary surface for a $1s$ atomic orbital.

electron in a diffuse region around a nucleus there is a distribution of charge around the nucleus, and atomic orbitals are represented as charge clouds having varying charge density. On this basis, the single electron in the $1s$ orbital of a hydrogen atom may be represented as shown in Fig. 26a, which tries to depict the variation in charge density. A rather simpler method, however, is more commonly used, and this maps out a boundary surface within which an electron might be said to exist; the boundary surface for a $1s$ orbital is shown in Fig. 26b.

This means that the electron exists somewhere within a sphere with the atomic nucleus at its centre. All electrons in s orbitals have similar spherical boundary surfaces, and s orbitals are said to be spherically symmetrical.

The charge cloud is not concentrated in any particular direction.

p orbitals, however, are dumb-bell in shape and have a marked directional character depending on whether a p_x, a p_y or a p_z orbital is being considered. The boundary surfaces of p orbitals are shown in Fig. 27. These p orbitals possess a nodal plane, i.e. a plane in which the probability of finding the electron is zero.

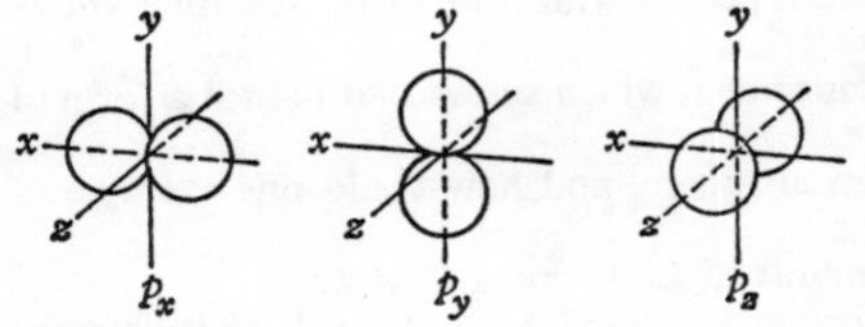

FIG. 27. The boundary surfaces of p_x, p_y and p_z orbitals.

The shape of the boundary surfaces of s and p orbitals, and of other orbitals obtained by combination of s and p orbitals, is of very great significance, in considering chemical combination between atoms, as will be seen in the next chapter.

13. The uncertainty principle. One of the fundamental differences between an electron and a larger particle is that the wave-like aspect of the electron is of much greater significance than the wave-like aspect of the larger particle (p. 118).

There is, too, a further difference. Both the position and the velocity of a large particle, e.g. a planet, at any one time can be measured with fair accuracy, but this cannot be done for anything so small as an electron. There is simply no way of measuring the velocity of an electron exactly and of locating it exactly at any one time. This is because any method of measurement affects the electron being measured. An electron might be detected, for example, by using very short wavelength X-rays, or γ-rays, if the electron would cause scattering of the rays, but the speed and direction of the electron would be affected by the rays, and electrical or mechanical methods would have the same result.

The information which can be obtained about an individual electron is, therefore, far from precise, and this is one example of the application of the Uncertainty Principle put forward by Heisenberg in 1927. This states that the more accurately the position of a particle is defined the less accurately is its velocity known, and the more accurately the velocity is defined the less accurately is its position known.

de Broglie's relationship (p. 117) indicates why this is so. A long wavelength can be measured with greater fractional accuracy than a short one. If, therefore, a particle has a small momentum, and a correspondingly large wavelength, the wavelength can be measured with some accuracy, but this is at the expense of a relatively inaccurate determination of the small momentum. Alternatively, if the momentum is large and can be measured with some accuracy, the wavelength will be small and will not be known with any accuracy.

Because it is impossible to know the position and the velocity of any one electron, the best that can be done is to try to determine the probable position and the probable velocity. Fortunately, wave mechanics enables the various probabilities to be calculated, but, in so doing, the simple physical picture of a single, particle-like electron moving with known velocity in a definite and precise orbit is lost.

QUESTIONS ON CHAPTER 11

1. How would you try to convince a person with little or no scientific knowledge that light is a form of energy?

2. Describe any experimental arrangement which shows that light is a form of energy.

3. What different types of spectra are there, and how could one example of each type be obtained?

4. What was the origin of the symbols K, L, M and s, p, d, f?

5. The value of the Rydberg constant is 1.097373×10^7 m^{-1}. Calculate the wavelength of the first lines in each of the four common spectral series of hydrogen.

6. The Planck constant can be measured by investigation of black body radiation or of the photochemical effect. Describe the general principle of these methods.

7. (a) What observations led to the original ideas of the quantum theory? (b) How does the quantum theory account for the variation of the specific heat capacity of a solid with temperature?

8. 'Planck's quantum theory is the foundation of modern chemistry.' Comment on this statement. To what extent would chemistry have developed without the quantum theory?

9. The value of the Planck constant is 6.6256×10^{-34} when expressed in J s. Express it in any three other sets of units.

10. What observations were mainly responsible for the original presentation of the quantum theory?

11. If the energy levels in a hydrogen atom corresponding with principal quantum numbers 1, 2, 3, 4, 5 and 6 are -217.9, -54.48, -24.2, -13.62, -8.7 and -6.05 joule $\times 10^{20}$ respectively, calculate the wavelengths of lines you would expect to observe in a hydrogen spectrum. What would be the ionisation energy for hydrogen?

12. The relationship $\lambda = h/mv$ shows that an electron of mass 9.1×10^{-31} kg and velocity 5.9×10^5 m s^{-1} has an associated wavelength of 1.234 nm. Calculate the wavelength associated with a golf ball of mass 45 gramme travelling at 30 m s^{-1}.

13. What is the importance of the Pauli principle?

14. Write short notes on the following: (i) stationary states, (ii) Planck's constant, (iii) electron spin, (iv) Thomsen.

15. What part did a study of X-ray spectra play in the development of chemistry?

16. Explain the relationship between the position of an element in the periodic table and the atomic structure of its atoms.

17. Give an account of the experimental observations which led to the discovery of the noble gases, and describe a chemical method of preparing crude argon from air. Comment on the importance of the inert gas group in the study of the periodic classification of the elements and the understanding of atomic structures. The atomic numbers of the noble gases are He, 2; Ne, 10; A, 18; Kr, 36; Xe, 54. (W.)

18. Explain concisely what is meant by the periodic classification of the elements, indicating, with one example of each, the significance of the terms 'group', 'short period' and 'transitional element'. How is the electronic structure of an element related to (a) its position in the periodic classification, (b) its valency?

Give the electronic structures of the following atoms: argon, carbon, chlorine, sodium. How is it that iron (atomic number 26) can have two valencies? (N.)

19. Justify the statement: 'The atomic number of an element gives more information regarding its atomic structure and is of greater usefulness in its classification than its relative atomic mass.' (N.)

20. Write an account on 'Particles and Waves'.

21. To what extent is it correct to refer to a β-particle as a β-ray or an α-particle as an α-ray. Could a γ-ray be referred to as a γ-particle? If so, why is it not more commonly done?

22. What is the uncertainty principle and how does it affect theories of atomic structure?

23. Comment on the statement that 'an observer must always be part of his experiment'.

Chapter 12
Radioactive disintegration and isotopes

A DETAILED investigation of the disintegration of naturally occurring radioactive elements was carried out, at the start of this century, by Rutherford, Soddy, Russell and others. The investigation involved consideration of the rate of radioactive disintegration or decay, the types of radioactive change which took place, and the nature of the products.

Perhaps the most important result was the discovery of isotopes. Many products of radioactive decay were discovered, and it was found that some of them were chemically identical, having atoms with the same atomic numbers but different relative atomic masses. Such elements were called isotopes, by Soddy, and, since then, many isotopes of non-radioactive elements have been isolated.

RADIOACTIVE DISINTEGRATION

1. Rate of disintegration. The natural disintegration of a radioactive element cannot be retarded or speeded up by any physical or chemical means, and the rate of a radioactive change is a characteristic of the atom of the element concerned. There are various ways in which the rate can be expressed.

The commonest method is to use the *half-life period*, which is the time required for one half of the element concerned to disintegrate, or the time for the radioactivity of an element to be reduced to half its initial value. The half-life period of radium, for example, is 1 620 years. This means that 1 kg of radium will be reduced to 0.5 kg in 1 620 years, to 0.25 kg in a further 1 620 years, and so on. The original amount makes no difference; 1 g would be reduced to 0.5 g in the same time as 1 kg is reduced to 0.5 kg. For other elements the half-life period may be a fraction of a second, or millions of years. It is convenient to remember that a time of 10 half-lives reduces the activity of a radioactive element about 1 000 times.

The rate of disintegration is directly proportional to the number of atoms present, and the rate gets slower and slower as more and more atoms break up. The rate can be expressed in a relationship similar to that used for first-order reactions (p. 337),

$$-\frac{\mathrm{d}N}{\mathrm{d}t} = \lambda N$$

where N is the number of undecayed atoms, t is the time in seconds and λ is a constant. If N_0 is the number of radioactive atoms present at zero time, then

$$N/N_0 = e^{-\lambda t}$$

The constant, λ, is called the *radioactive decay constant*; its units are s^{-1}.

After the half-life period, N will equal 0.5 N_0, so that

$$0.5 = e^{-\lambda t_{\frac{1}{2}}}$$

$$t_{\frac{1}{2}} = \frac{\ln 2}{\lambda} = \frac{0.693}{\lambda}$$

where $t_{\frac{1}{2}}$ is the half-life period in seconds. It will be seen that a large value of λ gives a small value of $t_{\frac{1}{2}}$. The rate of disintegration is expressed in terms of the number of disintegration per second, i.e. the number of atoms which split up in one second.

Another unit of disintegration rate known as *the curie* is also used. This was, originally, taken as the rate of disintegration of 1 g of radium, but it is now redefined as 37×10^9 disintegrations per second. The amount of a radioactive substance which undergoes 37×10^9 disintegrations per second is also called a curie. The millicurie (one thousandth of a curie) and the microcurie (one millionth of a curie) are also used.

The radioactive decay constant for ^{238}U, with a half-life of 4.5×10^9 years will be given by

$$\lambda = \frac{0.693}{4.5 \times 10^9 \times 365 \times 24 \times 3600} = 4.883 \times 10^{-18} \text{s}^{-1}$$

238 g of ^{238}U contain 6.02×10^{23} atoms. The rate of decay of 238 g of ^{238}U in disintegrations per second will, therefore, be $4.883 \times 10^{-18} \times 6.02 \times 10^{23}$, i.e. 2.939×10^6. The rate of decay of 1 g of ^{238}U will be $\frac{1}{238}$ of this.

As 1 curie is a rate of disintegration of 3.7×10^{10} s^{-1}, 238 g of ^{238}U represent $2.939 \times 10^6/3.7 \times 10^{10}$, i.e. 7.943×10^{-5} curie. 1 curie of ^{238}U will be 2.996 $\times 10^6$ g.

For cobalt-60, with a half-life of 5.2 years, 1 curie is equivalent to 8.73×10^{-4} g. It will be seen that there is a remarkable range of disintegration rates and that the mass of a radioactive substance representing 1 curie varies very greatly.

These calculations can be generalised into the statements

No. of grammes of isotope equivalent to 1 curie $= 8.872 \times 10^{-14} \times A \times t_{\frac{1}{2}}$

No. of curies per gramme of isotope $\qquad = \dfrac{1.127 \times 10^{13}}{A \times t_{\frac{1}{2}}}$

where A is the mass number and $t_{\frac{1}{2}}$ the half-life period of the isotope in seconds.

2. Types of disintegration. The commonest type of radioactive change is brought about by the loss of either an α- or a β-particle. γ-radiation may result from energy changes within an atom following the loss of an α- or β-particle. In artificial radioactivity, too, disintegration may take place by other processes (p. 146).

a. *Loss of an α-particle.* Disintegration by loss of an α-particle (a helium nucleus of mass 4 and charge $+2$) means a lowering of mass number of 4 and of atomic number of 2. Uranium, for example, with mass number 238 and atomic number 92 forms an element with a mass number of 234 and an atomic number of 90 when it loses an α-particle. The new element is called uranium X_1; it is an isotope of thorium. The change is conveniently represented as

$$^{238}_{92}U \longrightarrow UX_1(\text{or } ^{234}_{90}Th) + {}^{4}_{2}He$$

the superscripts showing the mass numbers and the subscripts showing atomic numbers.

b. *Loss of a β-particle.* Loss of a β-particle (an electron) causes no significant change in mass number but an increase of 1 in atomic number. This is because the loss of 1 negative charge is equivalent to the gain of 1 positive charge. Uranium X_1, for example, disintegrates by loss of β-particles, the product, uranium X_2, having a mass number of 234 and an atomic number of 91; it is an isotope of protoactinium. The change is written as

$$UX_1(\text{or } ^{234}_{90}Th) \longrightarrow UX_2(\text{or } ^{234}_{91}Pa) + {}^{0}_{-1}e$$

3. Disintegration series.

The final outcome of much research was that the disintegration of naturally occurring radioactive elements could be summarised in three radioactive disintegration series, known as the uranium, thorium and actinium series. The disintegration series of uranium is typical and the table on page 125 shows how the various changes take place and the nature of the elements involved.

It is fairly unusual for a radioactive change to take place in two alternative ways, but radium C can either lose an α- or a β-particle. 99.96 per cent of radium C′ is formed and only 0.04 per cent of radium C″. Uranium I and II (atomic number, 92), radium A, C′ and F (atomic number, 84) and radium B, D and G (atomic number 82) form groups of isotopes in this series.

The thorium and actinium series are shown, using modern notation, as follows:

a. Thorium Series

$$^{232}_{90}Th \xrightarrow{-\alpha} {}^{228}_{88}Ra \xrightarrow{-\beta} {}^{228}_{89}Ac \xrightarrow{-\beta} {}^{228}_{90}Th \xrightarrow{-\alpha} {}^{224}_{88}Ra \xrightarrow{-\alpha}$$

$$\longrightarrow {}^{220}_{86}Rn \xrightarrow{-\alpha} {}^{216}_{84}Po \xrightarrow{-\alpha} {}^{212}_{82}Pb \xrightarrow{-\beta} {}^{212}_{83}Bi \quad \begin{array}{c} \xrightarrow{-\beta} {}^{212}_{84}Po \xrightarrow{-\alpha} \\ \\ \xrightarrow{-\alpha} {}^{208}_{81}Tl \xrightarrow{-\beta} \end{array} {}^{208}_{82}Pb$$

b. Actinium Series

$$^{231}_{91}Pa \xrightarrow{-\alpha} {}^{227}_{89}Ac \xrightarrow{-\beta} {}^{227}_{90}Th \xrightarrow{-\alpha} {}^{223}_{88}Ra \xrightarrow{-\alpha} {}^{219}_{86}Rn \xrightarrow{-\alpha}$$

$$\longrightarrow {}^{215}_{84}Po \xrightarrow{-\alpha} {}^{211}_{82}Pb \xrightarrow{-\beta} {}^{211}_{83}Bi \quad \begin{array}{c} \xrightarrow{-\beta} {}^{211}_{84}Po \xrightarrow{-\alpha} \\ \\ \xrightarrow{-\alpha} {}^{207}_{81}Tl \xrightarrow{-\beta} \end{array} {}^{207}_{82}Pb$$

The position of the various isotopes involved in these series in the periodic table is governed by the fact that loss of an α-particle leads to the formation of an element situated two places to the left in the periodic table whilst loss of a β-particle gives an element situated one place to the right.

It will be seen that the stable end product of each of the three disintegration series is an isotope of lead.

Element	Modern symbol	Periodic table group no.	Half-life
Uranium I	$^{238}_{92}U$	6	4.5×10^9 year
$\downarrow -\alpha$			
Uranium X_1	$^{234}_{90}Th$	4	24.5 day
$\downarrow -\beta$			
Uranium X_2	$^{234}_{91}Pa$	5	1.14 min
$\downarrow -\beta$			
Uranium II	$^{234}_{92}U$	6	2.7×10^5 year
$\downarrow -\alpha$			
Ionium	$^{230}_{90}Th$	4	8.3×10^3 year
$\downarrow -\alpha$			
Radium	$^{226}_{88}Ra$	2	1620 year
$\downarrow -\alpha$			
Radon	$^{222}_{86}Rn$	0	3.8 day
$\downarrow -\alpha$			
Radium A	$^{218}_{84}Po$	6	3 min
$\downarrow -\alpha$			
Radium B	$^{214}_{82}Pb$	4	26.8 min
$\downarrow -\beta$			
Radium C	$^{214}_{83}Bi$	5	19.7 min
$\swarrow -\beta \qquad -\alpha \searrow$			
Radium C′	$^{214}_{84}Po$	6	1.5×10^{-4} s
$-\alpha \searrow$ Radium C″	$^{210}_{81}Tl$	3	1.32 min
$\searrow -\beta$			
Radium D	$^{210}_{82}Pb$	4	22 year
$\downarrow -\beta$			
Radium E	$^{210}_{83}Bi$	5	5 day
$\downarrow -\beta$			
Radium F	$^{210}_{84}Po$	6	140 day
$\downarrow -\alpha$			
Radium G	$^{206}_{82}Pb$	4	Stable

ISOTOPES

4. Positive rays. Investigation of radioactive disintegration series shows that more than one element can occupy the same position in the periodic table. Such elements must have the same atomic numbers, but they are not absolutely alike for they have different mass numbers. Soddy, as a result of his work on radioactivity, called such elements isotopes. They have the same atomic number, but different mass numbers, or, in other words, the same arrangement of electrons and the same number of protons, but different numbers of neutrons in the nucleus.

The existence of isotopes of non-radioactive elements was shown by positive ray analysis. Positive rays, first discovered by Goldstein in 1886, are formed in a discharge tube as well as cathode rays, and may be observed by using a tube with a perforated cathode. If this is done, the

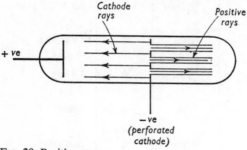

Fig. 28. Positive rays.

radiation passing through the holes in the cathode (Fig. 28) is found, by the effect of electric and magnetic fields, to consist of positively charged particles much heavier than the electrons of cathode rays.

Measurement of the charge/mass ratio for positive rays obtained in tubes containing different gases indicates that the rays are made up of positively charged atoms or molecules formed by loss of electrons from the atoms or molecules of the gas in the tube.

Application of magnetic and electric fields enables the various particles in the positive rays to be 'sorted out' according to their weight. Thomson first investigated positive rays from a tube containing neon, of relative atomic mass 20.183, and he found that neon was, in fact, made up of two isotopes, one of mass 20 and one of mass 22.

This method of positive ray analysis for detecting isotopes was further developed by Aston using an apparatus known as a mass spectrograph, and by Dempster, and others, using a mass spectrometer.

5. The mass spectrograph. In this apparatus, positive rays from a discharge tube are passed through two slits, A and B, in an evacuated vessel (Fig. 29).

The thin beam of rays is then subjected to an electric field between plates **P** and **O**, and this causes the beam to be deflected and spread out, because the slower positively charged particles are deflected more than the quicker ones. The spread-out beam is then passed through a diaphragm, **D**, and between the poles of a magnet, **M**. The magnetic field deflects the beam in the opposite direction to the deflection caused by the electric field and, once again, the slow particles are deflected most.

By adjusting the relative strengths of the electric and magnetic fields, all positively charged particles of the same mass, irrespective of their different velocities, can be focused on to the same line on a photographic plate at S.

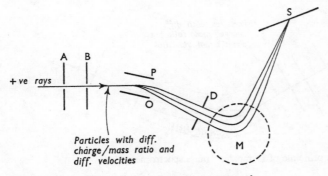

FIG. 29. The principle of Aston's mass spectrograph.

The result has the appearance of a line spectrum (p. 104), the lines on the plate each corresponding to a particle of definite mass in the positive rays; it is known as a mass spectrum. Moreover, the intensity of each line gives an indication of the relative amount of each particle present, though not very precisely.

As particles originally having widely different velocities are focused by the electrical and magnetic fields, Aston's method is known as velocity focusing. A mass spectrograph gives a photographic record, and is mainly intended for measuring isotopic masses.

6. The mass spectrometer. The relative abundance of each particle present in a beam of positive rays is best measured by a mass spectrometer. In this apparatus, a beam of positive rays is obtained by passing a slow stream of gas or vapour (from a heated filament), at very low pressure, through a beam of electrons. Collision with the electrons gives a beam of positive rays.

The beam consists of positively charged particles with different velocities and different charge/mass ratios. The beam is now passed through an electrical field between plates A and B (Fig. 30) and this produces a beam of particles all having about the same velocity, but still with different charge/mass ratios. On passing this beam through a magnetic field, particles with the same charge/mass ratio can be focused on to a conductor connected to an electrometer. Changing the magnetic field, or the electrical field between A and B, brings another group of particles with a different charge/mass ratio into focus on the conductor. The charge on the electrometer when different particles are in focus measures the relative abundance of the various particles.

A mass spectrometer differs from a mass spectrograph in three respects. The beam of positive rays is obtained differently, the particles in the beam

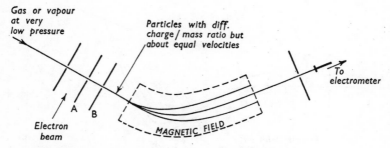

FIG. 30. The principle of Dempster's mass spectrometer.

are accelerated by an electrical field so as to have about the same velocity, and measurement of charge replaces the use of a photographic plate. The method of focusing is known as direction focusing, and the method has been applied, more recently, by Nier.

7. Examples of isotopes. Almost all elements have now been found to have stable, naturally occurring isotopes, and many have a very wide variety. Typical examples are given below.

Element	Isotope	Relative atomic mass of isotope	Relative abundance (per cent)
Carbon	$^{12}_{6}C$	12.000 00	98.89
	$^{13}_{6}C$	13.003 35	1.11
Chlorine	$^{35}_{17}Cl$	34.968 85	75.53
	$^{37}_{17}Cl$	36.965 34	24.47

a. *Isotopes of hydrogen.* Hydrogen has two main isotopes, known as hydrogen or protium and heavy hydrogen or deuterium. The atomic structures of the two isotopes are

| Hydrogen or   Heavy hydrogen |
| protium, ^1_1H or deuterium, ^2_1H or D |

$$\left(\text{1p}\right) \quad \text{1e} \qquad\qquad \left(\begin{array}{c}\text{1p}\\\text{1n}\end{array}\right) \quad \text{1e}$$

The existence of deuterium was discovered by Washburn and Urey, in 1931, by a careful examination of the line spectrum (p. 104) of the hydrogen resulting after several litres of liquid hydrogen had been evaporated.

This discovery was followed, in 1932, by the discovery of deuterium oxide, or heavy water, D_2O or ^2H_2O, by Washburn and Urey. They found that the residue remaining after water has been almost completely decomposed by electrolysis was rich in deuterium oxide, and almost pure deuterium oxide can now be obtained in large quantities by an extension of this method.

Normal hydrogen contains some deuterium (about 0.015 per cent) and ordinary water contains about the same proportion of heavy water.

The chemical properties of hydrogen and deuterium, or of water and deuterium oxide, are very similar, but the physical properties differ, as is seen in the following table:

	Hydrogen	*Deuterium*	*Water*	*Deuterium oxide*
Relative atomic mass .	1.007 825	2.0140		
f.p. or m.p. . . .	14 K	18.7 K	0°C	3.802°C
b.p.	20.4 K	23.5 K	100°C	101.42°C
Density (g cm^{-3}) at 20°C			0.998	1.106
Temp. of maximum density			4.0°C	11.6°C
Ionic product at 25°C .			1.0×10^{-14}	1.8×10^{-15}

A third isotope of hydrogen, known as tritium, can be made, though it is very doubtful if it exists naturally in ordinary hydrogen. Tritium is formed, for example, when the ^{6}Li isotope of lithium is bombarded by a stream of neutrons,

$$^6_3Li + ^1_0n \longrightarrow ^4_2He + ^3_1H$$

The structure of the tritium atom is

Tritium, ^3_1H

b. *Isotopes of uranium.* These isotopes are of particular significance in atomic energy work (p. 149). The structures of the isotopes are as follows:

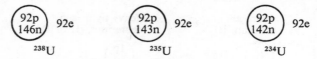

Natural uranium consists almost entirely of ^{238}U, with about 0.72 per cent (1 part in 140) of ^{235}U and 0.006 per cent of ^{234}U.

c. *Isotopes of oxygen.* Oxygen has three isotopes, with atomic structures

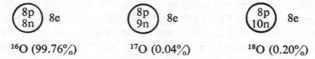

The percentage abundance of each isotope in ordinary oxygen is shown in brackets.

In conjunction with the three possible isotopes of hydrogen it is evident that these three isotopes of oxygen can form at least nine 'different' sorts of water formulated as $^1H_2{}^{16}O$, $^1H_2{}^{17}O$, $^1H_2{}^{18}O$, $^2H_2{}^{16}O$, $^2H_2{}^{17}O$, $^2H_2{}^{18}O$, $^3H_2{}^{16}O$, $^3H_2{}^{17}O$ and $^3H_2{}^{18}O$.

8. Physical relative atomic masses. The definition of the relative atomic mass of an element (p. 20) as

$$\frac{\text{Mass of 1 atom of the element}}{\frac{1}{16} \times \text{Mass of 1 atom of oxygen}}$$

has had to be adjusted since the discovery that there are three isotopes of oxygen.

Ordinary oxygen is a mixture of all three isotopes, and the mass of 1 atom of oxygen in the above definition now means the average mass of an atom in the mixture of isotopes existing in ordinary oxygen. Relative atomic masses measured by chemical methods, using this definition, are referred to as *chemical relative atomic masses.*

When, for example, the relative atomic mass of chlorine is given as 35.457, this is the chemical value. It does not mean that any single atom of chlorine has this relative atomic mass, but that this is the average value of the isotopes of which chlorine is made up. Chlorine does, in fact (p. 128), consist of two isotopes, ^{35}Cl and ^{37}Cl, and it is because these isotopes are present in the ratio of about 3:1 that the chemical relative atomic mass of chlorine is about 35.5.

Chemical relative atomic masses are perfectly satisfactory for many chemical purposes, but *physical relative atomic masses* give a better measure of the relative masses of individual isotopes. In physical relative atomic masses, oxygen in the definition given at the beginning of this section is taken to mean the ^{16}O isotope of oxygen.

$$\text{Physical relative atomic mass of an element} = \frac{\text{Mass of 1 atom of element}}{\frac{1}{16} \times \text{Mass of 1 atom of } ^{16}O}$$

On this scale, the relative atomic mass of ordinary oxygen has a value of 16.0043. This is the averaged atomic mass of ^{16}O, ^{17}O and ^{18}O isotopes, and it is the physical relative atomic masses, or isotopic masses, of individual isotopes which are of major significance. These can be measured by the mass spectrograph, a modern instrument giving an accuracy of 1 part in 100000. Such accurate values for individual isotopes are particularly important in work on nuclear reactions.

In 1961, the International Union of Pure and Applied Chemistry agreed to replace the ^{16}O isotope of oxygen by the ^{12}C isotope of carbon as the basis for atomic mass definition. This is because, nowadays, the determination of accurate isotopic masses in a mass spectrograph is done in relation to the mass of the ^{12}C isotope. This method is adopted because it is easy to obtain ions containing many carbon atoms in a mass spectrograph.

Thus

$$\text{Physical relative atomic mass of an element} = \frac{\text{Mass of 1 atom of element}}{\frac{1}{12} \times \text{Mass of 1 atom of } ^{12}C}$$

This new definition causes a lowering of relative atomic mass values as expressed on the ^{16}O scale. The values on that scale have to be multiplied by approximately 0.9997 to obtain the values on the ^{12}C scale.

9. Separation of isotopes. As isotopes have the same chemical properties, their separation must depend on physical methods making use of the slight mass difference between different isotopes. The following methods are the most important.

a. *Electromagnetic separation.* This method adopts the principle of the mass spectrometer (p. 127). Groups of different isotopes are deflected differently by a magnetic field and can be collected in containers placed in appropriate positions. The method is expensive if used for large quantities of material.

b. *Gaseous diffusion.* The rate of diffusion of a gas is inversely proportional to its density, according to Graham's law (p. 39). Thus, when a gas containing a mixture of two isotopes is allowed to diffuse through a porous partition, the lighter isotope passes through more rapidly than the heavier one.

A single stage diffusion through one partition gives only a very partial separation, and, in practice, several thousand stages, in a cascade arrangement, may be necessary to bring about a good separation. A cascade arrangement uses the same sort of principle as a fractionating column. The lighter gas mixture passing through a porous partition is taken on to the next stage, whilst the heavier gas mixture which has not diffused is returned to an earlier stage. The principle is illustrated in Fig. 31.

This method was adopted for separating ^{235}U and ^{238}U, a mixture of $^{235}UF_6$ and $^{238}UF_6$, which are gases, being used. The hexafluoride is one of the few gaseous compounds of uranium, and has the advantage that fluorine has no isotopes. It is, however, very corrosive.

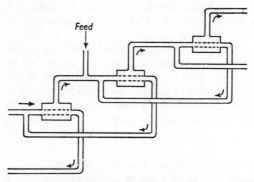

Fig. 31. Part of the arrangement used for separating isotopes by gaseous diffusion. The dotted lines represent porous partitions. The lighter fraction, passing through a partition, is passed on to the right. The heavier fraction is returned, for recycling, to the left. Pumps are used to maintain circulation, and several thousand stages may be required to give a good separation.

c. *Thermal diffusion.* A long, vertical, cylindrical tube, with an electrically heated wire running down its axis, is used in this method. If a gaseous mixture of isotopes is introduced into the tube, the lighter isotopes diffuse more rapidly towards the central, hot region where they are carried upwards by convection currents. The heavier isotopes, near the cold, outer surface of the tube, are carried downwards. Thus the lighter isotope collects at the top of the tube, and the heavier one at the bottom.

QUESTIONS ON CHAPTER 12

1. If the half-life of radium is 1 620 years, what is its radioactive decay constant? What will be the rate of disintegration of 1 g of radium in terms of atoms disintegrating per second? How many curies is this? How many grammes of radium represent 1 curie?

2. If the decay constant for radium is 1.356×10^{-11}, calculate the time required for (a) 10 per cent, (b) 90 per cent of a sample of radium to disintegrate.

3. If 1 curie of radium of isotopic weight 226 and half-life 1 620 year is 1.02 g,

how many grammes of cobalt 60, of half-life 5.2 year, will represent 1 curie? If 1 curie of potassium 40 is 146 kg, what is its half-life?

4. Radon has a half-life of 3.8 day. Plot a graph of the percentage of a sample of radon which has decayed against the time in days.

5. Prove that the mass of a radioactive isotope required to give 1 curie of radiation is equal to $8.87 \times 10^{-14} \times M \times t_{\frac{1}{2}}$, where M is the isotopic mass and $t_{\frac{1}{2}}$ the half-life period. Prove, also, that the number of curies per gramme is given by $1.13 \times 10^{15}/M \times t_{\frac{1}{2}}$.

6. If a sample of a radioactive isotope of half-life 3.11 hour has an activity of 1000 s^{-1} at a certain time, what will be its activity one hour later?

7. What is the percentage loss in the activity of a radioactive element after 1, 10 and 100 half-life periods?

8. The period of average life for a radioactive substance is the time within which the whole of the substance would disintegrate if the initial disintegration rate remained constant. How is the period of average life related to the disintegration constant?

9. Explain why it is that the determining factor for the detection of a radioactive substance by the measurement of its radiation is the product of the amount of substance and its disintegration constant.

10. The ratio by mass of radium to uranium I in old minerals is 3.3×10^{-7}. In such old minerals the uranium I is in equilibrium with the radium. If the disintegration constant for radium is $1.39 \times 10^{-11} \text{ s}^{-1}$, what is the value of the disintegration constant for uranium I?

11. Give an account of the early work on radioactivity of Rutherford, Russell and Soddy.

12. ^{238}U emits an α-particle to form UX_1. This, in turn, emits a β-particle to give UX_2. What are the atomic and mass numbers of UX_1 and UX_2?

13. $^{239}_{93}Np$ emits a β-particle to give an isotope which then decays to ^{235}U. What particle is emitted in the final decay?

14. ^{238}U decomposes by emitting (a) and forming ^{234}Th. This isotope undergoes decomposition with β-emission forming (b). What are (a) and (b)?

15. The thorium, uranium and actinium disintegration series are sometimes known as the $4n$, $4n + 2$, and $4n + 3$ series respectively. Why is this? What isotopes might exist in the $4n + 1$ series?

16. Write symbols showing the mass number and the atomic number for an electron, a proton, a neutron, a positron, a deuteron and an α-particle.

17. Write symbols showing the mass number and the atomic number of the atoms formed when (a) $^{226}_{88}Ra$ loses an α-particle, (b) $^{210}_{83}Bi$ loses a β-particle, (c) $^{234}_{92}U$ loses five α-particles, (d) $^{231}_{91}Pa$ loses an α- and a β-particle, (e) $^{226}_{88}Ra$ loses two β-particles.

18. What is (a) an α-particle, (b) a β-ray, (c) a γ-ray? Show how the position of an element in the periodic table is affected by (d) the loss of an α-particle, and (e) the emission of a β-ray from the nucleus.

Radium (atomic number 88, atomic weight 226) is placed in Group 2 of the periodic classification. The following is a consecutive series of elements produced by its radioactive disintegration together with their respective atomic weights: radium emanation, 222; radium A, 218; radium B, 214; radium C, 214; radium C', 214; radium D, 210; radium E, 210; radium F, 210; radium G, 206.

Construct a small table showing (i) their atomic numbers, (ii) their main group placings in the periodic table, (iii) the type of radiation emitted at each stage of their disintegration. Identify any sets of isotopes which occur in the series. (W. S.)

19. Three kinds of fundamental particles are of great importance in chemical theory. Name them and give their relative masses and charges.

What is meant by (*a*) the atomic number (*Z*) and (*b*) the relative atomic mass (*M*) of an element? For sodium $Z = 11$ and $M = 23$; how must the nucleus of this atom be built up? How is the fractional relative atomic mass of chlorine (Cl = 35.5) explained? How does the explanation conflict with Dalton's concept of the nature of atoms and which other of his postulates concerning elements are no longer acceptable? (N.)

20. Explain the essential differences between Aston's mass spectrograph and Dempster's mass spectrometer.

21. Explain why it is that the charge/mass ratio for cathode rays is independent of the nature of the gas in the discharge tube, whilst the charge/mass ratio for positive rays is smaller than that for cathode rays and depends on the nature of the gas in the tube.

22. If natural oxygen consists of 99.758 per cent of an isotope of mass 16, 0.0373 per cent of an isotope of mass 17.004 534 and 0.2039 per cent of an isotope of mass 18.004 855, calculate the average mass of an atom in natural oxygen.

23. Compare the functioning of Aston's mass spectrograph with that of an achromatic prism.

24. Give an account of some typical uses of isotopes.

25. It has been suggested sodium chloride obtained from the depths of the sea might be richer in ^{37}Cl than sodium chloride found on the earth's surface. Is there any validity in this?

Chapter 13
The atomic nucleus

1. Splitting the atom. In naturally occurring radioactive elements the nuclei of the atoms concerned are constantly splitting, and the process cannot be stopped. The nuclei of other atoms are more stable, but they can, nowadays, often be split by bombardment with particles such as α-particles, protons or neutrons.

The first artificial disintegration of a nucleus was observed by Rutherford, in 1919, when he passed α-particles from Radium C (p. 125), through nitrogen gas and discovered that protons were ejected. In 1925 Blackett demonstrated that this ejection of a proton was accompanied by the formation of the ^{17}O isotope of oxygen. The so-called nuclear reaction was represented as,

$$^{14}_{7}N + {}^{4}_{2}He \longrightarrow {}^{1}_{1}H + {}^{17}_{8}O$$

At this time, α-particles were the most convenient particles available for use as projectiles in bombarding nuclei. β-particles (electrons) could have been used, equally simply, but they are far too light to be effective. It would be like shooting peas at elephants. The α-particle is heavy, in comparison, but it is not a perfect bombarding particle because of its positive charge. The nucleus being bombarded is positively charged so that another positively charged particle must have sufficient energy to overcome the electrostatic repulsion if it is to 'score a hit'.

The developments in nuclear disintegration have centred round the introduction of new bombarding particles and of new methods of increasing their energy.

In 1932, Cockroft and Walton used an electrostatic accelerator to provide high-speed protons. By this means they converted atoms of lithium into atoms of helium,

$$^{1}_{1}H + {}^{7}_{3}Li \longrightarrow 2 \, {}^{4}_{2}He$$

Deuterons (deuterium nuclei) were also used; they have twice the mass of a proton but still suffer by being positively charged. Electrostatic accelerators were followed by cyclotrons, synchrotons and cosmotrons, all designed to provide charged particles with greater and greater energy. Protons with energy of about 1.6×10^{-9} J or 10 GeV, can be obtained.

Charged particles have their limitations, however, no matter what their energy may be, and the discovery of the neutron (p. 101) provided one of the best bombarding particles, for it is quite heavy and has no electrical charge. It can, therefore, approach a nucleus without being electrically repelled. Examples of its use are given in the following section.

2. Artificial radioactivity. Elements with atomic number greater than 83 are naturally radioactive, but other naturally occurring isotopes are not. It is possible, however, to make isotopes which are radioactive by nuclear reactions. The radioactivity is referred to as artificial or induced radioactivity. More than a thousand radioactive isotopes are now known, and many of them are extremely useful (see section **4**).

The first radioactive isotopes to be produced were obtained by M. and Mme Curie-Joliot, in 1933. They bombarded aluminium, beryllium and magnesium with α-particles and obtained radioactive isotopes of phosphorus, nitrogen and silicon respectively, e.g.

$$^{27}_{13}Al + {}^{4}_{2}He \longrightarrow {}^{30}_{15}P + {}^{1}_{0}n$$

Other nuclear reactions giving radioactive isotopes of sodium, carbon and sulphur respectively are represented by the following equations:

$$^{23}_{11}Na + {}^{1}_{0}n \longrightarrow {}^{24}_{11}Na$$
$$^{10}_{5}B + {}^{2}_{1}H \longrightarrow {}^{11}_{6}C + {}^{1}_{0}n$$
$$^{35}_{17}Cl + {}^{1}_{0}n \longrightarrow {}^{35}_{16}S + {}^{1}_{1}H$$

3. Detection and measurement of radiation. The detection and measurement of the various sorts of radiation associated with radioactivity is very important both for research purposes and in order to ensure that radiation does not build up to dangerous limits, for exposure to too much radiation is injurious.

Detection and measurement methods depend, in principle, on the detection of charged particles. X-rays and γ-rays, though not charged, can be detected because, when they pass through a gas, they eject electrons from the gas molecules and, therefore, form charged particles. Neutrons similarly, can be made to give rise to charged particles by collision with hydrogen atoms, for fast neutrons, or by collision with boron-10 or lithium-6, for slow neutrons.

a. *Cloud chambers.* It has already been mentioned (p. 91) that cathode rays cause the formation of a fog when passed through a supersaturated vapour by providing nuclei for condensation.

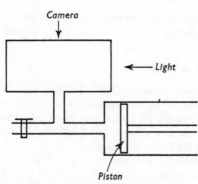

FIG. 32. Principle of the Wilson cloud chamber.

Wilson, in 1912, designed a very important piece of apparatus, known as a cloud chamber, for making use of this phenomenon to make the paths of charged particles visible.

The space inside the cloud chamber (Fig. 32) contains, initially, dust-free air, saturated with water vapour. Sudden expansion of this air, brought about by the rapid movement of a piston, causes cooling so that the air becomes supersaturated. The passage of charged particles into the chamber, at this stage, leaves visible tracks, which can be photographed, along the paths of the particles.

Not only can the particular paths be followed, but the ranges and energies of the particles can be measured. An α-particle, being heavy, normally travels in straight lines, but β-particles are more readily deviated by collision with air molecules.

Collisions between particles, resulting in the formation of new particles, can also be followed in the cloud chamber.

In recent modifications a cloud chamber can be made to operate automatically by the arrival of particular radiation at other suitably placed detectors. When used in conjunction with a magnetic field, provided by an external electromagnet, a cloud chamber shows the paths of charged particles as curves. The direction of the curvature in relation to the direction of the magnetic field shows the sign of the charge on the particle, and the amount of curvature enables the speed of the particle to be calculated if its mass and charge are known.

Finally, a continuous or diffusion cloud chamber can operate continuously. In such a cloud chamber a continuous region of supersaturated vapour is provided by allowing vapour to diffuse downwards from the top of the cloud chamber, which is kept hot, to the bottom which is kept cold.

b. *Bubble chambers.* In bubble chambers, developed by Glaser, since 1952, a superheated liquid is used instead of the supercooled vapour of the cloud chamber. Charged particles now provide nuclei for the formation of bubbles within the liquid and the tracks of bubbles are denser than the tracks formed in a cloud chamber.

c. *Geiger counters.* A Geiger–Müller counter consists, essentially, of a hollow cylinder with a thin, metal wire passing along its axis and insulated from the cylinder (Fig. 33). The cylinder is filled with a mixture of gases at

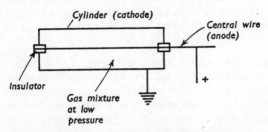

FIG. 33. Principle of the Geiger–Müller counter.

low pressure, the nature of the gases depending on the use of the counter. The cylinder, or cathode, can be earthed, whilst the central wire, or anode, can be raised to a potential of up to 1 kV.

Passage of a charged, or ionising, particle through the gas in the cylinder enables a pulse of current to flow between the wire and the cylinder. Such a pulse can be amplified so that it can be recorded on a loud-speaker as a sharp click or on an automatic recorder which counts the number of pulses.

Many different forms of Geiger counter are available commercially, some being able to deal with counts of about 10^5 per second.

d. *Scintillation counters.* The spinthariscope (p. 98) relied on the fact that fluorescence is caused when an α-particle falls on to zinc sulphide. Visual observation of the flashes of light from a zinc sulphide screen provided one of the earliest methods of detection.

Nowadays, precisely the same principle has been adopted in what is known as a scintillation counter. The flashes of light from a zinc sulphide screen, or some other phosphor, are allowed to fall on to a cathode with a photo-sensitive coating. The flashes of light release electrons from this cathode, by the photo-electric effect. These electrons are accelerated from one cathode to another, releasing, at each cathode, an increasing number of electrons, until a pulse of current suitable for amplification is obtained. The device is known as a photo-multiplier and radiations up to a million per second can be recorded.

e. *Photographic emulsion method.* The passage of charged particles through a photographic emulsion shows up as a track on development. This method has been particularly useful in the study of cosmic rays and is described further on page 146.

f. *Use of semi-conductors.* The conductivity of some semi-conductors (p. 197) is greatly increased by exposure to electromagnetic and ionising radiation (p. 348) and the current flowing through such a semi-conductor can be used to give a direct measure of the intensity of radiation. Cadmium sulphide crystals can be used, for example, to measure the intensity of γ-rays in this way.

4. Use of radioactive isotopes. Naturally occurring radium has been used for some time in radiotherapy, particularly in the treatment of cancer, and the γ-rays from radium or radon have also been used in taking radiographs. The present commercial availability of artificial radioactive isotopes has, however, greatly increased the research and industrial applications of radio techniques. The following selection of examples will illustrate the range of usage.

a. *Tracer techniques.* An object with a very small amount of a radioactive isotope attached to it, or incorporated in it, can readily be detected by picking up the radiation from it. A joint in a buried pipe or cable can be marked, for example, in this way, or the efficiency of a mixing process can be followed by adding a radioactive isotope to one of the ingredients before mixing and then observing how even the level of radiation is throughout the final mixture.

The wear inside an engine can be measured by making the various moving parts radioactive and measuring the resulting radioactivity in the circulating oil. The flow of a material, e.g. molten glass through a furnace, fertiliser through a drier, or gases in ventilation systems, can all, also, be controlled by tracer techniques.

There are, too, many biological and medical uses. The uptake of phosphorus by a plant from a phosphate fertiliser can be traced by using a fertiliser containing ^{32}P, and radioactive tracer studies, using ^{14}C, have been very valuable in elucidating the nature of photosynthesis. Introduction of radioactive ^{59}Fe into the blood stream has enabled the part played by iron in blood formation and use to be studied, and blood circulation is followed by injecting a saline solution containing a little ^{24}Na. ^{131}I has been used both in the diagnosis of thyroid diseases and in research on the functioning of the thyroid gland.

In more specifically chemical procesess, the diffusion of a metal containing radioactive isotopes into another metal, the solubility of water (made from tritium) in hydrocarbons, and the solubility of very sparingly soluble salts (made radioactive) in a solvent, can all be measured. Moreover, chemical reactions of the type,

$$AB + B' \longrightarrow AB' + B$$

where B and B' are the same element can be followed if non-radioactive B and radioactive B' are used. In this way exchange rates can be measured. The detailed behaviour of any element in other chemical reactions can be investigated, too, by using a radioactive isotope of the element.

b. *Direct use of radiation.* The β- and γ-rays emitted by radioactive isotopes can be very useful, particularly in measuring or controlling thicknesses of materials. The amount of radiation passing through a material will decrease as the material gets thicker. If a source of β- or γ-rays is placed on one side of the material and a detector on the other side, the scale reading on the detector will give a measure of the thickness of the material and can, in fact, be calibrated directly in thickness units. Such measurements can be made without touching the material concerned so that they can be applied, for example, to sheet steel coming through a high-speed rolling mill.

Typical uses are in controlling the thickness of paper, plastic sheeting and sheet metal, in checking the packing of tobacco in cigarettes, in measuring or monitoring the thickness of coating of one metal on another, and in checking the level to which a container is filled with a liquid or solid. β-rays can be used for material thicknesses up to about 1.5 mm of steel; γ-rays up to about 100 mm of steel.

γ-rays are also used in radiography, instead of X-rays. A radioactive isotope which gives γ-rays is much cheaper, much smaller, and more portable and manoeuvrable than X-ray equipment, but γ-ray photography is much slower than that using X-rays. A typical use is provided by the standard method of testing welds when laying a pipeline. ^{60}Co or ^{192}Ir

are used to provide γ-rays which enable a radiograph of the weld to be taken.

5. Types of nuclear reaction. It is conventional to represent a nuclear reaction by writing the target element, the bombarding particle, the emitted particle or radiation, and the new isotope formed. Thus the reaction,

$$^{35}_{17}Cl + ^1_0n \longrightarrow ^{35}_{16}S + ^1_1H$$

is written as $^{35}_{17}Cl$ (n, p) $^{35}_{16}S$, where n and p stand for neutron and proton respectively.

On this basis, possible nuclear reactions, with examples, may be summarised as follows:

Reaction type	Example
n, p	$^{14}_{7}N + ^1_0n \longrightarrow ^{14}_{6}C + ^1_1H$
n, α	$^6_3Li + ^1_0n \longrightarrow ^3_1H + ^4_2He$
n capture	$^{107}_{47}Ag + ^1_0n \longrightarrow ^{108}_{47}Ag$
p, n	$^{63}_{29}Cu + ^1_1H \longrightarrow ^{63}_{30}Zn + ^1_0n$
p, α	$^{19}_{9}F + ^1_1H \longrightarrow ^{16}_{8}O + ^4_2He$
d, n	$^{56}_{26}Fe + ^2_1H \longrightarrow ^{57}_{27}Co + ^1_0n$

Other nuclear reactions involving p capture; p, d; α, p; d, p; α, n; d, α; γ, n; and γ, p; where p stands for proton, d for deuteron, n for neutron, are also possible.

6. Trans-uranic elements. Uranium has the highest atomic number of any naturally occurring element and was, for a long time, placed at the end of the periodic classification of elements. Trans-uranic elements have, however, now been prepared by nuclear reactions.

a. *Neptunium and plutonium.* These elements were discovered by McMillan and Abelson, in 1940, by bombarding $^{238}_{92}U$ with neutrons. The first result is the capture of a neutron forming a $^{239}_{92}U$ isotope,

$$^{238}_{92}U + ^1_0n \longrightarrow ^{239}_{92}U$$

but this isotope is radioactive, losing β-particles in two stages to form neptunium and plutonium,

$$^{239}_{92}U \longrightarrow ^{239}_{93}Np + _{-1}^{0}e \qquad ^{239}_{93}Np \longrightarrow ^{239}_{94}Pu + _{-1}^{0}e$$

b. *Americium, curium, berkelium and californium.* These four elements were all discovered by Seaborg and his co-workers, the first two in 1944 and the last two in 1949 and 1950.

Americium was made from plutonium by neutron bombardment. Two neutrons were first captured and the unstable isotope produced then lost a β-particle,

$$^{239}_{94}Pu + 2^1_0n \longrightarrow ^{241}_{94}Pu \longrightarrow ^{241}_{95}Am + _{-1}^{0}e$$

Curium, berkelium and californium were made by bombarding plutonium americium and curium, respectively, with α-particles,

$$^{239}_{94}Pu + ^4_2He \longrightarrow ^{242}_{96}Cm + ^1_0n \qquad ^{241}_{95}Am + ^4_2He \longrightarrow ^{243}_{97}Bk + 2^1_0n$$
$$^{242}_{96}Cm + ^4_2He \longrightarrow ^{245}_{98}Cf + ^1_0n$$

c. *Einsteinium and fermium.* These elements were first isolated, in 1952, from the debris of a thermo-nuclear explosion. Their formation involved the capture of seventeen neutrons by a uranium atom, and the subsequent loss of seven or eight β-particles,

$$^{238}_{92}U + 17^{1}_{0}n \longrightarrow {}^{255}_{92}U \qquad {}^{255}_{92}U \longrightarrow {}^{255}_{99}E + 7_{-1}^{0}e \qquad {}^{255}_{92}U \longrightarrow {}^{255}_{100}Fm + 8_{-1}^{0}e$$

Einsteinium and fermium can also be made by bombarding plutonium in an atomic pile with a high intensity of neutrons, but the process is slow.

d. *Mendelevium.* This element was produced and identified, in 1955, by bombarding einsteinium with α-particles,

$$^{253}_{99}E + {}^{4}_{2}He \longrightarrow {}^{256}_{101}Mv + {}^{1}_{0}n$$

e. *Nobelium.* This was first made in 1958 by bombarding curium with accelerated carbon ions,

$$^{246}_{96}Cm + {}^{12}_{6}C \longrightarrow {}^{254}_{102}No + 4^{1}_{0}n$$

f. *Lawrencium.* This was first identified in 1961 and is formed by bombarding californium with accelerated boron ions,

$$^{252}_{98}Cf + {}^{11}_{5}B \longrightarrow {}^{257}_{103}Lw + 6^{1}_{0}n \qquad {}^{252}_{98}Cf + {}^{10}_{5}B \longrightarrow {}^{257}_{103}Lw + 5^{1}_{0}n$$

7. Nuclear stability. An atomic nucleus is made up of a number of protons equal to its atomic number, plus a sufficient number of neutrons to give the necessary mass number. The protons and neutrons in a nucleus are called nucleons.

Because the nucleus of an isotope must contain a whole number of protons and neutrons it was, for a while, thought that isotopic masses, i.e. the relative atomic masses of isotopes, might all be whole numbers. This revived a hypothesis first suggested by Prout, in 1815, in which he supposed that every atom was made up of hydrogen atoms.

Careful measurement, by mass spectrographic methods, show, however, that relative isotopic masses, on the ^{12}C scale, are not exactly whole numbers. Some illustrative figures are given below

^{1}H 1.007825	^{4}He 4.00260	^{12}C 12.00000
^{2}H 2.01409	^{7}Li 7.01600	^{35}Cl 34.96885

These figures show how much heavier one atom of a particular isotope is than $\frac{1}{12}$th of an atom of ^{12}C. The actual mass of a ^{12}C atom is readily obtained, for 12 gramme of the ^{12}C isotope contains 6.02252×10^{23} atoms. One atom of ^{12}C, therefore, has a mass of $12/6.02252 \times 10^{23}$ gramme, and $\frac{1}{12}$th of this is $1/6.02252 \times 10^{23}$, i.e. 1.66043×10^{-24} g or 1.66043×10^{-27} kg. This mass is often referred to as *an atomic mass unit*, the actual mass of any one atom being equal to its relative isotopic mass multiplied by the value of the atomic mass unit.

For an isotope, the nearest whole number to its relative isotopic mass is known as the *mass number*. The slight difference between the actual relative isotopic mass and the mass number, known as the *mass defect*, is due to the fact that the masses of protons and neutrons in a nucleus are not strictly additive because some part of their mass is converted into energy. This energy is the *binding energy* which holds the nucleus together; the greater it is, the more stable will the nucleus be.

For deuterium, with a simple nucleus of one proton and one neutron, the figures are as follows. The relative isotopic mass is 2.01409, giving an actual mass of 3.4426×10^{-27} kg. The mass of the nucleus will be 3.34426×10^{-27} less the mass of an electron (9.0191×10^{-31} kg), i.e. 3.34335×10^{-27} kg. The deuterium nucleus consists of one proton, of mass 1.67252×10^{-27} kg, and one neutron of mass 1.67482×10^{-27} kg, giving an additive mass of 3.34734×10^{-27} kg.

The mass converted into energy, for a deuterium nucleus, to be formed, must, therefore, be $(3.34734 - 3.34335) \times 10^{-27}$ kg, i.e. 0.00399×10^{-27} kg. This is equivalent to a bonding energy of 3.583×10^{-13} joule or 2.238 MeV, as explained in section 8.

In general, for a nucleus containing Z protons, and with a mass number of A, the number of neutrons will be $(A - Z)$. If the mass of the proton is p, that of the neutron n, and that of the actual nucleus M, the mass converted into energy will be given by

$$Zp + (A - Z)n - M$$

and this can be converted into energy units as explained below.

8. Note on units. When mass is converted into energy, the two are related by the equation (p. 3).

$$\underset{\text{(in joules)}}{E} = \underset{\text{(in kilogrammes)}}{m} \times \underset{(c \text{ in metres per second})}{c^2}$$

The value of c^2 is approximately 8.99×10^{16} so that

$$1 \text{ kg of mass} = 8.99 \times 10^{16} \text{ joule of energy}$$

One electron volt (1 eV), which is the energy acquired or lost by a particle of unit electronic charge on passing through a potential difference of 1 volt, is equal to 1.602×10^{-19} joule, so that

$$1 \text{ kg of mass} = (8.99 \times 10^{16})/(1.602 \times 10^{-19}) \text{ eV}$$
$$= 5.613 \times 10^{35} \text{ eV}$$
$$= 5.613 \times 10^{29} \text{ MeV}$$

The unit of atomic mass is 1.66043×10^{-27} kg so that

$$1 \text{ atomic mass unit (amu)} = 8.99 \times 10^{16} \times 1.66043 \times 10^{-27} \text{ J}$$
$$= 1.494 \times 10^{-10} \text{ J}$$
$$= 9.328 \times 10^{8} \text{ eV}$$
$$= 932.8 \text{ MeV}$$

The figures given here are only approximate to show the relationship between the various units. Where necessary, more accurate figures are used in the text.

9. Packing fraction and binding energy. The binding energy, or nuclear stability, of a nucleus is conventionally represented, on a comparative scale, as a packing fraction, this being defined by

$$\frac{\text{Packing}}{\text{fraction}} = \frac{\text{Isotopic mass} - \text{Mass number}}{\text{Mass number}}$$

Because the isotopic mass of ^{12}C is the same as its mass number, the packing fraction of the ^{12}C isotope is zero. The packing fractions of other

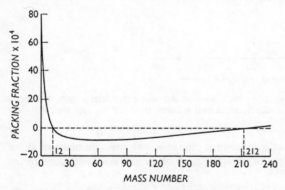

FIG. 34. The general way in which packing fractions vary with mass number. See, also, Fig. 36, p. 145.

isotopes give the gain or loss in mass per nucleon, as compared with the nuclear packing in ^{12}C. Because ^{12}C is, arbitrarily, given a packing fraction of zero, other isotopes may have positive or negative packing fractions.

Using the figures given on page 141, the packing fraction for ^{4}He is given by

$$\text{Packing fraction for }^{4}He = \frac{4.00260 - 4}{4} = 6.5 \times 10^{-4}$$

whilst that for ^{35}Cl is given by

$$\text{Packing fraction for }^{35}Cl = \frac{34.96885 - 35}{35} = -8.9 \times 10^{-4}$$

Plotting the packing fractions of isotopes against the corresponding mass numbers gives a curve of the general shape shown in Fig. 34.

The lower the packing fraction, the greater is the binding energy per nucleon, so that the curve of binding energy per nucleon against mass number is of the same shape, but upside down (Fig. 35).

The general shape of the curves shows that the elements between chromium and zinc lie at the minimum of the packing fraction curve and the maximum of the binding energy curve. They are, therefore, the most stable elements. If one of these elements could be converted into other elements with higher packing fractions a considerable amount of energy would be required. On the other hand, the heavier or lighter atoms could

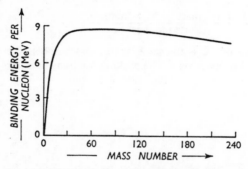

FIG. 35. The general way in which binding energies per nucleon vary with mass number. On a larger scale, maxima and minima show up for isotopes with low mass numbers.

undergo nuclear reactions, forming elements with mass numbers of about 60, i.e. with smaller packing fractions, which would be accompanied by the evolution of energy.

10. Nuclear structure. A more detailed plotting of packing fractions against mass number for the light isotopes shows (Fig. 36) successive minima for ^{4}He, ^{8}Be, ^{12}C, ^{16}O, ^{20}Ne and ^{24}Mg. The points occur as successive maxima on a large-scale binding energy curve. These isotopes all have nuclei which contain equal numbers of protons and neutrons, and they might all be regarded as made up of helium nuclei, containing 2 protons and 2 neutrons. The presence of 'pre-formed' helium nuclei (α-particles) within the nucleus of every atom could account for the frequency with which α-particles are produced in radioactive changes. An α-particle might have some unusual stability as a nuclear component. The minima in the packing fraction curve at ^{4}He, ^{8}Be, ^{12}C, ^{16}O, ^{20}Ne and ^{24}Mg, certainly suggest some regularity about the arrangement of protons and neutrons in atomic nuclei.

It is also noticeable that the great majority of known nuclei, which are stable, have even numbers of protons and neutrons. Indeed, only four stable nuclei, ^{2}H, ^{6}Li, ^{10}B and ^{14}N, have odd numbers of both protons and neutrons.

Furthermore, there is a fairly regular change in the number of neutrons in stable, non-radioactive nuclei as compared with the number of protons (Fig. 37). For light isotopes, the numbers of protons and neutrons are approximately equal, but, as the isotopic mass increases, the number of neutrons becomes gradually larger than the number of protons.

A detailed consideration of the neutron/proton ratio in all known isotopic nuclei also reveals some suggestive, but unexplained, regularities.

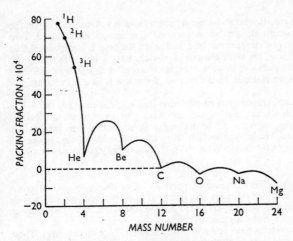

FIG. 36. Packing fractions of isotopes with low mass
numbers, showing various minima. See, also, Fig. 34, p. 143.

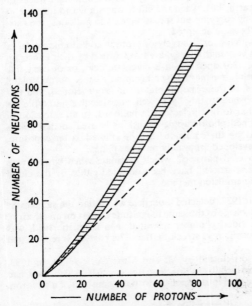

FIG. 37. Variation of number of neutrons with number of protons in stable,
non-radioactive nuclei. All the known nuclei lie within the shaded area. The dotted
line shows equal numbers of neutrons and protons.

Although the facts mentioned above would seem to have a direct bearing on the nature of nuclear structure the problems involved have not yet been solved. Two 'models' of an atomic nucleus seem to be possible. In the first, it is suggested that there is, within a nucleus, an arrangement of nuclear energy levels in which nucleons might exist in much the same way as electrons exist in orbits of different energy levels outside the nucleus. In the second, a nucleus is likened to a drop of liquid.

Whatever the actual arrangement of nucleons in a nucleus may be, the forces between them are very great, but, again, the nature of these nuclear forces is not understood. They seem to be something different from gravitational or electromagnetic forces. The most hopeful theory of nuclear forces, first suggested by Yukawa in 1935, involves mesons (p. 147), but this theory fails to account for all the known facts.

11. Subnuclear particles. The existence of particles other than protons, neutrons and electrons enters into any detailed discussion of nuclear structure, and a wide variety of such particles have been predicted and/or discovered. The part they all play in nuclear affairs is not fully understood.

Many of these discoveries have come about from a study of *cosmic rays*. These are rays, whose possible existence has been suspected for many years, emanating from outer space. Some sort of cosmic radiation was first suspected when it was found that gases in sealed vessels became very feebly ionised on standing. It was difficult to decide whether the effect was caused by γ-rays from small traces of radioactive material in the earth, but, when the effect was shown to increase at higher and higher altitudes, by carrying out measurements in balloons, the existence of cosmic rays slowly became accepted.

Nowadays, cosmic rays, which consist largely of protons and atomic nuclei of other light elements, all with very high energy, are investigated by their effects on a Wilson cloud chamber (p. 136) operated in a magnetic field, or on very fine photographic emulsions containing more silver bromide than a normal plate. The tracks made by the rays on these special plates can be developed, and this photographic method, established mainly by Powell, is particularly suitable for the high-altitude work that has to be carried out in balloons. In an extension of the method, a stack of stripped photographic emulsions is used so that tracks penetrating both into and along the emulsions can be studied. After exposure, the strips of emulsion are developed separately on glass plates.

Many of the newly discovered particles, which originate when cosmic rays collide with atoms and other particles, have been detected either by the cloud chamber or the photographic emulsion method.

a. *The positron.* Anderson, in 1932, detected a particle which had the same mass as an electron and a positive charge of the same magnitude as that on an electron. This particle, detected in a cloud chamber operated in a magnetic field, was thought to originate when cosmic rays struck matter. The particle is represented as $^0_1$e or e$^+$.

The same particle was also detected by M. and Mme Curie-Joliot, in 1933, when they found that the artificially radioactive isotope, $^{30}_{15}$P, which they had prepared (p. 136), rapidly changed into silicon-30 with the emission of a positron,

$$^{30}_{15}\text{P} \longrightarrow {}^{30}_{14}\text{Si} + {}^0_1\text{e}$$

Free positrons are thought to be stable, like free electrons, but, in the presence of any matter, a positron rapidly combines with an electron to give γ-radiation. The process is known as pair-annihilation. Conversely, γ-rays passed into a cloud chamber in a magnetic field, can be shown to give positrons and electrons, a process known as pair-production.

b. *Mesons.* There are a variety of particles, some negatively charged, some positively charged, and some neutral, with masses between those of an electron and a proton. They are known, collectively, as mesons.

The μ-meson, or muon, with a negative or positive charge, was the first to be detected, in 1937, by Anderson and Neddemeyer. The particle was detected in cosmic ray tracks in a cloud chamber operating in a magnetic field at an altitude of 4 000 m. It has a mass 207 times that of an electron, and is very unstable, disintegrating to give electrons or positrons, according to its charge, together, probably, with neutrinos and antineutrinos,

Negative muon $(\mu^-) \longrightarrow$ Electron + Neutrino (ν) + Antineutrino (ν^-)

Positive muon $(\mu^+) \longrightarrow$ Positron + Neutrino (ν) + Antineutrino (ν^-)

The discovery of muons was followed by Powell's discovery, in 1947, by photographic emulsion methods, of the π-meson, or pion (π). This particle has a mass 273 times that of an electron and may be negatively or positively charged, or neutral. Pions were later (1948) obtained when high energy protons collided with atomic nuclei. Pions are unstable and rapidly disintegrate into muons and neutrinos.

$$\pi^+ \longrightarrow \mu^+ + \nu$$
$$\pi^- \longrightarrow \mu^- + \nu$$

The π-meson is the particle theoretically predicted by Yukawa in his theory of nuclear forces (p. 146).

Other mesons, all about 1 000 times heavier than an electron are also known; they are called *K*-mesons.

c. *Hyperons.* These are particles similar to mesons but with masses greater than that of a proton.

d. *Anti-proton.* The detection of the positively charged electron (a positron) came a long time after the discovery of the electron. Similarly, the detection of a negatively charged proton, an anti-proton, did not come until 1955, when it was detected in the bombardment of copper by very high-speed protons.

e. *Neutrino and anti-neutrino.* The existence of these particles was first predicted to account for an apparent loss of energy when atoms lose β-particles. They have also been found in other radioactive changes and detected in the radiation from nuclear reactors. They have no charge and a mass smaller, even, than that of an electron.

(Questions on this Chapter will be found on p. 156.)

Chapter 14
Nuclear energy

1. Energy changes in nuclear reactions. The relationship between mass and energy given by Einstein's equation (p. 3),

$$E \quad = \quad m \quad \times \quad c^2$$
$$\text{(in joules)} \quad \text{(in kilogrammes)} \quad (c \text{ in m s}^{-1})$$

can be verified, experimentally, by measuring the energy changes in nuclear reactions. This can be done because the energy changes in these reactions are so very much greater than those in ordinary chemical reactions.

The Einstein equation shows that 1 kg is equal to 8.99×10^{16} joule of energy, and the relationships between other units used is discussed on page 142.

In the bombardment of lithium by high-speed protons (p. 135) the range of the α-particles produced, as shown in the Wilson cloud chamber, is such that their energy must be 8.6 MeV, and, as two α-particles are formed,

$$^1_1\text{H} + {}^7_3\text{Li} \longrightarrow 2\,{}^4_2\text{He}$$

the energy liberated in the reaction must be 2×8.6, i.e. 17.2 MeV. This is equivalent to 3.0643×10^{-29} kg (p. 142).

The atomic masses participating in the reaction are

$$^1_1\text{H} \quad + \quad {}^7_3\text{Li} \quad \longrightarrow \quad 2\,{}^4_2\text{He}$$
$$1.007825 \quad 7.01600 \quad 2 \times 4.00260$$

giving a decrease in mass of 3.0876×10^{-29} kg.

It will be seen that the decrease in mass expected from the amount of energy liberated is in good agreement with the actual decrease in mass. The agreement between decrease in mass and energy released in nuclear reactions is, in fact, so close that measurement of energy release can be used for finding mass values.

2. Nuclear fission. Before 1939 all nuclear reactions had consisted solely in the removal from a nucleus of a relatively small particle such as an α-particle, a proton or a neutron. In 1939, however, Hahn and Strassman realised that isotopes of medium mass number were obtained when uranium was irradiated with neutrons, and that this must represent a splitting of an atom into two, more or less similar, halves.

Fermi had been studying the irradiation of uranium with neutrons since 1934, but the various products found had been regarded as trans-uranic elements. Work in 1939 proved, however, that the fission products fell into two groups. Some were of mass numbers 85–100, others of mass numbers 128–150. Hahn and Strassman identified, originally, ^{139}Ba, but many other products are now known. Similar fission of thorium was also discovered about the same time.

To determine which isotope of uranium was undergoing fission, Nier separated ^{235}U and ^{238}U (p. 130), whereupon it was found that ^{235}U generally splits when bombarded by slow (thermal) or fast neutrons, whereas ^{238}U tends to absorb slow neutrons, with subsequent formation of neptunium and plutonium, and is only split by fast neutrons with high energy.

The main processes taking place can be summarised as follows:

a. *Fission of* ^{235}U *by slow or fast neutrons,*

$$^{235}_{92}U + ^{1}_{0}n \longrightarrow ^{144}_{56}Ba + ^{90}_{36}Kr + 2^{1}_{0}n$$

b. *Capture of slow neutrons by* ^{238}U,

$$^{238}_{92}U + ^{1}_{0}n \longrightarrow ^{239}_{92}U + \gamma\text{-rays}$$

The $^{239}_{92}U$ then disintegrates by β-emission,

$$^{239}_{92}U \longrightarrow ^{239}_{93}Np + ^{0}_{-1}e \qquad ^{239}_{93}Np \longrightarrow ^{239}_{94}Pu + ^{0}_{-1}e$$

c. *Fission of* ^{238}U *by fast neutrons,* which is relatively rare,

$$^{238}_{92}U + ^{1}_{0}n \longrightarrow ^{143}_{34}Xe + ^{94}_{38}Sr + 2^{1}_{0}n$$

In the fission of ^{235}U by slow neutrons, the mass change, summarised below,

$$\underbrace{^{235}_{92}U + ^{1}_{0}n}_{236.0526} \longrightarrow \underbrace{^{144}_{56}Ba + ^{90}_{36}Kr + 2^{1}_{0}n}_{235.8454}$$

$$\underset{235.0439 \quad 1.0087}{} \qquad \underset{143.881 \quad 89.947 \quad 2.0174}{}$$

is 0.2072. This represents a release of about 200 MeV of energy for every one atom of ^{235}U which undergoes fission. For the complete fission of all the atoms in 1 kg of ^{235}U the release of energy would be about 20 million kilowatt hours, which is equal to the energy obtainable from about 3×10^{6} kg of coal. Such a complete release of all the energy available from ^{235}U is not, of course, practicable. At the moment, the energy which can be obtained directly from 1 kg of natural uranium is equal to that available from about 20000 kg of coal. But there are possible ways of obtaining more energy from natural uranium by indirect methods (p. 153), and it should be possible, before very long, to get energy equivalent to 1×10^{6} kg of coal from 1 kg of natural uranium.

3. Nuclear chain reactions. When ^{235}U nuclei are split by neutrons, two or three other neutrons are released as a result of the fission process, and this enables the setting up of a chain reaction, which may be represented diagrammatically as follows:

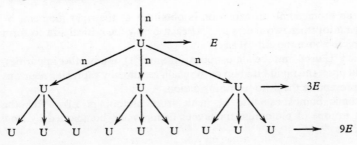

E is the amount of energy released per fission, but the total energy available from the chain reaction is enormous if the chain can be built up quickly and fully.

There are, in practice, two ways of making use of this chain reaction to produce energy. In an atomic bomb, the chain is uncontrolled and the enormous release of energy causes a vast explosion; in an atomic pile, the chain is controlled, and atomic energy is released for peaceful purposes.

4. Atomic bomb. If a piece of ^{235}U is bombarded by a stream of neutrons, at least one neutron will cause the fission of at least one ^{235}U nucleus. The release of energy for this single fission will be quite small, but, if three other neutrons (known as secondary neutrons) are also produced they will be

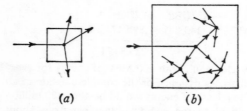

(a)   (b)

Fig. 38. Critical size. In (a), the secondary neutrons escape without causing fission. In (b), the secondary neutrons cause fission and a chain reaction can be set up.

able to participate, under the right conditions, in further nuclear fissions, and so on.

This will only occur if the piece of ^{235}U is large enough. If it is too small the secondary neutrons will escape from the ^{235}U without causing any further fission (Fig. 38). The chain will be broken. A small piece of ^{235}U will not, therefore, cause an explosion, and is safe, but a larger piece, from which the secondary neutrons cannot escape, will explode. There is, therefore, a critical size below which ^{235}U is safe but above which it is not safe.

In an atom bomb an explosion is obtained, at the right moment, by bringing together two pieces of ^{235}U each below the critical size to form one piece above the critical size.

The ^{235}U used must be almost pure, for any ^{238}U or any other impurities, would upset the rapid build-up of the chain reaction by capturing neutrons and preventing them from causing fission.

Atomic bombs can also be made from plutonium. Like ^{235}U, the ^{239}Pu isotope of plutonium undergoes fission when bombarded by slow neutrons.

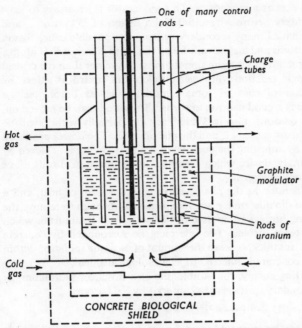

One of many control rods

Charge tubes

Hot gas

Graphite modulator

Rods of uranium

Cold gas

CONCRETE BIOLOGICAL SHIELD

FIG. 39. Diagram of Calder Hall type of reactor. At Calder Hall the moderator structure is 12 metre across and 9 metre high. It is made of 58 000 separate pieces of graphite. There are 1 700 vertical channels for the uranium rods. Each vertical rod of uranium is made up from six fuel elements, which are rods of uranium about 25 mm in diameter and 1 metre long in metal containers with spiral fins to facilitate heat transfer. Over 10 000 fuel elements can be loaded into the reactor through charge tubes, and removed, when necessary, through the same tubes.

5. Atomic piles or nuclear reactors. The number of neutrons resulting from a single fission is known as the *multiplication factor*. When this is less than 1 a chain reaction is impossible; when greater than 1 a chain reaction might take place. In an atomic pile, or a nuclear reactor, fissionable material is so arranged that the multiplication factor can be carefully controlled.

These piles, or reactors, liberate energy, and also provide beams of slow or fast neutrons for research and industrial purposes, such as the manufacture of radioactive isotopes.

a. *The Calder Hall, Magnox or Mark I reactor.* In the commonest type of reactor, first operated commercially at Calder Hall in 1956, rods of natural uranium containing both ^{235}U and ^{238}U atoms are used as fuel. They are placed in magnesium alloy (Magnox) cans within a block of very pure graphite, which acts as a moderator. This moderator is necessary to slow down the secondary neutrons resulting from fission of ^{235}U atoms, and thus to ensure that as many secondary neutrons as possible cause fission of further ^{235}U atoms and escape capture by the more plentiful ^{238}U atoms.

The overall amount of uranium present must be greater than the critical size or else a chain reaction will never develop, but control is gained by incorporating control rods of some material, such as boron-steel or cadmium, which is a good neutron absorber. These rods can be moved up, out of the pile, or down into it. When they are partially withdrawn the multiplication factor exceeds 1, and the pile operates. The power produced can be changed by adjusting the position of the rods. Safety is achieved by having an automatic device which ensures that all the control rods move into the pile in any emergency.

The energy liberated as the pile operates is in the form of heat, and a stream of carbon dioxide, under pressure, is passed through to remove the heat, the resulting hot gas being used to boil water, the steam from which can operate a steam turbine. The pile must be surrounded by concrete walls, several metres thick, to keep the amount of escaping radiation within safe limits. The concrete is referred to as a biological shield.

Besides liberating energy, the atomic pile also produces plutonium, as some of the secondary neutrons are captured by ^{238}U atoms,

$$^{238}_{92}U + ^{1}_{0}n \longrightarrow ^{239}_{92}U \longrightarrow ^{239}_{94}Pu + 2\,^{0}_{-1}e$$

This plutonium can be separated from the uranium by chemical means. It is required for defence purposes, in making bombs, and can also be used in fast reactors as described in d. below.

Since the original Calder Hall reactors were constructed, ten other power stations using improved Magnox designs have been built at Chapelcross, Berkeley, Bradwell, Hunterston, Trawsfyndd, Hinkley Point, Dungeness, Sizewell, Oldbury and Wylfa. They provide plutonium and more than 5000 megawatts of electricity.

The advantage of the Magnox design is that it uses natural uranium as the fuel. It is, however, big, partially because of the necessity for a graphite moderator. Moreover, the efficiency of the reactor is not very high. This is partly because the fission products which build up lower the efficiency, and this means that the fuel elements have to be replaced periodically. This is a lengthy process and can only be carried out when the reactor is fully shut down.

b. *Advanced gas-cooled reactor (AGR or Mark II)*. The next stage in the development of Magnox reactors into more economic units is represented by the advanced gas-cooled reactor (AGR). Such a reactor has been in operation, at Windscale, since 1962. It has a graphite moderator and a carbon dioxide coolant, but the fuel elements consist of uranium oxide enriched with ^{235}U. The reactor operates at a higher temperature than the Magnox type, and the fuel life is longer.

AGR stations are at present being built at Dungeness, Hinkley Point, Hunterston and Hartlepool and these are planned to provide 5000 megawatts of electricity by 1974. Further stations at Heysham and Connah's Quay are under consideration.

c. *Steam generating heavy water reactor (SGHWR)*. In this type of reactor, uranium oxide pellets are placed in zirconium-alloy cans which are surrounded by heavy water as a moderator. Ordinary water is used as the coolant and some of it boils in the cooling process. The resulting steam is used to drive a turbine.

The first SGHWR was opened at Winfrith in 1969.

d. *Fast reactors*. Magnox, advanced gas-cooled and SGHWR reactors are called thermal reactors because they use a moderator to slow down the neutrons. The slow neutrons are necessary to cause fission of further ^{235}U atoms, but some plutonium is also formed by neutron capture by ^{238}U atoms.

The plutonium formed can itself be fissioned but only by fast neutrons. If plutonium is used as the fuel in a reactor, then, it is not necessary to use a moderator and the resulting reactor can be much smaller in size.

The first fast reactor was built at Dounreay and has been in operation since 1959. A central core of plutonium, from which the energy is liberated by fission, is surrounded by ^{238}U. This uranium captures any neutrons that escape from the central core, and is converted into plutonium. It can be arranged, by careful design, that a rather larger number of plutonium atoms are formed in the surrounding uranium than are used up in the central core. This type of reactor, then, produces more 'fuel' than it actually uses. It is known as a *fast* or *fast fission breeder reactor*, because it relies on fast neutrons and because it 'breeds' plutonium from uranium-238.

The central core of plutonium produces energy more rapidly than a uranium-type pile, and the heat produced is removed by circulating a liquid metal, such as an alloy of sodium and potassium, or molten sodium. As with uranium piles, biological shielding is necessary, and there is also a large metal sphere surrounding the whole to prevent escape of radioactive material in case of fire.

A second fast reactor is being built at Dounreay and is designed to produce 250 megawatts of electricity when at full power.

6. Nuclear fusion. When a uranium atom is split into two or more atoms of medium relative atomic mass, there is a decrease in mass and a corresponding release of energy. The packing fraction curve (p. 143) shows that there would be a similar result if two light nuclei could be combined together to give a heavier one.

Combination, or fusion, of two nuclei is not, however, at all easy. They are both positively charged and repel each other. At high temperatures, however, nuclei may have sufficient energy to enable them to overcome the repulsive forces and to unite.

a. *Stellar energy.* Von Weizsacker and Bethe proposed, in 1938, that the vast amount of energy emitted by the sun and by stars might originate from a process involving nuclear fusion.

The modern view is that there are two cycles, one producing the energy in very hot stars, and the other taking place in cooler stars, such as the sun.

In very hot stars, the cycle of changes thought to take place is as follows:

$$^{12}_{6}C + {}^{1}_{1}H \longrightarrow {}^{13}_{7}N + \gamma$$
$$^{13}_{7}N \longrightarrow {}^{13}_{6}C + {}^{0}_{1}e$$
$$^{13}_{6}C + {}^{1}_{1}H \longrightarrow {}^{14}_{7}N + \gamma$$
$$^{14}_{7}N + {}^{1}_{1}H \longrightarrow {}^{15}_{8}O + \gamma$$
$$^{15}_{8}O \longrightarrow {}^{15}_{7}N + {}^{0}_{1}e$$
$$^{15}_{7}N + {}^{1}_{1}H \longrightarrow {}^{12}_{6}C + {}^{4}_{2}He$$

The overall result is that four protons combine to give one helium nucleus, with the emission of γ-rays and two positrons (p. 146).

$$4\,{}^{1}_{1}H \longrightarrow {}^{4}_{2}He + 2\,{}^{0}_{1}e + \gamma$$

^{12}C appears to act in the nature of a catalyst. The energy liberated by such a cycle of changes is in agreement with the energy liberated by hot stars, and each stage of the cycle can be simulated in the laboratory.

In cooler stars, such as the sun, the cycle of changes is thought to be,

$$4\,{}^{1}_{1}H \longrightarrow 2\,{}^{2}_{1}H + 2\,{}^{0}_{1}e$$
$$2\,{}^{1}_{1}H + 2\,{}^{2}_{1}H \longrightarrow 2\,{}^{3}_{2}He + \gamma$$
$$2\,{}^{3}_{2}He \longrightarrow {}^{4}_{2}He + 2\,{}^{1}_{1}H$$

the overall result being the same as for the first cycle of changes, though with a smaller liberation of energy.

b. *The hydrogen bomb.* The temperature attained when ^{235}U or plutonium atoms undergo fission in an explosive chain reaction are high enough to initiate fusion, or thermonuclear, reactions involving hydrogen isotopes, and such reactions have been brought about in hydrogen bombs. In effect, a ^{235}U or plutonium fission explosion acts as a detonator for a fusion reaction.

The precise nature of the fusion reactions used in practice is a military secret, but various reactions, involving $_1^1H$, $_1^2H$ (D), or $_1^3H$ (T) (p. 129), all capable of liberating a lot of energy, are possible. For example:

a	$_1^2H + _1^1H \longrightarrow _2^3He + 5 \text{ MeV}$
b	$_1^2H + _1^2H \longrightarrow _2^3He + _0^1n + 3.3 \text{ MeV}$
c	$_1^2H + _1^3H \longrightarrow _2^4He + _0^1n + 17.6 \text{ MeV}$
d	$_1^3H + _1^1H \longrightarrow _2^4He + 20 \text{ MeV}$
e	$_1^3H + _1^3H \longrightarrow _2^4He + 2 \, _0^1n + 11 \text{ MeV}$

Mixtures containing tritium, ^3H, are the easiest to explode but suffer from the disadvantage that tritium, which has to be manufactured by neutron bombardment of lithium-6,

$$_3^6Li + _0^1n \longrightarrow _2^4He + _1^3H$$

is not stable. Deuterium is, therefore, probably the main component of a hydrogen bomb. If reaction b could be made to take place completely, 1 kg of deuterium would provide as much energy as 2 865 000 kg of coal.

The advantage, if that be the right word, of a hydrogen bomb is that unlimited amounts of hydrogen isotope mixtures can be detonated as the size of a single bomb is not limited by critical-size restrictions as it is in a uranium bomb. Bombs much more powerful than uranium bombs have, in fact, been exploded.

c. *Controlled thermonuclear fusions.* Efforts to control a useful thermonuclear fusion reaction have not yet been successful. If they ever are, a new source of energy will become available.

Bombardment of heavy water with accelerated deuterons does lead to the reaction,

$$_1^2H + _1^2H \longrightarrow _2^3He + _0^1n + 3.3 \text{ MeV}$$

for neutrons are emitted, but very few fusions take place, and more energy has to be put into the accelerated deuterons than is obtained from the reaction.

For successful fusion reactions, which would liberate more energy than is needed to initiate them, it seems likely that temperatures of the order of 10^8 K will have to be achieved, and no constructional material will withstand such temperatures. The problem may not, however, be insoluble, and several hopeful lines are being investigated.

In one approach a current is passed through a gas in a tube, squeezing the gas into a narrow column along the axis of the tube. This happens because magnetic forces pinch the gas into the centre. An additional magnetic field along the axis of the tube, provided by an external electromagnet, also helps to keep the gas on the axis of the tube, and away from the walls. It is hoped, in this way, to be able to heat deuterium gas, out of contact with any material which might melt, by passing very high currents through it. None of the experiments have yet been fully successful, but if they can be developed, or some alternative method of heating devised, deuterium, which is relatively cheap and relatively abundant, may be the fuel of the future.

QUESTIONS ON CHAPTERS 13 AND 14

1. Write full equations for the nuclear reactions summarised as follows:
(a) $^{23}_{11}Na$ (α, p) $^{26}_{12}Mg$, (b) ^{12}C (d, n) ^{13}N, (c) ^{26}Mg (n, α) ^{23}Ne, (d) $^{26}_{12}Mg$ (p, n) $^{26}_{13}Al$.

2. What does X stand for in the following equations:

i $^{14}_7N + ^4_2He \longrightarrow ^{17}_8O + X.$

ii $^1_1H + ^7_3Li \longrightarrow 2X.$

iii $2^2_1D \longrightarrow ^3_2He + X.$

iv $^8_4Be + ^4_2He \longrightarrow X.$

3. Write down, in words, the changes represented by the following:

a $^{27}_{14}Si \longrightarrow ^{27}_{13}Al + \beta^+$

b ^{55}Mn (n, γ) ^{56}Mn.

c $^4_2He + ^9_4Be \longrightarrow ^{12}_6C + ^1_0n.$

d $^1_1H + ^1_0n \longrightarrow ^2_1H + \gamma.$

e $^{235}_{92}U + ^1_0n \longrightarrow 2^1_0n + \gamma + ^{147}_{60}Nd + ^{87}_{32}Ge.$

4. What does Y stand for in the following: (a) ^7_3Li (p, n) Y, (b) $^{43}_{20}Ca$ (α, p) Y, (c) ^7_3Li (Y, n) ^7_4Be, (d) $^{59}_{27}Co$ (Y, α) $^{56}_{25}Mn$?

5. Mass spectrographic investigation of natural cadmium shows the presence of isotopes of mass 110, 111, 112, 113, 114 and 116. After the same sample of natural cadmium has been exposed to a beam of slow neutrons for some time, ^{113}Cd was found to be almost absent. Account for this observation.

6. What conclusions can be drawn about the values of the neutron/proton ratio for stable isotopes?

7. Helmholtz said, in 1881, that 'the very mightiest among the chemical forces are of electric origin'. Comment on this statement.

8. What energy, in electron-volts, corresponds to radiation of wavelength (i) 1 μm and (ii) 200 nm? What wavelength, in nanometres, corresponds to an energy of 2.48 eV?

9. Explain what is meant by the following relationships: (a) $1 \text{ cm}^{-1} = 11.97$ $J \text{ mol}^{-1} = 1.23958 \times 10^{-4}$ eV atom^{-1}, (b) 1 eV $= 99.16 \text{ kJ mol}^{-1}$.

10. ^{55}Mn is one of the stablest nuclei. If its physical relative atomic mass on the ^{12}C scale is 54.956, determine its binding energy. Take the relative masses of the proton and neutron as 1.0076 and 1.009 respectively.

11. Calculate the binding energy of ^{12}C, taking the masses of the proton, the neutron and ^{12}C as 1.0078, 1.0089 and 12.0000 respectively.

12. The radius of the nucleus of an atom is given, approximately, by the relationship,

$$r = r_0 A^{\frac{1}{3}}$$

where r is the radius in cm, r_0 equal 1.5×10^{-13} cm, and A is the mass number of the nucleus. Calculate the approximate radius of (i) a neutron, and (ii) a proton. Calculate also the approximate radius of a ^{64}Cu nucleus. What will be the approximate density of the nucleus in g cm^{-3}, assuming it to be spherical and uniformly packed?

13. An atomic nucleus has been likened to a drop of incompressible liquid. Discuss this liquid-drop model.

14. Compare the repulsive force between two protons separated by 10^{-10} m, as in a hydrogen molecule, with that when the distance of separation is only 10^{-15} m, as in an atomic nucleus. Comment on the significance of your answer.

15. Deuterium nuclei of atomic mass 2.014 can be split up into protons of mass 1.0076 and neutrons by γ-radiation of energy 2.26·MeV. Calculate the mass of the neutron from this data in atomic mass units.

16. The proton bombardment of the isotope ^{11}B yields three alpha particles for each proton and each ^{11}B atom. Each alpha particle has energy of 2.85 MeV. If the atomic mass of the alpha particle is 4.0034 and that of the proton is 1.0081 calculate the atomic mass of ^{11}B.

17. The relative atomic mass of helium is 4.0039 and that of deuterium is 2.0147. If it was possible to make 1 mole of helium by fusing 2 mol of deuterium together, how many joules would be released?

18. ^{13}C can be converted into nitrogen by capture of a proton. If the atomic masses are 13.0075, 1.0081 and 14.0075, respectively, what is the heat of the reaction?

19. If a ^{235}U atom fissions, after absorption of a neutron, according to the equation,

$$^{235}_{92}U + ^{1}_{0}n \longrightarrow ^{148}_{57}La + ^{85}_{33}Br + 3^{1}_{0}n$$

and if the relative isotopic masses concerned are U = 235.124, n = 1.009, La = 147.961, and Br = 84.938, calculate the energy given out from one fissioned atom in MeV.

Chapter 15
The chemical bond

THE precise nature of the chemical bond is one of the most fundamental problems in chemistry, for it is the formation of bonds between atoms which leads to compound formation, and, in a chemical reaction, bonds are first broken and then reformed. The more information that can be gleaned about bonds between atoms the more fully will chemical properties be understood. The electrovalent bond, the covalent bond and the dative bond are considered in this chapter and, in more detail, in the three following ones. Metallic bonding, hydrogen bonding and van der Waals' bonding are dealt with in Chapter 19.

1. The electrovalent or ionic bond. The first detailed mode of forming a chemical bond was suggested, by Kossel, in 1916. He regarded the arrangements of electrons in the noble gas atoms (p. 116) as being particularly stable arrangements which accounted for the inability of the noble gases to enter into chemical combination.

His suggestion was that elements placed just before a noble gas in the periodic table might attain a noble gas structure by gaining electrons and forming negatively charged ions, whilst elements placed just after a noble gas might attain a noble gas structure by losing electrons and forming positively charged ions.

Thus, chlorine, with a structure, 2.8.7, could become a chloride ion, Cl^-, with a structure, 2.8.8 (argon structure), by gaining an electron. Sodium, 2.8.1, could form a Na^+ ion, with a structure 2.8 (neon structure), by losing an electron.

Both the chloride and sodium ions would have stable structures, and, by combining, would form sodium chloride. For two ions, sometimes known as an *ion-pair*, the state of affairs could be represented, showing only the outermost, or valency, electrons as

$$Na\bullet \quad + \quad {}_{\times}^{\times\times}\!Cl\,{}_{\times}^{\times} \quad \longrightarrow \quad [Na]^+ \left[\bullet\,{}_{\times\times}^{\times\times}\!Cl\,{}_{\times}^{\times}\right]^-$$

Other similar examples are provided by calcium bromide and potassium sulphide,

$$[Ca]^{2+} \quad \begin{array}{c} \left[{}_{\times}^{\bullet}{}_{\times\times}^{\times\times}\!Br\,{}_{\times}^{\times}\right]^- \\[10pt] \left[{}_{\times}^{\bullet}{}_{\times\times}^{\times\times}\!Br\,{}_{\times}^{\times}\right]^- \end{array} \qquad \begin{array}{c} [K]^+ \\[20pt] [K]^+ \end{array} \quad \left[{}_{\times}^{\bullet}{}_{\times\times}^{\times\times}\!S\,{}_{\times}^{\times}\right]^{2-}$$

Calcium bromide,
$CaBr_2$

Potassium sulphide,
K_2S

Such bonding between ions is due to electrostatic attraction between oppositely charged ions. The resulting bonds are known as electrovalent or ionic bonds, and compounds containing such bonds are known as electrovalent or ionic compounds.

When, as in practice, many ions are involved, they are held together, by electrostatic forces, within a crystal, and crystals which are made up of ions are known as *ionic crystals* (p. 168). The arrangement of ions within such crystals, i.e. the crystal structure, is mainly determined by the charges and sizes of the ions. Typical structures are shown on pp. 170–2.

Many compounds can be satisfactorily formulated with ionic bonds, and they have certain characteristic properties, summarised as follows:

i Individual molecules of ionic compounds do not exist because the compounds are made up of an interlocking structure of ions. The formula of an ionic compound simply shows the relative numbers of each ion.

ii Ionic compounds are invariably electrolytes, for, in the presence of an ionising solvent, such as water, the forces between the ions are so greatly reduced that the ions 'fall apart', and the free ions, in the resulting solution, are able to move under the influence of an electrical field as in electrolysis.

iii Ionic compounds are often hard solids because the inter-ionic forces within an ionic crystal are usually strong.

iv Ionic compounds generally have high melting points because a lot of thermal energy is required to break down the inter-ionic forces and form a liquid. Once an ionic compound has been melted, the melt can undergo electrolysis because it contains free ions. A high melting point is associated, too, with a high boiling point for ionic compounds.

v Ionic compounds are commonly soluble in water, or other ionising solvents, and insoluble in benzene or other organic solvents.

The commonest and most stable ionic compounds are formed between elements preceding, and elements following, a noble gas in the periodic table. Elements preceding a noble gas form negatively charged ions (anions) by gaining electrons; such elements are said to be *electronegative*. Elements following a noble gas, and readily losing electrons to form positively charged ions, are said to be *electropositive*.

Further details of ionic compounds are given in Chapter 16.

2. The covalent bond. There are many substances which cannot be formulated with ionic bonds, either because they are non-electrolytes, e.g. tetrachloro-methane, or because the atoms bonded together are the same, so that neither would be expected to transfer an electron to the other, e.g. chlorine, Cl_2.

To account for the formation of such molecules, Lewis, in 1916, suggested that atoms might gain noble gas structures, not by complete transference of electrons, as in ionic bonding, but by sharing electrons.

On this idea the chlorine and hydrogen molecules are represented as

$$\overset{\times\times}{\underset{\times\times}{\times}} Cl \overset{}{\times} \overset{\bullet\bullet}{\underset{\bullet\bullet}{Cl}} \colon \quad \text{and} \quad H \overset{\times}{\bullet} H$$

one electron from each atom being held in common by both. The shared pair of electrons constitutes what is known as a covalent bond.

In tetrachloromethane (carbon tetrachloride) and methane the molecules are represented as

$$\colon \overset{\bullet\bullet}{Cl} \colon$$
$$\colon \overset{\bullet\bullet}{Cl} \overset{\times}{\underset{\times}{\times}} C \overset{\times}{\underset{\times}{\times}} \overset{\bullet\bullet}{Cl} \colon$$
$$\colon \overset{}{\underset{\bullet\bullet}{Cl}} \colon$$

Tetrachloromethane

$$H$$
$$H \overset{\times}{\underset{\bullet}{\colon}} C \overset{\times}{\underset{\bullet}{\colon}} H$$
$$H$$

Methane

and the double and triple bonds of organic compounds, as exemplified by ethene (ethylene) and ethyne (acetylene), are written as

$$\underset{H}{\overset{H}{}} \overset{\bullet}{\underset{\times}{C}} \overset{\times\times}{\underset{\circ\circ}{\circ}} C \overset{}{\underset{}{}} \underset{H}{\overset{H}{}}$$

Ethene

$$H \overset{\times}{\colon} C \overset{\times\circ}{\underset{\circ}{\circ}} C \overset{}{\colon} H$$

Ethyne

A shared pair of electrons represents one valency bond, and it is convenient to denote such a single covalent bond by a single line. Ammonia and hydrogen chloride, for example, are written as

$$\begin{array}{c} H - N - H \\ | \\ H \end{array} \quad \text{or} \quad H \overset{\times\times}{\underset{\bullet\times}{\colon}} N \overset{}{\colon} H \qquad H - Cl \quad \text{or} \quad H \overset{\times\times}{\underset{\times\times}{\colon}} Cl \overset{}{\times}$$

The molecules of oxygen and nitrogen, with double and triple bonds, are

$$\overset{\times}{\underset{\times}{\times}} O \overset{\times}{\underset{\circ}{\circ}} O \overset{\circ}{\underset{\circ}{\circ}}$$

O=O

$$\overset{\times}{\underset{\times}{\times}} N \overset{\circ}{\underset{\circ}{\circ}} N \overset{\circ}{\underset{\circ}{\circ}}$$

N≡N

A covalent bond is directed in space so that the atoms in a covalent compound are linked in a definite position in relation to each other and the molecules formed may exist as distinct particles.

The atoms in a water molecule, for example, are arranged as shown,

$$\overset{\times\times\times}{\underset{\times}{}} O \overset{H}{\underset{H}{\bullet}}$$

or

$$O \overset{H}{\underset{H}{\nwarrow\nearrow}} \text{104° 31'}$$

with a bond angle of 104° 31′. Similarly, when carbon forms four single covalent bonds the bonds are arranged tetrahedrally. The molecule of methane, therefore, although commonly represented, for convenience, as flat, is really three-dimensional with the four hydrogen atoms at the corners, and the carbon atom at the centre, of the tetrahedron. The fact that many molecules are three-dimensional must not be forgotten. They are often portrayed as flat simply because it is so much more difficult to show their spatial arrangement on paper.

a. *Molecular crystals.* When a substance is made up of a large number of individual, covalent molecules, the solid form of the substance is either amorphous or exists in crystal structures known as molecular crystals. The atoms within each separate molecule are united by covalent bonds, but the molecules are held together, within the crystal, by weak intermolecular forces. It is the weakness of these forces that explains why such substances commonly exist as gases or liquids at room temperature. If they are solid, they are generally soft.

Such substances have certain general properties, listed below, which are opposite to those of ionic compounds.

i The substances are made up of individual covalent molecules, with weak inter-molecular forces.

ii The substances are non-electrolytes. Electrolysis involves ions, and covalent compounds do not contain ions. Most organic compounds are non-electrolytes, and the covalent bond is of great significance in a study of organic chemistry. The ways in which carbon can form single, double and triple covalent bonds is particularly important, some typical examples being given below:

Ethanal
(Acetaldehyde)

Ethanonitrile
(Methylcyanide)

Methylamine

iii The substances are gases, liquids or soft solids at room temperature, because of the weak inter-molecular forces.

iv The substances have low melting and boiling points because little energy is needed to break down the weak inter-molecular forces.

v The substances are commonly soluble in organic solvents, but insoluble in water or other ionising solvents. Covalent substances may, however, appear to be soluble in water because they react with it.

A comparison of the properties of sodium chloride and tetrachloromethane shows, in a rather exaggerated way, the typical differences between an electrovalent compound forming ionic crystals and a covalent compound forming molecular crystals.

	Sodium chloride	Tetrachloromethane
i	Composed of Na^+ and Cl^- ions in an ionic crystal	Composed of individual CCl_4 molecules with weak intermolecular forces. Forms a molecular crystal when solid
ii	Electrolyte	Non-electrolyte
iii	Hard solid at room temperature	Liquid at room temperature. Soft, when solid
iv	m.p. $= 803°C$ b.p. $= 1430°C$	m.p. $= -28°C$ b.p. $= 77°C$
v	Soluble in water. Insoluble in benzene	Soluble in benzene. Insoluble in water

b. *Atomic crystals.* In substances which form molecular crystals, covalent bonds hold atoms together in discrete molecules. Covalent bonds can, however, hold atoms together in inter-locking structures known as atomic crystals.

Diamond, with carbon atoms linked to each other, and silica, with alternating silicon and oxygen atoms, are simple examples. The crystal structure of diamond is shown on p. 183.

No single molecules of such substances exist; they are sometimes said to be made up of *giant molecules*. Moreover, the covalent bonds holding the whole crystal together are strong so that such substances are hard solids, and have high melting and boiling points. Like all covalent substances, they are non-electrolytes.

3. The dative bond. The shared pair of electrons of a covalent bond may also be formed by one of the two bonded atoms providing both electrons. If so the bond is sometimes called a dative bond, but as it is just like a covalent bond, once formed, the two are not always distinguished.

The atom providing the two electrons to make up the dative bond is known as the *donor*. It must, of course, have an 'unused' pair of electrons available, and such a pair is referred to as a *lone pair*. The atom sharing the pair of electrons from the donor is known as the *acceptor*.

When it is not necessary to distinguish between a dative bond and a covalent bond the — symbol is used for both. Two other symbolisms to represent a dative bond are, however, in use and have certain points in their favour. The first shows a dative bond between atoms A and B as

$A \longrightarrow B$, A being the donor and B the acceptor. This indicates, in a convenient way, the origin of the electrons making up the bond.

The second shows an $A \longrightarrow B$ bond as $A^{\oplus}\text{—}B^{\ominus}$. This method indicates the electrical charges which develop on atoms A and B as a result of dative bond formation. A, the donor, develops a positive charge by partly transferring two electrons to B; B, the acceptor, develops a corresponding negative charge. On this basis, the dative bond can be regarded as a covalent bond with a certain amount of ionic character, and the term coionic, instead of dative, is intended to describe this state of affairs. Other terms which have been used are co-ordinate bond, semi-polar bond or semi-polar double bond. The use of dative is preferred because it was the term finally used by Sidgwick, who first suggested the formation of this type of bond.

The following structural formulae show the various ways in which dative bonds can occur in typical compounds.

a. *Nitromethane*, $CH_3.NO_2$,

b. *Ethanoisonitrile* (*Methyl isocyanide*), $CH_3.NC$,

c. *The ammonium ion*, $NH_4{}^+$,

d. *Aluminium chloride*, Al_2Cl_6,

(Questions on the contents of this Chapter will be found on pp. 205-7.)

Chapter 16
The electrovalent or ionic bond

1. Stable ionic structures. So far as simple, negatively charged ions (anions) are concerned the only known ions have noble gas structures, the commonest simple anions being,

O^{2-}	2.8	F^-	2.8	Br^-	2.8.18.8
S^{2-}	2.8.8	Cl^-	2.8.8	I^-	2.8.18.18.8

Many simple cations e.g. Li^+, Na^+, K^+, Rb^+, Cs^+, Mg^{2+}, Ca^{2+}, Sr^{2+}, Ba^{2+} and Al^{3+}, also have noble gas structures.

Cations which do not have noble gas structures are, however, also fairly common. Many metals in the B sub-groups of the periodic table can form ions. Zinc, 2.8.18.2, for example, forms a zinc ion, Zn^{2+}, with a structure 2.8.18, and copper, 2.8.18.1, forms the copper(ɪ) ion, Cu^+, with the same 2.8.18 structure. The stability of an ion with such a structure is not, generally, so great as that of an ion with a noble gas structure, and this explains the greater ease of formation of, for instance, a calcium ion, Ca^{2+} as compared with a zinc ion, Zn^{2+}.

Moreover, an arrangement of 18 electrons in the outermost orbit of an ion is not very stable in elements in which the group of 18 electrons has only just filled up. One or more of the 18 electrons can be lost quite easily to form an ion with a greater charge. Copper, silver and gold, for instance, form ions as below:

Cu	2.8.18.1	Ag	2.8.18.18.1	Au	2.8.18.32.18.1
Cu^+	2.8.18	Ag^+	2.8.18.18	Au^+	2.8.18.32.18
Cu^{2+}	2.8.17	Ag^{2+}	2.8.18.17	Au^{3+}	2.8.18.32.16

the stabler ion in each case being underlined.

In ions formed from transitional elements, any number of electrons, from 9 to 17, can occur in the outermost orbit, and the small differences in stability between two or more alternative structures explains why transitional elements often have variable valency and form two or more ions. For example, iron, 2.8.14.2, forms both iron(ɪɪ) ions, 2.8.14, and iron(ɪɪɪ) ions, 2.8.13, and cobalt, 2.8.15.2, forms both cobalt(ɪɪ) ions, 2.8.15, and cobalt(ɪɪɪ) ions, 2.8.14.

2. Limitations to formation of ions. Whether or not an atom will form an ion depends to some extent on the stability of the ionic structure which it might form. As ionisation depends on the gaining (anion formation) or losing (cation formation) of electrons, however, both the size of the atom and its atomic number are also important. The size of the atom determines the distance of the valency electrons from the nucleus. The atomic number determines the positive charge on the nucleus, and it is this charge which holds the electrons in position.

The farther an electron is from the nucleus, the less firmly is it held and the more easily can it be lost. Thus in the group lithium, sodium, potassium, rubidium and caesium the last atom, which is the largest, forms an ion most easily. The loss of an electron from a caesium atom is so easy that caesium shows a very strong photoelectric effect (p. 98). Similarly, barium ionises more readily than strontium, calcium or magnesium.

Once an electron has been lost from an atom, the remaining ones are held more firmly and are not lost so easily. It is in this way that the formation of cations is limited to those with a charge of four units, and such highly charged ions are rare and only formed by large atoms. Tin and lead do form M^{4+} ions, but the smaller atoms of carbon and silicon do not. Similarly, aluminium forms Al^{3+}, but there is no B^{3+} ion.

In the formation of anions, the positive charge on the nucleus of an atom may be able to hold one extra electron, and can sometimes hold two or three, but three is the limit. The smaller the anion, the more easily can the nuclear charge hold the extra electrons, for, in a small ion, they are nearer to the nucleus than in a larger one. As a result, simple anions are limited to those of hydrogen (in salt-like hydrides such as lithium hydride, LiH), the halogens, oxygen, sulphur, selenium and tellurium, and nitrogen and phosphorus (in some nitrides and phosphides). In the halogen series, fluorine, with the smallest atom, most easily forms an anion. Thus mercury(II) aluminium and tin(IV) fluorides are ionic compounds, whereas the corresponding chlorides are covalent.

3. Fajans's rules. From the general considerations outlined in the preceding section, it is possible to summarise the conditions favouring the formation of an ion. An ion will be formed most easily:

a if the electronic structure of the ion is stable,

b if the charge on the ion is small, and

c if the atom from which the ion is formed is small for an anion, or large for a cation.

These rules, in a different form, were first suggested by Fajans, in 1924, and are usually known as Fajans's rules.

Fajans used the conception of the atomic volume of an element, i.e. the relative atomic mass of the element divided by its density. This gives an approximate measure of the size of an atom of an element, and it follows, from what has been said above, that anions will be formed most easily by elements with low atomic volume, and cations most easily by elements with large atomic volume.

The so-called diagonal relationships in the periodic table shown below,

Li   Be   B   C

Na   Mg   Al   Si

are probably due to the comparative ease of ionisation of the elements joined by arrows. Increase in ionic charge in moving from left to right will lead to greater difficulty in ionisation, but increase in atomic size in passing down a group will lead to greater ease of ionisation.

4. Measurement of ease of formation of ions. More quantitative measurements of the ease with which various atoms will form ions can be made in three different ways.

a. *Electrode potentials.* The standard electrode potential of an element (p. 426) is the potential difference between the element and a solution containing the ions formed from the element in a concentration of 1 mol of ion in 1 dm^3. The electrode potential of hydrogen is arbitrarily chosen as zero, and other values are measured on this scale. The values for some common elements are given on p. 428.

An element with a high negative electrode potential is one which readily loses electrons and forms positive ions. A large positive electrode potential means a readiness to form negative ions by gain of electrons.

The resulting list of electrode potentials is sometimes known as the electrochemical series or the reactivity series.

This series is a useful guide to the readiness with which an element will ionise, but it is based on measurements made on ions in solution, and, under such conditions, hydration of the ions may take place and affect their ease of formation.

b. *Ionisation energy.* A direct measurement of the ease with which an atom can lose an electron, with no hydration complications, is provided by the ionisation energy of an element. This is the energy required to withdraw an electron from an atom against the attraction of the nuclear charge. In other words, it is the heat of reaction of the change

$$\text{atom} + \text{energy} \longrightarrow \text{cation} + 1 \text{ electron}$$

Ionisation energies can be measured by spectroscopic methods, or by measuring the current passing through a discharge tube containing gas or vapour of the element under consideration as the applied voltage is gradually increased. At certain voltages there are marked rises in the current passing and these correspond to the points at which atoms of the element concerned lose one, two, three or more electrons.

Some typical ionisation energies are given below. The values given are in kJ mol^{-1}; they can be converted into eV by dividing by 96.48. They show quantitatively some of the effects which have already been mentioned qualitatively.

H $\longrightarrow$ H$^+$ 1322	Cu $\longrightarrow$ Cu$^+$ 740.6	Be $\longrightarrow$ Be^{2+} 2644	Zn $\longrightarrow$ Zn^{2+} 2623
Li $\longrightarrow$ Li$^+$ 518.8	Cu $\longrightarrow$ Cu^{2+} 2682	Mg $\longrightarrow$ Mg^{2+} 2176	Cd $\longrightarrow$ Cd^{2+} 2485
Na $\longrightarrow$ Na$^+$ 497.9	Ag $\longrightarrow$ Ag$^+$ 723.8	Ca $\longrightarrow$ Ca^{2+} 1724	Hg $\longrightarrow$ Hg^{2+} 2799
K $\longrightarrow$ K$^+$ 422.6	Au $\longrightarrow$ Au^{3+} 895.4	Sr $\longrightarrow$ Sr^{2+} 1602	B $\longrightarrow$ B^{3+} 6832
Rb $\longrightarrow$ Rb$^+$ 401.7		Ba $\longrightarrow$ Ba^{2+} 1460	Al $\longrightarrow$ Al^{3+} 5099
Cs $\longrightarrow$ Cs$^+$ 380.7			

IONISATION ENERGIES. (Values in kJ mol^{-1}.)

c. *Electron affinity*. The electron affinity of an atom is the energy given out when an extra electron is taken up by the atom. It is the heat of reaction of the change

$$\text{atom} + 1 \text{ electron} \longrightarrow \text{anion} + \text{energy}$$

Electron affinities cannot be measured very easily. Spectroscopic methods can be used, and one main method depends on a study of the change of equilibrium of the above reaction with temperature. Some values of electron affinities, in kJ mol^{-1}, are given below,

$$\underset{68.62}{\text{H} \longrightarrow \text{H}^-} \quad \underset{414.2}{\text{F} \longrightarrow \text{F}^-} \quad \underset{389.1}{\text{Cl} \longrightarrow \text{Cl}^-} \quad \underset{364.0}{\text{Br} \longrightarrow \text{Br}^-} \quad \underset{330.5}{\text{I} \longrightarrow \text{I}^-}$$

5. Ionic radii. X-ray analysis of ionic crystals (p. 215), and other methods, give values for the equilibrium distance between two adjacent ions in a crystal. The values are very small and are generally expressed in nanometres (nm), 1 nm being equal to 10^{-9} m. An older unit, 1 Ångström unit (1 Å), equal to 10^{-10} m is also still used to some extent. The internuclear distance for the halides of sodium and potassium are given below in nm:

	KF 0.266	KCl 0.314	KBr 0.329	KI 0.353
	NaF 0.231	NaCl 0.281	NaBr 0.298	NaI 0.323
Difference	0.035	0.033	0.031	0.030

If the ions in a crystal are regarded as spheres, the internuclear distance between two ions will be made up of the sum of the ionic radii of the two ions. Moreover, the constancy of the difference between the internuclear distances of sodium and potassium halides, as shown, indicates that the ionic radii of anions and cations must be reasonably constant in a series of such compounds.

The actual internuclear distance does not give a value for an ionic radius until any one ionic radius is decided by some other method. By adopting ionic radii for fluoride and oxide ions of 0.136 and 0.140 nm, Pauling has drawn up tables of ionic radii. Typical values are given below.

Li^+ 0.060		Be^{2+} 0.031						O^{2-} 0.140	F^- 0.136
Na^+ 0.095		Mg^{2+} 0.065		Al^{3+} 0.050				S^{2-} 0.184	Cl^- 0.181
K^+ 0.133	Cu^+ 0.096	Ca^{2+} 0.099	Zn^{2+} 0.074	Sc^{3+} 0.081	Ga^{3+} 0.062	Ti^{4+} 0.068		Se^{2-} 0.198	Br^- 0.195
Rb^+ 0.148	Ag^+ 0.126	Sr^{2+} 0.113	Cd^{2+} 0.097	Y^{3+} 0.093	In^{3+} 0.081	Zr^{4+} 0.080	Sn^{4+} 0.071	Te^{2-} 0.221	I^- 0.216
Cs^+ 0.169	Au^+ 0.137	Ba^{2+} 0.135	Hg^{2+} 0.110	La^{3+} 0.115	Tl^{3+} 0.095		Pb^{4+} 0.084		

IONIC RADII. (Values in nanometres.)

The figures illustrate a number of conclusions which would be expected on qualitative lines.

a. The ions in any one vertical group of the periodic table increase in size as the relative atomic mass of the element increases.

b. For a series of ions with the same arrangement of electrons as in neon, e.g. F^-, Na^+, Mg^{2+}, Al^{3+}, the size of the ion decreases as the atomic number increases. Increasing atomic number means increasing nuclear charge and a correspondingly greater inwards attraction on the electrons.

c. When an element forms two positively charged ions the ion with the lower charge is larger than the more highly charged ion. Pb^{2+}, for example, with radius of 0.121 nm is larger than Pb^{4+}, with radius of 0.084 nm.

d. Cations are smaller than the atom from which they are formed, whilst anions are larger.

6. Ionic crystals. The electrostatic forces holding the ions together in an ionic crystal are non-directional and the arrangement of ions within the crystal is mainly controlled by the sizes and the charges of the ions concerned.

a. Sodium chloride crystal structure. The way in which Na^+ and Cl^- ions build up into a sodium chloride crystal structure can be followed in stages. An ion-pair, shown in Fig. 40, has a strong residual field and will attract a

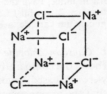

Fig. 42. Four ion-pairs of sodium chloride. The comparative sizes of the ions are not shown in this figure, but they are the same as in Fig. 41.

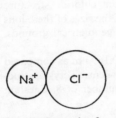

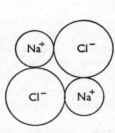

Fig. 40. Ion-pair of sodium chloride.

Fig. 41. Two ion-pairs of sodium chloride.

second ion-pair, just as two magnets would, as shown in Fig. 41. Four ion-pairs will arrange themselves as in Fig. 42, and a larger number will take up the arrangement shown in Fig. 43, which gives the crystal structure of sodium chloride.

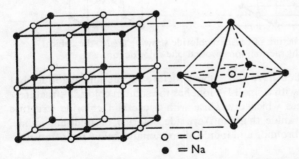

o $= Cl$
● $= Na$

Fig. 43. Crystal structure of sodium chloride, showing (*right*) the octahedral arrangement of six sodium ions around one chloride ion.

In a sodium chloride crystal, each sodium ion is surrounded octahedrally by six chloride ions, and each chloride ion by six sodium ions. The *co-ordination number*, the number of nearest neighbours of each ion in the structure, is 6.

Other compounds, made up of simple ions, which have the same crystal structure as sodium chloride include the halides of lithium, sodium, potassium, rubidium and caesium (except the chloride, bromide and iodide of caesium), the oxides and sulphides of magnesium, calcium, strontium and barium, the chloride, bromide and iodide of silver, the monoxides of cadmium, iron(II), cobalt(II), nickel(II) and manganese(II), and monosulphides of manganese(II) and lead(II).

One, or both, of the simple ions in a sodium chloride crystal may, also, be replaced by a charged group of atoms. Ammonium chloride, bromide and iodide, for instance, have the sodium chloride structure above their transition temperatures (p. 218) whilst Ca^{2+} and Co_3^{2-} ions in calcite, or Ca^{2+} and $(C\equiv C)^{2-}$ ions in calcium carbide, are also arranged with the sodium chloride structure.

b. *Caesium chloride structure.* A plan of the sodium chloride structure shows the comparative sizes of the ions concerned. If the sizes of these ions were increased beyond certain limits it is clear, on geometrical grounds, that the arrangement of ions could not exist.

Thus, caesium chloride, with a Cs^+ ion of radius 0.169 nm as compared with a radius of 0.095 nm for the Na^+ ion, has a body-centred crystal structure (Fig. 44) and a corresponding co-ordination number of 8. Caesium

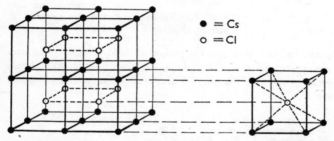

$\bullet = Cs$
$\circ = Cl$

FIG. 44. Crystal structure of caesium chloride showing (*right*) the cubical arrangement of eight caesium ions around one chloride ion.

bromide and iodide have the same structure as caesium chloride, but caesium fluoride, with a small F^- ion, has the sodium chloride structure.

Other substances which crystallise with a caesium chloride structure include caesium cyanide, thallium(I) cyanide, chloride, bromide and iodide, and the chloride, bromide and iodide of ammonium below their transition points (p. 218).

7. Structures of CA_2 ionic crystals. When the charge on the cation is twice that on the anion the crystal can be represented as CA_2. The two commonest crystal structures for ionic compounds of this type are the *fluorite*, CaF_2 and the *rutile*, TiO_2, *structures*.

In the fluorite structure (Fig. 45) of CaF_2, each Ca^{2+} ion is surrounded by eight F^- ions at the corners of a cube, and each F^- ion is surrounded by four Ca^{2+} ions arranged tetrahedrally. The coordination numbers are 8 and 4.

In the rutile structure of TiO_2 (Fig. 46) each Ti^{4+} ion is surrounded by six O^{2-} ions arranged octahedrally, and each O^{2-} ion is surrounded by three Ti^{4+} ions arranged triangularly. The co-ordination numbers are 6 and 3.

A third closely related structure is the *anti-fluorite structure* which is the normal fluorite structure with the anions and cations interchanged.

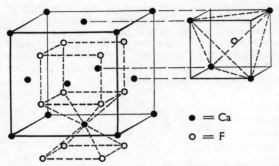

FIG. 45. Fluorite structure showing (*right*) the tetrahedral arrangement of four calcium ions around one fluoride ion, and (*below*) the cubical arrangement of eight fluoride ions around one calcium ion.

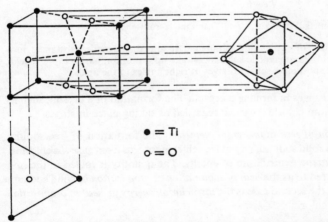

FIG. 46. Rutile structure showing (*right*) the octahedral arrangement of six oxide ions around one titanium ion and (*below*) the triangular arrangement of three titanium ions around one oxide ion.

8. The magnetic moments of ions. Much information has been obtained about the arrangement of electrons in ions from magnetic measurements.

Substances may be classified as *paramagnetic* or *diamagnetic*. Paramagnetic substances are drawn into a strong magnetic field, i.e. they take up a position parallel to a magnetic field (Fig. 47a). Diamagnetic substances tend to be drawn out of a magnetic field, i.e. they set themselves at right angles to the field (Fig. 47b). Substances normally regarded as magnetic, e.g. iron, steel, cobalt, nickel and magnetic alloys, are paramagnetic, but the degree of magnetism possessed by such substances is much greater than that of any others. They are said to be *ferromagnetic*.

An ion containing an unpaired electron is paramagnetic, whilst one in which all the electrons are paired is diamagnetic. A spinning electron is equivalent to an electric current in a circular conductor and, as such, behaves as a magnet. The moment of such a magnet is measured in magnetons and the magnetic moment of a paramagnetic substance containing n unpaired electrons can be shown to be $\sqrt{n(n + 2)}$ magnetons. This enables the number of unpaired electrons in any ion to be determined by measurement of its magnetic moment.

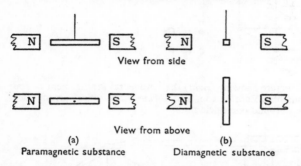

FIG. 47. Effect of suspending paramagnetic and diamagnetic substances in a magnetic field. Paramagnetic substances take up a position parallel to the field; diamagnetic substances set themselves at right angles to the field.

9. Energy changes in forming a crystal. The formation of a crystal, such as that of sodium chloride, can be regarded as taking place in stages.

a. *Formation of free atoms from elements.* The formation of free sodium atoms from solid sodium, and of free chlorine atoms from gaseous chlorine, both involve the expenditure of energy. The quantity involved in the first case is referred to as the *heat of sublimation* or *atomisation* of solid sodium, S_{Na}, that in the second case is the *dissociation energy* or *heat of atomisation* of gaseous chlorine, D_{Cl}.

b. *Formation of an ion from a free atom.* The energies involved in the formation of an Na^+ ion from an Na atom, or a Cl^- ion from a Cl atom, are the ionisation energy of sodium, I_{Na}, (p. 166) and the electron affinity of chlorine, E_{Cl}, (p. 167) respectively.

c. *Formation of an ion pair.* There will be an attractive force between an Na^+ and a Cl^- ion. If the two ions are at a distance r apart the force will be equal to e^2/r^2, and if they come closer together by a distance dr there will be a release of energy equal to $e^2 \times dr/r^2$. In forming an ionic bond in an ionic pair, it is a matter of bringing the two ions to their interionic distance, d, and the energy released in doing this will be given by

$$\int_{\infty}^{d} \frac{e^2}{r^2} \cdot dr \qquad \text{or} \qquad \frac{e^2}{d}$$

The formation of an ionic bond will only take place if the bonded ions are stabler than the free atoms, i.e. if $(I_{Na} - E_{Cl} - e^2/d)$ is negative.

d. *Crystal or lattice energy.* The simple ion pair formed in c. has a strong residual field so that many similar ion pairs attract each other and build up into an ionic crystal. This involves further energy changes, and the total energy change when free ions form together into a crystal is known as the crystal or lattice energy. Alternatively, this energy can be regarded as that required to break down a crystal structure into free ions. The lattice energy can be calculated for a given crystal, e.g. C_{NaCl}, in terms of the forces acting within the crystal.

e. *The Born–Haber cycle.* The heat of formation of crystalline sodium chloride, i.e. the heat of reaction of

$$Na(s) + \tfrac{1}{2}Cl_2(g) \longrightarrow NaCl(s) + H \text{ (heat of reaction)}$$

is, therefore, made up of five different energy terms. All the energy changes involved are interrelated in the Born–Haber cycle which can be summarised as follows:

$$
\begin{array}{ccc}
Na(s) + \tfrac{1}{2}Cl_2(g) & \xrightarrow{\ -H\ } & NaCl(s) \\
{\scriptstyle +S_{Na}}\downarrow \quad {\scriptstyle +D_{Cl}}\downarrow & & \uparrow {\scriptstyle -C_{Na\,Cl}} \\
\underset{\text{(free atoms)}}{Na + Cl} & \xrightarrow{\ I_{Na}-E_{Cl}\ } & \underset{\text{(free ions)}}{Na^+ + Cl^-}
\end{array}
$$

By Hess's law the energy change in the formation of crystalline sodium chloride must be the same however it is formed and it therefore follows that

$$H = -S_{Na} - D_{Cl} - I_{Na} + E_{Cl} + C_{NaCl}$$

If any five of the quantities concerned can be measured or calculated the sixth value can be obtained. The relationship is useful for finding values of crystal energies from measured values of H, S_{Na}, D_{Cl}, I_{Na}, and E_{Cl}, or for obtaining values of E_{Cl} from measured values of H, S_{Na}, D_{Cl}, I_{Na} and calculated values of C_{NaCl}. Similar considerations apply to other crystals.

(Questions on the contents of this Chapter will be found on pp. 205–7.)

Chapter 17
The covalent bond

1. Molecular orbitals. It is clear that ions are held together by electrostatic attraction in an electrovalent or ionic bond, but it is less easy to understand why a shared pair of electrons should hold two atoms together. There are two ways of viewing the problem. One involves the use of the conception of resonance and is deferred until Chapter 18; the other line of approach is afforded by the method of molecular orbitals.

In this method, a *molecule* is supposed to have electronic orbitals associated with it in much the same way as a single atom has (p. 108). All, or nearly all, of the electrons associated with the component atoms of a molecule are supposed to enter the various molecular orbitals, which fill up according to certain rules just as atomic orbitals do (p. 111). The Pauli principle (p. 109) applies to molecular orbitals, as to atomic orbitals, and no molecular orbital can contain two precisely similar electrons. This means that any particular molecular orbital can only contain two electrons, and that these two must have different spins.

The nomenclature s, p and d used for atomic orbitals is replaced by that of σ, π and δ for molecular orbitals, σ and π being the most important. Electrons in some of these molecular orbitals contribute towards the binding together of the atoms in a molecule; such orbitals are known as *bonding orbitals*. Others cause repulsion between atoms and are known as anti-bonding orbitals.

The nomenclature most commonly used indicates (*a*) the nature of the molecular orbital, i.e. σ or π, occupied by electrons in a molecule, (*b*) the atomic orbital from which the electrons originated, and (*c*) whether the molecular orbital is bonding or anti-bonding. A $\sigma 1s$ orbital, for example, is a σ orbital made up by interaction of $1s$ atomic orbitals; it is a bonding orbital. A $\sigma^* 1s$ orbital is the corresponding anti-bonding orbital.

2. The hydrogen molecule. In a hydrogen molecule, H_2, the two $1s$ electrons, one from each of the two hydrogen atoms concerned, are present in a $\sigma 1s$ molecular orbital. This is a bonding orbital and constitutes the single covalent bond in the hydrogen molecule. Since the orbital can only hold two electrons if they have different spins, it is clear that a molecule will only be formed from two hydrogen atoms containing electrons with opposed spins. This is an important point. A covalent bond is not simply a shared pair; *it is a shared pair of electrons with opposed spins*. This means that only single, unpaired electrons, in atoms, can participate in covalent bond formation.

An atomic orbital can be represented as a charge cloud of varying density, or, more conveniently, by mapping out the boundary surface within which the electron might be said to exist (p. 119). The same procedure can be adopted for molecular orbitals, as is illustrated for the hydrogen molecule in Fig. 48.

It is the accumulation of negative charge between the two positively charged atomic nuclei which is responsible for holding them together. Such an accumulation of charge occurs when two s orbitals containing electrons with opposed spins overlap.

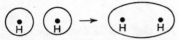

Two 1s atomic or- A σ1s molecular
bitals; electrons orbital. A bond-
with different ing orbital form-
spins. ing a σ-bond.

FIG. 48. Formation of a σ-bond from a pair of s electrons with different spins.

If electrons with the same spin are involved, an anti-bonding molecular orbital is formed (Fig. 49).

The bonding molecular orbital formed from any pair of s electrons with different spins is plum-shaped, as shown in the hydrogen molecule. It is symmetrical about the line joining the two nuclei and has no nodal plane,

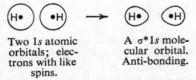

Two 1s atomic A σ*1s mole-
orbitals; elec- cular orbital.
trons with like Anti-bonding.
spins.

FIG. 49. Formation of anti-bonding molecular orbital from a pair of s electrons with like spins.

i.e. a plane in which the probability of finding an electron is zero (p. 119). The bond which it forms is referred to as a σ-bond, and the electrons which occupy a σ orbital are referred to as σ-electrons.

The bonding orbital formed from any pair of p electrons with opposed spins is different. It is known as a π orbital and its boundary surface is shown in Fig. 50.

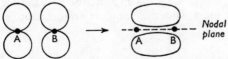

Two p atomic orbitals. A π-molecular orbital.

FIG. 50. Formation of a π-molecular orbital from a pair of p electrons with different spins. A π-bond.

π-orbitals, like p atomic orbitals from which they are formed (p. 119), have a nodal plane. The bonds which they form are known as π-type.or π-bonds and electrons occupying a π orbital are known as π-electrons.

There are, then, really two kinds of covalent bond. A σ-bond holds two atoms together by an accumulation of charge between their nuclei. A π-bond holds two atoms together by accumulation of charge alongside the two nuclei.

3. Directional nature of covalent bonds. The electrical field surrounding an ion extends equally in all directions and, to that extent, the bond formed by an ion has no directional characteristics. That covalent bonds can be directed in space is shown, however, both by the existence of stereoisomerism in covalent compounds and by the actual measurement of the angles between covalent bonds.

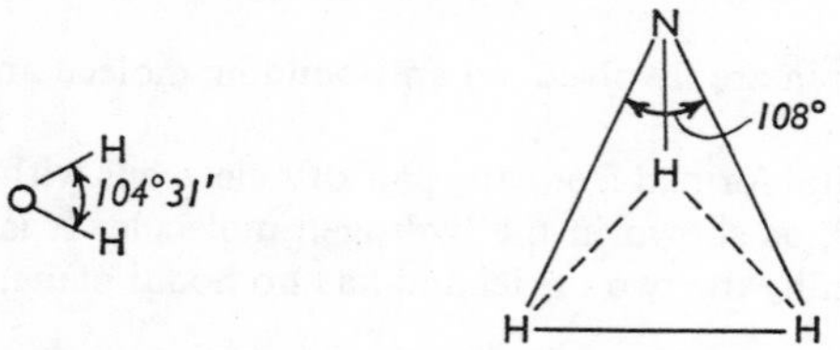

FIG. 51. Bond angles in water and ammonia molecules.

The atoms in a water molecule, for instance, are not in a straight line, and in the ammonia molecule the atoms are in a pyramidal arrangement with bond angles between adjacent N—H bonds of 108° (Fig. 51).

The directional nature of some covalent bonds is explained in terms of the various atomic orbitals participating in bond formation. Only those orbitals containing an unpaired electron will, it must be remembered, be available for contributing to a shared pair. s orbitals are said to be spherically symmetrical, and they have no directional character. A p orbital, however, is dumb-bell in shape and has, therefore, a directional nature. Moreover, the three p orbitals in any one shell, i.e. the p_x, p_y and p_z orbitals, are directed mutually at right angles to each other (p. 119).

In a water molecule each hydrogen atom has an s electron available for bond formation. The oxygen atom, with an electronic arrangement (p. 114) of,

$$O \quad \begin{array}{ccccc} 1s & 2s & 2p_x & 2p_y & 2p_z \\ \uparrow\downarrow & \uparrow\downarrow & \uparrow\downarrow & \uparrow & \uparrow \end{array}$$

has two 2p electrons available, the two 2p orbitals being at right angles to each other. The two O—H bonds will be s—p bonds, and the molecule of water might well be expected to be represented in some such way as in Fig. 52, with a bond angle of 90°. The fact that the measured bond angle is 104° 31′ is, possibly, due to the ionic character (p. 187) of the O—H bond. This gives positive charges to the two hydrogen atoms in a water molecule and causes them to repel each other to some extent making the bond angle greater than the expected 90°. For hydrogen sulphide, where the ionic character is not so great, the observed angle is 92° 20′.

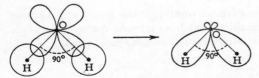

FIG. 52. The way in which two s orbitals from hydrogen atoms might be expected to interact with p orbitals from an oxygen atom to form a water molecule with a bond angle of 90°.

In ammonia the nitrogen atom, with a structure of

	$1s$	$2s$	$2p_x$	$2p_y$	$2p_z$
N	↿⇂	↿⇂	↿	↿	↿

has three $2p$ electrons available for bond formation. The ammonia molecule therefore contains three s—p bonds and it would be expected that these would be at right angles to each other. The actual bond angles are each 108°, and the divergence from 90° is again probably due to the partial ionic character of the bonds. The bond angles in phosphine, PH_3, are about 95°.

4. Hybridisation. The arrangement of extra-nuclear electrons in carbon atoms,

	$1s$	$2s$	$2p_x$	$2p_y$	$2p_z$
C	↿⇂	↿⇂	↿	↿	

shows only two unpaired electrons, and this would indicate that carbon ought to be 2-valent. To account for the quadrivalency of carbon it must be assumed that the arrangement of electrons in the carbon atom is changed, prior to reaction, in such a way as to provide four unpaired electrons. The change which takes place provides one $2s$ and three $2p$ electrons,

	$1s$	$2s$	$2p_x$	$2p_y$	$2p_z$
C	↿⇂	↿	↿	↿	↿

If four such unpaired electrons took part in the formation of four covalent bonds with, say, hydrogen it might be expected that one of the four bonds would be an s—s bond with no directional character, whilst the other three would be s—p bonds directed at right angles to each other. This is not in accordance with the known facts, for carbon's four valency bonds are arranged tetrahedrally.

The matter is resolved by supposing that one s and three p orbitals can

combine together into four different orbitals, and it can be shown, theoretically, by combining the wave-functions of the one s and three p orbitals,
that the result will be four equivalent orbitals directed towards the corners
of a regular tetrahedron.

This process of combining orbitals together in this sort of way is known
as hybridisation and the resulting orbitals are referred to as *hybrid orbitals*.
Bonds formed from hybrid orbitals are known as *hybrid bonds*.

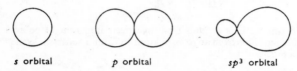

s orbital   p orbital   sp³ orbital

FIG. 53. The boundary surface of an sp^3 orbital formed by hybridisation of s and p
orbitals. With a carbon atom, four similar sp^3 orbitals are arranged
tetrahedrally.

The hybrid orbitals of carbon referred to above are labelled as sp^3
orbitals, and their shape is shown in Fig. 53. The directional arrangement of
the four sp^3 orbitals towards the corners of a regular tetrahedron accounts
for the tetrahedral arrangement of carbon's four valencies, first suggested
by le Bel and van't Hoff.

FIG. 54. Three sp^2
orbitals in the same
plane but at angles
of 120° to each
other.

5. The nature of the C=C bond. The type of hybridisation described in the preceding section, sometimes
known as tetrahedral hybridisation, is not the only
way in which s and p orbitals can be hybridised.

It is possible, for instance, for one 2s and two 2p, or
for one 2s and one 2p orbitals to be hybridised. With
one 2s and two 2p orbitals the resulting hybrid orbitals
are all equivalent and are coplanar at an angle of
120° to each other (Fig. 54). The hybrid orbitals are
known as sp^2 orbitals. One 2p orbital remains unchanged (Fig. 55).

This type of hybridisation (trigonal hybridisation) occurs in the formation of a C=C bond. Each carbon atom linked by the bond has three
equivalent sp^2 orbitals. In the molecule of ethene (ethylene), two of these
orbitals from each carbon atom form covalent bonds (σ-bonds) with
hydrogen atoms. The remaining sp^2 orbital of each carbon atom forms a
σ-bond between the carbon atoms. The two carbon, and four hydrogen,
atoms are all in the same plane.

At right angles to this plane there remains the unchanged 2p orbital of
each carbon atom and these two 2p orbitals interact to form a π-bond
between the two carbon atoms (Fig. 56).

A double bond between carbon atoms consists, therefore, of a σ-bond and a π-bond. The two bonds are not of equal strength, for a σ-bond is stronger than a π-bond. Two atoms linked by a σ-bond can rotate freely

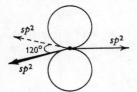

FIG. 55. The 2p orbital remaining after the formation of three coplanar sp^2 orbitals.

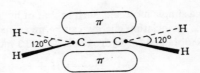

FIG. 56. The bonding in an ethene molecule. Four σ-bonds between carbon and hydrogen atoms. One σ-bond and one π-bond between carbon atoms.

about the bond, unless there is some steric interference, but free rotation is prevented when two atoms are linked by a π-bond.

6. The nature of the C≡C bond.

FIG. 57. Two colinear sp orbitals.

Hybridisation of one $2s$ and one $2p$ orbitals leads to two equivalent orbitals, which are colinear (Fig. 57). This type of hybridisation, known as digonal hybridisation, and forming sp orbitals, occurs in the formation of a C≡C bond. Two $2p$ orbitals remain unchanged on each carbon atom.

Each carbon atom linked by the triple bond has two equivalent, coplanar, sp orbitals. In the molecule of ethyne (acetylene), one of these orbitals from each carbon atom forms a σ-bond with hydrogen atoms, and the remaining sp orbital of each carbon atom forms a σ-bond between the two carbon atoms. The four atoms are all colinear.

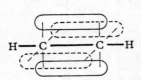

FIG. 58. The bonding in an ethyne molecule. Two σ-bonds between carbon and hydrogen atoms. One σ-bond and two π-bonds (in planes at right angles to each other) between carbon atoms.

The two remaining $2p$ orbitals of each carbon atom interact to form two π-bonds between the carbon atoms, these two bonds being in planes at right angles to each other (Fig. 58).

The C≡C bond consists, therefore, of a σ-bond and two π-bonds.

7. Bond energies.

Determination of the bond energy in a diatomic molecule containing a covalent bond involves the measurement of the heat of formation of the molecule from its free atoms, or, alternatively, the heat of dissociation of the molecule into its free atoms. The bond energy of a bond A—B is equal to U, where

$$\begin{array}{ccccccc} A & + & B & \longrightarrow & A\text{—}B & + & U \\ \text{(atom)} & & \text{(atom)} & & \text{(molecule)} & & \text{(energy)} \end{array}$$

In a simple molecule, such as that of hydrogen, the heat of dissociation, or the dissociation energy, can be measured by spectroscopic means. The value obtained for hydrogen is 432.6 kJ mol^{-1}, i.e.

$$H_2 \longrightarrow H + H - 432.6 \text{ kJ}$$

and this value is equal to the bond energy of the H—H bond.

With polyatomic molecules the situation is more complicated but bond energies can be obtained by thermo-chemical measurements as described on p. 302.

Some bond energies for common bonds are given below.

H—H 432.6	C—C 341.4	N—N 160.7	O—O 1 46.0	S—S 266.9	F—F 139.3	F—H 554.0
	C—N 290.0	N—H 388.7	O—H 461.2	S—H 366.1	Cl—Cl 241.8	Cl—H 432.6
	C—O 341.0				Br—Br 192.9	Br—H 365.3
	C—H 413.4	Si—Si 177.8			I—I 151.5	I—H 298.7

BOND ENERGIES OF SINGLE BONDS. (Values in kJ mol^{-1})

C=C 611.3	C=N 564.8	N=N 408.4	C=O 723.8	O=O 401.7
	C≡C 803.7	C≡O 1071	N≡N 941.4	

BOND ENERGIES OF MULTIPLE BONDS. (Values in kJ mol^{-1}.)

These values show how much energy is required to rupture a particular bond and they therefore indicate the relative stability of various bonds. Notice, for instance, the wide range of stable bonds formed by carbon. They also enable the heats of formation of some compounds to be calculated. The heat of formation of ethanol, for example, can be obtained by adding up the bond energies of five C—H, one C—C, one C—O and one O—H bonds. The value obtained is 3 210.6, which compares well with the experimental value of 3 210.

Heats of reactions may also be calculated from bond energies. The reduction of an alkene to an alkane, for instance, involves bond changes represented as follows:

$$C{=}C \ + \ H{-}H \ \longrightarrow \ C{-}C \ + \ 2(C{-}H)$$

$$\underbrace{611.3 \qquad\quad 432.6}_{1\,043.9} \qquad \underbrace{341.4 \qquad 2 \times 413.4}_{1\,168.2}$$

The reduction therefore involves an evolution of 124.3 kJ and this is in very good agreement with experimental values.

It will be seen, however, that such uses of bond energies do not always give figures in agreement with the experimentally measured values, and this has important consequences (p. 191).

8. Bond lengths and covalent radii. The bond length of a covalent bond is the distance between the nuclei of the two bonded atoms. The distances involved are very small and they are usually expressed in nanometres (p. 167).

Bond lengths can be measured by X-ray analysis of crystals, by the diffraction of X-rays, electrons or neutrons by gases or vapours, or by spectroscopic methods. The results of measurements on a wide variety of compounds shows that the inter-atomic distance between two atoms A and B linked by a covalent bond is very nearly constant, and is independent of the varied nature of the molecules in which the bond length is measured. The C—Cl distance, for instance, is 0.176 nm in mono-, di-, tri- and tetra-chloromethane. Similarly the C—C distance is 0.154 nm whether the bond occurs in diamond, propane, 1,3,5-tri-methylbenzene, ethanal trimer, or many other compounds.

It is found, moreover, that the interatomic distance A—B is equal to the arithmetic mean of the A—A and B—B distances, i.e.

$$A{-}B = \frac{A{-}A + B{-}B}{2}$$

and, because of this simple relationship, it is possible to allot what are called covalent radii to the elements such that $r_A + r_B = A{-}B$. The covalent radii of these elements which normally form covalent bonds are summarised below.

As is to be expected, the heavier elements in any one group of the periodic table have the larger radii; the heavier elements contain more electrons. For elements in the same horizontal periods those with higher atomic numbers have the smaller radii; these elements have their outer electrons in the same orbits, but the electrons are more strongly attracted by increasing positive charges on the nuclei.

H 0.030	B 0.088	C 0.077	N 0.070	O 0.066	F 0.064
		Si 0.117	P 0.110	S 0.104	Cl 0.099
			As 0.121	Se 0.117	Br 0.114
			Sb 0.141	Te 0.137	I 0.133

SINGLE-BOND COVALENT RADII. (Values in nanometres.)

Similar multiple-bond covalent radii for double and triple bonds, summarised below, show that the double-bond radius of an atom is about 13 per cent, and the triple-bond radius about 22 per cent, less than the corresponding single-bond radius.

B= 0.076	C= 0.067	Si= 0.107	N= 0.060	O= 0.055	S= 0.094
B≡ 0.068	C≡ 0.060		N≡ 0.055	O≡ 0.050	

MULTIPLE BOND COVALENT RADII. (Values in nanometres.)

The values for both single-bond and multiple-bond covalent radii are found to disagree with the measured bond lengths in some molecules, an effect which is explained by resonance (p. 191).

9. Atomic or covalent crystals. In atomic or covalent crystals, the structural units are atoms, and the bonding between the atoms is covalent, or predominantly covalent. Atomic crystals are not so common as ionic crystals (p. 169). The three main structures involved in atomic crystals are described below.

a. *The diamond structure.* In a diamond crystal (Fig. 59) each carbon atom is surrounded by four other carbon atoms arranged tetrahedrally. The atoms are linked by covalent bonds, which are sp^3 hybrid bonds (p. 178), and the interlocking network of atoms accounts for the extreme hardness of diamond.

Germanium, silicon and grey-tin also have a diamond structure, and it is common for elements in group 4 to exhibit allotropy. Carbon atoms, for example, can be linked together in a different structure, as in graphite (p. 204).

b. *The zinc-blende structure.* Zinc blende is one form of zinc sulphide. The crystal structure (Fig. 60) is related to that of diamond, the only difference being that adjacent atoms are different. Thus each zinc atom is

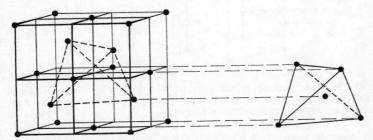

FIG. 59. The crystal structure of diamond showing (*right*) the tetrahedral arrangement of four carbon atoms around a central carbon atom.

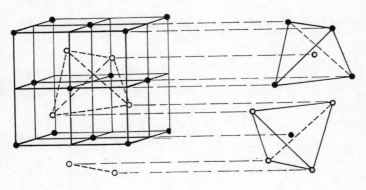

● = S

○ = Zn

FIG. 60. The crystal structure of zinc blende showing (*upper right*) the tetrahedral arrangement of four sulphur atoms about a zinc atom and (*lower right*) the tetrahedral arrangement of four zinc atoms about a sulphur atom. Compare the crystal structure of diamond (Fig. 59).

surrounded by four sulphur atoms arranged tetrahedrally, and each sulphur atom by four zinc atoms similarly arranged.

Other common substances which crystallise with a zinc-blende structure are copper(I) chloride, bromide and iodide, aluminium phosphide, silicon carbide, and the sulphides of beryllium, cadmium and mercury(II).

c. *The wurtzite structure.* Wurtzite is another form of zinc sulphide, and the structure (Fig. 61) is closely related to the zinc-blende structure above.

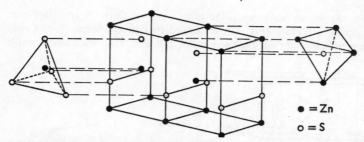

FIG. 61. The crystal structure of wurtzite showing (*right*) the tetrahedral arrangement of four zinc atoms around a sulphur atom, and (*left*) the tetrahedral arrangement of four sulphur atoms around a zinc atom.

Each atom is surrounded by four nearest neighbours, arranged tetrahedrally, as in zinc blende. It is only when the second nearest neighbours are considered that the wurtzite structure differs from that of zinc blende.

(Questions on the contents of this Chapter will be found on pp. 205–7.)

Chapter 18
Resonance

1. Resonance hybrids. In the preceding chapters the various different bonds have been regarded as distinct types, but there is plenty of evidence that the actual chemical bonds occurring in some compounds are of intermediate type.

The actual arrangement of electrons in a particular molecule cannot always be satisfactorily represented, by using accepted symbols, in terms solely of simple ionic or covalent bonds. No single structural formula which can be written for benzene, for example, accounts for all the known properties of benzene, and this is so for many other compounds.

The actual electronic arrangement in such compounds, which cannot be represented adequately by accepted symbols, has, therefore, to be represented in terms of other possible, though non-existent, structures which can be formulated using accepted symbols. The actual structure which must be visualised does not consist of a mixture of the various possible structures. It is a single structure of its own, but as it cannot be written down simply on paper, it is necessary to think of it in terms of structures which can.

A carbon dioxide molecule, for instance, can be represented by the three possible structures shown below.

I.   O=C=O   or   O × C O O

II.   O←C≝O   or   O : C : O

III.   O≝C→O   or   O : C : O

The actual structure of carbon dioxide which best accounts for all its properties, particularly for the measured bond lengths and heat of formation (p. 191), must be considered as something closely related to all the three possible structures, but something which is different from all of them.

The actual structure is said to resonate between the structures I, II and III, or to be a resonance hybrid of the three structures. Alternatively, using a terminology introduced by Ingold, the term mesomeric forms or mesomeric structures are used and the conception is known as mesomerism ('between the parts').

Various analogies have been suggested to facilitate the building up of a visual picture of what resonance means. Perhaps the best is the idea of describing a rhinoceros in terms of a unicorn and a dragon. The rhinoceros, which has an actual existence, is thought of, as a sort of resonance hybrid, in terms of unicorns and dragons, which do not exist.

It must be emphasised that all the molecules in a resonance hybrid are alike, and that it is not just a mixture of different molecules. It is particularly important to distinguish between resonance and tautomerism, for they are easily confused. The latter may be regarded as the existence of two or more forms of a substance having different arrangements of *atoms*; the forms can sometimes be isolated. The possible structures contributing to a resonance hybrid have the same arrangement of atoms but different arrangements of *electrons*.

2. Resonance energy. One of the most important points connected with resonance is that the resonance hybrid is a more stable structure than any of the structures contributing to it.

The increased stability is accounted for in the following way. If it is supposed that a resonance hybrid has two resonating structures, I and II, they can each be represented by wave functions, in simple cases, which give the

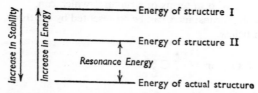

FIG. 62. Illustration of the increased stability of the actual structure caused by resonance between two possible structures, I and II.

energies of the structures they represent. By combining the two wave functions, it is found that a third wave function will also represent the system and that this function corresponds to a lower energy value, i.e. a higher stability.

The resonance energy of a substance is the extra stability of the resonance hybrid as compared with the most stable of the resonating structures (Fig. 62). It will be seen that the resonance hybrid is more stable than any of the resonating structures. It is, therefore, wrong to speak of the resonance hybrid being intermediate between the resonating structures so far as energy is concerned.

For carbon dioxide, the resonance energy is 154.4 kJ. This value is obtained by subtracting the calculated heat of formation of the $O{=}C{=}O$ structure (1 447.6 kJ, p. 180) from the observed value of the heat of formation (1602 kJ). In general

$$\begin{array}{c} \text{Resonance} \\ \text{energy} \end{array} = \begin{array}{c} \text{Observed heat of} \\ \text{formation} \end{array} - \begin{array}{c} \text{Calculated heat} \\ \text{of formation} \end{array}$$

3. Ionic character of covalent bonds. A covalent bond between an atom A and an atom B involves sharing of electrons and is represented as A⋮B.

An ionic bond involves complete transfer of electrons and is represented as $(A)^+(\overset{\cdot}{\underset{\times}{\cdot}}B)^-$.

In a covalent bond between unlike atoms the pair of shared electrons will not necessarily be shared equally by both atoms, for if, in a bond A—B, the atom B has a stronger attraction for electrons than A the shared pair will be attracted towards B and away from A. Any permanent displacement of electrons of this sort in a covalent bond will give the bond some ionic character, and the actual bond will have to be represented as a resonance hybrid between covalent and ionic forms,

$$A—B \quad \longleftrightarrow \quad A^+B^-$$

In an alternative method of describing the same state of affairs an *inductive effect* is said to exist in the bond; this is symbolised as A $\rightarrow$ B.

The extent of the ionic character, or the inductive effect, in a covalent bond will depend on the relative attraction for electrons of the bonded atoms. A covalent bond between like atoms will have no ionic character or inductive effect; it might be regarded as a true covalent bond. But in a bond between an electropositive atom, A, and an electronegative atom, B, the ionic structure A^+B^- may have a similar stability to the covalent structure A—B. If so, the ionic character of the actual bond will be high.

The stability of the A^+B^- form will clearly depend on the relative 'affinities' of A and B for electrons. Various terms are used to describe this, and atoms with a strong 'affinity' for electrons are said to be strongly electrophilic, to be electronegative, or to have a high electronegativity. The greater the difference between the electronegativities of the atoms A and B the stabler A^+B^- or A^-B^+ will be and the more ionic in character will the bond between A and B be.

4. Electronegativity scales. It is possible to give quantitative values to the electronegativities of atoms, and to draw up electronegativity scales.

Pauling has based his scale on the estimated contribution of the A^+B^- structure to the actual bond existing between A and B. This is done by making the following measurements:

Actual bond energy measured experimentally $= H$
Bond energy if the bond was truly covalent $= Q$
Resonance energy caused by ionic character of bond $= H - Q$

Values of H can be measured experimentally, but a truly covalent bond between unlike atoms is hypothetical, and therefore, the value of Q can only be obtained indirectly. Pauling has obtained values of Q by taking the bond energy of the truly covalent bond A—B(E_{AB}) as either the arithmetic or the geometric mean of the bond energies of the bonds A—A(E_{AA}) and B—B(E_{BB}), i.e.

$$E_{AB} = \frac{E_{AA} + E_{BB}}{2} \quad \text{or} \quad E_{AB} = \sqrt{E_{AA} \times E_{BB}}$$

The difference between the values of H and Q, obtained in these ways, is then taken as the ionic resonance energy of the bond A—B. Typical values obtained are 267.8 kJ for H—F and 6.69 kJ for H—I. The smaller value for HI is to be expected, for iodine is well known to be less electronegative than fluorine, and the higher ionic resonance energies would be expected the greater is the difference between the electronegatives of the two atoms linked together.

Pauling has used the ionic resonance energies as a means of estimating this difference in electronegativity and has drawn up an electronegativity scale, given below, on the basis that the electronegativity, x, of an element is such that $(x_B - x_A)$ is approximately equal to the square root of the ionic resonance energy of the bond A—B expressed in eV (p. 142).

			H 2.1			
Li 1.0	Be 1.5	B 2.0	C 2.5	N 3.0	O 3.5	F 4.0
Na 0.9	Mg 1.2	Al 1.5	Si 1.8	P 2.1	S 2.5	Cl 3.0
						Br 2.8
						I 2.5

ELECTRONEGATIVITY VALUES.

Mulliken has arrived at an electronegativity scale in close agreement with that of Pauling by taking the electronegativity of an atom as the arithmetic mean of its ionisation energy (p. 166) and its electron affinity (p. 167). These values, when scaled down, give electronegativity values very similar to those of Pauling's scale.

The greater the difference, $(x_B - x_A)$, between the electronegativities of two atoms the greater the ionic character of the bond A—B. An $(x_B - x_A)$ value of 1.7 leads to 50 per cent ionic character; a value of 2.3 to 73 per cent. The percentage ionic character of typical bonds is given below:

C—H 4%	N—H 19%	O—H 39%	F—H 60%
C—F 43%	C—Cl 11%	C—Br 3%	C—I 0%

5. Effect of ionic character. A covalent bond with some ionic character differs from a truly covalent bond in three main ways.

a. *Bond length.* The sum of the single-bond covalent radii of two atoms A and B only gives the bond length, A—B, when the bond is purely covalent. A covalent bond with some ionic character will have a slightly different length, and there are various empirical formulae relating the difference in length to the electronegativities of the atoms concerned.

b. *Chemical reactivity.* A covalent bond with some ionic character has electrical charges associated with it. Such charges render the bond more liable to attack by other charged atoms or groups, and, therefore, affect chemical reactivity and the detailed mechanism of chemical reactions.

c. *Dipole moments.* A magnet has a magnetic moment equal to ml, where m is the pole strength of the magnet and l the distance between the poles. In a similar way, two equal and opposite, but separated, electrical charges constitute an electrical dipole moment measured by the charge multiplied by the distance between the two charges.

Dipole moments are expressed in terms of coulomb metre units.* All dipole moments are of the order 10^{-30} C m, and this is to be expected for the charge on an electron is of the order 10^{-19} C and molecular distances are of the order 10^{-10} m.

Any bond which has any degree of polarity will have a corresponding dipole moment, though it does not follow that compounds containing such bonds will have dipole moments, for the polarity of the molecule as a whole is the vector sum of the individual bond moments. The C—Cl bond, for instance, has definite polarity, C→–Cl or $C^{\oplus}$—$Cl^{\ominus}$, and a definite dipole moment. Tetrachloromethane, however, has no dipole moment for the resultant of the four C—Cl moments is zero. By comparison, mono-, di- and tri-chloromethane all have dipole moments.

The positive and negative ends of a dipole are conveniently represented by an arrow over the bond or compound concerned, e.g. $\overrightarrow{A—B}$. The arrow head indicates the negative end and the tail the positive end. The methods of dipole moment measurement are outside the scope of this book, but such measurements provide a mass of data which can be used in a variety of ways to solve chemical problems.

The ionic character of a bond, for instance, can be estimated from the values of its dipole moment and its bond length. The dipole moment for HCl, for example, is 3.436×10^{-30} C m. If the bond was fully ionic the expected dipole moment would be equal to the charge on the electron multiplied by the bond distance, i.e. $(1.602 \times 10^{-19}) \times (1.29 \times 10^{-10})$ or 20.67×10^{-30}. The actual ionic character of the H—Cl bond may, therefore, be taken as $\frac{3.436}{20.67} \times 100$ per cent, i.e. approximately 17 per cent. This result is in good agreement with that obtained from the electronegativities of hydrogen and chlorine (p. 188).

* An older unit, the debye (D) is still used; it is equal to $3.335\ 640 \times 10^{-30}$ C m.

6. Bond order. The preceding sections have been concerned with the possibilities of resonance between a purely covalent and an ionic bond. Resonance may also occur between single and multiple covalent bonds.

The actual bonds between carbon atoms in a benzene molecule (p. 193) are neither C—C nor C=C bonds. They are something intermediate, a fact which can be expressed in terms of resonance,

$$C—C \quad \longleftrightarrow \quad C=C$$

A single bond, C—C, is said to be of order 1; a double bond, C=C, of order 2; and a triple bond, C≡C, of order 3. An actual bond intermediate between a C—C and a C=C bond would have a bond order between 1 and 2. Many examples are known where actual bonds in molecules are of these intermediate types.

7. Conditions necessary for resonance. The several structures which may contribute to a resonance hybrid can only do so under certain conditions.

a. They must all have their various atoms in the same relative positions. Different arrangements of atoms leads to tautomerism.

b. The several structures must each be reasonably stable. If very unstable the contribution they could make to the resonance hybrid would be so small as to be negligible.

c. The number of unpaired electrons must be the same in all the several structures to allow a continuous change from one bond type to another.

8. Evidence for resonance. There are four main lines of evidence which suggest that a particular substance may exist as a resonance hybrid.

a. *Chemical evidence.* When one structural formula fails to account for the known properties of a substance, as with benzene, the formula is probably only one of other possible formulae contributing to a resonance hybrid.

b. *Bond length.* When the measured bond lengths in a molecule do not agree with the bond lengths calculated from covalent radii it is probable that the molecule ought to be represented as a resonance hybrid.

c. *Bond energies.* When the experimentally measured values of the heats of formation or hydrogenation of a compound are greater than the heats of formation or hydrogenation calculated from the bond energies of the bonds in a structure purporting to represent the compound, then that single structure is probably only one of many contributing to a resonance hybrid.

d. *Dipole moments.* A single structure for the molecule of a compound may not be able to account for the observed dipole moment.

9. Examples of resonance. Typical examples are given below.

a. *Carbon dioxide*. The formula of carbon dioxide was, for a long time, thought to be $O{=}C{=}O$. If this was the correct formula the bond distances in carbon dioxide ought to be equal to the sum of the double-bond covalent radii of carbon (0.067 nm) and oxygen (0.055 nm). This calculated bond length of 0.122 nm is not in agreement with the measured value of 0.115 nm.

Moreover, the bond energy of the $C{=}O$ bond is 723.8 kJ so that the calculated heat of formation of carbon dioxide, if its structure is $O{=}C{=}O$, would be 1 447.6 kJ. The actual measured value is 1 602 kJ.

It is, therefore, much better to represent carbon dioxide as a resonance hybrid between the possible structures given on page 185.

The structures II and III must contribute equally for they would, individually, have dipole moments whereas carbon dioxide has no dipole moment. The resonance energy (p. 186) of carbon dioxide is 154.4 kJ, i.e. the difference between the measured and calculated heats of formation.

b. *Carbon monoxide*. The electronic structure to be allotted to carbon monoxide, with carbon normally four-valent and oxygen two-valent, has always provided something of a problem. It is best represented as

$$\underset{\text{I}}{C{=}O} \quad \longleftrightarrow \quad \underset{\text{II}}{C{\lessgtr}O}$$

The calculated bond distances for structures I and II are 0.122 nm and 0.110 nm respectively, whereas the measured bond length is 0.113 nm.

Carbon monoxide has a dipole moment which is very nearly zero whereas both structure I and structure II would have large dipole moments if they existed individually.

The heat of formation of the $C{=}O$ bond is 723.8 kJ, whereas the observed heat of formation for carbon monoxide is 1 071 kJ, the resonance energy being 347.2 kJ.

c. *Dinitrogen oxide*. A resonance hybrid between

$$\underset{\text{I}}{N{\lessgtr}N{=}O} \quad \longleftrightarrow \quad \underset{\text{II}}{N{\equiv}N{\rightarrow}O}$$

The calculated bond lengths for the various bonds involved in these structures are $N{=}N$, 0.120 nm; $N{\equiv}N$, 0.110 nm; $N{=}O$, 0.115 nm; and $N{-}O$, 0.136 nm. The length of the molecule, which is linear, is 0.231 nm, the probable bond lengths being $N{-}N$, 0.112 nm and $N{-}O$, 0.119 nm.

The dipole moment of dinitrogen oxide is very small so that the two structures must contribute almost equally.

d. *Nitrogen oxide*. A resonance hybrid between

$$\overset{x}{\underset{x}{N}}{}^{x}_{x} : \overset{\cdot\cdot}{O} : \quad \longleftrightarrow \quad \overset{x}{\underset{x}{N}}{}^{x\cdot}_{x} : \overset{\cdot}{O} :$$

The observed bond distance is 0.114 nm. The bond length for the $N{=}O$ bond would be expected to be 0.115 nm and for the $N{\equiv}O$ bond, 0.105 nm.

e. *The nitrate*(v) *ion.* A resonance hybrid between

$$O{=}N\!\!\begin{array}{c}{}^{\nearrow O}\\{}_{\searrow O^-}\end{array} \quad\longleftrightarrow\quad {}^-O{-}N\!\!\begin{array}{c}{}^{\nearrow O}\\{}_{\searrow O}\end{array} \quad\longleftrightarrow\quad O{\leftarrow}N\!\!\begin{array}{c}{}^{\nearrow O^-}\\{}_{\searrow O}\end{array}$$

The observed bond length is 0.121 nm. Calculated bond lengths are $N{-}O$, 0.136 nm and $N{=}O$, 0.115 nm. The resonance energy is 188 kJ.

f. *The nitro group.* A resonance hybrid between

$$-N\!\!\begin{array}{c}{}^{\nearrow O}\\{}_{\searrow O}\end{array} \quad\longleftrightarrow\quad -N\!\!\begin{array}{c}{}^{\nearrow O}\\{}_{\searrow O}\end{array}$$

The $N{-}O$ bond lengths in a variety of nitro-compounds vary between 0.121 and 0.123 nm. A single bond between nitrogen and oxygen atoms would give a bond length of 0.136 nm; a double bond would give 0.115 nm.

1,4-dinitrobenzene has no dipole moment, which indicates that the dipole moments in the two nitro groups must be equal in magnitude, but opposite in direction. The dipole moment of each nitro group must, in fact, act in a direction which bisects the ONO angle. The dipole moments of each of the above structures would act along the direction of the dative bonds.

g. *The carbonate ion.* A resonance hybrid between

$$O{=}C\!\!\begin{array}{c}{}^{\nearrow O^-}\\{}_{\searrow O^-}\end{array} \quad\longleftrightarrow\quad {}^-O{-}C\!\!\begin{array}{c}{}^{\nearrow O}\\{}_{\searrow O^-}\end{array} \quad\longleftrightarrow\quad {}^-O{-}C\!\!\begin{array}{c}{}^{\nearrow O^-}\\{}_{\searrow O}\end{array}$$

The calculated bond length for the $C{=}O$ bond is 0.122 nm and for the $C{-}O$ bond, 0.143 nm. The observed bond length in the carbonate ion is 0.131 nm. The resonance energy is 176 kJ.

h. *The sulphate*(vi) *ion.* In the sulphate(vi) ion, the four oxygen atoms are arranged around the central sulphur atom almost tetrahedrally. The measured bond distances are all equal to 0.151 nm. The calculated bond distance for single bonds between sulphur and oxygen atoms is 0.170 nm, and for double bonds, 0.149 nm.

It is clear that there must be some double-bond character about the bonds in the ion, and it is best represented as a resonance hybrid between many various structures of which some typical ones are shown:

$$\left\{\begin{array}{c}O\\\uparrow\\O{\leftarrow}S{\rightarrow}O\\\downarrow\\O\end{array}\right\}^{2-} \qquad \left\{\begin{array}{c}O\\\|\\O{-}S{\rightarrow}O\\\|\\O\end{array}\right\}^{2-} \qquad \left\{\begin{array}{c}O\\\|\\O{-}S{=}O\\\|\\O\end{array}\right\}^{2-}$$

The last of these structures is probably the most important, but there are many similar ones because the double bonds may have alternative positions.

10. The structure of benzene. Numerous suggestions have been made regarding the structure of benzene. It is, nowadays, regarded as a resonance hybrid or as a conjugated system with a ratherspecial type of π-bonding.

a. *Resonance in benzene.* The properties of benzene can be accounted for, reasonably satisfactorily, by regarding it as a resonance hybrid between the two well-known Kekulé structures with three possible Dewar structures making a small contribution.

I	II	ONE OF THREE
KEKULÉ STRUCTURES		DEWAR STRUCTURES

The bond length of a C—C bond would be expected to be 0.154 nm, and that of a C=C bond, 0.134 nm. The actual bonds between carbon atoms in benzene are all alike, with a bond length of 0.139 nm.

The actual measured heat of formation of benzene is 5 501 kJ, but the calculated heat of formation for structures I or II is only 5 339 kJ. The resonance energy of benzene is, therefore, 162 kJ. A similar figure is obtained, too, from heats of hydrogenation. The expected heat of hydrogenation for structures I or II would be 359.1 kJ, i.e. three times the heat of hydrogenation of a C=C bond, given on page 303 as 119.7 kJ. The measured heat of hydrogenation of benzene is, however, only 208.4 kJ. The resonance energy, from these figures, is 150.7 kJ.

Representation of benzene as a resonance hybrid does not postulate the existence of true C=C bonds in a benzene molecule, and this accounts for the lack of true unsaturation properties such as those shown by ethene.

b. *Benzene as a conjugated system.* The benzene molecule may also be considered from the point of view of molecular orbitals. This method envisages the electrons of each carbon atom in a benzene molecule as existing in a state of sp^2 hybrid orbitals, as in ethene (ethylene) (p. 178). Each carbon atom will have three coplanar sp^2 orbitals directed at angles of 120° to each other and a p-orbital at right angles to the plane of the three sp^2 orbitals.

For each carbon atom, one of the sp^2 orbitals forms a σ-bond with a hydrogen atom, and the other two form σ-bonds with adjacent carbon atoms (Fig. 63). Each carbon atom still holds a *p*-orbital with an axis at right angles to the plane of the carbon atoms (Fig. 64). These six *p*-orbitals, one from each carbon atom, are pictured as interacting all round the ring and not just participating in any one particular bond.

The resulting π-orbitals, pictured as charge-clouds both above and below the plane of the ring (Fig. 65), are said to be non-localised. The

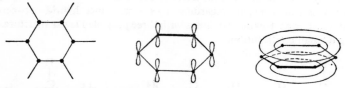

Fig. 63 (*left*). The σ-bonds in a benzene molecule.
Fig. 64 (*centre*). The six *p*-orbitals at right angles to the plane of the carbon atoms in a benzene molecule.
Fig. 65 (*right*), Formation of non-localised π-orbitals all round the benzene ring.

bonding between carbon atoms, on this basis, consists of a σ-bond with some π-bonding.

The interaction of π-bonds separated by σ-bonds is known as conjugation, and a system containing a —C=C—C=C— arrangement of bonding is known as a *conjugated system*. In all such systems, whether they be ring systems as in benzene or straight-chain systems as in buta-1,3-diene, the formation of non-localised π-orbitals affects the bond lengths, the bond energies and the chemical properties of the molecules concerned.

11. The nature of the covalent bond. A picture of covalent bonding in terms of molecular orbitals, involving σ- and π-bonding, has already been given. From the point of view of resonance, the nature of a covalent bond is best illustrated by a consideration of the covalent bond in a hydrogen molecule.

The hydrogen molecule consists of two nuclei which will be distinguished as A and B, and two electrons, which will be called 1 and 2. Electron 1 may be associated with nucleus A or with nucleus B, and similarly for electron 2. When the two nuclei are far apart they may be represented as

Structure I $H_A^{•1}$ $_B^2H$ Structure II $H_A^{•2}$ $_B^1H$

and a wave function can be calculated for each structure.

If the two nuclei in structure I are brought near to each other, the electrons associated with each nucleus will interact and the corresponding energy change can be calculated from the wave functions. On plotting the energy of the system against the internuclear distance, line 2 in Fig. 66 is obtained. The same line shows the energy changes when the two nuclei in structure II are brought together.[*]

If, however, the two possible structures are combined together, i.e. resonate to give a resonance hybrid, and the wave functions are treated accordingly, the change in energy of the system as the internuclear distance is varied is given by either line 1 or line 3 in Fig. 66. Line 1 results if the spins of electrons 1 and 2 are similar; line 3, if the spins are opposed.

Only line 3 shows a pronounced minimum and this corresponds to the forma-

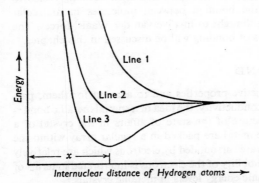

FIG. 66. The energy charges as two hydrogen atoms come together.

tion of a stable hydrogen molecule. The atoms are linked by a pair of electrons, *with opposed spins*, at an internuclear distance of x.

The energy of the covalent bond in a hydrogen molecule is mainly due to the resonance energy between the two structures I and II, and this is so for any other covalent bond.

(Questions on the contents of this Chapter will be found on pp. 205-7.)

Chapter 19
Metallic, hydrogen and van der Waals' bonding

THE nature of the bonding in the majority of chemicals can be described in terms of ionic or covalent bonds. The situation in metals requires, however, the postulation of a special type of metallic bond. There are, also, a number of hydrogen compounds in which hydrogen bonding seems to play a large part, and the bonding between molecules in molecular crystals, and in liquids, is thought to involve van der Waals' forces. The nature of these three types of bonding will be discussed in this chapter.

THE METALLIC BOND

Metals have very distinctive properties and to account for them, particularly for the electrical conductivity, the idea of a special metallic bond is necessary. The general picture of the state of affairs in the crystal of a metal is that atoms of the metal are packed in a regular array within the crystal and that the atoms are surrounded by electrons which are relatively mobile. The cause of the mobility of the electrons is interpreted in terms of the band model of electronic energy levels in metallic crystals.

1. The band model. The arrangement of electrons in a single atom of sodium is

$1s$	$2s$	$2p$	$3s$
2	2	6	1

each electron being in a particular energy level. If two sodium atoms come close together, they will interact with each other so that the energy levels in each atom will be slightly affected. In a crystalline array of close-packed atoms, this interaction will cause the single, discrete energy levels of the single atoms to be replaced by bands of closely related energy levels. For two atoms there would be two levels within a band; for n atoms there are n possible levels. The bands must be regarded as belonging to the crystal as a whole and not to any one individual atom.

In a sodium crystal, then, n levels within the $1s$ band can be postulated. As each of the n atoms has two $1s$ electrons, all the n levels will be full, and this will be so, too, for all the levels within the $2s$ and $2p$ bands. For the n possible levels in the $3s$ band, however, there are only n electrons, whereas the band could hold $2n$ electrons. All the levels in the $3s$ band are not, therefore, full. As the energy difference between the levels in the $3s$ band are very small, these $3s$ electrons can move about within the $3s$ band very easily. In a crystal under ordinary conditions, equal numbers of $3s$ electrons will move in all directions, but, under an applied potential difference, the electrons will move, preferentially, in one direction and a current will flow. It is this movement of electrons, within what is sometimes known as a conduction band, which accounts for electrical conductivity.

For a metal with an even number of valency electrons, e.g. Mg, 2.8.2, it might be expected that all the levels in the $3s$ band would be full because each atom contributes two $3s$ electrons. Magnesium is, however, still a good conductor and this is because some of the levels in the $3p$ band overlap those in the $3s$ band and thus provide opportunities for movement of electrons.

Electrical conductivity is associated, then, with either partially filled bands or with the overlapping of an unfilled band with a full one.

It might be expected that increase in temperature would cause more electrons to be 'promoted' into conducting bands with a consequent increase in conductivity or decrease in resistance. In general, however, the conductivity of a metal decreases with increase in temperature, i.e. the resistance increases. This is because increasing temperature produces increased thermal vibration within a metal crystal. This upsets the regularity within the crystal and interferes with the ease of movement of electrons through the crystal. It is rather like comparing movement through the ranks of a battalion of soldiers on parade with that through a London crowd. Similarly, the introduction of an impurity may upset the regular array and so cause increased resistance. The resistance of copper, for example, is greatly increased by even traces of impurities.

2. Semi-conductors. Some pure and impure metals have rather low conductivities which increase with temperature. Such conductivity comes about in three different ways.

a. *Intrinsic semi-conductors.* In magnesium there is an overlap between the $3s$ and $3p$ bands. In substances like germanium or grey-tin there is no actual overlap of bands but the energy difference between the highest filled band and the next empty one is very small. It might, therefore, be possible for an electron to gain enough energy to pass from the full to the empty band; conductivity would result. The chance of an electron in the full band being able to pass into the empty band would increase as the temperature increased, i.e. as the energy of the electrons increased. With these substances, called intrinsic semi-conductors, the conductivity therefore increases with increase in temperature.

b. *n-type semi-conductors.* The addition of an impurity to a metal will provide additional energy levels, and, if these levels are correctly related to the bands within the pure metal, conductivity may result.

If the impurity contains a full energy level which is just below that of an empty band in the pure metal, the electrons from the impurity might have enough energy to pass into the empty conducting band in the pure metal. This happens when arsenic or antimony are added to germanium. The passage of electrons from an energy level in the arsenic or antimony into one in the germanium causes the germanium to become negatively charged. It is, therefore, known as *n*-type germanium.

c. *p-type semi-conductors.* If the impurity contains an empty energy level just above that of a full band in the pure metal, the electrons from the full band in the pure metal might be able to pass into the empty level of the impurity. This happens when gallium or indium are added to germanium. Passage of electrons from the germanium to the gallium or indium causes the germanium to become positively charged. It is called p-type germanium.

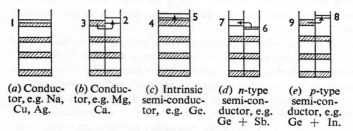

(a) Conduc-   (b) Conduc-   (c) Intrinsic   (d) n-type   (e) p-type
tor, e.g. Na,   tor, e.g. Mg,   semi-conduc-   semi-con-   semi-con-
Cu, Ag.   Ca.   tor, e.g. Ge.   ductor, e.g.   ductor, e.g.
                                                   Ge + Sb.   Ge + In.

FIG. 67. Electrical conductivity. Full bands are shaded.
(a) Conductivity due to partially filled band, 1.
(b) Conductivity due to overlap of empty band, 2, with full band, 3.
(c) Conductivity due to narrow gap between full band, 4, and empty band, 5.
(d) Conductivity due to full band in impurity, 6, just below empty band in pure metal, 7.
(e) Conductivity due to empty band in impurity, 8, just above full band in pure metal, 9.

Semi-conductors of the n- and p-type have a conductivity which, like that of germanium, increases with temperature. They are of great importance in making transistors.

The various types of conductivity are summarised in Fig. 67.

THE HYDROGEN BOND

3. Introduction. A hydrogen atom is normally monovalent and forms only one bond, but in some compounds it appears to form two bonds.

Hydrogen fluoride, for example, is known, from relative molecular mass measurements, to be associated, i.e. to exist as $(HF)_n$, and the acid salt, potassium hydrogendifluoride, KHF_2, is also well known and must be derived from the acid H_2F_2.

Originally, $(HF)_n$ and KHF_2 were formulated by assuming that hydrogen could act as an acceptor and form a dative bond with fluorine acting as the donor:

$$(H-F \longrightarrow H-F \longrightarrow H-F)_n \qquad K^+(F \longrightarrow H-F)^-$$

There is, however, no reason to assume that a hydrogen atom can act as an acceptor in this way for it has only one stable orbit ($1s$) which can only hold two electrons, and the formulae written above give hydrogen four electrons.

The linkage previously represented as a dative bond is now regarded as a special type of bond known as a hydrogen bond. The mechanism of its formation is thought to be electrostatic. The HF_2^- ion, for instance, is envisaged as two negatively charged fluoride ions linked together by a positively charged hydrogen ion (proton). The proton is thought to be able to exert a sufficiently strong electrostatic attraction to do this because of its small size. Hydrogen bonding is, indeed, sometimes called proton bonding.

To distinguish a hydrogen bond it is best to write it as a dotted line so

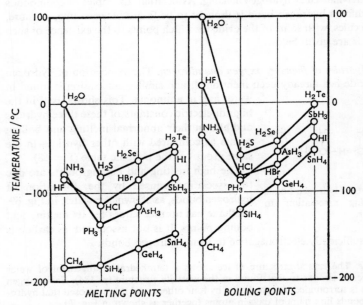

FIG. 68. The abnormal melting and boiling points of water, hydrogen fluoride and ammonia.

that the HF_2^- ion becomes $(F\cdots H-F)^-$, and, in general, when a hydrogen bond links two atoms, A and B, the structure is represented as $A-H\cdots B$ or as a resonance hybrid between $A-H\cdots B$ and $A\cdots H-B$.

That the electrostatic mechanism for the formation of a hydrogen bond is probably correct is shown by the fact that a hydrogen bond $A-H\cdots B$ is formed most easily if A and B have high electronegativities. Thus the tendency of an $A-H$ bond to form a hydrogen bond with another atom, B, increases rapidly from $C-H$ through $N-H$ and $O-H$ to $F-H$, and it decreases in passing from $O-H$ to $S-H$ or from $F-H$ to $Cl-H$. This shows that the bond $A-H$ has the greatest tendency to form hydrogen bonds when the ionic character of the bond is greatest, i.e. when the bond has the greatest polar character, $A^{\delta-}$—$H^{\delta+}$.

Fluorine, with the highest electronegativity, forms the strongest hydrogen bonds, and by far the greatest number of hydrogen bonds known are those which unite two oxygen atoms.

The hydrogen bond is a weak bond, the strength of the strongest being about 20–40 kJ as compared with strengths of 200–400 kJ for normal covalent bonds (p. 180).

4. Inter-molecular hydrogen bonding. Association. Examples of compounds in which a hydrogen bond is thought to be found between two, or more, molecules, with some of the evidence which points to the existence of such bonds are given below.

a. *Hydrides of fluorine, oxygen and nitrogen.* The association of hydrogen fluoride has already been mentioned, and similar association is found in water and in ammonia. This shows itself in the high di-electric constants of these three hydrides and also in their abnormal melting and boiling points as compared with other hydrides in the same groups of the periodic table (Fig. 68).

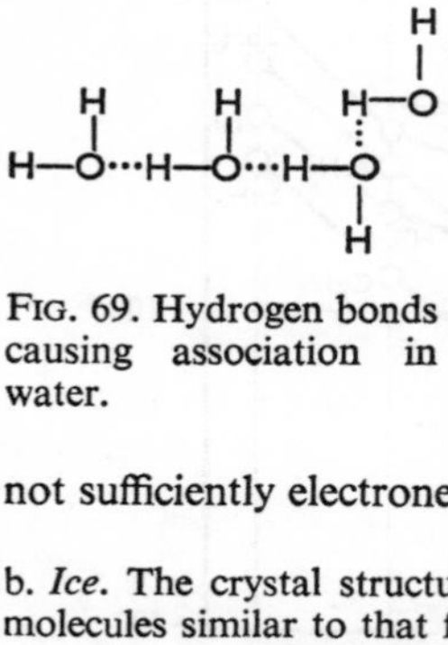

The high melting and boiling points are due to association caused by the formation of hydrogen bonds, as shown, for water, in Fig. 69. Methane has normal values for its melting and boiling points. It is not associated, as carbon is not sufficiently electronegative to form hydrogen bonds.

FIG. 69. Hydrogen bonds causing association in water.

b. *Ice.* The crystal structure of ice shows a tetrahedral arrangement of water molecules similar to that found in the wurtzite structure (p. 184). Each oxygen atom is surrounded, tetrahedrally, by four others, and it is supposed that hydrogen bonds link pairs of oxygen atoms together as shown in Fig. 70.

The distance between adjacent oxygen atoms is 0.276 nm and this suggests that the hydrogen atom linking the two oxygen atoms together is not midway between them, for the O—H distance in water vapour is 0.096 nm and not half 0.276 nm. Distance measurements on other compounds, too, indicate that the hydrogen atom in a hydrogen bonded pair of atoms is not equidistant from the two atoms.

The arrangement of the water molecules in ice is a very open structure and this explains the low density of ice. When ice melts, the structure breaks down and the molecules pack more closely together so that water has a higher density. This breaking down process is not complete until a temperature of 4°C is reached and it is on these lines that the abnormal behaviour of water is explained.

c. *Alcohols.* There is a marked difference between the boiling points of alcohols and the corresponding sulphur analogues (thiols or mercaptans), e.g.

Boiling point	CH_3OH 64.5°C	C_2H_5OH 78°C	C_3H_7OH 97°C	C_4H_9OH 117°C
Boiling point	CH_3SH 5.8°C	C_2H_5SH 37°C	C_3H_7SH 67°C	C_4H_9SH 97°C

Alcohols are, therefore, thought to be associated in much the same way as water.

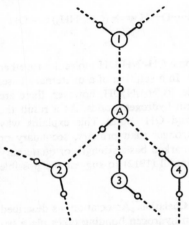

FIG. 70. The crystal structure of ice. The central oxygen atom, A, is surrounded tetrahedrally by the oxygen atoms, 1, 2, 3 and 4. All other oxygen atoms are arranged similarly. The hydrogen atoms are shown as small circles, and the dotted lines indicate the hydrogen bonds.

d. *Carboxylic acids.* Some carboxylic acids associate into dimers, i.e. two molecules link together, both in the vapour state and in certain solvents. Partition coefficient measurements of the distribution of ethanoic (acetic) acid between water and benzene show, for instance, that the acid is present as a dimer in the organic solvent. This dimer is written as

$$H_3C-C\begin{matrix} O-H\cdots O \\ \diagdown O\cdots H-O \end{matrix}C-CH_3$$

and the presence of an eight-membered ring is confirmed by electron diffraction studies. Relative density measurements also show the presence of double molecules of ethanoic acid in the vapour state.

In aqueous solution, the molecules of a carboxylic acid link up with water molecules rather than form dimers.

e. *Amines.* The association and basic strength of amines are both explained in terms of hydrogen bonding.

Primary and secondary amines are associated to some extent, though not greatly because nitrogen does not form hydrogen bonds very readily. Tertiary amines are not associated at all for they have no hydrogen atom capable of forming hydrogen bonds. Thus trimethylamine, which is not associated, has a lower boiling point (4°C) than dimethylamine (7°C) even though it has a higher relative molecular mass.

In aqueous solution, amines react with water molecules as shown,

$$CH_3{\cdot}NH_2 + H_2O \rightleftharpoons CH_3{\cdot}\overset{\overset{\displaystyle H}{|}}{\underset{\underset{\displaystyle H}{|}}{N}}{-}H{\cdots}O{-}H \rightleftharpoons (CH_3{\cdot}NH_3)^+ + OH^-$$

The resulting solution will contain some $CH_3{\cdot}NH_3OH$ molecules together with some $CH_3NH_3{}^+$ and OH^- ions. In a solution of a quaternary base, e.g. tetramethylammonium hydroxide, $[(CH_3)_4N]OH$, however, there are no hydrogen atoms which could form hydrogen bonds. As a result the solution contains only $(CH_3)_4N^+$ and OH^- ions. This explains why quarternary bases are very much stronger than primary, secondary or tertiary amines. It was, in fact, the marked basic strength of quaternary bases which first led Moore and Winmill (1912) to suggest the possible existence of hydrogen bonds.

5. Intra-molecular hydrogen bonding. Chelation. Association, as described in the preceding section, occurs when hydrogen bonding takes place between two or more molecules, i.e. when there is inter-molecular hydrogen bonding. Hydrogen bonding may, however, also take place within a single molecule; this is known as intra-molecular bonding. It may lead to the linking of two groups to form a ring structure. Such an effect is known as chelation, though the term is used in a wider sense (p. 481) and this is but one kind of chelation. Examples are provided by the substances mentioned below.

a. 1,2-*substituted benzene compounds.* 2-nitrobenzenol (*o*-nitrophenol) boils at 214°C as compared with 290°C for 3- and 279°C for 4-nitro benzenol. 2-nitrobenzenol is, moreover, volatile in steam, and more soluble in water than the other two isomers.

All these facts can be accounted for on the assumption that 2-nitro benzenol contains an internal hydrogen bond represented as

This intra-molecular hydrogen bonding prevents inter-molecular bonding between two or more molecules. But in the 3- and 4-isomers, intra-molecular bonding is not possible because of the size of the ring which would have to be formed. Inter-molecular bonding therefore takes place, and this causes some degree of association, which accounts for the higher boiling points of the 3- and 4-isomers.

The low solubility of 2-nitrobenzenol may be explained in two ways. The formation of an internal hydrogen bond 'suppresses' the hydroxylic character of the compound and this causes a lowering of solubility in water. In other words, the formation of an internal hydrogen bond prevents hydrogen bonding between 2-nitrobenzenol and water and this results in reduced solubility.

The effect of hydrogen bonding in 2-nitrobenzenol is also shown spectroscopically. A normal —OH group is found to give rise to a particular line in the infra-red absorption spectrum of the substance concerned. The spectra of 3- and 4-nitrobenzenol show this line, but that of 2-nitrobenzenol does not. There is not the same difference between the three methyl ethers of the nitro benzenols because hydrogen bonding cannot take place in these ethers.

Other compounds in which intra-molecular hydrogen bonding plays the same part as in 2-nitrobenzenol include 2-hydroxybenzaldehyde, 2-chloro-benzenol and 2-hydroxybenzoic acid.

b. *Ethyl 3-oxobutanoate*. Ethyl 3-oxobutanoate (ethyl acetoacetate or aceto-acetic ester) exists in two tautomeric forms known as the keto- and enol-forms.

$$CH_3 . \underset{\underset{O}{\|}}{C}\cdot CH_2 \cdot COOC_2H_5 \longrightarrow CH_3 \cdot \underset{\underset{OH}{|}}{C}{=}CH\cdot COOC_2H_5$$

keto-form   enol-form

Meyer, in 1920, succeeded in separating these two forms by fractional distillation under reduced pressure in specially cleaned quartz apparatus (aseptic distillation).

Alcohols have, in general, higher boiling points than the corresponding ketones. Compare, for example, propan-2-ol (iso-propyl alcohol) (82°C) and propan-2-one (acetone) (56°C). But the enol-form of ethyl 3-oxobutanoate has a lower boiling point than the keto-form. This is probably due to intra-molecular hydrogen bonding in the enol-form which prevents inter-molecular bonding and thus prevents association which would raise the boiling point. This intra-molecular hydrogen bonding is represented as,

$$\begin{array}{c} H \\ \| \\ C \\ H_3C{-}C \qquad C{-}OC_2H_5 \\ O{-}H\cdots\cdots O \end{array}$$

The presence of such a bond is supported by the fact that the enol-form is less soluble in water, and more soluble in cyclohexane, than the keto-form. This indicates a suppression of hydroxylic character in the enol-form.

The hydrogen bond may be a weak one, but it is important and probably plays a major role in many biological processes.

VAN DER WAALS' FORCES

Electrostatic attraction holds the ions together in an ionic crystal, and covalent bonds hold the atoms together in an atomic or covalent crystal. The nature of the forces holding the molecules together in a molecular crystal are, however, of a different nature, as are the forces holding molecules together in a liquid. These forces are known as van der Waals' forces after the man who first took them into account in modifying the gas equation (p. 73).

6. Nature of the forces. When the molecules held together in a liquid or a molecular crystal are polar in nature there will be a dipole–dipole attraction between the molecules when they are correctly orientated, and this is probably the main contribution to van der Waals' forces in such cases. The forces in the gaseous state will be very small because the molecules are far apart. Nevertheless, the deviation of gases from ideal behaviour (p. 72) is at least partially due to such forces in the gaseous state. In the solid and liquid states the molecules are closer together and the dipole–dipole forces become strong enough to give some cohesion to the liquid or solid.

With non-polar substances, such as nitrogen, oxygen, the noble gases, carbon tetrachloride and benzene, there must still be some binding forces in the liquid and solid state. The molecules concerned are non-polar because they are symmetrical, but slight displacement of the nuclei or the electrons in a molecule will give rise to an electrical dipole. It is thought that such a slight displacement is constantly occurring within a molecule or atom. The displacement is temporary and random, so that the molecule or atom as a whole, over a period of time, has no observable dipole. But if it has a dipole at any one instant it will induce another dipole in a neighbouring molecule or atom and an attractive force will be established between the two. The binding forces in non-polar liquids and solids are thought to originate from this cause. The forces concerned are relatively small, being ten to twenty times less than those involved in ionic or covalent bonding.

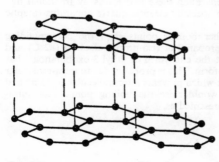

Graphite.

7. Graphite. Layer lattice structures. Graphite (p. 221) provides a simple example of a crystal involving van der Waals' forces. The crystal contains layers of hexagonally arranged carbon atoms, and, within the layers, the carbon atoms are linked by covalent bonds like those in benzene (p. 193). The C—C distance in one of the graphite layers is 0.142 nm as compared with 0.335 nm in benzene.

Adjacent layers in graphite are held together, 6.335 nm apart, by van der Waals' forces. The weakness of these forces allows the layers to slide over each other, which is why graphite is soft and acts as a lubricant.

In cadmium chloride, Cd^{2+} ions are surrounded octahedrally by six Cl^- ions and the $CdCl_6$ groups are linked together into layers. Such layers are held together by van der Waals' forces, as in graphite. Similar arrangements occur in many di- and tri-chlorides, bromides and iodides and in some sulphides and hydroxides.

In the hydroxides of zinc, beryllium, aluminium and iron(III), layers are held together by hydrogen bonds (p. 198) instead of van der Waals' forces.

QUESTIONS ON CHAPTERS 15-19

1. Cite briefly two pieces of evidence which help to indicate the real existence in matter of discrete entities described as atoms and molecules.
Explain by reference to the theory of atomic structure what is meant by the terms (i) atomic number, (ii) isotope, (iii) neutron.
Discuss how far the properties of the following compounds may be inferred from the modes of linkage of their component atoms: (a) magnesium oxide, (b) tetrachloromethane, (c) ammonium chloride, (d) copper(II) sulphate(VI)-5-water. (N. S.)

2. In what ways are a sodium ion and a neon atom alike? How do they differ?

3. Discuss the arrangement of valency electrons in the oxides of (i) nitrogen, (ii) sulphur, (iii) carbon.

4. What is the arrangement of electrons in the following ions; Ag^+, Ca^{2+}, Fe^{2+}, Hg^{2+}, Al^{3+}, Fe^{3+}, Bi^{3+}? Which of these ions would you expect to be most readily formed from its atoms?

5. Give the arrangement of electrons in the following ions: F^-, Cl^-, Br^-, I^-, O^{2-} and S^{2-}. Which of these ions is the strongest reducing agent?

6. Hydrogen can form both an anion and a cation. Explain why this is so, and give examples of compounds containing hydrogen anions and of those containing hydrogen cations.

7. Show the arrangement of valency electrons in the following compounds: ethane, ammonium bromide, potassium hexacyanoferrate(II), trichloromethane and nitric(V) acid.

8. The number of water molecules attached to a hydrated cation in aqueous solution increases as the size of the cation, and its charge, increase. Illustrate this statement.

9. Suggest a reasonable arrangement of bonds in the following ions: CO_3^{2-}, SiO_4^{4-}, NO_3^-, PO_4^{3-}, SO_4^{2-}, SO_3^{2-}, $S_2O_3^{2-}$, ClO_3^-, ClO_4^-.

10. Show the probable arrangement of bonds in the following: BF_3, NH_3, SiH_4, ClF, $Ag(NH_3)_2^+$, HCN, $CH_3 \cdot NC$, $C_6H_5 \cdot NO_2$.

11. What is the arrangement of valency electrons in the following: H_2S, PH_3, BCl_3, NH_4^+, the nitronium ion NO_2^+, the peroxide ion O_2^{2-}, the carbide ion?

12. What is the probable arrangement of electrons in the following compounds of iodine: HI, NaI, ICl_3, IF_5, IF_7, I_3^-, $CH_3 \cdot I$ and NI_3?

13. Discuss briefly in electronic terms (a) the formation of ions of the type MO^{n-} where n is 1, 2 or 3, (b) the existence of a stable compound $BCl_3 \cdot NH_3$, and (c) the valency types shown in ammonium carbonate.

14. What experiments would you carry out in order to try to decide whether a pure white solid was an ionic or covalent compound?

15. Explain what is meant by the terms covalence, electrovalence and co-ordinate valence. Illustrate your answer with one concrete example of each type.

Indicate the distribution of electrons in the electron 'shells' of the atoms of the elements X, Y and Z of atomic numbers 7, 19 and 35 respectively. Write electronic formulae for the following: (a) the hydrides of X, Y and Z; (b) the reaction between the hydrides of X and Z; (c) the reaction between the hydride of Y and water; (d) the state of the hydride of Z in aqueous solution; (e) the compound of Y and Z; (f) the compound YXO_3. (W.)

16. Examine the ease of formation and stability of the Cu^+ and Zn^{2+} ions from the point of view of Fajans's rules. Why do you think it is that the Cu^+ ion is found so rarely?

17. State Fajans's rules and illustrate their application.

18. Fajans's rules suggest that electrovalency is favoured by small ionic charge and by large cation radius. If Z is the charge on a cation, and r its radius, examine the statement that the oxide of a metal is basic if $\sqrt{Z}/r$ is less than 2.2, and acidic if $\sqrt{Z}/r$ is greater than 3.2.

19. Show how the electronic theory of valency provides an explanation of the regular variation of the valency of the typical elements in passing from group to group of the periodic classification of the elements.

20. The first, second and third ionisation energies of sodium, magnesium and aluminium are tabulated below in kJ mol^{-1}.

Na	494	4 561	6 920
Mg	732	1 443	7 680
Al	577	1 812	2 745

Comment on any significant points.

21. Explain and illustrate the meaning of the terms electropositive and electronegative.

22. Calculate the expected volume occupied by a water molecule from (a) the bond angle and bond distances in a water molecule, (b) the fact that the density of water at 4°C is 1 g cm^{-3}. Comment on the results.

23. Account for the non-existence of the pentachlorides of nitrogen and bismuth.

24. Certain elements of high relative atomic mass exhibit a valency two less than that which would be expected from their position in the periodic table. This is known as the inert pair effect. Give varied examples of its occurrence.

25. Suggest reasons for the following: (a) magnesium oxide has a much higher melting point than sodium fluoride, (b) anhydrous aluminium chloride is soluble in benzene whereas the hydrated form is insoluble, (c) liquefied hydrogen chloride is non-conducting, whereas a solution of hydrogen chloride in water conducts electricity.

26. The oxy-acids exist as resonance hybrids. Give the arrangement of bonds in any one important canonical form of H_2CO_3, HNO_3, H_2SiO_3, H_3PO_4, H_2SO_4, H_2SO_3, $HClO_3$ and $HClO_4$.

27. What experimental evidence would you try to obtain before deciding to regard a binary compound as a resonance hybrid between the two forms $A{=}B$ and $A{\equiv}B$?

28. Give the main resonating structures of carbon dioxide, carbon monoxide, dinitrogen oxide, the nitrate(III) ion, ozone O_3, the nitride ion and benzene.

29. Give varied examples to show that it is an over-simplification to say that atoms combine in order to attain an octet of electrons in their outermost orbit.

30. By considering the orbitals available for bond formation predict the geometrical arrangement of atoms in the following: NH_4^+, CH_4, NH_3, H_2S and BCl_3.

31. 'The tendency of electrons of like spin to avoid each other does more than any other single factor to determine the shapes and properties of molecules.' Comment.

32. The C—H bond lengths in ethyne, ethene and methane are 0.1057, 0.1079 and 0.1094 nm respectively. Suggest a reason for these changes.

33. Comment on the statement that all compounds of carbon, nitrogen, oxygen and fluorine, containing only σ-bonds, can be regarded as tetrahedral molecules if lone pairs be considered as bonding electrons.

34. If the first ionisation energy of hydrogen is taken as being 1 unit the values for the elements from helium to argon, inclusive, are 1.81, 0.40, 0.69, 0.61, 0.83, 1.07, 1.00, 1.37, 1.59, 0.38, 0.56, 0.44, 0.60, 0.80, 0.76, 0.96 and 1.15. Plot these figures against atomic number, and comment on any points of significance.

35. What type of bond hybridisation would you expect in simple compounds of (a) boron, and (b) beryllium? What effect would the hybridisation have on the geometry of the molecules concerned?

36. What is meant by the statement that the ions of transitional metals only differ from the inert gas type of configuration in possessing some additional d electrons in excess of the configuration ns^2np^6? Illustrate your answer with examples.

37. The fact that *trans*-butenedioic (fumaric) acid has a higher melting point than *cis*-butenedioic (maleic) acid, with which it is isomeric, can be accounted for in terms of hydrogen bond formation. How might this be done?

38. *m*- and *p*-compounds differ markedly from the corresponding *o*-compounds in being (a) less volatile, (b) more miscible with water, and (c) less miscible with benzene. Illustrate and explain this statement.

39. How would you expect intermolecular hydrogen bonding to affect (i) the molecular weight, (ii) the parachor, (iii) the boiling point, (iv) the vapour pressure (v) the surface tension of a compound? Use these criteria to discuss the possibility of hydrogen bonding in water.

40. How do you think hydrogen bonds might be formed in (a) chlorobenzenol, (b) the enol form of ethyl 3-oxobutanoate, (c) 2-hydroxybenzylaldehyde, (d) 2, 2, 2-trichloroethane-1, 1-diol or chloral hydrate, (e) α-acetylaminoacetamide?

41. Illustrate the meaning of hybridisation by considering the bonding in the following substances: $BeCl_2$, BF_3, CF_4, PF_5, SF_6 and IF_7.

42. Suggest reasons why the nitrate(v) and carbonate ions are planar whilst the chlorate(v) ion is pyramidal.

43. Compare the nature of the bonding in (a) benzene and borazole, $B_3N_3H_6$, (b) graphite and boron nitride, (c) sulphur dioxide, nitrogen dioxide and carbon dioxide, (d) sulphur dichloride oxide or thionyl chloride, sulphur dichloride dioxide or sulphuryl chloride and phosphorus trichloride oxide or phosphorus oxychloride.

44. What part does electronegativity play in deciding oxidation numbers?

45. The measured dipole moments of the hydrogen halides are given below in C m units

Hydrogen halide	HF	HCl	HBr	HI
Dipole moment ($\times 10^{30}$)	6.371	3.436	2.602	1.268

Comment on these figures. Taking the charge on an electron as 1.602×10^{-19} C and the bond distance in hydrogen chloride as 1.29×10^{-10} m, estimate the ionic character of the bond in hydrogen chloride.

46. Examine the statement that bonds are to be regarded as ionic when the electronegativity difference between the two bonded elements is greater than 1.7.

Chapter 20
Crystal structure

1. Crystal systems. The crystals of a given substance have plane surfaces, known as faces, and the angle between the faces is constant, however irregularly the crystal may have grown. Different crystals of the same substance may not look alike, because different faces can grow at different rates, but the corresponding faces always intersect at the same angle. Moreover, cleavage or splitting of a crystal occurs along definite planes so that the same interfacial angles are found even when a crystal is split.

The number of different crystal shapes is large, and some substances can crystallise in two or more different forms. Sulphur, for example, forms crystals of α- and β-sulphur, under different conditions, and sodium chloride, which usually forms cubic crystals, gives octahedral crystals in the presence of carbamide (urea). Crystalline shape may, too, be affected by temperature. Ammonium chloride, for example, gives one type of crystal below its transition temperature, and another above. This occurrence of the same chemical substance in different crystalline forms is known as polymorphism or allotropy if only elements are concerned (p. 217).

The wide variety of crystals may be classified into seven crystal systems, according to the set of axes which must be used to characterise the crystal faces. A crystal which is a perfect cube, for example, can be characterised by three axes, at right angles to each other, and with $a = b = c$. On this basis, the seven crystal systems may be summarised as follows, where α is the angle between b and c, β that between a and c, and γ that between a and b.

System	Axes	Angles	Maximum no. of planes of symmetry	Example
Cubic	$a = b = c$	$\alpha = \beta = \gamma = 90°$	9	Sodium chloride
Tetragonal	$a = b \neq c$	$\alpha = \beta = \gamma = 90°$	5	White tin
Orthorhombic	$a \neq b \neq c$	$\alpha = \beta = \gamma = 90°$	3	Rhombic sulphur
Monoclinic	$a \neq b \neq c$	$\alpha = \gamma = 90°; \beta \neq 90°$	1	Monoclinic sulphur
Rhombohedral or trigonal	$a = b = c$	$\alpha = \beta = \gamma \neq 90°$	3	Calcite
Hexagonal	$a = b \neq c$	$\alpha = \beta = 90°; \gamma = 120°$	7	Quartz
Triclinic	$a \neq b \neq c$	$\alpha \neq \beta \neq \gamma \neq 90°$	0	Copper(II) sulphate

Each system includes a number of forms or classes. The cubic system, for example, includes the cube and the regular octahedron. Crystals in each system have a maximum number of planes of symmetry, and other symmetrical features based on axes and centres of symmetry. The detailed symmetry of the forms or classes within one system vary, and there are, altogether, thirty-two crystal forms or classes.

2. Space lattices and unit cells. The external shape of a crystal, and its symmetry, reflect an orderly array of particles within the crystal. These particles may be ions, atoms or molecules, and the arrangement of these structural units within the crystal is represented by a space lattice or a unit cell.

A space lattice is a regular pattern of points, the points representing the positions of the structural units which, when in position, make up the crystal structure. The space lattice extends in all directions, but it is only necessary to consider a specimen portion of it, which is representative of the whole. Such a portion is known as a unit cell. It is defined as *the smallest portion of a space lattice, which, by moving a distance equal to its own dimensions in various directions, can generate the complete space lattice.*

The simplest unit cells (Fig. 71) are based on a cube as shown below:

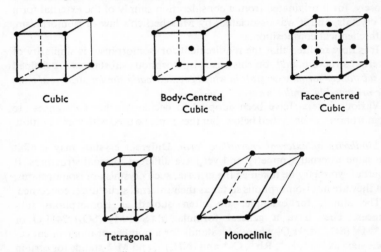

Cubic   Body-Centred Cubic   Face-Centred Cubic

Tetragonal   Monoclinic

FIG. 71. Simple unit cells.

a. *Simple cubic.* This unit cell has points at each corner of a cube.

b. *Body-centred cube.* This has points at each corner of a cube and a point at the centre of the cube.

c. *Face-centred cube.* This has points at the corner of a cube and at the centres of each face of the cube.

Other unit cells may be regarded as derived from cubic cells by modification of edge lengths and/or by skewing. Simple tetragonal and monoclinic unit cells are shown in Fig. 70.

3. Binding forces in crystals. The structural units which occupy the points in a space lattice of a crystal are held together by different kinds of forces in different substances.

Crystals made up of ions and held together by electrostatic forces are called *ionic crystals* (p. 169). Crystals made up of molecules held together by weak, van der Waals' forces, are called *molecular crystals* (p. 161). Crystals made up of atoms held together by covalent forces are called *atomic or covalent crystals* (p. 182).

Metallic bonding (p. 196) occurs in crystals of metals, and hydrogen bonding is also found in some crystals (p. 201).

4. Isomorphism. Substances which from very closely related crystals are said to be isomorphous, but it is not always very easy to decide when two substances are isomorphous. The term, moreover, is used somewhat loosely, for it originated from a consideration simply of the external form of crystals, and it was associated, by Mitscherlich's law of isomorphism, with chemical composition.

It is now realised that the requirement for isomorphism is similarity of internal structure, and, on this basis, isomorphous substances are defined as *substances forming crystals in which geometrically similar structural units are arranged in similar ways.*

Various criteria have been adopted to decide whether substances are isomorphous, as described below, but they must be used with some caution.

a. *Similarity of external crystalline form.* Different crystals may exhibit the same external symmetry and yet have different internal structures. If external symmetry be taken as the criteria, such crystals are isomorphous, but they are not isomorphous so far as their internal structure is concerned.

The alums, for example, are often quoted as isomorphous substances. They have a general formula $M_2^+SO_4 \cdot M_2^{3+}(SO_4)_3 \cdot 24H_2O$ or $M^+M^{3+}(SO_4)_2 \cdot 12H_2O$, where M^+ stands for a number of monovalent cations such as Na^+, K^+, Rb^+, Cs^+ and $(NH_4)^+$, and M^{3+} stands for certain trivalent cations such as Al^{3+}, Cr^{3+} and Fe^{3+}.

The crystals of different alums are all very much alike, and exhibit the same external symmetry, but the interfacial angles do vary a little. This is because the ions in different alum crystals are of different sizes, and the slight distortion caused means that there are really three groups of alums with slightly different internal structures. The four commonest alums, listed below, are, however, all in the same group, so that they are definitely isomorphous.

$K_2SO_4 \cdot Al_2(SO_4)_3 \cdot 24H_2O$	$K_2SO_4 \cdot Cr_2(SO_4)_3 \cdot 24H_2O$
POTASH ALUM	CHROME ALUM
$(NH_4)_2SO_4 \cdot Al_2(SO_4)_3 \cdot 24H_2O$	$(NH_4)_2SO_4 \cdot Fe_2(SO_4)_3 \cdot 24H_2O$
AMMONIUM ALUM	IRON ALUM

Similarly, although the interfacial angles of calcium carbonate (calcite), iron(II) carbonate (chalybite) and manganese(II) carbonate (dialogite) vary a little, the crystals have the same external symmetry, and the same internal arrangement, and are, therefore, isomorphous.

b. *Similarity of chemical constitution.* Mitscherlich's law of isomorphism (1819) stated that *substances which have similar chemical compositions are isomorphous.*

This is true in such substances as the alums and the carbonates listed in a., and also in the two series of simple sulphates(VI):

$$K_2SO_4 \qquad Rb_2SO_4 \qquad Cs_2SO_4 \qquad (NH_4)_2SO_4$$
$$ZnSO_4 \cdot 7H_2O \qquad MgSO_4 \cdot 7H_2O \qquad NiSO_4 \cdot 7H_2O$$

but there is not necessarily any relation between chemical composition and isomorphism.

Very similar chemicals need not be isomorphous, e.g. caesium chloride and rubidium chloride; and compounds with no real chemical resemblances may be isomorphous. Calcium carbonate and sodium nitrate(v), or barium sulphate(VI) and potassium manganate(VII) (permanganate), for example, are isomorphous because their crystals are made up of positive and negative ions of the same geometrical shape and comparable sizes. Similarly all substances with a sodium chloride crystal structure are isomorphous and these include such compounds as lead(II) sulphide, calcium oxide, calcium carbide, scandium nitride, and silver chloride as well as the halides of lithium, sodium, potassium, rubidium and caesium (excepting the chloride bromide and iodide of caesium).

c. *Formation of overgrowths.* A crystal of potash alum can be overgrown on top of a crystal of chrome alum, and vice versa. Similarly, sodium nitrate(v) can be grown on top of a calcite (calcium carbonate) crystal.

It is, in fact, fairly common for isomorphous substances to give overgrowths, but the formation of such overgrowths will only occur within certain structural limits. Potassium sulphate and caesium sulphate, though isomorphous, will not form overgrowths because the Cs^+ ion is so much bigger than the K^+ ion (p. 168). But rubidium and caesium sulphates, or rubidium and potassium sulphates, will form overgrowths.

d. *Formation of mixed crystals or solid solutions.* The crystals formed from a solution containing both potash and chrome alums will contain Cr^{3+}, Al^{3+} and SO_4^{2-} ions within a single crystal structure, the proportion of Cr^{3+} or Al^{3+} being variable over a very wide range depending on the composition of the original solution.

The crystals are known as mixed crystals, though this is something of a misnomer. There is only a single crystal containing a random arrangement of Cr^{3+} or Al^{3+} ions, and not a mixture of crystals of potash alum with crystals of chrome alum. The term solid solution is, therefore, a useful alternative.

In the mixed crystals, Cr^{3+} and Al^{3+} ions are interchangeable. When two substances crystallise with identical space-lattices it is to be expected that points in the lattice might be occupied by alternative structural units from the two substances so long as the geometrical difference between the structural units is not too great. One unit can simply replace another similar one without affecting the crystal structure.

Thus, potassium chloride and potassium bromide form a continuous series of mixed crystals, i.e. they are miscible in all proportions, as the Cl^- and Br^- ions only differ in size by about 8 per cent. Potassium chloride and potassium iodide, however, have a difference in size between Cl^- and I^- ions of about 21 per cent, and only partial solid solution occurs.

Isomorphous substances commonly form mixed crystals, but they do not necessarily do so. Potassium chloride, for instance, will not form mixed crystals with lead(II) sulphide. Moreover, substances which form mixed crystals, e.g. calcium fluoride and yttrium fluoride or silver bromide and silver iodide, are not necessarily isomorphous, or even chemically similar.

5. Other types of mixed crystals. The random replacement of one ion by another in the mixed crystals formed from potassium chloride and potassium bromide gives rise to what are sometimes referred to as *substitutional* solid solutions.

Other crystal structures can be formed in which atoms or ions get 'trapped' in spaces between the structural units of a regular crystal lattice. This gives rise to *interstitial* solid solutions. It is also possible to obtain crystals in which some of the sites for structural units are vacant.

Some examples of these, and related, types of crystal are given below. Some of the products which can be formed are very surprising.

a. *Metallic hydrides.* Many transitional metals can absorb large amounts of hydrogen, and the absorbed gas can generally be liberated by pumping at a high enough temperature. During the absorption, the crystal lattice of the metal is expanded but not greatly distorted and it is thought that the small hydrogen atoms are situated between the metallic atoms in interstitial compounds.

The amount of hydrogen absorbed varies with the conditions but hydrides with 'formulae' such as $TiH_{1.73}$, $TaH_{0.76}$, $CeH_{2.8}$, $LaH_{2.8}$ and $VH_{0.6}$ have been reported.

b. *Iron(II) sulphide and iron(II) oxide.* Iron(II) sulphide, commonly used as an example to illustrate the validity of the law of constant composition (p. 4), has, in fact, a variable composition. Chemical analysis of different crystalline specimens show that they vary in composition from FeS to $FeS_{1.14}$. Such 'formulae' suggest that the crystals contain excess sulphur and that they may be interstitial in nature, but density measurements show quite clearly that the crystals are really deficient in iron so that the formulae variation of FeS to $Fe_{0.88}S$ is perhaps preferable. To maintain an electrical balance within the crystal, some of the Fe^{2+} ions are converted into Fe^{3+} ions.

Similarly, crystalline iron(II) oxide shows a composition varying from $FeO_{1.055}$ to $FeO_{1.19}$, again because of a deficiency of iron. Pure iron(II) oxide, with exact composition represented by FeO, is not known.

c. *Graphitic compounds.* Graphite, with its layer-lattice structure (p. 204), can form a number of 'compounds' in which other atoms or molecules take up a position between the layers of carbon atoms in the graphite structure.

Graphite absorbs liquid potassium, for example, to form 'alloys' with detectable compositions represented by KC_8, KC_{16}, KC_{24} and KC_{40}. Graphite also reacts with strong oxidising agents, such as nitric acid or potassium chlorate, to form graphitic oxides with compositions varying from $C_{2.9}O$ to $C_{3.5}O$. Similar 'compounds' are formed between graphite and fluorine, e.g. $(CF)_n$, and between graphite and acids such as sulphuric acid, e.g. $C_{24}HSO_4 \cdot 2H_2SO_4$.

d. *Clathrates.* Benzene-1, 4-diol, quinol, $C_6H_4(OH)_2$, crystallises from an aqueous solution, in the presence of argon at 40 atmospheres, with a structure in which an atom of argon is 'trapped' inside a 'cage' of hydrogen-bonded quinol molecules. The argon can be liberated by melting or dissolving the crystalline product. The 'compound', known as a clathrate, has a fomula $[C_6H_4(OH)_2]_3A$.

Xenon, krypton, hydrogen sulphide, hydrogen chloride, hydrogen cyanide, sulphur dioxide and carbon dioxide, all with molecules of the correct size to fit inside the benzene-1, 4-diol, 'cage', form similar clathrates.

Benzene molecules can also be 'trapped' by shaking with an ammoniacal solution of nickel(II) cyanide. The resulting clathrate compound has a formula $Ni(CN)_2 \cdot NH_3 \cdot C_6H_6$.

X-RAY ANALYSIS OF CRYSTALS

6. Diffraction of light. When a beam of light is passed through a diffraction grating, which consists of a large number of very fine opaque lines, parallel to each other and of equal width, drawn on a piece of glass, a series of spectra can be observed on either side of the original path of light. If monochromatic light is used, the spectra are replaced by a series of bright images on a dark background.

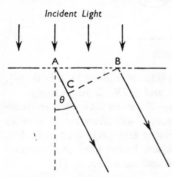

FIG. 72. A small portion of a diffraction grating.

A small portion of a grating is shown in Fig. 72. If monochromatic light of wavelength λ is incident perpendicularly on the upper face of the grating all the clear spaces act as secondary sources of light and emit rays in all directions. The rays in any one direction from each of the spaces will interfere with each other. In simple terms, where a crest of one wave coincides with a crest of another the resulting displacement will be increased, i.e. the light will be brighter, Similarly, when a crest of one wave coincides with a trough of another there will be no resultant displacement, i.e. there will be no visible light. The conditions leading to maximum or zero displacements are summarised in Figs. 73 and 74.

The ray from A will reinforce the ray from B to give a maximum displacement only if the distance AC is equal to $n\lambda$, where n has any integral value. As AC is equal to AB$\sin \theta$, and AB is equal to the spacing of the grating, d, it follows that, for reinforcement,

$$d\sin \theta = n\lambda$$

If the grating has 550 lines to the mm, then AB(d) is equal to approximately 1.81×10^{-3} mm. If the wavelength of light used is 5.8×10^{-4} mm

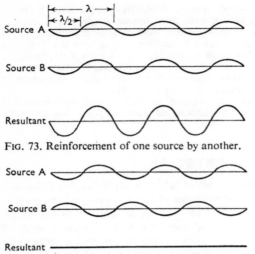

FIG. 73. Reinforcement of one source by another.

FIG. 74. Neutralisation of one source by another.

(yellow light), then $\sin \theta = 0.3197n$ the values of θ given by $n = 0$, ± 1, ± 2 and ± 3 being $0°$, $\pm 18° \, 39'$, $\pm 39° \, 45'$ and $\pm 73° \, 33'$ respectively.

The light emerging from such a grating, if viewed through a movable telescope, will show up as a series of bright and dark lines as in Fig. 75 where bright lines are indicated by B and dark ones by D.

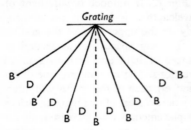

FIG. 75.

7. Diffraction by X-rays. In 1912, Laue suggested that X-rays should be diffracted in the same way as light waves, if they were electromagnetic waves of short wavelength, so long as a diffraction grating with a small enough spacing could be found. A crystal with a regular array of lattice planes was considered to be capable of acting as such a grating, and Friedrich and Knipping found that this was so. They passed a beam of X-rays through a thin section of a zinc blende crystal, and on photographing the emergent rays they found that the plate showed a bright central spot surrounded symmetrically by other bright spots caused by diffraction of the X-rays.

This original diffraction of X-rays was a diffraction of transmitted rays, but it is rather easier, both theoretically and experimentally, to treat the diffraction of reflected X-rays.

The atoms or ions in a crystal are arranged in a series of planes, and the lines KL and MN in Fig. 76 may be taken as representing two such parallel planes. When a beam of X-rays is incident at a glancing angle of θ some reflection takes place at each plane and diffraction is caused by the interference of the reflected rays. The path difference between the reflected rays from KL and those from MN is equal to AYB. Taking the distance between the two planes as d, it follows that AY is equal to $d\sin\theta$. For the reflected

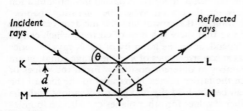

FIG. 76.

rays to reinforce, the necessary condition is that AYB (the path difference) must be equal to an integral number of wavelengths. If, therefore, the wavelength of the X-rays which are used is λ, the condition for reinforcement is that

$$n\lambda = 2d\sin\theta$$

This is the basic relationship of X-ray analysis of crystals. It is generally referred to as the *Bragg equation*.

8. Outline of experimental methods. The general idea behind the X-ray analysis of crystals is easy enough to understand. By reflecting X-rays of known wavelength from parallel lattice planes, and by measuring the values of the glancing angles, and the values of n, which give rise to maxima in the intensity of the reflected X-rays, it is possible to measure values of d, i.e. the distance between the parallel lattice planes. This can be done for different lattice planes so that the internal dimensions of the crystal can be obtained.

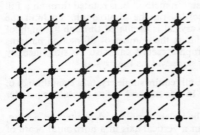

FIG. 77. Some of the parallel planes in a typical crystal structure.

The various experimental methods for putting these principles into practice are outlined below. In no case is the interpretation of the results at all easy. In particular, the problem is complicated by the presence in a crystal of many parallel lattice planes.

A two-dimensional representation of a crystal structure as shown, for example, in Fig. 77, has, amongst others, the three sets of parallel planes shown. In interpreting the experimental data, however, some assistance is provided by the fact that a plane which is thickly populated with ions or atoms is much more effective in reflecting X-rays than one in which the atoms or ions are more thinly spread. The intensity of the reflected X-ray beams can, therefore, assist in the interpretation of the experimental results.

Laue's method of passing X-rays through a crystal has been largely superseded by reflection methods, summarised below:

a. *Bragg's X-ray spectrometer.* In this instrument, which looks very much like an optical spectrometer, a narrow beam of monochromatic X-rays is allowed to fall on a single crystal so mounted that the X-ray beam is incident on one of the important crystal faces. The reflected beam of X-rays is passed into an ionisation chamber which can be rotated around the crystal. When the ionisation chamber is in the correct position in relation to the crystal and the incident beam, the reflected X-rays cause ionisation so that there is a flow of current, which can be measured on an electrometer, dependent on the intensity of the X-rays. A series of current maxima is obtained as successive values of n in the Bragg equation are attained with the ionisation chamber in different positions. From the values of n, λ and θ, the distance between the lattice planes, d, parallel to the face of the crystal exposed to the X-rays can be obtained. By making observations with the crystal in different positions, other d values for other planes can also be found.

Bragg's method provides a lot of detailed information about the internal dimensions of a crystal, but a large number of readings have to be taken and it requires the use of a fairly large crystal with well-defined faces. Other methods overcome these difficulties.

b. *Rotating crystal method.* In this method, a small crystal is mounted so that it can be rotated about a vertical axis parallel to one of the crystal axes. A narrow, horizontal beam of monochromatic X-rays is allowed to fall on the rotating crystal. For certain positions of the crystal, the relationship $n\lambda = 2d\sin\theta$ will be satisfied, and corresponding spots can be recorded on a circular photographic film. By taking three photographs with the crystal rotating about three different crystal axes, the crystal structure can be obtained, but the interpretation of the photographs is difficult as a number of diffraction patterns from the rotating crystal are recorded simultaneously.

This is, however, the commonest technique used for investigating crystal structure.

c. *Oscillating crystal method.* If the crystal in method b. is rotated through a full 360° a large number of spots are recorded on the photograph. To limit the number, and facilitate interpretation, a crystal is oscillated through an angle of 10° or 20° about a vertical axis. This limits the number of diffraction patterns formed. Interpretation of the photographic record can also be simplified by oscillating the photographic film in the same way as the crystal.

d. *Powder method.* Methods a., b. and c. all require single crystals, even though b. and c. can be used with small crystals. In this method, a fine crystalline powder can be used.

The powder, contained in a thin-walled glass capillary, deposited on a fibre, or moulded into a wire, is rotated about a vertical axis in a horizontal beam of monochromatic X-rays. The crystals in the powder are orientated in all directions, but there will always be some crystals at the correct angle to the incident X-rays for diffraction for each set of lattice planes. The different diffraction patterns are recorded on a surrounding, cylindrical strip of film.

POLYMORPHISM AND ALLOTROPY

9. Polymorphism. Substances which can crystallise in more than one form are said to be polymorphic. If the number of crystalline forms is two, they are *dimorphic*; if three, *trimorphic*; and so on. The term polymorphic applies to crystalline forms of both elements and compounds, but the term *allotropy* is also used to describe different forms of the same element, in the same state, whether they be crystalline forms or not. Polymorphism, then, includes all cases of allotropy caused by variation in crystalline form, but some examples of allotropy, e.g. ozone and oxygen, are caused by different arrangements of atoms in molecules and do not involve crystal structure at all.

The crystals formed by polymorphic substances have structures which can readily change to another type, particularly if the temperature is changed. In some cases the change takes place at a definite temperature known as the *transition temperature*.

When the change can take place reversibly, i.e. when one form of the substance is stable above the transition temperature and the other form is stable below it, the type of polymorphism is known as *enantiotropy*, from the Greek meaning opposite change. When the change is not reversible and can only take place in one direction the polymorphism is known as *monotropy*, from the Greek meaning one change. In monotropic substances there is no definite transition temperature, and one form of the substance is more stable than the other. The unstable form will change into the stable form at all temperatures, but the change may be very slow. So slow, in fact, that the unstable form can sometimes be kept for a very considerable time in what is known as a métastable condition (p. 526).

The distinction between enantiotropy and monotropy is made clearer by a consideration of vapour pressure-temperature curves as given for sulphur and phosphorus on pp. 219–20.

10. Examples of polymorphic compounds. The following are some examples of common polymorphic substances.

a. *Mercury(II) iodide*. Red mercury(II) iodide turns yellow on heating, the transition point being 126°C. The yellow form does not immediately revert to the red form on cooling for it can be kept in a metastable state, which passes over into the red form slowly, on standing, or quickly if disturbed by touching. The yellow crystals of mercury(II) iodide are rhombic; the red ones are tetragonal.

b. *Zinc oxide*. Zinc oxide, which can form crystals with the zinc blende or wurtzite structures (p. 183) is white at room temperature but changes into a yellow form above 250°C.

c. *Ammonium chloride*. Ammonium chloride forms crystals with a caesium chloride structure (p. 171) below the transition point of 184.3°C, and crystals with the sodium chloride structure (p. 170) above the transition point.

d. *Ammonium nitrate*(v). This salt has five polymorphic forms with transition temperatures as shown,

$$V \underset{}{\overset{-18°C}{\rightleftharpoons}} IV \underset{}{\overset{32.5°C}{\rightleftharpoons}} III \underset{}{\overset{84.2°C}{\rightleftharpoons}} II \underset{}{\overset{125°C}{\rightleftharpoons}} I$$

The polymorphism of ammonium compounds is due to the rotation of the ions in the crystal which enables them to take up different positions in relation to each other.

e. *Silicon* (IV)*oxide, silicon dioxide, or silica*. This can exist in three crystalline forms,

$$Quartz \underset{}{\overset{870°C}{\rightleftharpoons}} Tridymite \underset{}{\overset{1470°C}{\rightleftharpoons}} Cristobalite$$

All three forms are stable and all occur naturally.

Other polymorphic compounds include ammonium bromide, iodide and sulphate, caesium chloride (which changes to a sodium chloride structure at 460°C), calcium carbonate (forming calcite and aragonite structures), and zinc sulphide (with zinc blende and wurtzite structures).

11. Allotropy. *When an element exists in two or more forms, in the same state, it is said to exhibit allotropy.* When the forms have different crystal structures the elements are, really, polymorphic, and the allotropy they exhibit can be either enantiotropy or monotropy (p. 217). The commonest examples are:

Enantiotropic elements: Sulphur and Tin
Monotropic elements: Phosphorus and Carbon

Allotropy may also be caused by differences in the arrangement of atoms in molecules, as with oxygen and ozone (trioxygen).

a. *Enantiotropy of sulphur*. Two crystalline forms of sulphur, rhombic(α) and monoclinic(β), are known to exist, and they are both stable under certain conditions of temperature and pressure. At the transition point, 95.5°C, both rhombic and monoclinic sulphur can co-exist, but at higher temperatures all rhombic sulphur changes into monoclinic sulphur, and at lower temperatures all monoclinic sulphur changes into the rhombic form. The changes are not, however, very rapid.

$$Rhombic\ (\alpha)\ S \underset{}{\overset{95.5°C}{\rightleftharpoons}} Monoclinic\ (\beta)\ S \underset{}{\overset{120°C}{\rightleftharpoons}} Liquid\ S$$

The complete vapour pressure-temperature diagram for sulphur is given on p. 526. In considering the allotropy, only the vapour pressure-temperature curves of rhombic, monoclinic and liquid sulphur need be considered, and these are given in Fig. 78.

Slow heating of rhombic sulphur will change it to monoclinic sulphur at the transition temperature, and further heating will convert the monoclinic sulphur into liquid sulphur at the melting point of monoclinic sulphur, T_1. On slowly cooling liquid sulphur, the reverse changes will occur.

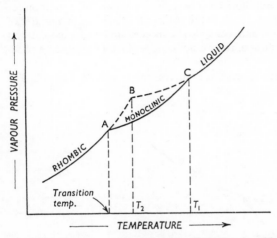

FIG. 78. Vapour pressure-temperature curves for rhombic, monoclinic and liquid sulphur, typical of an enantiotropic system. Compare Fig. 79, p. 220 and Fig. 171, p. 526.

If rhombic sulphur is rapidly heated, however, the dotted curve, ABC, will be followed, as the transformation to monoclinic sulphur will not take place quickly. Rhombic sulphur will then be found to melt at T_2. Similarly, rapid cooling of liquid sulphur may cause the curve, CBA, to be followed, with rhombic sulphur being formed direct from the liquid form.

It will be seen that the transition temperature is below the melting points of both allotropes. This is always so for substances exhibiting enantiotropy, and the pressure-temperature curves given for sulphur are typical of those for any enantiotropic system.

b. *Enantiotropy of tin.* Tin has three solid allotropes, with two transition points, as summarised below:

$$\text{Grey tin} \xrightleftharpoons{13°C} \text{White tin} \xrightleftharpoons{161°C} \text{Rhombic tin} \xrightleftharpoons{232°C} \text{Liquid tin}$$

The main point of interest is that the density of grey tin is smaller than that of white tin. When, therefore, white tin changes into grey tin the increase in volume causes the metal to expand and crumble in places. The effect is known as tin plague. Grey tin is formed only in the severest persistent winter conditions as the change from white tin is slow and as white tin can exist below the transition point in a metastable condition.

c. *Monotropy of phosphorus.* The allotropes of phosphorus are red and white phosphorus, but only red phosphorus is stable. White phosphorus slowly changes into red phosphorus at any temperature, so there is no transition point. At room temperatures the change from white to red is so very slow that white phosphorus can be kept in a metastable condition for a considerable time. The change to the red form takes place in a matter of days, however, at 250°C, and it can also be speeded up by adding a catalyst such as iodine. Red phosphorus cannot change directly into white phosphorus.

The vapour pressure-temperature curves are shown in Fig. 79. The

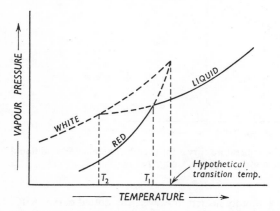

Fig. 79. Vapour pressure-temperature curves for white, red and liquid phosphorus, typical of a monotropic system. Compare Fig. 78.

curve for white phosphorus lies above that for red. This means that the vapour pressure of white phosphorus at any temperature is greater than that of red phosphorus.

If red phosphorus is heated it will melt at T_1 into liquid phosphorus. If white phosphorus is heated it will melt at T_2 to form liquid phosphorus. If liquid phosphorus is rapidly cooled, white phosphorus will be formed. Thus, to convert red phosphorus into white it is necessary to melt, or vaporise, the red form and to cool the resulting liquid or vapour quickly.

The transition point between red and white phosphorus has no real meaning for it lies at a temperature above the melting points of the two allotropes, i.e. at a point when they have both ceased to be solid allotropes at all. This is typical of all monotropic systems, which also have vapour pressure-temperature curves like those of phosphorus.

As red phosphorus is the stable form it has a lower energy content than white phosphorus, i.e. energy has to be put into red phosphorus to convert it into the white form. The amount of energy can be obtained from measurements of the heats of combustion of the red and white allotropes, as follows:

$$P_4 + 5O_2 \longrightarrow P_4O_{10} \quad \Delta H = -3012 \text{ kJ}$$
(white)
$$P_4 + 5O_2 \longrightarrow P_4O_{10} \quad \Delta H = -2939 \text{ kJ}$$
(red)
$$P_4 \longrightarrow P_4 \quad\quad\quad \Delta H = -73.64 \text{ kJ}$$
(white)   (red)

It is characteristic of the unstable form, i.e. white phosphorus, (a) to have a higher energy content, (b) to exert a higher vapour pressure at any temperature, and (c) to be more soluble.

d. *Monotropy of carbon.* The allotropes of carbon are diamond and graphite, with graphite being just the stable form, at least at low temperatures. Thus graphite has a lower energy content than diamond,

$$C + O_2 \longrightarrow CO_2 \quad \Delta H = -393.4 \text{ kJ}$$
(graphite)
$$C + O_2 \longrightarrow CO_2 \quad \Delta H = -395.5 \text{ kJ}$$
(diamond)
$$C \longrightarrow C \quad\quad \Delta H = -2.1 \text{ kJ}$$
(diamond)   (graphite)

Diamond might, therefore, be expected to change at room temperature into graphite, but, perhaps fortunately, such a change is infinitesimally slow and has never been observed. At very high temperatures and pressures, however, conversion of graphite into diamond seems possible.

e. *Allotropy dependent on different molecular species.* The allotropy of oxygen is not dependent on crystal structure. It depends, instead, on the possible existence of two different molecules, O_2 for dioxygen and O_3 for trioxygen (ozone).

Dioxygen is the stable allotrope,

$$3O_2 \longrightarrow 2O_3 \quad\quad \Delta H = 284.5 \text{ kJ}$$

and it has to be highly energised, in an ozoniser, to convert it into trioxygen. Once formed, trioxygen, as the unstable allotrope, reverts at all temperatures to oxygen.

Liquid sulphur contains at least three allotropes, known as $S\lambda$, S_μ and S_π. It is probable that they differ in molecular constitution, $S\lambda$ probably being S_8 and S_π, S_4, but the molecular formula of S_μ is not known. At any one temperature they exist in equilibrium, and the equilibrium position changes with temperature. It has been estimated that there is 0 per cent S_μ, 3.7 per cent S_π and 96.3 per cent $S\lambda$, at 120°C, and 37 per cent S_μ, 4 per cent S_π and 59 per cent $S\lambda$ at the boiling point of sulphur.

The dynamic equilibrium existing between S_λ, S_π and S_μ provides an example of what is known as *dynamic allotropy*.

12. Measurement of transition temperatures. The change from one polymorphic form to another may result in a change of colour, density or solubility, or in an absorption or evolution of heat, and investigation of such properties can be used for measuring transition temperatures. Similar methods are also available for measuring the transition temperatures between different salt hydrates.

a. *Colour change.* If a little mercury(II) iodide is placed in a melting point tube attached to a thermometer immersed in a heating bath, it is possible to

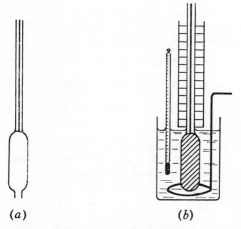

(a)   *(b)*

FIG. 80. A dilatometer. (*a*) Before filling. (*b*) Set up for making measurements.

record the temperature at which the red mercury(II) iodide changes into the yellow form.

Alternatively, copper(I) tetraiodomercurate(II) can be used. It changes very sharply from a red to a black form on heating.

b. *Density change.* Two polymorphic forms of the same substance have different densities so that there is a change in volume at the transition point, and the temperature at which this change occurs can be measured by using a *dilatometer*. This consists (Fig. 80) of a glass bulb connected to a length of capillary tubing which is either calibrated or placed against a scale. Initially, the bottom of the bulb is open, so that it may be filled.

The method is suitable for measuring the transition point of sulphur. Some powdered rhombic sulphur is placed in the bulb, which is then sealed. Some inert liquid, such as liquid paraffin, is then introduced into the bulb, through the capillary tube, until there are no air bubbles in the bulb or tube and until the liquid level is at the lower end of the capillary. The dilatometer is then immersed in a heating bath and the temperature is slowly raised, the liquid level in the capillary tube being recorded at regular temperature intervals.

At first, the liquid level rises steadily as the contents of the bulb expand. When the transition temperature is reached, however, the rhombic sulphur begins to undergo a volume change, as it is converted into monoclinic sulphur. This causes a marked change in the rate of rise of the liquid level. When all the rhombic sulphur has changed to monoclinic, the liquid level again rises steadily, as at the beginning. A plot of liquid level against temperature gives a curve as in Fig. 81, line (a).

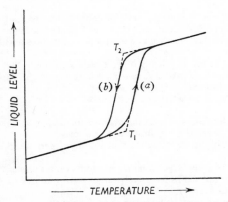

FIG. 81. Volume change at transition point as measured on a dilatometer.

On slow cooling of the dilatometer, the reverse changes take place, but, because of thermal lag, the plot of liquid level against temperature gives a curve as in Fig. 81, line (b). The transition temperature is taken as the mean of temperatures T_1 and T_2.

In a more accurate application of the dilatometer a mixture of rhombic and monoclinic sulphur is placed in the bulb. The dilatometer is then immersed in a bath at constant temperature. After the dilatometer has attained the temperature of the bath, the liquid level is observed. If the bath temperature is equal to the transition temperature there will be no change in the liquid level, and the transition temperature can be measured by finding what bath temperature will give no change in liquid level.

c. *Solubility change.* Two forms of the same substance have different solubilities if they are both soluble in the same solvent. Thus the solubility curve of a polymorphic substance shows distinct breaks at the transition points, for it is really made up of two or more solubility curves.

This method is widely used to measure the transition points of salt hydrates, solubility curves such as those shown in Fig. 83 on p. 230 being obtained.

d. *Evolution or absorption of heat.* The cooling curve of a single substance will be regular, unless some change of state occurs, but the cooling curve of a polymorphic substance will show distinct breaks at the transition temperature. Such temperatures can therefore be measured by plotting cooling point curves, but the method can only be used for substances which can be well stirred or which are good conductors of heat. It is suitable for obtaining the transition temperatures of some salt hydrates, e.g. anhydrous and hydrated sodium sulphate(vi), or of different forms of a metal.

Other physical properties such as viscosity, electrode potential, electrical or thermal conductivity and vapour pressure can also be used to measure transition temperatures.

QUESTIONS ON CHAPTER 20

1. Explain the relationship between (a) cubic and hexagonal close-packed structure, (b) cubic close-packed structures and the structures found in sodium chloride, zinc blende and fluorite, and (c) hexagonal close-packed structures and the wurtzite structure.

2. By considering structures made up of spheres of unit radius calculate the relative sizes of the spheres which would just fit into (a) a triangular site, (b) a tetrahedral site, and (c) an octahedral site.

3. What is meant by saying that a tetrahedral site is smaller than an octahedral site?

4. Is it true to say that the building units in a crystal arrange themselves so as to occupy the minimum volume of space? Give illustrative examples.

5. What percentage of the available space is occupied in a cubic close-packed structure?

6. Would the location of a successively larger number of atoms on a number of spheres all with the same centre but with successively larger radii give rise to a crystal structure? Explain your answer.

7. Why do you think it is that most crystals, however carefully grown, are imperfect in some way or other? What are the main types of imperfection?

8. What result would you expect if you were to heat a crystal of sodium chloride in an atmosphere of sodium vapour?

9. Comment on the statement that the hydrate of a substance with the greatest number of molecules of water of crystallisation crystallises from aqueous solution at the lowest temperature.

10. Draw diagrams to show that a tetrahedron results (a) when half the corners of a cube are connected, and (b) when half the faces of an octahedron are extended.

11. Describe what you would do to make a single crystal of potash alum such as would be suitable for displaying its regular external shape.

12. How would you attempt to establish, in a school laboratory, whether two substances were isomorphous?

13. Discuss the various criteria which can be used to decide whether two substances are isomorphous.

14. Apply Le Chatelier's principle in a discussion of the effect of change of pressure on the changes (a) water to ice, (b) rhombic to monoclinic sulphur, and (c) grey to white tin.

15. Calculate the heats of combustion of diamond and graphite and red and white phosphorus in ozone.

16. Explain what is meant by the statement that 'polymorphism of the elements is known as allotropy?' Do you think there is any point in maintaining the use of the word allotropy? Give your reasons.

17. What are ortho- and para-hydrogen? Are they to be regarded as allotropes of hydrogen?

18. Mixed crystals are sometimes called solid solutions. What arguments can you put forward in favour of the separate terms?

19. How would you demonstrate, experimentally, that sodium and potassium chlorides can form mixed crystals?

20. Give an account of interstitial compounds.

21. Write an account on one of the following topics: (i) mica, (ii) naturally occurring silicates, (iii) the piezoelectric effect, (iv) polymorphism, and (v) types of symmetry.

22. Discuss (a) any one example of allotropy, and (b) any one example of polymorphism.

23. How is the fact that sulphur can appear to have two different melting points explained?

24. How would you investigate the formation of mixed crystals by potassium chloride and potassium bromide?

25. Write notes on (a) the law of constant composition, (b) clathrates, (c) silica, and (d) the allotropy of tin.

26. Compare and contrast the allotropy of (i) dioxygen and trioxygen with that of (ii) diamond and graphite.

Chapter 21
Solutions of solids in liquids

1. General terms used. The most familiar type of solution is formed when a solid, e.g. common salt, dissolves in a liquid, e.g. water, but many other types are also known, and *a solution may be defined as a perfectly homogeneous mixture.*

The components mixed together in a solution may be solids and/or liquids and/or gases, giving six types of solution containing two components only.

The composition of any solution may be variable over a wide or narrow range. Gases, for instance, are usually miscible in all proportions, but water will only dissolve a limited amount of common salt at any one temperature, and mercury and water are completely immiscible.

The terms *solvent* and *solute* are commonly used in discussing solutions, the solvent being the substance present in excess. In simple language, the solvent does the dissolving, and the solute is dissolved. The relative proportions of solvent and solute in a solution can be expressed in a number of ways.

a. *Percentage by weight.* This is the number of grammes of solute in 100 g of solution. Similarly, the number of grammes of solute in 100 g of solvent may be used.

b. *Grammes per cubic decimetre.* This is the number of grammes of solute in 1 cubic decimetre of the solution. It is sometimes referred to as the *mass concentration* and other units such as $kg\ m^{-3}$ can be used.

c. *Moles per cubic decimetre.* This is the number of moles of solute in 1 cubic decimetre of solution. Alternatively, $mol\ m^{-3}$ units can be used. This is the commonest method of expressing concentration and the concentration of a substance, x, in such units is written as $[x]$. If 1 mol of solute is dissolved in 1 cubic decimetre of solution, the solution is said to be 1M or to have a concentration of $1\ mol\ dm^{-3}$. If one-tenth of a mol of solute is dissolved in 1 cubic decimetre of solution, the solution is M/10 or has a concentration of $0.1\ mol\ dm^{-3}$.*

This method of expressing concentration is important because 1 mol of any substance contains the same number of molecules. Equal volumes of M solutions always, therefore, contain the same number of solute molecules.

d. *Normality.* A normal, N, solution contains 1 gramme-equivalent of solute per cubic decimetre of solution. A 2N solution of sulphuric acid, for example, is 1M; it contains $98\ g\ dm^{-3}$ and the equivalent mass of the acid is 49 whilst its relative molecular mass is 98.

---

* M and 0.1M solutions used to be called molar and deci-molar respectively. Such terms may, in fact, persist, but should, strictly, no longer be used. The term 'molar' is now to be used as implying 'divided by amount of substance' so that the term molar solution becomes meaningless.

Normalities used to be used extensively in expressing concentrations of solutions but their use, nowadays, is limited.

e. *Molal concentration*. This is the number of moles of solute in 1 kg of solvent.

f. *Mole fraction*. This is the number of moles of solute divided by the total number of moles of solute and solvent. The mole fraction is proportional to the actual number of solute molecules present in the solution. If the mole fraction is 0.5, for example, it means that half the molecules in a solution are solute molecules and half are solvent molecules. The mole fractions of solute and solvent will always add up to 1.

Moles per cubic decimetre, moles per cubic metre and mole fraction are probably the most important and general methods of expressing concentration, but all the other ways are in use, and it is important to differentiate clearly between them.

SOLUTIONS OF SOLIDS IN LIQUIDS

2. Introduction. A solid which will dissolve in a liquid at a particular temperature is said to be soluble in the liquid at that temperature, but it may be more, or less, soluble at a different temperature or in a different liquid. Solubility depends, in fact, on the nature of the solid and liquid concerned, and on the temperature.

In general, a solid will dissolve in a liquid which is chemically similar to it, or 'like dissolves like'. Organic (covalent) compounds will usually dissolve in organic (covalent) solvents such as benzene, ethoxyethane (ether), propan-2-one (acetone) or tetrachloromethane (carbon tetrachloride), but not in ionising solvents such as water. On the other hand, ionic compounds are more likely to dissolve in ionising solvents, such as water, than in organic solvents.

There are, however, many exceptions to such general statements, and the precise mechanism of the formation of a solution is not known. Certainly the liquid causes the solid to split up into dispersed particles of very small size, and for ionic solids the dielectric constant of the liquid is an important matter. Organic solvents have low dielectric constants whereas ionising solvents have high ones. Water is the commonest ionising solvent but liquid hydrogen fluoride, liquid ammonia and liquid sulphur dioxide are also good solvents for ionic compounds.

In the process of dispersing the particles of a solid, energy changes take place, and heat is usually absorbed when a solid dissolves in a liquid. It is for this reason, and as a consequence of Le Chatelier's principle (p. 321), that the solubilities of most solids increase with temperature. Those solids whose solubility in water decreases with temperature, e.g. anhydrous sodium sulphate(VI) (p. 230) and sodium carbonate monohydrate, evolve heat on dissolving.

3. Saturated and supersaturated solutions. If a small amount of a soluble solid is added to a liquid in which it will dissolve, a solution is formed. But if more and more solid is added to the same amount of solvent, at the same temperature, a point will be reached at which no more solid will dissolve. At this point, a saturated solution is said to have been formed. Addition of more solid causes no increase in the concentration of the saturated solution. An equilibrium is set up between the saturated solution and undissolved solid, solid depositing from the solution at the same rate as solid passes into solution from the undissolved solid present. *A saturated solution can*, therefore, *be defined as a solution which is in equilibrium with undissolved solid, at a particular temperature.* If the temperature be changed, the equilibrium will be upset and either more solid will pass into solution or more solid will deposit from the solution.

Under rather peculiar conditions, supersaturated solutions can be obtained. These are solutions which contain more dissolved solute than is required to form a saturated solution at the same temperature. Supersaturated solutions are, theoretically, unstable, but they can be obtained, and kept, in some cases; they are said to be metastable.

They are most easily made from some hydrated salts, e.g. sodium sulphate(VI)–10–water or sodium thiosulphate(VI)–5–water. If crystals of hypo, for example, are carefully heated, a solution of sodium thiosulphate(VI) in its own water of crystallisation is obtained. If this solution is allowed to cool, without shaking or admission of dust particles, a supersaturated solution results. Shaking, or addition of any small particles of dust or a crystal of hypo, causes the supersaturated solution to crystallise very quickly into a solid mass and with a considerable evolution of heat.

4. Measurement of solubility. The solubility of a solid in a liquid is usually defined as *the maximum number of grammes of the solid which will dissolve in 100 g of liquid at the temperature concerned and in the presence of excess, undissolved solid.* Other units such as kg of solute for 100 kg of solvent or mol kg^{-1} may be used.

Measurement of the solubility of a solid which is reasonably soluble requires, first, a known mass of a saturated solution of the solid. This is then analysed to find the mass of solute it contains. The mass of solvent is found by subtraction, so that the solubility can be calculated in terms of grammes of solute per 100 g of solvent.

A saturated solution is made by stirring the chosen solvent with excess of the solute in a container immersed in a thermostat at the temperature required.

After undissolved solid has had time to settle, a portion of the supernatant saturated solution is withdrawn and placed in a container of known mass. The total mass is then measured to give the mass of the saturated solution. In the withdrawal of this sample of saturated solution it is important not to include any solid. This can be done, with care, using a pipette, the tip being covered with a plug of glass wool if necessary. The pipette used must be at the same temperature as the saturated solution.

Analysis of the known mass of saturated solution is best done volumetrically if suitable reagents are available. Standard silver nitrate(v) can be used, for example, if the solute is a chloride, or standard potassium manganate(vii) (permanganate) can be used for an ethanedioic (oxalic) acid solute. If volumetric methods cannot be adopted, careful evaporation to dryness must be used.

The measurement of the solubility of sparingly soluble substances requires special methods mentioned on pp. 139 and 386.

5. Solubility curves. The solubility of a solid in a liquid varies with temperature, and it is convenient to represent the variation on a graph plotting

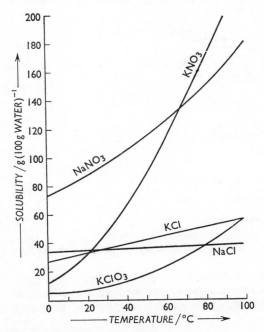

Fig. 82. Continuous solubility curves.

solubility against temperature. Such a graph is known as a solubility curve. As will be seen (Fig. 82) the solubility of the majority of solids increases with increasing temperature. Solubility curves may, also, show marked changes of direction due either to a polymorphic change (p. 223) or a change in hydrate of the solute concerned (Fig. 83).

The discontinuous curve for ammonium nitrate(v) is caused by a polymorphic change from the β-rhombic to the α-rhombic form at the transition temperature of 32°C. Discontinuous curves caused by changes in hydration are also shown. Two or more separate solubility curves for different forms of the same solid are really being plotted on one diagram,

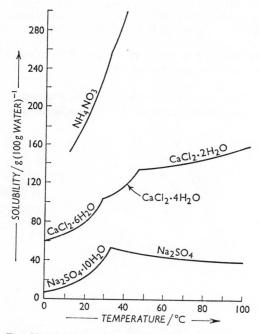

FIG. 83. Discontinuous solubility curves.

different polymorphs or different hydrates being stable over certain temperature ranges.

The solubility curves shown in Figs. 82 and 83 are limited to a temperature range of 0–100°C. A more complete representation of a solid–liquid system can be given by choosing a wider range of temperature; this is done on p. 257.

6. Crystallisation from solution. The mass of crystals obtained by cooling a hot saturated solution of known concentration to a particular temperature can readily be obtained from a solubility curve.

100 g of water, for example, will dissolve 172 g of potassium nitrate(v)

at 80°C, and only 20 g at 10°C (Fig. 82). If, therefore, a saturated solution of potassium nitrate(v) at 80°C containing 172 g is cooled at 10°C, 152 g of solid will crystallise out. If the cooling is done rapidly a mass of small, badly shaped crystals will form, but slow cooling will produce fewer, larger and better shaped crystals.

A mixture of two solids can be separated by making use of their different solubilities in the same solvent. The process is known as *fractional crystallisation*. It depends on differences in solubility and differences in the shapes of solubility curves. Fractional crystallisation is very widely used for separating mixtures of solids. It can be applied in a number of ways, as illustrated in the following examples.

a. *Separation of potassium nitrate(v) and potassium chlorate(v)*. The solubility curves for these two substances are given in Fig. 82, and it will be seen that potassium nitrate(v) is more soluble than potassium chlorate(v). At 20°C and 50°C, the solubilities in gramme per 100 g of water are

	KNO_3	$KClO_3$
20°C	31	8
50°C	86	18

Consider a mixture of 20 g of potassium nitrate(v) and 18 g of potassium chlorate(v). If this mixture is dissolved in 100 g of water at 50°C the solution will be just saturated with potassium chlorate(v) but not with potassium nitrate(v), neglecting any effect the solubility of one salt may have on that of the other. On cooling to 20°C, 10 g of potassium chlorate(v) will crystallise out, but no crystals of potassium nitrate(v) will form for, even at 20°C, the solution is still not saturated with potassium nitrate(v). Thus pure potassium chlorate(v) is obtained from the mixture.

The conditions in this particular example are extremely favourable. Both potassium chlorate(v) and potassium nitrate(v) have 'steep' solubility curves, and there is a wide difference, at all temperatures, in their solubilities. Moreover, favourable numerical figures were chosen to illustrate the point. The same principle can, nevertheless, be applied in many other cases. Typical examples are provided by the separation of potassium chloride from carnallite, $KCl \cdot MgCl_2 \cdot 6H_2O$, and the separation of potassium bromate(v) from potassium bromide in the mixture obtained by reaction between bromine and hot potassium hydroxide solution.

b. *Recrystallisation in organic chemistry*. In organic chemistry, the purification of a substance generally involves the removal of only a relatively small amount of impurity. 47.5 g of A, for example, may be contaminated with 2.5 g of B, i.e. 5 per cent impurity. Assume that both A and B are soluble in water, the solubilities at 20°C being 10 g and 4 g per 100 g of water respectively. If 50 g of the mixture of A and B is dissolved in 100 g of hot water and the resulting solution is allowed to cool to 20°C, 37.5 g of pure A will crystallise out, leaving 10 g of A and all the B in the mother liquor.

If the original mixture of A and B had contained 10 per cent impurity, i.e. 45 g of A with 5 g of B, similar treatment would have given crystals containing 35 g of A and 1 g of B. The impurity would have been reduced from 10 per cent to about 3 per cent, but this first crop of crystals would have to be recrystallised for further purification. It may not, therefore, be possible to achieve complete purification by a single crystalisation.

c. *Separation of substances with similar solubilities.* The main difficulties of fractional crystallisation arise when it is necessary to deal with very

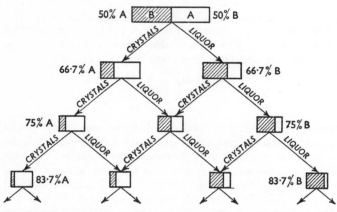

FIG. 84. Diagrammatic representation of fractional crystallisation of a mixture of A and B. A is slightly less soluble than B, and the relative amounts of B and A are shown by the shaded and non-shaded areas.

small quantities of material and/or when the two components to be separated are so chemically similar that they have very similar solubilities in all solvents.

Fractional crystallisation, under such conditions, involves a lot of careful but rather tedious work. Mme Curie, for example, took about four years to separate radium bromide from barium bromide in her isolation of radium, and the separation of mixtures of salts of the rare-earths by fractional crystallisation can be lengthy.

The general principle adopted is outlined in Fig. 84. The original mixture of A and B is dissolved in a hot solvent in which A is slightly less soluble than B. Quantities are taken such that, on cooling, about half the dissolved solid will crystallise out. The crystals resulting will be enriched in the less soluble A, whilst the mother liquor will be enriched in the more soluble B.

The enriched crystals are redissolved and the solution is recrystallised, yielding a crop of crystals still richer in A, and a mother liquor (C) still richer in B. The original mother liquor gives, on evaporation and recrystallisation, another B-rich liquid and an A-rich solid (D). C and D are mixed together as shown on Fig. 84.

By repeated treatment, as summarised in Fig. 84, it is possible to obtain pure A and pure B, but the crystallisation process may have to be repeated very many times.

d. *Crystallisation from a mixture of potassium chloride and sodium nitrate(v).* A solution containing two salts may deposit crystals of a third salt under certain conditions. A solution made by dissolving potassium chloride and sodium nitrate(v) in water contains K^+, Na^+, Cl^- and NO_3^- ions. When such a solution is hot, sodium chloride is the least soluble salt which can be formed from the ions present, and it is the first substance to crystallise out on cooling. Removal of the sodium chloride by filtration and further cooling of the filtrate yields crystals of potassium nitrate(v), and such a method is used in the manufacture of this salt.

7. Distribution of a solid between two immiscible solvents. If a solid or liquid is added to a mixture of two immiscible liquids, in both of which it is soluble, it will distribute itself between the two liquids according to the distribution or partition law. This law states that a liquid or solid distributes itself between two immiscible solvents in such a way that the ratio of the concentrations in the two solvents is constant at a fixed temperature, provided that the solute is in the same molecular state in both solvents, i.e.

$$\frac{\text{Concentration of } X \text{ in solvent A}}{\text{Concentration of } X \text{ in solvent B}} = K \text{ (a constant)}$$

This constant is generally known as the partition coefficient. It is only a constant for given substances, and it varies with temperature.

The validity of the partition law can readily be demonstrated by shaking iodine with benzene and water, or butanedioic (succinic) acid with ethoxyethane (ether) and water, at a fixed temperature. On standing, two layers separate out, and the concentration in each layer can be measured by pipetting off a known volume and titrating. Different original amounts of solute and solvents can be used.

It is important to realise that it is the ratio of the *concentrations* which matters and not the ratio of the total masses of solute. Clearly, too, the same units of concentration must be used for each solvent.   ·

8. The partition law and Henry's law. The generalised statement of Henry's law (p. 240) states that the concentrations of any single molecular species in two phases in equilibrium bear a constant ratio to each other at a fixed temperature.

The partition law is simply an application of this general law, the two phases being the two immiscible solvents, and the single molecular species being the distributed solute.

Alternatively, Henry's law as applied to gases dissolved in liquids may be regarded as an application of the partition law. For a gas dissolved in a liquid, the partition law would indicate that

$$\frac{\text{Concentration of gas in gaseous layer}}{\text{Concentration of gas in solution}} = K \text{ (a constant)}$$

This can be rewritten as

$$\frac{\text{Pressure of gas}}{\text{Mass dissolved per cm}^3} = K' \text{ (a constant)}$$

which is an expression of Henry's law.

9. Effect of association and dissociation on partition law. Henry's law and the partition law only hold if the same molecular species is considered in each phase. If a solute associates or dissociates in one or both of the solvents concerned, the partition law must be modified, though it still holds so long as identical molecular species are considered. A full consideration of the matter is lengthy for association and/or dissociation might take place in one or both solvents and, moreover, the association or dissociation may only be partial. Two comparatively simple cases will be considered.

a. *Dissociation in one solvent.* Consider a solute, AB, which is dissociated, to a degree α, in solvent 2 but not in solvent 1. If the total concentration of AB, as measured by analysis, in solvent 2 is c_2 the concentration of undissociated AB molecules in solvent 2 must be $c_2(1 - \alpha)$,

$$\text{AB} \quad \rightleftharpoons \quad \text{A}^+ + \text{B}^-$$
$$c_2(1 - \alpha)$$

If the concentration of AB in solvent 1 is c_1, then

$$\frac{c_1}{c_2(1 - \alpha)} = K$$

b. *Association in one solvent.* Consider a solute, AB, which is associated, to a degree α, into double molecules in solvent 2 but is not associated in solvent 1. If the total concentration of solute, as measured by analysis, in solvent 2 is c_2, the concentration of unassociated molecules will be $c_2(1 - \alpha)$, and that of associated molecules will be $\frac{1}{2} c_2 \alpha$,

$$2\text{AB} \quad \rightleftharpoons \quad (\text{AB})_2$$
$$c_2(1 - \alpha) \qquad \frac{1}{2} c_2 \alpha$$

Application of the equilibrium law (p. 309) shows that

$$\frac{[(\text{AB})_2]}{[\text{AB}]^2} = K = \frac{c_2 \alpha}{2[c_2(1 - \alpha)]^2}$$

The concentration of unassociated molecules in solvent 2, $c_2(1 - \alpha)$, s, therefore, equal to $\sqrt{c_2\alpha/K2}$.

If the concentration of unassociated molecules in solvent 1 is c_1, then

$$\frac{c_1}{\sqrt{c_2\alpha/K2}} = K' \quad \text{or} \quad \frac{c_1}{\sqrt{c_2\alpha}} = K''$$

When α is 1, i.e. when there is complete association into double molecules in solvent 2,

$$\frac{c_1}{\sqrt{c_2}} = K$$

10. Solvent extraction. Organic compounds are generally much more soluble in such organic solvents as ethoxyethane (ether) or benzene than in water, and these organic solvents are also immiscible with water. An organic compound can, then, often be extracted from an aqueous solution or suspension by adding ethoxyethane or benzene, shaking, separating the two layers in a separating funnel, and, finally, distilling off the ethoxyethane or benzene to leave the purified compound required.

In such a solvent extraction it is advantageous to use a given volume of organic solvent in small lots rather than in one whole.

Suppose that 100 cm³ of benzene is available for extracting a solute, X, dissolved in 100 cm³ of water, and that the partition coefficient of X between benzene and water is 5, i.e.

$$\frac{\text{Concentration of } X \text{ in benzene}}{\text{Concentration of } X \text{ in water}} = 5$$

If the whole of the benzene is added to the 100 cm³ of solution, X will distribute itself between the benzene and the water. If W_B gramme pass into the benzene layer, and W_W remain in the water, then

$$\frac{W_B/100}{W_W/100} = 5 \quad \text{or} \quad \frac{W_B}{W_W} = 5 \quad \text{or} \quad \frac{W_B}{W_B + W_W} = \frac{5}{6}$$

This means that the 100 cm³ of benzene has extracted $\frac{5}{6}$ths of the total amount of X originally present in the water.

Using only 50 cm³ of benzene

$$\frac{W_B/50}{W_W/100} = 5 \quad \text{or} \quad \frac{W_B}{W_W} = \frac{5}{2} \quad \text{or} \quad \frac{W_B}{W_B + W_W} = \frac{5}{7}$$

This means that 50 cm³ of benzene will extract $\frac{5}{7}$ths of the original amount of X. $\frac{2}{7}$ths of X will, therefore, remain in the water, but addition of a second 50 cm³ portion of benzene will extract $\frac{5}{7}$ths of the X which is still present, i.e. $\frac{5}{7}$ths of $\frac{2}{7}$ths or $\frac{10}{49}$ths of the original amount of X.

The first 50 cm³ of benzene, therefore, extracts $\frac{5}{7}$ths and the second 50 cm³ portion extracts a further $\frac{10}{49}$ths of the original amount of X. The total amount of X extracted by two 50 cm³ portions of benzene will be $\frac{45}{49}$ths, and this is more than the $\frac{5}{6}$ths extracted by one 100 cm³ portion of benzene.

A still greater proportion could be extracted by using four 25 cm³ portions of benzene.

QUESTIONS ON CHAPTER 21

1. What is the mole fraction of ethanol ethyl alcohol in a mixture of 10 g of it with 10 g of water? If the mole fraction of ethanol in a mixture of it with water is 0.5, what is the percentage of ethanol by mass?

2. A solution of ethanoic (acetic) acid containing 80.8 g dm⁻³ of ethanoic acid of molecular weight 60.1 has a density of 1.0097 g cm⁻³. Calculate the mole fraction of ethanoic acid in the solution.

3. The following figures give the solubilities of barium ethanoate (acetate) in water at different temperatures,

Temp. °C	0	8	18	25	35	40	50	60	70	80
Solub./g(100 g water)⁻¹	58	62	70	77	76	78	77	75	74	74

Express these results graphically, and interpret the form of the curve, noting that the crystals which separate from the saturated solution contain the following percentages of barium, viz. below 24°C, 44.3 per cent; between 25°C and 40°C, 50.2 per cent; above 41°C, 53.7 per cent. (O. & C.)

4. Define 'saturated solution', 'supersaturated solution' and 'solubility' as applied to a solid dissolved in a liquid. How may the solubility of ethanedioic (oxalic) acid in water be determined at 35°C?

50.42 g of a saturated solution of a certain sulphate(VI) (formula weight = 159.6) saturated at 20°C were made up to 250 cm³. 25 cm³ of this diluted solution give 1.265 g of barium sulphate(VI) when treated with barium chloride solution. Determine the solubility of the original sulphate(VI) at 20°C. At 40°C the solubility is 28.5 g per 100 g water. Deduce whether the process of solution in this case is exothermic. (N.)

5. Describe, with some experimental detail, how you would plot the solubility curve of potassium chloride.

6. What form does the solubility curve of iron(III) chloride take? Comment on any points of interest.

7. Describe, as precisely as possible, what happens when sodium chloride dissolves in water. What energy changes take place?

8. When common salt is added to water it dissolves. When it is heated, it melts. What have these two processes in common and how do they differ? In what ways do the dissolving and melting of (a) sucrose, and (b) oxygen differ from common salt?

9. Make a list of actual chemical processes in which fractional crystallisation is used.

10. Potassium salts often have 'steeper' solubility curves than the corresponding sodium salts. To what extent is this true, and what bearing does it have on the fact that potassium salts are commonly used in the laboratory in preference to the corresponding sodium salt, even though the sodium salt may be cheaper.

11. Describe, with experimental detail, how you would obtain a sample of pure benzenecarboxylic (benzoic) acid from an impure sample by recrystallisation.

12. To what extent is it true to say that the first crop of crystals from a solution containing two solutes is enriched in the least soluble solute?

13. What is meant by solubility and partition coefficient? Explain the principles of (*a*) recrystallisation, and (*b*) extraction with ether. (Oxf. Prelims.)

14. Benzenecarboxylic (benzoic) acid is distributed between water and benzene according to the following figures:

| Conc. in water | 0.0150 | 0.0195 | 0.0302 |
| Conc. in benzene | 0.2420 | 0.4120 | 0.9700 |

What conclusions can you draw?

15. What laws describe the distribution of a substance between two non-miscible solvents? A solution of 10 g of A in 100 g of water is shaken with (*a*) 100 g of ethoxyethane (ether), (*b*) five successive quantities of 20 g of ethoxyethane. Calculate the total amount of A removed from the aqueous solution in the two cases. The distribution ratio of ether: water of A is 5.

16. The solubility of iodine in water is 0.7 per cent of that of iodine in carbon disulphide. An aqueous solution of iodine containing 0.1 g of iodine per 100 cm³ is shaken with carbon disulphide. To what value does the concentration of the aqueous solution sink (i) when 1 dm³ of it is shaken with 50 cm³ of carbon disulphide, and (ii) when 1 dm³ of it is shaken successively with five separate quantities of carbon disulphide of 10 cm³ each? (C. Schol.)

17. The ratio of the solubility of octadecanoic (stearic) acid in *n*-heptane to that in ethanoic acid is 4.95. How many extractions of 10 cm³ of a solution of octadecanoic acid in ethanoic acid with successive 10 cm³ portions of *n*-heptane are needed to reduce the octadecanoic acid content of the ethanoic acid layer to less than 0.5 per cent of its original value?

18. The partition coefficient of a solute between water and ethoxyethane is D, the solute being more soluble in ethoxyethane. If the solute is extracted from an aqueous solution using an equal volume of ethoxyethane in two successive portions prove that the fraction of the solute extracted is $\dfrac{D(D+4)}{(D+2)^2}$.

19. X is a weak tribasic acid ($M_r = 210$) which is soluble in benzene and in water. After 2.800 g of X had been shaken with 100 cm³ of benzene and 50 cm³ of water until completely dissolved, it was found that 25 cm³ of the aqueous layer required 14.50 cm³ of normal sodium hydroxide solution for neutralisation. Calculate the ratio of the concentrations of X which are contained in equal volumes of benzene and water. (N.)

Chapter 22
Solutions of gases in liquids

1. Measurement of the solubility of a gas in a liquid. Some confusion has been caused by the use of various definitions and various units for the solubility of a gas, and care must be taken in using any table of solubility figures. Two definitions are most commonly used.

a. *The solubility of a gas in a liquid is the volume of gas in cubic centimetres which will just saturate 1 cm³ of liquid, the gas volume being measured at the temperature and pressure at which the measurement of solubility is made.*

b. *The absorption coefficient of a gas in a liquid is the volume of the gas in cubic centimetres which will just saturate 1 cm³ of liquid, the gas volume being measured at s.t.p.*

The values obtained for gases differ so widely that two experimental methods are required, one for very soluble gases which are those which react with, or ionise in, the solvent, e.g. ammonia, hydrogen chloride and sulphur dioxide, and another for sparingly soluble gases. Typical values for absorption coefficients of some common gases at 101.325 kN m^{-2} pressure and different temperatures are summarised below.

	0°C	20°C	50°C
Nitrogen	0.0235	0.0161	0.0109
Oxygen	0.0489	0.0310	0.0209
Hydrogen	0.0215	0.0184	0.0161
Carbon dioxide . . .	1.713	0.878	0.436
Hydrogen chloride . . .	506	442	362
Ammonia	1 300	710	—

a. *Sparingly soluble gases,* e.g. oxygen in water. The apparatus used is shown in Fig. 85, C and E being three-way taps. A is first filled with mercury, by raising B, and oxygen is then passed in, from X, through C. Its volume is measured, at atmospheric pressure, by levelling the mercury in A and B. The mass of the absorption vessel, D, is measured, first empty and then full of water. The full absorption vessel is connected to the gas burette and the connecting tube is filled with oxygen by passing a stream of the gas in at X and out at Y. Alternatively, a connecting tube with a very narrow bore, whose volume can be neglected, is used.

Some gas from A is transferred into D by raising B and allowing a measured mass of water to run out of D. The absorption vessel is shaken until no more oxygen will dissolve in the water in D, i.e. until the mercury level in A is constant. At this point, the volume of gas remaining in A is recorded, again with the levels in A and B equal.

The volume of water which has been used as solvent is equal to the original mass of water in D less the mass which was run out. The volume of gas which has dissolved in this volume of water to give a saturated solution is equal to the initial volume of gas in A minus the final volume in A and minus the volume of water run out of D. This volume of gas which has dissolved has been measured at room temperature and at atmospheric pressure. To calculate the absorption coefficient this volume must be converted to s.t.p. taking into account the water vapour pressure at the temperature concerned.

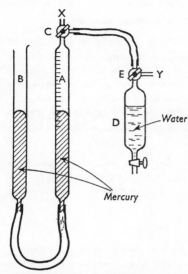

For measurements at temperatures other than room temperature, the absorption vessel must be placed in a thermostat.

Fig. 85. Apparatus for measuring the solubility of a sparingly soluble gas.

b. *Very soluble gases.* Solutions of all the very soluble gases can be estimated by volumetric analysis and this method is adopted for measuring their solubility. It is simply a matter of obtaining a known mass of saturated solution of the gas at a particular temperature. If the mass of the gas in this solution is then determined volumetrically the mass of solvent will be equal to the mass of solution minus the mass of gas so that the solubility can be calculated.

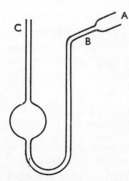

The mass of an empty, thin-walled, glass pyknometer (Fig. 86) is first measured and it is then about one-third filled with, say, water. Ammonia gas, say, is passed in, through A, until the water is saturated, and then the pyknometer is sealed by a blow-pipe flame, first at B and then at C. The mass of the sealed pyknometer is measured so that the mass of saturated solution can be found.

The whole pyknometer is then immersed in a known volume (excess) of standard acid, and broken. The mass of ammonia in the saturated solution is measured by back titration of the un-neutralised acid.

Fig. 86. A pyknometer.

For measurements at temperatures other than room temperature the pyknometer must be immersed in a thermostat whilst the saturated solution is being made.

2. Effect of temperature and pressure on the solubility of a gas.

a. *Temperature*. When gases dissolve there is generally an evolution of heat, and it would therefore be expected, on the basis of Le Chatelier's principle (p. 321) that the solubility of a gas would decrease with increase in temperature. For most gases, this is found to be so, though hydrogen and the noble gases behave differently over some ranges of temperature.

On heating a solution of a gas, the gas is usually expelled, though in some cases, e.g. a solution of hydrogen chloride in water, a constant boiling mixture (p. 249) may be formed.

b. *Pressure*. The variation in the solubility of a gas with pressure is summarised in Henry's Law (1803) which states that *the mass of a gas dissolved by a given volume of liquid at a constant temperature is proportional to the pressure of the gas*, i.e.

$$\text{Mass dissolved} \propto \text{Pressure}$$

This can also be expressed in the statement that *the volume of gas dissolved, measured at the pressure used, is independent of the pressure*. Both statements are only true for gases which do not ionise or associate in, or react with, the solvent.

That the second statement follows from the first can easily be shown: Let x g be dissolved by V cm³ of liquid at pressure, p. Then $2x$ g will be dissolved by V cm³ at pressure, $2p$. But the volume of $2x$ g of gas measured at pressure, $2p$, is equal to the volume of x g measured at pressure p. The volume of gas dissolved, measured at the pressure used, is therefore constant.

Henry's law can be expressed in a third, and more general form. If a unit volume of liquid is considered, the mass of gas dissolved in it will be equal to the mass concentration of the solution. Moreover, the pressure of a gas is only a convenient way of expressing its concentration. The proportionality between the mass dissolved and the pressure can, therefore, be rewritten as

$$\begin{array}{ccc} \text{Concentration of gas} & & \text{Concentration of gas} \\ \text{in solution} & \propto & \text{above the solution} \end{array}$$

or in the statement that *the concentrations of any single molecular species in two phases in equilibrium bear a constant ratio to each other at a fixed temperature*. This is a very generalised statement of Henry's law and applies not only to the solution of a gas in a liquid but, also, to the distribution of a solid between two immiscible liquids (p. 233).

Like other gas laws, Henry's law is only accurately true for ideal gases

(p. 35). Compound formation between gas and liquid, and association or dissociation of the gas in the solution, cause deviations. The law does not apply to a solution of ammonia in water (compound formation) or to one of hydrogen chloride in water (ionic dissociation).

3. Solution of two or more gases in the same liquid. Henry's law applies separately to a mixture of two gases dissolved in the same liquid, so long as the partial pressure of each gas is used in considering the gas.

A solution of air in water may be regarded, in an over-simplified way, as a solution of a mixture of 20 per cent by volume of oxygen and 80 per cent of nitrogen. The absorption coefficient of oxygen is approximately 0.05 at 15°C and 101.325 kN m^{-2} pressure, and that of nitrogen, under the same conditions, is approximately 0.025. The solubility of oxygen is, in fact, about twice that of nitrogen.

If air is dissolved in water at 15°C and 101.325 kN m^{-2} pressure, the partial pressure of the oxygen will be 20/100 $\times$ 101.325 kN m^{-2}, neglecting any water vapour pressure. At this pressure, 1 cm^3 of water will dissolve

$$0.05 \times \frac{20}{100} \times \frac{101.325}{101.325} \text{ cm}^3$$

of oxygen, this volume being measured at s.t.p. Similarly, the volume of nitrogen dissolved by 1 cm^3 of water will be

$$0.025 \times \frac{80}{100} \times \frac{101.325}{101.325} \text{ cm}^3.$$

The proportion, by volume, of oxygen to nitrogen is therefore

$$\frac{\text{Volume of oxygen}}{\text{Volume of nitrogen}} = \frac{0.05}{0.025} \times \frac{20}{80} = \frac{1}{2}$$

A solution of air in water, at 15°C and 101.325 kN m^{-2}, therefore contains approximately 33$\frac{1}{3}$ per cent of oxygen and 66$\frac{2}{3}$ per cent of nitrogen by volume, and a gas of this composition will be obtained on boiling the solution. Gas mixtures with still higher oxygen content can be obtained by further dissolving and boiling-off procedures.

QUESTIONS ON CHAPTER 22

1. Describe the experimental method of measuring the solubility, at room temperature, of either (i) nitrogen, (ii) hydrogen chloride or (iii) sulphur dioxide.

2. (*a*) Describe, with adequate practical details, how you would determine the solubility of (i) ammonium ethanedioate (oxalate) in water at 100°C, (ii) nitrogen in water at room temperature. Why is sodium chloride soluble in water but almost insoluble in benzene?

(*b*) A sample of water gas has the following composition by volume: 45 per cent hydrogen; 44 per cent carbon monoxide; 4 per cent carbon dioxide; 7 per cent nitrogen. If water is saturated at 15°C and 101.325 kN m^{-2} pressure with this water gas, what is the percentage of carbon dioxide in the gas mixture which would be completely expelled from the solution by boiling it? One volume of water at 15°C dissolves one volume of carbon dioxide and 0.02 volume of each of the other gases. The vapour pressure of water at 15°C is 1.733 kN m^{-2}. (W.)

3. Henry's law can be expressed in the form $p = C \times K$ where p is the gas pressure in N m^{-2}, C is the mole fraction of gas in the solution and K is a constant. The value of K for carbon dioxide in water at 25°C is 1.25×10^6. Calculate the solubility of carbon dioxide in water at 101.325 N m^{-2} pressure, expressing the answer in mol dm^{-3}.

4. Discuss the effect of temperature, pressure, the nature of the gas, and the presence of other gases on the solubility of a gas in a liquid. Describe how you would determine the solubility of ammonia in water at the temperature of the laboratory, being given a supply of aqueous ammonia. (S.)

5. By dissolving air in water and boiling off the dissolved gas a mixture enriched in oxygen is obtained. This mixture can, theoretically, be still further enriched in oxygen by redissolving and subsequent boiling off. How many times must this dissolving–boiling off process be repeated before the oxygen content exceeds 90 per cent?

6. Taking air as containing 78 per cent nitrogen, 21 per cent oxygen and 1 per cent argon by volume calculate the percentage composition of the gas boiled off from a saturated solution of air in water. The absorption coefficients of nitrogen, oxygen and argon are 0.0239, 0.0489 and 0.0530 respectively.

7. Would you expect any correlation between the solubilities of those gases which dissolve without chemical reaction and their boiling points? Examine whether such a correlation does exist for some common gases.

8. What happens when a lump of sugar is added to a fizzy drink? Can you account for your observation satisfactorily? Can you suggest any further experiments you could carry out to test your hypothesis? Do you think an equal weight of common salt would have a greater or smaller effect than the sugar? Why?

9. Find out anything you can about the importance of the solubility of various gases in blood so far as deep-sea diving is concerned. Write a short account on the topic.

Chapter 23
Solutions of liquids in liquids

S OME pairs of liquids (and this discussion will be limited to pairs) are completely miscible in all proportions, others are completely immiscible, and others are partially miscible. Some examples of each type are considered in this chapter.

COMPLETELY MISCIBLE LIQUIDS

1. Ideal or perfect solutions. An ideal or perfect gas (p. 35) would have no cohesive forces between its molecules. In an ideal or perfect solution, the cohesive forces would be just the same as those existing in the separate components of the solution. A solution made from A and B would only be ideal if the forces existing in the solution of A and B were just the same as those existing in pure A and pure B. Ideal solutions are rare and most solutions deviate considerably from the ideal.

2. Vapour pressure of ideal solution of two liquids. There is, in a solution, as in a pure liquid (p. 79), a tendency for some molecules to escape from the solution and pass into the vapour phase. This escaping tendency, which is known as the fugacity in a completely ideal system, manifests itself in the

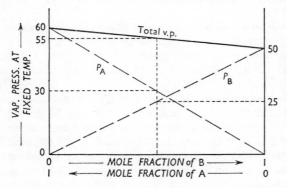

FIG. 87. The vapour pressure–composition diagram for an ideal solution.

measurable vapour pressure above a solution. This vapour pressure varies with temperature and also with the composition of the solution. The change with composition, for ideal solutions, is controlled by Raoult's law (p. 263) which states that *the partial vapour pressure of A in a solution, at a given temperature, is equal to the vapour pressure of pure A, at the same temperature, multiplied by the mole fraction* (p. 227) *of A in the solution.*

In an ideal solution, components A and B will have just the same tendency to pass into the vapour phase as they have in pure A and pure B, because the internal forces within the liquids are alike. There will, however, be relatively fewer particles of A in a solution containing both A and B than in pure A so that the partial vapour pressure of A above the solution might be expected, ideally, to be proportional to the mole fraction (p. 227) of A in the solution. Similarly the partial vapour pressure of B above the solution would be proportional to the mole fraction of B. The total vapour pressure above the solution would be equal to the sum of the partial vapour pressures of A and B.

This is illustrated in Fig. 87. The vapour pressure of pure B is 50, but it is only 25 when the mole fraction of B, in a solution with A, is 0.5. Similarly the vapour pressure of pure A is 60, but only 30 at a mole fraction of 0.5. The total vapour pressure of a mixture of A and B at a mole fraction of 0.5 will, therefore, be 25 plus 30, i.e. 55.

Numerical results of this type are given only by ideal solutions, e.g. *n*-hexane and *n*-heptane or bromoethane and iodoethane, both at 30°C.

3. Vapour pressure of non-ideal solutions of two liquids. The straight line relationship between vapour pressure and mole fraction (composition) for an ideal solution of two liquids becomes curved when the liquids deviate from ideal behaviour. Four different examples are shown in Fig. 88.

a. *Negative deviation from Raoult's Law.* Liquids which give curves as in (*a*) or (*b*) in Fig. 88 are said to have a negative deviation from Raoult's law. The total vapour pressure above the liquids is less than it would be if the liquids were ideal. There is less tendency for molecules to escape from the solution than from the pure liquids.

This must indicate stronger attractive forces between the molecules in solution than between those in the pure liquids, which may be due to association of one, or both, of the components in solution or to some degree of compound formation between the components of the solution. It is related to a contraction in volume and an evolution of heat in making the solution, whereas there is no volume, or heat, change in making an ideal solution.

Water and nitric(v) acid provide an example of a pair of liquids showing negative deviation from Raoult's law. Other pairs are given on p. 250.

b. *Positive deviation from Raoult's Law.* Liquids giving vapour pressure-composition curves as in (*c*) or (*d*) in Fig. 88, are said to have a positive deviation from Raoult's law. The total vapour pressure is greater than would be expected for ideal liquids, the molecules in the solution having a greater tendency to escape from the solution than from the pure liquid. This must be due to weaker attractive forces between the molecules in solution than between those in the pure liquids. There is, in simple language, some reluctance to mix completely, and this is associated with an increase in volume and an absorption of heat on mixing.

Ethanol and water show a positive deviation from Raoult's law, and other pairs are mentioned on p. 250.

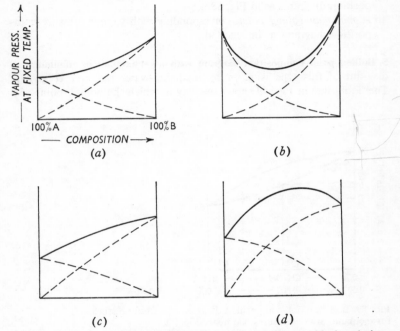

FIG. 88. Vapour pressure–composition diagrams for non-ideal solutions of A and B. (a) and (b) are for liquids showing a negative deviation from Raoult's law. (c) and (d) show a positive deviation. The dotted lines show the partial pressures of A and B.

4. Boiling-point–composition diagrams. The way in which the total vapour pressure of a mixture of two liquids varies with composition is of importance because of its bearing on the possibility of separating the components of a mixture of two liquids by fractional distillation.

For this purpose, it is most convenient to use boiling point–composition diagrams, at a fixed pressure, instead of vapour pressure–composition diagrams, at a fixed temperature. As a high vapour pressure corresponds to a low boiling point, and vice versa, the boiling point–composition diagram, at a fixed pressure, for a pair of liquids can readily be obtained from the corresponding vapour pressure–composition diagram at fixed temperatures. The two diagrams are similar in shape except that they are inverted. Compare, for example, the diagrams in Figs. 89 and 90.

There are three important types of boiling point–composition diagram:

i *No maximum or minimum.* This type corresponds with the vapour pressure–composition diagrams in Figs. 87, 88a and 88c.

ii *A maximum boiling point.* corresponding with vapour pressure–composition diagrams as in Fig. 88b.

iii *A minimum boiling point,* corresponding with vapour pressure–composition diagrams as in Fig. 88d.

5. Boiling point–composition diagram with no maximum or minimum. A diagram of this type is given by methanol-water mixtures (Fig. 89). The liquid line in Fig. 89 shows the way in which the boiling point of a

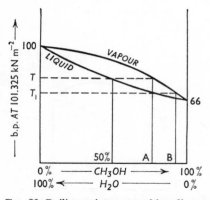

Fig. 89. Boiling point–composition diagram at a fixed pressure for methanol–water mixtures. Compare Fig. 90.

methanol–water mixture varies with composition, at a fixed pressure.

For a liquid mixture of any one composition, the vapour with which it is in equilibrium will be richer in the more volatile component, i.e. in methanol. The liquid line has, therefore, an associated vapour line. The vapour pressure–composition diagram corresponding with this boiling point-composition diagram is shown in Fig. 90.

When a mixture of methanol and water containing 50 per cent of each is boiled, it will boil at temperature T. The vapour coming from it will have a composition represented by A, and, on condensing, this vapour will form a liquid with the same composition. If this liquid is again boiled, it will now boil at temperature T_1, giving a vapour of composition B, and this vapour will condense into a liquid whose composition is also B. By repeating this boiling–condensing–boiling process pure methanol could be obtained, but the method would be tedious, and the same result can be obtained in one operation by fractional distillation using a fractionating column (Fig. 91).

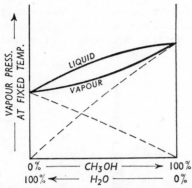

FIG. 90. Vapour pressure–composition diagram at a fixed temperature for methanol–water mixtures.

6. Fractional distillation. Mixtures with a boiling point–composition diagram of Type 1 can be separated into their component parts by fractional distillation, and this is most commonly done by using a fractionating column.

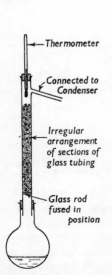

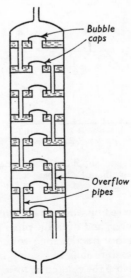

FIG. 91. A fraction-
ating column.

FIG. 92. A fractionating tower.

A simple and effective column for laboratory use consists of a long glass tube packed with short lengths of glass tubing, glass beads or specially made porcelain rings (Fig. 91). The aim is to obtain a large surface area, and there are many patent designs of column. Industrially, a fractionating tower is used. Such a tower is divided into a number of compartments by means of trays set one above the other (Fig. 92). These trays contain

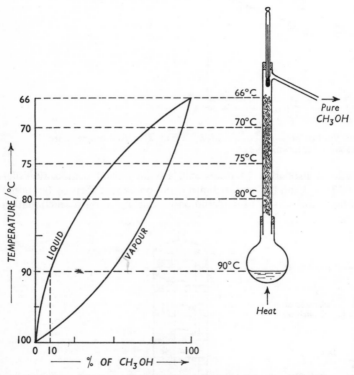

FIG. 93. Idealised and simplified representation of the fractional distillation of a mixture of methanol (10 per cent) and water (90 per cent) using a fractionating column.

central holes, covered by what are known as bubble caps, to allow vapour to pass up the tower, and overflow pipes to allow liquid to drop down.

In both laboratory fractionating columns and industrial fractionating towers the temperature falls in passing from the bottom to the top. As vapour passes up into the column or tower it condenses, first, in the lowest part. As more hot vapour ascends, however, this condensed liquid is boiled again, giving a vapour which condenses higher up the column or tower.

This liquid, in its turn, is again heated and boiled by ascending vapour so that the composition of the mixture is constantly being enriched in the more volatile component on passing up the column or tower. Finally, the more volatile liquid emerges in a pure form from the top of the tower.

At each point in a column, or at each plate in a tower, an equilibrium between liquid and vapour is set up and this is facilitated by (*a*) an upward flow of vapour and downward flow of liquid, (*b*) a large surface area, (*c*) slow distillation. It is also preferable to maintain the various levels of the column or tower at a steady temperature so that external lagging or electrical heating jacket is often used.

The state of affairs existing in an idealised and simplified distillation of a mixture of methanol and water, containing 10 per cent by mass of methanol, is shown in Fig. 93. The figure shows five liquid–vapour equilibria which are set up at different temperatures in the fractionating column; in reality the liquid–vapour equilibria change continuously in passing up the column. The purpose of the fractionating column is to facilitate the setting up of these equilibria.

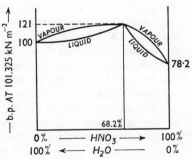

Mixtures of varied composition can be drawn off from different points on the column or tower as is done, for instance, in the fractional distillation of crude oil in a refinery.

7. Boiling point–composition diagram with a maximum. The vapour pressure–composition diagram for nitric(v) acid–water mixtures shows a minimum (as in Fig. 88b) and the corresponding boiling point–composition diagram, with a maximum, is shown in Fig. 94.

FIG. 94. Boiling point–composition diagram at a fixed pressure for nitric(v) acid–water mixtures.

On distilling a mixture of nitric(v) acid and water containing less than 68.2 per cent nitric(v) acid, the distillate will consist of pure water and the mixture in the flask will become more and more concentrated until it contains 68.2 per cent nitric(v) acid. At this stage, the liquid mixture will boil at a constant temperature because the liquid and the vapour in equilibrium with it have the same composition, i.e. 68.2 per cent nitric(v) acid.

Mixtures containing more than 68.2 per cent nitric(v) acid will give a distillate of pure nitric(v) acid until the residue in the flask reaches the 68.2 per cent nitric(v) acid composition. Thereafter, the distillate will be 68.2 per cent nitric(v) acid as before.

A mixture with this type of boiling point–composition curve cannot be completely separated by fractional distillation. It can only be separated into one component and what is known as the constant boiling mixture, maximum boiling point mixture, or azeotropic (ζειν, to boil; ατροπος, unchanging) mixture.

Maximum boiling point mixtures are also obtained from mixtures of water with hydrofluoric, hydrochloric, hydrobromic, hydriodic, sulphuric(VI) and methanoic (formic) acids; and from propanone (acetone) and trichloromethane (chloroform) or benzenol (phenol) and phenylamine (aniline).

8. Boiling point–composition diagram with a minimum. Ethanol and water give a vapour pressure–composition diagram with a maximum, as in Fig. 88d, and the corresponding boiling point–composition diagram, with a minimum, is shown in Fig. 95.

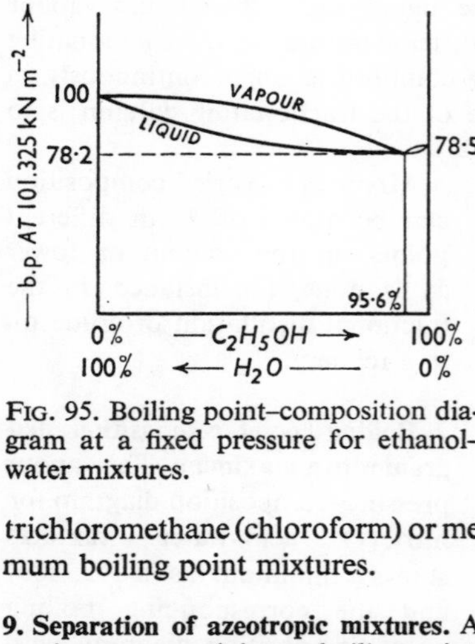

FIG. 95. Boiling point–composition diagram at a fixed pressure for ethanol-water mixtures.

It is not possible to get a complete separation of ethanol and water by fractional distillation. A mixture containing more than 95.6 per cent ethanol can be separated into pure ethanol and a minimum boiling point mixture, with a composition of 95.6 per cent ethanol. A mixture containing less than 95.6 per cent ethanol can be separated into pure water and the same minimum boiling point mixture.

Water with propanol or butanol or pyridine, and ethanol with trichloromethane (chloroform) or methyl benzene (toluene), also give minimum boiling point mixtures.

9. Separation of azeotropic mixtures. An azeotropic mixture may have either a maximum or a minimum boiling point but, at any one pressure, it has a fixed composition. It is unusual for this composition to correspond with that of any simple chemical formula for the mixture, and there is definitely no compound formation because the composition of the mixture does depend on pressure and, moreover, the mixture can be separated into its component parts fairly easily. Such separation can be brought about by the following methods:

a. *By distillation with a third component.* The azeotropic mixture of ethanol and water contains 95.6 per cent of alcohol at normal atmospheric pressure. If benzene is added, distillation yields, first, a ternary azeotropic mixture of ethanol water and benzene, then a binary azeotropic mixture of ethanol and benzene, and finally, absolute ethanol.

b. *By chemical methods.* Quicklime may be used to remove the water from an azeotropic mixture of ethanol and water, or concentrated sulphuric(VI) acid will remove aromatic or unsaturated hydrocarbons from mixtures with saturated hydrocarbons in the refining of petrols and oils.

c. *Adsorption*. Charcoal or silica gel may adsorb one of the components.

d. *Solvent extraction*. One component can be extracted by a solvent.

IMMISCIBLE LIQUIDS

10. Vapour pressure of immiscible liquids. Some liquids will not mix at all and simply separate out into two distinct layers according to their densities. Examples are provided by mercury and water or by paraffin and water. Water floats on top of mercury; paraffin floats on top of water. Such liquids, which will not mix, must have strongly repulsive forces between their molecules.

The lack of any sort of interaction between two immiscible liquids, excepting purely surface effects (p. 492), means that each liquid behaves almost independently of the other. Thus *the vapour pressure above an agitated mixture of two immiscible liquids, at any temperature, is equal to the sum of the vapour pressures of the individual liquids at the same temperature.* Agitation of the mixture is necessary to enable each liquid to establish its vapour phase. Moreover, the vapour pressure above the mixture will be independent of the amount of each liquid present, so long as there is enough to give a saturated vapour.

As the temperature is increased, the vapour pressure of each liquid will rise, so that the total vapour pressure above the mixture will also rise. When this total vapour pressure becomes equal to the external pressure, the mixture will boil, and this boiling point will be lower than the boiling points of either of the two separate liquids.

A mixture of phenylamine (aniline) (b.p. $= 184°C$) and water (b.p. $=$ $100°C$), for example, will boil at $98°C$, under a pressure of 101.325 kN m^{-2} (760 mm). At $98°C$, the vapour pressure of phenylamine is 7.065 kN m^{-2}, and that of water is 94.260 kN m^{-2}. The combined vapour pressure is, therefore, 101.325 kN m^{-2}, and the mixture will continue to boil at $98°C$ so long as both phenylamine and water are present in it. The vapour coming from the boiling mixture will contain both phenylamine vapour and water vapour, in volumes proportional to their partial vapour pressures.

The relative masses of phenylamine and water in the vapour may be calculated as follows, A standing for phenylamine and W for water.

$$\frac{\text{Vol. of } A \text{ vapour}}{\text{Vol. of } W \text{ vapour}} = \frac{\text{No. of molecules of } A}{\text{No. of molecules of } W} = \frac{\text{Vapour pressure of } A}{\text{Vapour pressure of } W}$$

$$\frac{\text{Mass of } A}{\text{Mass of } W} = \frac{\text{Relative density of } A \times \text{Vol. of } A \text{ vapour}}{\text{Relative density of } W \times \text{Vol. of } W \text{ vapour}}$$

$$\therefore \frac{\text{Mass of } A}{\text{Mass of } W} = \frac{\text{Relative density of } A \times \text{Vapour pressure of } A}{\text{Relative density of } W \times \text{Vapour pressure of } W}$$

$$\therefore \frac{\text{Mass of } A}{\text{Mass of } W} = \frac{46.5 \times 7.065}{9 \times 94.260} = \frac{1}{2.5} \text{ (approx.)}$$

11. Steam distillation. Steam distillation depends on the properties of immiscible liquids and is a useful process for separating a liquid or a solid from a mixture. It is most successful when the liquid or the solid to be separated (*a*) is immiscible with or insoluble in water, (*b*) has a high relative molecular mass, and (*c*) exerts a high vapour pressure at about 100°C. Any impurities present must be non-volatile under the conditions used

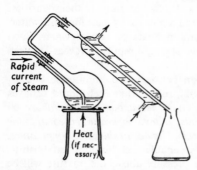

Steam distillation is particularly useful for the purification of substances that decompose at temperatures near their normal boiling points. For, in steam distillation, a substance is distilled at a temperature considerably below its normal boiling point.

The method is to pass steam through the mixture in an apparatus as shown in Fig. 96. The distillate, collected in the receiver, consists of water and the solid or

FIG. 96. Apparatus for steam distillation.

liquid required. The solid or liquid can be isolated by filtration, by extraction with a solvent (p. 235), or by using a separating funnel and drying with a drying agent.

If any mixture of a substance, *X*, with water and other non-volatile impurities is subjected to steam distillation, the argument in section **10**, shows that the distillate collected will contain *X* and water in the proportion

$$\frac{\text{Mass of } X}{\text{Mass of water}} = \frac{\text{Relative density of } X \times \text{Vapour pressure of } X}{\text{Relative density of water} \times \text{Vapour pressure of water}}$$

It is desirable to get as much *X* as possible in the distillate, and that is why a high relative density (or relative molecular mass) and a high vapour pressure for *X* is helpful. The use of superheated steam, i.e. steam at high pressure, also helps by increasing the vapour pressure of *X* in relation to that of water, and the low relative density of water is an important factor.

By measuring the relative masses of *X* and water in the distillate, the relative density, and hence the relative molecular mass of *X* can be calculated, if the vapour pressures of *X* and water at the temperature concerned and the relative density of water are known. This provides a method of relative molecular mass measurement, though it is not of any practical significance.

PARTIALLY MISCIBLE LIQUIDS

Ethoxyethane (ether) will dissolve a little water (about 1.2 per cent at room temperature) to form a homogeneous solution, and water will also dissolve a little ethoxyethane (about 6.5 per cent at room temperature) to form a similar solution. Within these limits ethoxyethane and water are completely miscible. As their mutual solubilities are limited, however, ethoxyethane and water are only partially miscible. If equal volumes of the two liquids are shaken together, for instance, two layers will form, one a saturated solution of ethoxyethane in water and the other a saturated solution of water in ethoxyethane. These two solutions are described as conjugate solutions.

12. Critical solution temperature. Some mixtures only form similar, immiscible, conjugate solutions within certain temperature ranges. Equal weights of phenol and water, for example, give two conjugate solutions, existing in two layers, at temperatures up to 60°C, but are completely miscible above that temperature. The effect of composition and temperature is shown in a temperature–composition diagram (Fig. 97a).

The temperature above which phenol and water are miscible in any proportion, i.e. 66°C, is known as the upper critical solution temperature or the upper consolute temperature. There may be complete miscibility below that temperature but it will be dependent on the composition of the mixture. Above the curve there is always complete miscibility, i.e. one layer. Thus 80 per cent of phenol and 20 per cent of water, or 2 per cent of phenol and 98 per cent of water, will be completely miscible at 50°C. Below the curve, two layers will always form, and the curve will give the compositions of the two conjugate solutions making up the two layers. A mixture of 50 per cent phenol and 50 per cent water, for example, at 50°C, will form two layers whose compositions are given by A and B. The line YZ is known as a tie-line. The ratio YX/YZ is equal to the ratio of the mass of the phenol layer (of composition B) to that of the mass of the aqueous layer (of composition A).

The complete miscibility of phenol and water with increasing temperature comes about because their mutual solubilities increase as the temperature does. The curve in Fig. 97a can, in fact, be regarded as made up of two halves, the one being the solubility curve of water in phenol and the other the solubility curve of phenol in water. With triethylamine and water the mutual solubilities decrease as the temperature is increased. This leads to a temperature–composition diagram with a lower critical solution (or consolute) temperature of 18.5°C (Fig. 97b). A 50–50 mixture will be completely miscible at 10°C but will separate into two layers, with compositions C and D, at 50°C.

Mixtures of nicotine and water are very unusual as they have both an upper (208°C) and a lower (61°C) critical solution temperature (Fig. 97c). A 50–50 mixture is completely miscible at 60°C or 250°C, but it will form two layers, with compositions shown by E and F, at 150°C.

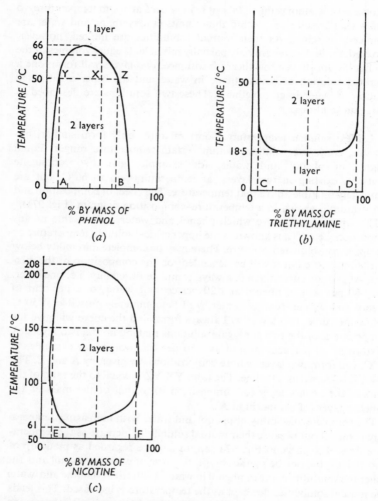

FIG. 97. Temperature–composition diagrams for
(a) phenol and water, (b) triethylamine and water,
(c) nicotine and water.

QUESTIONS ON CHAPTER 23

1. What are the main characteristics of (i) an ideal solution of one liquid in another, (ii) an ideal gas?

2. If water has a vapour pressure of x N m^{-2} at a certain temperature, what will be the vapour pressures above (a) a mixture of 18 g of water and 0.1 mol of

sucrose, (b) a mixture of 36 g of water and 5.85 g of sodium chloride, (c) a mixture of 18 g of water and 0.1 mol of a liquid with a vapour pressure of y mm, at the same temperature? Assume all solutions are ideal.

3. A liquid, A, with a vapour pressure of 16 kN m⁻² is mixed with an equimolecular portion of another liquid, B. The vapour pressure of the mixture of A and B is found to be 101.325 kN m⁻². What is the vapour pressure of pure B?

4. The vapour pressures of methanol and ethanol at 20°C are 12.530 and 5.866 kN m⁻² respectively. If 20 g of methanol give an ideal solution when mixed with 100 g of ethanol calculate (a) the partial pressure exerted by each component of the mixture, (b) the total vapour pressure above the mixture, and (c) the composition of the vapour.

5. What will be the composition of the vapour over an ideal solution of A in B, if the mole fraction of A is 0.25 and the vapour pressures of pure A and pure B are 8 and 13.332 kN m⁻² respectively?

6. The fact that nitric(v) acid–water and hydrochloric acid–water systems give constant boiling mixtures can be used to obtain standard solutions of these two acids. How would this be done in practice? At 101.325 kN m⁻² pressure, the constant boiling mixture from hydrochloric acid and water contains 20.22 per cent of hydrogen chloride by mass and has a density of 1.0962 g cm⁻³. What volume of this constant boiling mixture would be required to make 1 dm³ of a 1M solution of hydrochloric acid?

7. Draw approximate boiling point–composition diagrams for the following systems. A and B give the components together with their separate boiling points, C gives the percentage by weight of B in the constant boiling mixture, and D gives the boiling point of the constant boiling mixture:

A H_2O (100°C)	H_2O (100°C)	C_2H_5OH (78°C)
B HCl (−80°C)	C_3H_7OH (97.2°C)	C_6H_6 (80°C)
C 20.2%	71.7%	67.6%
D 108.6°C	87.7°C	68.2°C

8. How do an azeotropic mixture and a eutectic mixture (i) resemble, and (ii) differ from each other?

9. How would you obtain, experimentally, the data which would enable you to draw Fig. 89?

10. Explain two different methods for obtaining absolute ethanol.

11. The constant boiling mixture of hydrochloric acid and water has a composition approximating to the formula $HCl \cdot 8H_2O$. How would you try to convince a sceptic that the mixture was not a hydrate of hydrochloric acid?

12. Draw typical equilibrium diagrams showing the variation of the percentage composition of the vapour and liquid phases with boiling temperatures (at constant pressure) for mixtures of two completely miscible liquids A and B, which do not form a constant boiling mixture and of which A has the higher boiling point. Use the diagram to describe the changes which occur in the composition of the residual liquid and of the distillate when a 50 per cent mixture of A and B is distilled. Describe one useful practical application of constant boiling mixture formation. (W.)

13. Explain with the aid of diagrams how two liquids may be separated by distillation. What is the advantage of a fractionating column? In what circumstances is it not possible to separate a mixture of two liquids by fractional distillation. (O. Schol.)

14. Write short notes on (a) azeotropic mixtures, (b) consolute temperatures, (c) theoretical plate, (d) drip point.

15. Give examples of the use of steam distillation.

Chapter 24
Solidification of solutions. Eutectics

WHEN a single substance, in the liquid state, is cooled, it solidifies (freezes) into a single solid, and there are very few complications, but a solution contains at least two components and there are various possibilities on cooling a solution until it wholly or partially freezes.

1. Solidification of aqueous solutions of solids. On cooling a hot solution of sodium nitrate(v) in water, crystals of sodium nitrate(v) may separate out above 0°C, if the solution is concentrated enough ever to become saturated, but what will happen if the solution be cooled below 0°C, the freezing point of water?

If a dilute solution of sodium nitrate(v) is cooled below 0°C, pure ice will form when the freezing point of the solution is reached. This freezing point will be less than 0°C, because the added solute lowers the freezing point of the water (p. 273). As ice is formed, the solution becomes more concentrated so that its freezing point becomes lower still, but further cooling will deposit more ice. Eventually, at −17.5°C, the solution remaining will become saturated, and any further cooling will deposit a mixture of ice and solid sodium nitrate(v), the temperature remaining constant at −17.5°C until the whole system solidifies.

If a concentrated solution of sodium nitrate(v) is cooled, crystals of pure sodium nitrate(v) will be deposited until, at a temperature of −17.5°C, a mixture of ice and sodium nitrate(v) will again crystallise out at a constant temperature.

Such results, together with other useful information, can be summarised in a temperature–composition diagram (Fig. 98). The line BE represents the solubility curve of sodium nitrate(v) in water (compare Fig. 82, p. 229); the line AE shows the way in which the freezing point of water is lowered as more and more sodium nitrate(v) is added to it.

Point E is known as the *eutectic point*; it is the lowest temperature which can be reached before the whole system solidifies. It is also the only point at which ice, solid sodium nitrate(v) and saturated solution of sodium nitrate(v) in water are in equilibrium. At the eutectic point the crystals deposited from the solution have the same composition as the solution. This composition is given by C, i.e. 38.62 per cent by weight of sodium nitrate(v), and a mixture of this composition is known as the *eutectic* or the *eutectic mixture*.

The terms *cryohydric point* and *cryohydrate* were originally used instead of eutectic point and eutectic mixture. This was because it was originally supposed that the eutectic mixture was a definite hydrate, and so the solid obtained by freezing it was called a cryohydrate.

It is now known that the eutectic mixture is not a compound because (a) microscopic examination shows the presence of two kinds of crystal, particularly if coloured salts are involved, (b) only by chance does a eutectic mixture have a composition corresponding to a simple chemical formula, and (c) the composition and melting poiht of a eutectic mixture change with pressure.

The temperature–composition diagram for the sodium nitrate(v)–water system is divided into different areas. In area 1, below the eutectic point, only ice and solid sodium nitrate(v) can co-exist; everything is solid. In area 2, solid sodium nitrate(v) is in equilibrium with a saturated solution. In area 3, ice is in equilibrium with a solution. In area 4, a solution exists, with no solid present.

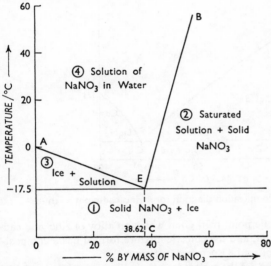

FIG. 98. Temperature–composition diagram for the sodium nitrate(v)–water system.

2. Solidification of mixtures of two liquids without compound formation. A

mixture of two liquids very often forms a eutectic on freezing in the same way as an aqueous solution of a solid, and the matter is particularly important in a study of alloy systems.

Pure zinc will melt on heating and resolidify, at its freezing point of 419°C, on cooling; pure, molten cadmium will resolidify at its freezing point of 321°C.

Addition of a little cadmium to molten zinc will lower the freezing point, but, on cooling, pure zinc will still separate out. The freezing point will continue to be lowered as more and more cadmium is added, or as more and more zinc separates out, until, eventually, a solid mixture of zinc and cadmium will separate. Similarly, addition of zinc to molten cadmium lowers the freezing point, and when sufficient zinc has been added, a solid mixture of zinc and cadmium again forms on cooling.

The temperature–composition diagram for a zinc–cadmium mixture is shown in Fig. 99. Point A shows the freezing point of pure zinc, and point B that of pure cadmium. The line AE shows the way in which the freezing point of zinc is lowered by adding cadmium; line BE shows the lowering of the freezing point of cadmium on adding zinc.

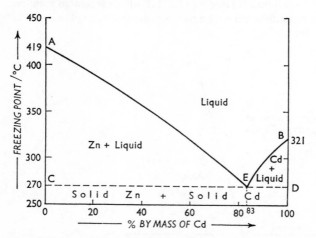

FIG. 99. Temperature–composition diagram for a zinc–cadmium mixture.

Point E is the eutectic point (270°C) at which an alloy of zinc and cadmium first separates out, and the eutectic mixture formed has a composition of 17 per cent zinc and 83 per cent cadmium.

At all points above AEB the system is entirely liquid; in the area ACE, solid zinc is present with molten mixture; in BDE, solid cadmium is present with molten mixture; below CED the system is entirely solid.

Other mixtures which form eutectics in the same way, include tin and lead, antimony and lead, gold and thallium, potassium and silver chlorides, bromomethane and benzene, and camphor and naphthalene.

3. Solidification of mixtures of two liquids with compound formation. If two liquids form a solid compound on cooling, the temperature–composition diagram is of the type as shown for a magnesium–zinc mixture in Fig. 100.

The maximum in the curve indicates the formation of a compound with a composition represented by C, and a corresponding formula of $MgZn_2$. The diagram is most simply regarded as made up of two halves. The left-hand half, ABCD, shows the formation of a normal eutectic mixture from a mixture of magnesium and the compound $MgZn_2$; the eutectic point is at X. The right-hand half, CDEF, shows a eutectic at Y, formed from a mixture of zinc and the compound $MgZn_2$.

The diagram can be divided up into various areas within which different phase relationships hold, as indicated.

The rounded maximum in the diagram indicates that the $MgZn_2$ compound is not very stable; it tends to dissociate into magnesium and zinc. If a stabler compound is formed the maximum is a sharper peak.

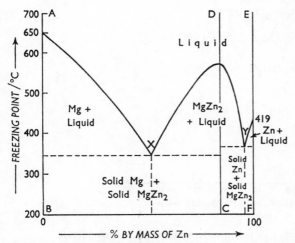

FIG. 100. Temperature–composition diagram for a magnesium–zinc mixture.

The formation of intermetallic compounds, and of compounds between pairs of organic substances, e.g. benzenol and aniline, is not uncommon, though the plotting of a temperature–composition diagram is often the only practical method of detecting the formation of such a compound.

More than one compound can be formed and this will be shown by more than one maximum in the temperature–composition diagram. Iron(III) chloride, for example, forms many different hydrates with water as shown in Fig. 101.

4. Solidification of mixtures of two liquids to form solid solutions. Pairs of substances which form eutectic mixtures do so because the substances concerned are not completely soluble in each other in the solid state. If two substances are soluble in the solid state they will form true solid solutions. Silver and gold provide an example, with a temperature–composition diagram as in Fig. 102.

The diagram is made up of two lines. One, known as the liquidus, shows the composition of the liquid phase and the way in which the freezing point of the liquid mixture varies with composition. The other, known as the solidus, shows the composition of the solid phase and the way in which the melting point of the solid mixture varies with composition.

For a pure substance, the melting point and freezing point are identical if super-cooling or super-heating (p. 270) are avoided, but this is not necessarily so for a mixture. In fact, the majority of mixtures do not have sharp melting or

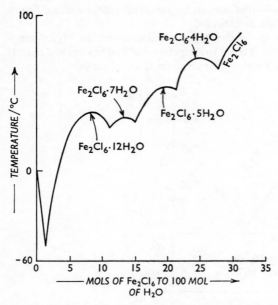

FIG. 101. The formation of four distinct hydrates of iron(III) chloride.

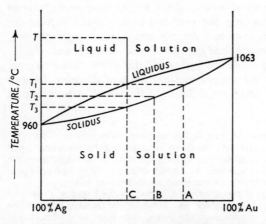

FIG. 102. Temperature–composition diagram for silver–gold mixtures which form solid solutions.

freezing points; candle-wax, butter, chocolate and solder are common examples. Eutectic mixtures are examples of mixtures which do have sharp melting and freezing points.

Solid solutions do not melt or freeze completely at one temperature. If a mixture containing silver and gold, of composition C, at temperature T is cooled, a solid of composition A will begin to separate out at temperature T_1. On cooling to temperature T_2, a solid of composition B will separate out. Solidification will only be complete at T_3. The liquid has, in fact, been depositing solid between T_1 and T_3. Similarly, if a solid of composition C was heated it would begin to melt at T_3, but would not be completely molten until T_1.

This type of temperature–composition curve is also given by tin and bismuth, mixtures of alums, and copper and nickel. The well-known copper–nickel alloys, Constantan and Monel are, in fact, solid solutions. The temperature–composition curve is analogous to that for a methanol–water mixture (p. 246), the only difference being that solid–liquid equilibria are concerned instead of liquid–equilibria.

Similarly, mixtures of solids which form solid solutions can give temperature–composition curves with maxima and minima, just as liquid mixtures can (p. 249). Very few mixtures of solids give curves with a maximum, but fusion mixture (equal parts of sodium and potassium carbonate) provides a good example of a mixture of solids which give a minimum. The melting points of sodium and potassium carbonate are 820° and 860°C respectively; that of fusion mixture is 690°C.

QUESTIONS ON CHAPTER 24

1. The following data applies to a mixture of gold and tellurium:

% mass of Te	0	10	20	30	40	42	50	56.4	60	70	82.5	90	100
f.p./ °C	1063	940	855	710	480	447	458	464	460	448	416	425	453

Plot the phase diagram for the system. Sketch the cooling curve which would be obtained from a melt containing 50 per cent by mass of tellurium.

2. The freezing point and the melting point of a pure substance are equal, but this is not generally so for most mixtures of two pure substances. Why is this? For what mixtures are the freezing point and melting point equal?

3. The boiling point of a liquid can be defined as the temperature at which the vapour pressure of the liquid becomes equal to the external pressure on the liquid. What would be the comparable definition for the melting point of a solid?

4. Every cryohydrate is a eutectic, but not every eutectic is a cryohydrate. Explain.

5. A mixture of salt and sand is scattered on frozen roads in winter months. What does it do?

6. The data below represent the temperatures at which solid begins to separate from fused alloys of zinc and antimony of the compositions shown:

% Sb	0	2	5	10	20	30	40	50	60	70	75	80	90	100
°C		443	420	428	456	502	535	553	565	566	530	507	5ſ0	561	632

Plot the figures on squared paper. What can you infer from the graph? What would happen if an alloy with 65 per cent antimony were gradually heated from 400° to 600°? (O. Schol.)

7. The freezing points of molten mixtures of bismuth and cadmium are as follows:

+ Cd	90	80	70	60	50	40	30	20	10	0
Temp./°C	300	280	255	225	195	145	175	205	235	270

Draw and interpret the freezing point curve. Describe and explain what will be observed on cooling each of the following from 400°C to 100°C: (a) pure Cd, (b) a mixture of 60 per cent Bi and 40 per cent Cd, (c) a mixture of 75 per cent Bi and 25 per cent Cd. (N.)

8. Two metals, A and B, melt at 650°C and 225°C respectively. They form an intermetallic compound, C, containing 75 per cent of A which melts at 800°C. Mixtures containing 20 per cent and 90 per cent of A melt at constant temperatures at 150°C and 500°C respectively. Draw an approximate equilibrium diagram and label each area to indicate the equilibrium involved. Explain what happens on cooling liquid mixtures containing (a) 10 per cent, (b) 50 per cent, (c) 90 per cent of A. Describe the effect of melting a mixture of zinc in a lead–silver alloy. Comment on the practical application which is made of this process. (W.)

9. Use the following data to draw a temperature–composition diagram for the water–sulphur trioxide system:

f.p./°C	−20	−70	−60	−28	−28	−50	−48	2	4	−26	−22	6	6
% of SO₃	10	30	36	44	48	57	61	67	71	75	77	80	83

Comment on the nature of the diagram.

10. Solid solutions may be either substitutional or interstitial. Explain, with illustrative examples, what this means.

11. What are the criteria which decide whether two metals will form solid solutions?

12. What are Hume-Rothery's rules?

13. Write an account of 'metallic alloys'.

Chapter 25
Vapour pressure, boiling point and freezing point of solutions with non-volatile solutes

THERE are certain properties which are common to all solutions with *non-volatile* solutes and which vary with the composition of the solution in the same sort of way. The most important of these properties are the vapour pressure, the boiling point, the freezing point and the osmotic pressure. Each of these can be measured for solutions of different concentration and the results obtained depend, essentially, on the total number of solute particles present in a fixed amount of solvent. For this reason the properties are sometimes known as *colligative properties*.

The laws governing these colligative properties are generally known as *the laws of dilute solutions*. The laws only hold at all accurately for dilute solutions because it is only at low concentrations that the solutions approximate to ideal solutions (p. 243).

The vapour pressure, boiling point and freezing point of dilute solutions are considered in this chapter. Osmotic pressure is considered in Chapter 26.

VAPOUR PRESSURE OF SOLUTIONS

1. Raoult's law. The way in which the vapour pressure of an ideal solution (p. 244) changes with composition of the solution is summarised in Raoult's law (p. 243) first put forward as a result of experimental measurements in 1886. The law can be stated in three ways.

First, *the partial vapour pressure of A in a solution, at a given temperature, is equal to the vapour pressure of pure A, at the same temperature, multiplied by the mole fraction* (p. 227) *of A in the solution.*

For solutions made up of volatile solvents and non-volatile solutes, the total vapour pressure of the solution will be equal to the partial vapour pressure of the solvent, for the solute will not contribute to the vapour pressure. Thus

$$p_{soln} = p_{solv} \times \text{Mole fraction of solvent}$$

or
$$1 - \frac{p_{soln}}{p_{solv}} = 1 - \text{Mole fraction of solvent}$$

or
$$\frac{p_{solv} - p_{soln}}{p_{solv}} = 1 - \frac{\text{Mole fraction}}{\text{of solvent}} = \frac{\text{Mole fraction}}{\text{of solute}}$$

This is the second form in which Raoult's law can be expressed. In words, *the relative lowering of the vapour pressure of a solution containing a non-volatile solute is equal to the mole fraction of the solute in the solution.*

If n is the number of molecules, or moles, of solute in a solvent, and N the number of molecules, or moles, of solvent, Raoult's law is expressed as

$$\frac{p_{solv} - p_{soln}}{p_{solv}} = \frac{n}{n + N}$$

For dilute solutions, n will be small as compared with N, so that $(n + N)$

will be approximately equal to N. Moreover, for any given solvent, p_{solv} will be constant, so that

$$p_{solv} - p_{soln} \propto \frac{n}{N}$$

and, for a given quantity of solvent, N will also be constant so that

$$p_{solv} - p_{soln} \propto n$$

This means that *equimolecular quantities of* **any** *non-volatile solute dissolved in the same quantity of the same solvent will produce the same lowering of vapour pressure*, and this is the third statement of Raoult's law. ·

This only applies for dilute solutions containing non-volatile solutes. Moreover, it does not apply for solutes which dissociate or associate in solution. If a solute dissociates, more particles will be present in solution than if it did not dissociate, and the lowering of vapour pressure will be proportionately greater. Association of a solute would give fewer particles in solution with a proportionately smaller lowering of vapour pressure.

If a solute with molecules AB caused a lowering of vapour pressure of x without dissociating or associating it would cause a lowering of $2x$ if it completely ionised into A^+ and B^- or a lowering of $x/2$ if it completely associated into $(AB)_2$ particles. The lowering of vapour pressure is really proportional to the number of dissolved particles in a solvent, and these particles can be molecules, ions or associated molecules, or a mixture of all three.

2. Derivation of Raoult's law from Henry's law or the partition law. The generalised form of Henry's law states (p. 240) that the concentrations of any single molecular species in two phases in equilibrium bear a constant ratio to each other at a fixed temperature.

Consider a solution containing n mol of solute and N mol of solvent. The mole fraction of solvent will be $N/(n + N)$, and this is a measure of the concentration of solvent in the solution. The vapour pressure of the solution (p_{soln}) is proportional to the concentration of solvent molecules in the gaseous phase above the solution. Therefore,

$$\frac{p_{soln}}{N/(n + N)} = \text{a constant}$$

For the pure solvent, the mole fraction of solvent, i.e. the concentration of solvent, will be 1 or N/N. The vapour pressure of the solvent (p_{solv}) will be proportional to the concentration of solvent in the gaseous phase above the solvent. Therefore,

$$\frac{p_{solv}}{N/N} = \text{the same constant}$$

Therefore,

$$\frac{p_{soln}}{N/(n + N)} = \frac{p_{solv}}{N/N}$$

or

$$\frac{p_{solv}}{p_{soln}} = \frac{N + n}{N}$$

or

$$\frac{p_{solv} - p_{soln}}{p_{solv}} = \frac{n}{N + n}$$

which is the expression of Raoult's law.

3. Determination of relative molecular masses from vapour pressure measurements. The expression of Raoult's law

$$\frac{p_{solv} - p_{soln}}{p_{solv}} = \frac{n}{N + n} \backsimeq \frac{n}{N}$$

can be written as

$$\frac{p_{solv} - p_{soln}}{p_{solv}} = \frac{w/m_r}{w/m_r + W/M_r} \backsimeq \frac{w/m_r}{W/M_r}$$

where w is the number of grammes of solute dissolved in W gramme of solvent, and m_r and M_r are the relative molecular masses of the solute and solvent respectively.

If the relative lowering of the vapour pressure be measured for a known concentration of solution, and if the relative molecular mass of the solvent is known, then the relative molecular mass of the solute, m, can be calculated as this will be the only unknown in the above expression. The correct value for the relative molecular mass would only be obtained if there was no dissociation or association.

The method is of very little practical significance because experimental measurement of the depression of freezing point or the elevation of boiling point of a solution give the same result more easily and accurately.

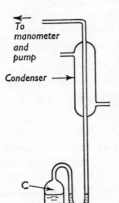

To manometer and pump

Condenser ⟶

C.

A   B

FIG. 103. An isoteniscope.

4. Measurement of vapour pressure.

a. *Barometric method.* Raoult measured vapour pressures by introducing a little of a liquid or a solution into the Torricellian vacuum of a barometer tube and measuring the depression of the mercury level. The method is not, however, very accurate or very convenient.

b. *The isoteniscope.* One of the best methods for measuring the vapour pressure of a liquid or a solution makes use of an isoteniscope (Fig. 103). Some liquid or solution is introduced into the small bulb. A, and the adjacent small U-tube, B. The liquid is boiled until all the air in the space, C, has been displaced. The bulb and U-tube are then placed in a thermostat. The pressure in the space, C, will be equal to the vapour pressure of the liquid or solution concerned, at the temperature of the thermostat.

The external pressure is then adjusted until the liquid levels in the two limbs of the U-tube are equal. When this is so, the external pressure, measured on a manometer, is equal to the vapour pressure in the space, C.

c. *Gas saturation method.* If a measured volume, V cm³, of dry air at a pressure P, and a temperature T, is slowly passed through a liquid at temperature T, so that the air becomes saturated with vapour, the loss in mass of the liquid will give the mass of vapour contained in the known volume of saturated air. The volume occupied by this mass of vapour (v cm³) at P and T can be calculated as it is known that 1 mol of the vapour would occupy 22400 cm³ at s.t.p. In the saturated air there is, then, v cm³ of vapour and V cm³ of air. The vapour pressure of the liquid concerned must, therefore, be equal to P multiplied by $v/(V + v)$.

5. Measurement of relative vapour pressure lowering. The relative difference between the vapour pressure of a solution and that of the pure solvent can be measured without, necessarily, measuring each vapour pressure separately. The method of Ostwald and Walker is the commonest (Fig. 104).

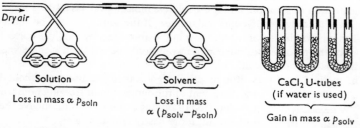

Fig. 104. Ostwald–Walker method of measuring the relative lowering of vapour pressure.

A slow stream of dry air is passed through a weighed series of bulbs containing the solution. The air absorbs a mass of vapour proportional to the vapour pressure of the solution, so that

$$\text{Loss in mass of 'solution' bulbs } (w_1) = kp_{\text{soln}}$$

The air is then passed through a similar series of bulbs, of known mass, containing pure solvent. The air takes up more vapour as the vapour pressure of the solvent is greater than that of the solution. The extra mass of vapour absorbed will be proportional to ($p_{\text{solv}} - p_{\text{soln}}$), so that

$$\text{Loss in mass of 'solvent' bulbs } (w_2) = k(p_{\text{solv}} - p_{\text{soln}})$$

It follows that

$$w_2 = kp_{\text{solv}} - w_1 \qquad \text{or} \qquad p_{\text{solv}} = \frac{w_2 + w_1}{k}$$

so that the relative lowering of vapour pressure is given by

$$\frac{p_{\text{solv}} - p_{\text{soln}}}{p_{\text{solv}}} = \frac{w_2}{w_2 + w_1}$$

If aqueous solutions are being used the air can be passed through weighed U-tubes containing anhydrous calcium chloride after passing through the aqueous solution and pure water. The increase in mass of the U-tubes will then give the value of $(w_2 + w_1)$ directly.

The temperature of the air, the solution and the solvent must be kept constant throughout, for vapour pressures change with temperature.

BOILING POINT OF SOLUTIONS

6. Relationship between boiling point and vapour pressure. The vapour pressure of a pure solvent rises with increasing temperature, as shown in Fig. 105. When the vapour pressure reaches the external pressure the

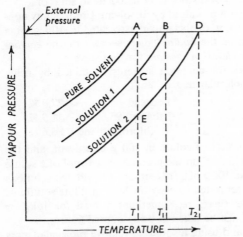

Fig. 105. Relationship between lowering of vapour pressure and elevation of boiling point.

solvent boils, so that the boiling point of the pure solvent is T. A solute dissolved in the solvent gives a solution, (1), with a lower vapour pressure, which will increase with rise of temperature as the vapour pressure of the pure solvent does. The boiling point of the solution will be T_1, which is higher than the boiling point, T, of the pure liquid. *Dissolving a non-volatile solid in a liquid causes an* **elevation** *of boiling point.*

Addition of more solute to form solution 2, will give a still further lowering of vapour pressure and raising of boiling point (T_2).

In dilute solutions the areas ABC and ADE are approximately similar triangles and

$$\frac{AB}{AD} = \frac{AC}{AE}$$

or $\dfrac{\text{Elevation of b.p. in 1}}{\text{Elevation of b.p. in 2}} = \dfrac{\text{Lowering of v.p. in 1}}{\text{Lowering of v.p. in 2}}$

The elevation of boiling point of a dilute solution is, therefore, *proportional to the lowering of the vapour pressure of the same solution.*

7. Determination of relative molecular masses from boiling point measurements. Because equimolecular quantities of any non-volatile solute dissolved in a given quantity of solvent produce the same lowering of vapour pressure and because lowering of vapour pressure and elevation of boiling point are proportional, it follows that equimolecular quantities of any non-volatile solute dissolved in a given quantity of solvent produce the same elevation of boiling point. As with vapour pressure lowering, this is only true provided the solute does not dissociate or associate in solution.

The elevation of boiling point produced by dissolving 1 mol of any non-volatile solute, which does not dissociate or associate, in 100 g of a particular solvent is known as the *boiling point constant* or the *ebullioscopic constant* or the *molar elevation constant* for that solvent.

Some experimental values for common solvents, obtained by using solutes of known relative molecular masses, are

Water	5.2 K	Benzene	27.0 K
Propan-2-one	17.0 K	Ethanol (95.6%)	11.5 K
Ethoxyethane	21.0 K	Trichloromethane	36.6 K

These constants are for 1 mol of solute in 100 g of solvent, and the use of such figures can be criticised on the grounds that 1 mol of a solute will not, generally, dissolve in 100 g of a solvent, or, if it will, the solutions obtained will be far too concentrated to obey the laws of dilute solutions. Boiling point constants are, therefore, sometimes quoted for 1 kg or 1 dm^3 of solvent, and care must be taken in using quoted figures.

Taking water as solvent, and using a solute of relative molecular mass M_r, it follows that

Mass of solute	Mass of solvent	Elevation of b.p.
1 mol	100 g	5.2 K
M_r g	100 g	5.2 K
M_r g	W g	$5.2 \times \dfrac{100}{W}$ K
x g	W g	$5.2 \times \dfrac{100}{W} \times \dfrac{x}{M_r}$ K

If T K is the measured elevation of boiling point for a solution of x gramme of a solute of relative molecular mass, M_r dissolved in W gramme of water,

$$T = 5.2 \times \frac{100}{W} \times \frac{x}{M_r}$$

or

$$M_r = \frac{5.2 \times 100 \times x}{T \times W}$$

For a solvent with a boiling point constant of B K

$$M_r = \frac{B \times 100 \times x}{T \times W}$$

By measuring T for a known concentration of solute in a solvent whose boiling point constant is known the relative molecular mass of the solute can be obtained.

It must be emphasised that correct values will only be obtained if the solute does not dissociate or associate, and if dilute solutions are used. Moreover, there must be no chemical reaction between the solute and solvent, and the solute used must be non-volatile.

8. Measurement of elevation of boiling point. The apparatus used for measuring boiling points or boiling point elevations, is simple, in principle, but complicated in detail because of the precautions which have to be taken to ensure steady and uniform heating and to avoid superheating and loss of solvent.

a. *Beckmann's method*. The apparatus is shown in Fig. 106. A is a glass tube which contains the liquid or solution under test. A side arm, B, is

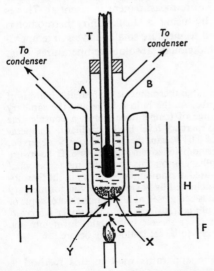

FIG. 106. Beckmann's apparatus for measuring boiling points.

connected to a condenser so that there will be no loss whilst the liquid or solution is boiling in A. The actual boiling point is measured on a thermometer, T. A contains some glass beads, or fragments of unglazed porcelain,

X, and has a short length of platinum wire, Y, sealed through it, to mini-mise superheating.

The liquid or solution in A is not heated directly. It is surrounded by a glass jacket, D, containing the same liquid as A and also fitted with a con-denser to avoid loss when the liquid is boiling. The whole is mounted on an asbestos tray, F, which has a small hole in the centre covered with fine wire gauze, G. Heat is provided by a small Bunsen burner flame below the wire gauze. The asbestos tray has two chimneys fitted, H, to conduct hot air away.

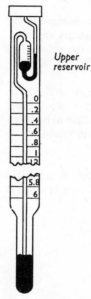

In use, a known mass of pure solvent is placed in A and its boiling point is measured. A pellet of solute, of known mass, is then added through the side arm, B, and the higher boiling point of the resulting solution is measured. Further pellets, of known mass, are added so that readings for solutions of higher concentrations may be obtained. Plotting the measured boiling point elevations against the total mass of solute added en-ables the different experimental readings to be averaged.

To obtain good results great care is required and accurate temperature measurement is essential. This is best achieved by using a Beckmann thermometer, which is designed to measure small changes in tempera-ture and not individual, absolute temperatures (Fig. 107).

FIG. 107. A Beck-mann thermo-meter.

It consists of a large thermometer bulb at the bottom of a fine capillary tube. As the bulb is large and the capillary fine, a small change of temperature causes a considerable movement of the mercury level in the capillary. Tempera-ture differences of 0.01 K can, therefore, easily be observed.

The whole scale of a Beckmann thermometer covers only about 6 K. To ensure that the level of mercury in the capillary is on this scale, no matter at what temperature the thermometer is being used, the amount of mercury in the thermo-meter bulb has to be variable. This is achieved by having a reservoir for mercury at the top of the thermometer. When it is being used at high temperatures some of the mercury from the thermometer bulb is transferred into the upper reservoir. At low temperatures, mercury from the reservoir is brought down into the thermometer bulb. The thermometer has to be set in this way before use.

b. *Landsberger–Walker method*. The apparatus used in this method is shown in Fig. 108. Pure solvent is first placed in the graduated glass tube, A, and vapour from the same solvent, boiling in a separate vessel, is passed in through the tube, B. The vapour eventually causes the solvent in A to boil, and also provides a thermal jacket around A, passing out of A through a small hole at the top.

When the solvent in A is boiling its boiling point is recorded on a thermometer, T, graduated in tenths of a degree. A pellet of solute, of known mass, is then added to the solvent in A, solvent vapour is again passed through the resulting solution until it boils, and the boiling point is recorded. At this point, the passage of solvent vapour is stopped so that the volume of solution in A can be recorded.

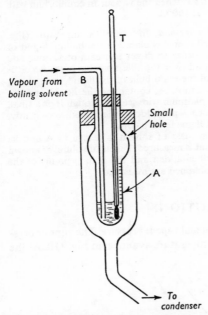

FIG. 108. Landsberger–Walker apparatus.

More vapour is passed into A until the solution is sufficiently diluted for a second boiling point and volume reading to be taken. Third and fourth sets of readings must also be taken, and the results averaged.

As it is the volume of solution which is measured in this experiment the readings must be converted into masses by multiplying by the density of the solvent at its boiling point. Alternatively, boiling point constants expressed in terms of 100 cm³ of solvent and not 100 g can be used. In both cases the density of the dilute solution is taken as being equal to the density of the pure solvent.

The method is quicker and simpler, but less accurate than Beckmann's method. Super-heating is avoided by using solvent vapour as the source of heat.

It is not readily apparent why it is possible to boil a solution by passing in the vapour of a solvent with a lower boiling point than the solution. Consider an aqueous solution with a boiling point, say, of 102°C, which is at 100°C. The water vapour above this solution will be at 98°C, say. If steam, at 100°C, is passed into the solution, it cannot pass through the solution, for, if it did, it would have to be in equilibrium with water vapour at 98°C. The steam, therefore, condenses and gives up its latent heat to the solution at 100°C. In theory the temperature of the solution is raised to its boiling point of 102°C when the vapour in equilibrium with it is steam at 100°C.

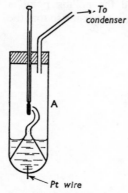

c. Cottrell's method. In Cottrell's apparatus (Fig. 109) a small funnel is placed in the boiling liquid or solution in order to direct a stream of vapour and boiling liquid over the thermometer bulb, which is itself out of the main bulk of the liquid.

The glass tube, A, containing the liquid or solution has a platinum wire sealed through it or a small piece of porous pot added. It is heated, very gently, by a small flame from below.

A known mass of solvent is placed in A and its boiling point is measured. Pellets of solute, of known mass, are then added and the boiling points of the solutions obtained are measured.

FIG. 109. Cottrell's apparatus.

FREEZING POINT OF SOLUTIONS

9. Relationship between freezing point and vapour pressure. The vapour pressure of a pure liquid changes with temperature as shown in Fig. 110. At the

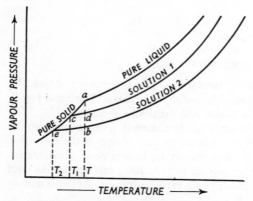

FIG. 110. Relationship between lowering of vapour pressure and lowering of freezing point.

freezing point of the liquid, T, there is a sharp change in the curve, the vapour pressure–temperature curve below T being that for a solid. If a non-volatile solute is dissolved in the liquid, the vapour pressure will be lowered, and the vapour pressure–temperature curve will be as shown in Fig. 110. The freezing point of the solution 1 will be T_1. *Dissolving a non-volatile solute in a liquid causes a* **lowering** *of freezing point.*

Addition of more solute to form solution 2 causes a further lowering of vapour pressure, and the freezing point falls to T_2.

For dilute solutions the areas abe and adc are approximately similar triangles so that, as for the elevation of boiling point (p. 268), *the lowering of freezing point of a dilute solution is proportional to the lowering of the vapour pressure of the same solution.*

Notice that this will only apply to a solution which, on freezing, deposits pure solid solvent.

10. Determination of relative molecular mass by freezing point measurements. Lowering of vapour pressure, elevation of boiling point, and lowering of freezing point are all proportional to each other, and all depend on the total number of solute particles in a solution. Thus 1 mol of any non-volatile solute, which does not dissociate or associate, dissolved in 100 g of any particular solid will cause the same lowering of vapour pressure, the same elevation of boiling point, and the same lowering of freezing point.

The freezing point lowering caused by 1 mol of such a solute in 100 g of solvent is called the *freezing point constant,* the *cryoscopic constant* or the *molar depression constant* for the solvent. Some experimental values are

Water	18.6 K	Benzene	50.0 K
Ethanoic acid	39.0 K	Camphor	400 K

As for boiling constants, figures may be quoted in terms of 1 kg or 1 dm³ of solvent instead of 100 g, and care must be taken to use the appropriate figures.

Freezing point measurements can be used in just the same way as boiling point measurements to determine relative molecular masses. Using a solvent with a cryoscopic constant of F K, 1 mol of any solute dissolved in 100 g of the solvent will give a freezing point depression of F K. x gramme of solute, of relative molecular mass M_r, dissolved in W gramme of solvent will give a depression of

$$F \times \frac{100}{W} \times \frac{x}{M_r} \text{ K}$$

If, therefore, the measured freezing point depression for such a solution is T K,

$$T = F \times \frac{100}{W} \times \frac{x}{M_r}$$

or

$$M_r = \frac{F \times 100 \times x}{T \times W}$$

Correct values for relative molecular mass will only be obtained if (*a*) the solute used is non-volatile, (*b*) the solute does not dissociate or associate, (*c*) dilute solutions are used, (*d*) there is no chemical reaction between solute and solvent, and (*e*) the solution, on freezing, deposits pure, solid solvent.

11. Measurement of lowering of freezing point. The lowering of freezing point can be measured more accurately, and with less trouble, than the elevation of boiling point. The main difficulty is in avoiding super-cooling.

a. *Beckmann's method* (Fig. 111). An inner glass tube, A, with a side arm, B, is used to contain the liquid or solution. The tube is fitted with a thermometer, which, for accurate work, must be a Beckmann thermometer (p. 270), and a stirrer passing through a short glass collar in the stopper. A fits into an outer tube, C, which provides an air jacket. The whole can be immersed in a freezing mixture which also contains a stirrer.

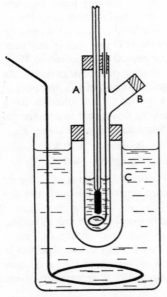

A known mass of solvent is placed in A, and, to save time, the solvent is first just solidified by immersing A directly in the freezing mixture. A is then warmed by holding in the hand whilst the solvent is well stirred. When all the solvent has melted, A is placed in the outer tube, C, and the whole is placed in the freezing mixture. The solvent is stirred steadily, and the freezing mixture is also stirred. Slow, steady cooling of the solvent is necessary.

The temperature recorded on the thermometer drops steadily and, because of some supercooling, which cannot be avoided, it reaches a level below that of the true freezing point of the solvent. As soon as crystallisation occurs, however,

Fig. 111. Beckmann's apparatus for measuring freezing points.

the temperature rises a little to the correct freezing point level and remains steady for a while. This steady temperature is recorded. In case too much super-cooling has taken place three readings must be taken.

A pellet of solute, of known mass, is then added to the solvent in A through the side arm, B, and is allowed to dissolve. The freezing point of the resulting solution is measured in the same way as for the pure solvent. Further pellets, of known mass, are also added to give readings for solutions of higher concentrations.

b. *Richards' method.* This method is simple, but lengthy, for aqueous solutions which can easily be analysed. It is difficult to carry out for non-aqueous solutions.

The solution concerned is mixed with ice in a vacuum flask. After some hours, equilibrium is established between the solution and the ice, at the freezing point of the solution. At this stage, the temperature is recorded and a sample of the solution is withdrawn for analysis. The freezing point of a solution of known concentration is, therefore, found.

c. *Rast's method.* This method makes use of the very high (400°C) freezing point constant of camphor. It can, however, only be used for solutes which dissolve in molten camphor. The resulting solutions have freezing points so much lower than that of pure camphor that the difference can be measured quite accurately using an ordinary thermometer and the ordinary method of measuring melting points using a melting point tube immersed in a suitable heating bath.

The melting point of camphor is first measured. The freezing point constant for that particular sample of camphor is then found by measuring the melting point of a mixture of a known mass of the camphor with a known mass of a solute of known relative molecular mass. The mixture is made by melting the two solids together and powdering the mass remaining on cooling.

Once the freezing point constant has been measured for the camphor sample it can be mixed with a solute of unknown relative molecular mass, in measured masses, and the melting point of the solution can be measured.

12. Boiling or freezing point constants and specific latent heat. Because of the relationship between evaporation and vapour pressure (p. 79), it is not surprising that there is also a relationship between the specific latent heat of a liquid and its boiling point or freezing point constants.

The boiling point constant is related to the specific latent heat of vaporisation; the freezing point constant to the specific latent heat of fusion. The relationship, in both cases is

$$K = \frac{0.0832T^2}{L}$$

where K is the boiling point or freezing point constant, T the absolute temperature, and L the specific latent heat of evaporation or fusion. The number 0.0832 is derived from R (the gas constant) divided by 100 (p. 37).

For water, with $T = 273$ K and specific latent heat of fusion $= 334.8$ J g^{-1}, the calculated freezing point constant is 18.6 K.

DISSOCIATION AND ASSOCIATION

13. Introduction. The colligative properties of solutions, i.e. the lowering of vapour pressure, the elevation of boiling point, the depression of freezing point and the osmotic pressure, all depend on the total number of solute particles present in a solution.

A solute which dissociates in solution by splitting up, either completely or partially, into ions will provide *more* particles than would otherwise be present and it will cause an increased effect. One which associates, by the linking together, either completely or partially, of solute molecules, will provide *fewer* particles and a decreased effect.

A relative molecular mass calculated from a measured effect which, on account of ionisation, is too high, will be too low. A relative molecular mass calculated from a lowered effect, due to association, will be too high.

Such anomalous results were first noticed by van't Hoff who introduced an empirical factor, i, to right matters. Such a factor, used before the cause of the anomalous results was understood, is not, now, of much significance.

14. Degree of ionisation. If a molecule, AB, ionises completely into A^+ and B^- ions, one molecule will produce two ions,

$$AB \longrightarrow A^+ + B^-$$

	AB	$A^+ + B^-$
Before ionisation	1 particle	
After complete ionisation	0 particles	2 particles

A solution made from AB molecules would, therefore, produce twice the effect than an equal amount of solute which did not dissociate. 1 mol of sodium chloride (NaCl), for example, dissolved in 100 g of water would lower the freezing point by 18.6°C if sodium chloride did not dissociate. If it is assumed that sodium chloride dissociates completely into Na^+ and Cl^- ions the observed depression of freezing point will be 37.2°C.

A molecule, e.g. AB_2, ionising completely into A^{2+} and two B^- ions will provide three ions, giving three times the effect, and so on.

When the ionisation is not complete but only partial, it is measured by what is known as the degree of ionisation. This can be expressed as a percentage or as a fraction. Thus, a degree of ionisation of 50 per cent means that 50 per cent, or half, of all the molecules originally present split up into ions. If 1 mol of solute was originally present, 0.5 mol will form ions and 0.5 mol will not.

For 1 mol of a solute, AB, giving A^+ and B^- ions, the situation, if the degree of ionisation is α, will be as follows:

	AB	$\longrightarrow$	A^+	+	B^-
Original amount before ionisation	1 mol		0		0
Amount after ionisation to a degree, α	$(1 - \alpha)$ mol		α mol		α mol

The number of particles present is proportional to the number of moles, so that ionisation causes a proportional change in the number of particles from 1 to $(1 + \alpha)$. The ratio of the expected effect, i.e. that resulting if there was no ionisation, to the actual effect caused by ionisation will be $1:(1 + \alpha)$;

$$\frac{\text{Expected effect if no ionisation}}{\text{Actual effect caused by ionisation}} = \frac{1}{1 + \alpha}$$

For a solute, AB_2, giving A^{2+} and two B^- ions, the situation will be:

	AB_2	$\longrightarrow$	A^{2+}	+	$2B^-$
Original amount before ionisation	1 mol		0		0
Amount after ionisation to a degree, α	$(1 - \alpha)$ mol		α mol		2α mol

In such a case,

$$\frac{\text{Expected effect}}{\text{Actual effect}} = \frac{1}{1 + 2\alpha}$$

Such relationships apply to any of the four colligative effects—the lowering of vapour pressure, the elevation of boiling point, the depression of freezing point, and osmotic pressure. If any one of these four effects is measured, and a relative molecular mass calculated from the measurements, then

$$\frac{\text{Real relative molecular mass}}{\text{Apparent relative molecular mass}} = \frac{1 + \alpha}{1} \text{ (for AB)}$$

$$\text{or} = \frac{1 + 2\alpha}{1} \text{ (for AB}_2\text{)}$$

Alternatively, the degree of ionisation of a solute whose real relative molecular mass is known can be obtained by measuring its apparent relative molecular mass.

QUESTIONS ON CHAPTER 25

1. The lowering of vapour pressure of a solution of 108.2 g of a substance X in 1 kg of water at 20°C is 24.79 N m^{-2}. The vapour pressure of water at 20°C is 2.338 kN m^{-2}. Calculate the relative molecular mass of X.

2. Explain why a liquid at a given temperature has a definite vapour pressure. How would you attempt to measure the vapour pressure of benzene at 60°C? (O. Schol.)

3. How would you determine the vapour pressure of water at various temperatures between 20°C and 100°C. Explain what use can be made of measurements of vapour pressures of pure liquids, solutions, and mixtures. (O. Schol.)

4. What is Raoult's law? An acidic substance was made up into two solutions, A and B, A containing 15.0 g of the substance in 100 g of water, and B containing 6.9 g of the substance in 100 g of benzene. These solutions, A and B, had vapour pressures of 99.13 and 99.02 kN m^{-2} respectively at the boiling points of the pure solvents at 101.325 kN m^{-2} pressure. Calculate the apparent relative molecular mass of the substance in each case. Suggest a reason or reasons why the results differ in the two solvents, and outline briefly an experiment to confirm your suggestions. (C. Schol.)

5. 10 g of a non-volatile substance, X, is dissolved in 100 g of benzene. The loss in mass of the solution on bubbling a slow stream of air through is 1.205 g. The same volume of air passed through pure benzene under the same conditions caused a loss of 1.273 g. What is the probable relative molecular mass of X?

6. Air was drawn very slowly over a solution of 38.89 g of a sugar in 100 g of water, then over distilled water at the same temperature and then brought through tubes of anhydrous calcium chloride of known mass. The loss in mass of the distilled water was 0.0921 g and the gain in mass of the calcium chloride tubes was 5.163 g. Calculate the relative molecular mass of the sugar.

7. Show, with the aid of a vapour pressure–temperature diagram, how the changes of vapour pressure, boiling point, and freezing point caused by dissolving a non-volatile solute in water are interrelated, and state how these properties of a solution vary with concentration.

Two solutions containing respectively 4.45 g of anthracene (molecular weight = 178) and 6.42 g of sulphur in 1 kg of carbon disulphide have the same vapour pressure at the same temperature. What is (a) the molecular formula of the dissolved sulphur, (b) the relative lowering of the vapour pressure? (W.)

8. The boiling point of ethanol is 78°C and its molar elevation constant per 100 g is 11.5 K. A solution of 0.56 g of camphor in 16 g of ethanol had a boiling point of 78.278°C. Calculate the relative molecular mass of camphor.

9. You are given a clinical thermometer graduated at intervals of $\frac{1}{5}$th from 94°F to 110°F. Devise a simple small-scale apparatus, using the thermometer and the ordinary materials of a laboratory, for determination of the relative molecular mass of a substance soluble in ethoxyethane. Ethoxyethane boils at 95°F and has a molecular elevation (K) of 2.11°C for 1 kg (or 3°C per dm³). What mass of phenylamine (C_6H_7N) when dissolved in 20 cm³ of ethoxyethane would raise the boiling point by 6°F? (Army and Navy.)

10. A solution of 0.06 g of a non-electrolyte AB, of high boiling point, in 100 g of water had a freezing point 0.006 K below that of pure water. A 0.1 g sample of AB, vaporised at atmospheric pressure, occupied a volume of 36.52 cm³ at 200°C and 45.74 cm³ at 240°C. What can you deduce from these data? (C. Schol.)

11. If a volatile solute is added to a solvent what effects may be noticed in the total vapour pressures of the mixture? What difference would there be if the solute were non-volatile? Explain the importance of these observations either on the separation of two volatile liquids by distillation or on the determination of relative molecular mass in solution. (O. & C. S.)

12. (a) Draw a diagram to show the effect on the vapour pressure of a volatile liquid of adding successive quantities of a non-volatile solute. (b) Draw diagrams to show the possible variations in vapour pressure at, say, 15°C when two volatile liquids (completely miscible) are mixed in varying proportions. Why is it not possible to tell from these diagrams whether the two liquids will form a constant boiling mixture or not? (c) Draw the temperature–composition diagram for two volatile liquids showing a constant boiling maximum. What is the pressure represented on this diagram? (O. & C. S.)

13. Solution of 0.142 g of naphthalene in 20.25 g of benzene caused a lowering of freezing point of 0.284 K. The molar depression constant per 100 g of benzene is 51.2 K. Calculate the relative molecular mass of naphthalene.

14. (a) The melting point of camphor is 177.5°C, whilst that of a mixture of 1 g of naphthalene (M.W. = 128) and 10 g of camphor is 147°C. What is the cryoscopic constant for camphor? The melting point of a mixture of 1 g of N-phenylethanamide (acetanilide) and 10 g of camphor is 148.5°C. What is the relative molecular mass of N-phenylethanamide? (b) What do you think are the advantages and disadvantages of Rast's method for molecular mass measurement?

15. If you had to determine the relative molecular mass of an organic non-electrolyte, what considerations would influence your choice between the boiling and freezing point methods?

16. Describe an experimental method for finding the relative molecular mass of cane sugar by studying its effect on the freezing point of water.

Two solutions, one containing 10 g of potassium chloride per kg water and the other 10 g of cane sugar, $C_{12}H_{22}O_{11}$, per kg water, are allowed to cool in the same freezing bath. (a) From which solution, and at what temperature, will ice separate; and (b) how much ice will have separated from that solution by the time it first appears in the other? Super-cooling should be neglected in this calculation and the salt should be regarded as completely dissociated. The freezing point constant for 1 kg water is 1.86 K mol⁻¹.

17. The freezing point of a solution of 20 g of mercury(II) chloride in 1 kg of water was $-0.138°C$. Would you expect this solution to conduct an electric current? Give reasons for your answer.

18. What advantages and disadvantages has ethane-1, 2-diol (ethylene glycol) over methanol as an anti-freeze for use in car radiators?

19. A car radiator contains 12 dm³ of water. Addition of 5 kg of ethane–1,2-diol (ethylene glycol) lowers the freezing point by 12 K. What mass of (a) propane–1,2,3–triol or glycerol, and (b) methanol would be required to give the same protection against freezing?

Chapter 26
The osmotic pressure of solutions

1. Diffusion. Gases diffuse quite rapidly (p. 38), but liquids and solutions, and even solids, will also diffuse, if more slowly.

Two miscible liquids, for instance, diffuse throughout the whole mixture until it is homogeneous. A solution will also diffuse into another solution, or into pure solvent, so long as the pairs of substances used are miscible.

It is possible, with care, to pour a saturated aqueous solution of copper(II) sulphate(VI) through a tube into a vessel containing water so that the blue saturated solution, with a higher density than water, forms a well-defined lower layer. On standing, the two layers slowly merge, until a homogeneous mixture of uniform concentration is obtained. Water diffuses into

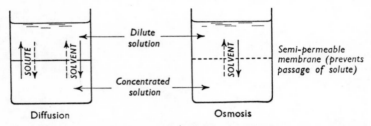

Diffusion   Osmosis

FIG. 112. Diffusion and osmosis. The bold arrows indicate quicker rates of diffusion than the dotted arrows. In diffusion, there is an overall passage of solvent into the concentrated solution, and of solute into the dilute solution. In osmosis, there is an overall passage of solvent into the concentrated solution. Passage of solute is prevented by the semi-permeable membrane.

the copper(II) sulphate(VI) solution, and copper(II) sulphate(VI) diffuses into the water layer.

It is not easy to make accurate measurements on such a system, but it can be shown that the rate of diffusion, measured as the mass of solute passing in unit time from solution to solvent, increases rapidly with rise of temperature and is directly proportional to the concentration of the solution used. Such results would be expected from the kinetic theory.

Similar diffusion will take place if two solutions of unequal concentration are in contact. Solvent will pass from the dilute to the concentrated solution, and solute will pass from the concentrated to the dilute, until equality of concentration is achieved. This is illustrated in Fig. 112.

2. Osmosis. If solutions of unequal concentration, or a solution and pure solvent, could be separated by a membrane which allowed solvent molecules to pass but was impervious to solute molecules, only solvent molecules could diffuse from the solvent or dilute solution into the more concentrated solution. Such membranes, which allow solvent to pass, but not solute, are called *semi-permeable membranes. Osmosis is the name given to the diffusion of solvent through a semi-permeable membrane from* *a solvent into a solution or from a dilute solution into a more concentrated one.* Osmosis is illustrated in Fig. 112.

It must be understood that the passage of solvent through a semi-permeable membrane is not simply one-way, but the diffusion from solvent to solution or from dilute to concentrated solution is more rapid than in the opposite direction. The resultant direction of the flow of solvent is from low to high concentration.

3. Examples of osmosis. Many natural membranes act as semi-permeable membranes, and various substances, especially copper(II) hexacyano-ferrate(II), $Cu_2Fe(CN)_6$, parchment paper and cellophane are also semi-permeable, if not ideally so. The following examples illustrate the action of semi-permeable membranes in osmosis.

a. If an animal bladder, e.g. a pig's bladder, is filled with ethanol and tied at the neck, it will swell when immersed in water, and finally burst. This is because water diffuses through the animal bladder much more rapidly than ethanol diffuses out. On filling the bladder with water and immersing in ethanol, the bladder will shrink. These observations were first made in 1784 by the Abbé Nollet.

b. A dry prune will swell when placed in water because water will pass through its semi-permeable skin. If the swollen prune is then placed in a concentrated solution of sugar it will shrink again, because osmosis now takes place from the weaker solution inside the prune to the more concentrated one outside.

c. The hard, outer shell of an egg can be removed by dissolving it in hydrochloric acid. If two eggs are treated in this way, one placed in water will swell, whilst the other placed in strong brine will shrink. The membrane beneath the outer shell of the eggs acts as a semi-permeable membrane. Slices of beetroot or carrot, or a segment of an orange behave similarly.

d. Crystals of copper(II) sulphate will form weird 'growths' when placed in a solution of potassium hexacyanoferrate(II). This is because the crystals become coated with a semi-permeable layer of copper(II) hexacyano-ferrate(II). Water then passes through this layer, stretching it until it bursts. As soon as it does burst, more copper(II) hexocyanoferrate(II) layer is formed, and so on.

e. Weird 'growths', as in d. are also obtained if crystals of many salts, e.g. iron(II) sulphate, nickel(II) chloride, cobalt(II) nitrate and iron(III) chloride, are placed in a solution of water glass (sodium silicate). The layers of metallic silicates formed by double decomposition reactions are semi-permeable. The growths resulting form what is commonly called a chemical garden.

f. If a beaker containing a concentrated solution of sugar in water and one containing pure water are placed under a bell-jar, water will pass, through the vapour phase, from the pure water to the concentrated solution. Air, in this instance, acts as the semi-permeable membrane.

g. A layer of dilute calcium nitrate(v) solution can be separated from a layer of concentrated calcium nitrate(v) solution by a layer of benzenol

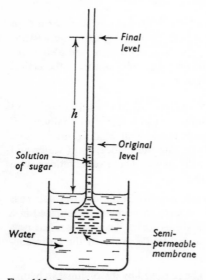

FIG. 113. Osmotic pressure. Osmosis takes place from the water into the sugar solution until the hydrostatic pressure due to the column of sugar solution becomes equal to the osmotic pressure of the solution, when osmosis ceases.

(phenol) solution. The benzenol solution acts as a semi-permeable membrane so that water passes from the dilute to the concentrated calcium nitrate(v) solution, and the passage can be seen by the rising of the benzenol layer (Fig. 114).

h. A thistle funnel can be sealed by cellophane so that it will hold a solution of sugar. If such a funnel is immersed in water (Fig. 113), water will pass through the semi-permeable cellophane so that the level of liquid in the thistle funnel will rise.

i. More striking results than in h. can be obtained by using a porous pot fitted with a long tube, instead of a thistle funnel. The porous pot must be made into a semi-permeable membrane by first filling it with potassium hexocyanoferrate(II) solution and standing it in copper(II) sulphate(VI) solution; the two solutions react together to form copper(II) hexacyanoferrate(II) in the pores of the pot. Alternatively the pot can be treated with a warm solution containing gelatine and propane-1,2,3-triol (glycerine).

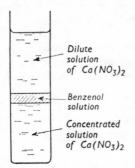

On standing, a gel (p. 508) forms in the pores and provides a very good semi-permeable membrane.

Once the porous pot is prepared, it is fitted with a stopper and a long tube and filled with a very concentrated solution of sugar, conveniently made by mixing equal volumes of water and golden syrup. On standing the pot in water the level of liquid in the long tube rises quite remarkably.

FIG. 114. Osmosis from dilute to concentrated calcium nitrate(v) solution causes the benzenol layer to rise.

4. Osmotic pressure. The apparatus described in g. in the preceding section, and illustrated in Fig. 113, can be used to illustrate the meaning of the term osmotic pressure, and to measure actual osmotic pressures approximately.

Water will pass through the semi-permeable membrane into the sugar solution so that the liquid level in the thistle funnel will rise. It will not, however, go on rising indefinitely. As the liquid level does rise, a hydrostatic pressure is built up, and when this reaches a certain value osmosis will cease. The value of this pressure is known as the osmotic pressure of the solution in the thistle funnel. It will be equal to the height of liquid in the thistle funnel multiplied by the density of the solution in the funnel.

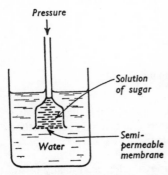

When the liquid in the thistle funnel has reached its final level osmosis stops, but not diffusion. Solvent will still be diffusing through the semi-permeable membrane but equally in both directions.

Osmosis from a solvent into a solution can be prevented not only by allowing a hydrostatic pressure to build up but also by applying an external pressure large enough to prevent osmosis. The minimum external pressure which will do this is known as the osmotic pressure of the solution concerned. Thus, referring to Fig. 115 the external pressure which would have to be applied down the thistle funnel to maintain the initial level of liquid steady would be the osmotic pressure of the solution in the thistle funnel.

FIG. 115. Osmotic pressure. The minimum pressure required to prevent osmosis into the sugar solution is equal to the osmotic pressure of the solution.

The term osmotic pressure is misleading if it is not understood, for a solution, by itself, cannot exert any pressure due to osmosis. *The osmotic pressure of a solution is the pressure required to prevent osmosis when the solution is separated from pure solvent by a semi-permeable membrane.*

5. Experimental measurement of osmotic pressure. There are two main difficulties involved in measuring the osmotic pressure of a solution. First, it is not easy to make a perfect semi-permeable membrane, which will with-

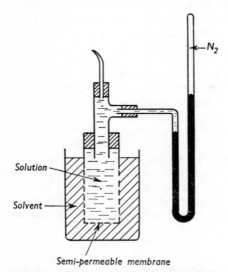

Fig. 116. Pfeffer's apparatus for measuring osmotic pressure. Morse and Fraser used the same principle.

stand the pressures that may be involved. Secondly, the high pressures which can be set up may not be easy to measure and can only be withstood by carefully designed joints in the apparatus used.

a. *Method of Pfeffer and of Morse and Frazer.* Pfeffer made many of the earliest measurements of osmotic pressure. He made semi-permeable membranes by allowing solutions of copper(II) sulphate(VI) and potassium hexacyanoferrate(II) to diffuse into the pores of a porous pot from opposite sides. Within the pores they reacted to form copper(II) hexacyanoferrate(II).

The pot was then fitted with a closed manometer containing mercury and nitrogen. The pot, and the space up to the mercury, was filled with the solution whose osmotic pressure was required. This filling was done through a tube which was finally sealed off. The pot was then placed in pure solvent maintained at a constant temperature. The solvent passed through the semi-permeable membrane into the solution and the highest pressure developed was recorded on the manometer. A diagrammatic representation of the apparatus is given in Fig. 116.

Morse and Frazer used essentially the same method, but deposited the copper(II) hexacyanoferrate(II) in the pores of a porous pot electrolytically. A solution of copper(II) sulphate(VI), containing a platinum anode, was placed inside the pot, whilst the pot was standing in a solution of potassium

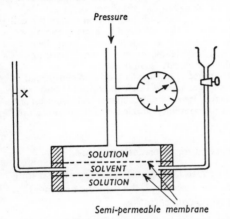

FIG. 117. Berkeley and Hartley's apparatus for measuring osmotic pressure.

hexacyanoferrate(II) containing a platinum cathode. Morse and Frazer were able to measure osmotic pressures up to about 30×10^3 kN m^{-2} (300 atmospheres). To measure such pressures they did not use a manometer but used, for example, measurements on the change of the refractive index of water on compression.

b. *Berkeley and Hartley's method.* In this method, the pressure necessary to prevent osmosis is measured. The apparatus used is shown diagrammatically in Fig. 117, but the detail of the complex jointing systems employed is omitted.

The pressure which had to be applied to the solution to maintain a steady liquid level at X was measured on a pressure gauge. The semi-permeable membrane was made by depositing copper(II) hexacyanoferrate(II) in the pores of a porous cylinder.

c. *Comparison of osmotic pressures.* Direct measurement of osmotic pressure by the methods of a. or b. is difficult, but it is easy to get a rough comparison of the osmotic pressure of two solutions by a method originated by de Vries.

If a plant or animal cell is placed in a solution it will shrink (undergo plasmolysis) or swell, unless the osmotic pressure of the external solution is equal to that of the solution inside the cell. The shrinking or swelling can be observed under a microscope.

De Vries placed the same plant cells, e.g. the epidermal cells of the leaves of tradescantia discolor, in two different solutions and was able to compare the osmotic pressures of the two solutions by measuring how much they had to be diluted to give osmotic pressures equal to that of the solution within the cells, i.e. to give conditions under which the plant cell did not shrink or swell.

Under such conditions, the external solution and the solution within the cell are said to be isotonic. If the membrane of the cell acted as a perfect semi-permeable membrane the two solutions would also be isosmotic, i.e. have the same osmotic pressures, but imperfect semi-permeable action of plant membranes means that isotonic solutions are not necessarily isosmotic.

Hamburger has used red blood corpuscles in the same way. When placed in solutions with lower osmotic pressures than the cell contents, the cells swell and burst, and the solution becomes red. In solutions of higher osmotic pressure, the cells shrink, fall to the bottom of the container, and leave the solution colourless.

6. Effect of concentration and temperature on osmotic pressure.

a. *Concentration*. At a given temperature, in dilute solution, and so long as the solute is neither dissociated nor associated, the osmotic pressure of a solution is directly proportional to its concentration,

$$\text{Osmotic pressure } (\Pi) \propto \text{Concentration of solution}$$

b. *Temperature*. Under the same conditions as in a., but for a given concentration, the osmotic pressure of a solution is proportional to the absolute temperature,

$$\text{Osmotic pressure } (\Pi) \propto \text{Absolute temperature}$$

c. *Relation to gas laws*. van't Hoff first drew attention, in 1877, to the fact that concentration and temperature had the same effect on the pressure of a gas as on the osmotic pressure of a solution.

The relationship

$$\text{Osmotic pressure } (\Pi) \propto \text{Concentration}$$

can be written as

$$\text{Osmotic pressure } (\Pi) \propto \frac{1}{\text{Volume of solution}} \text{ (for a given mass of solute)}$$

This is analogous to the statement of Boyle's law,

$$\text{Gas pressure} \propto \frac{1}{\text{Volume of gas}} \text{ (for a given mass of gas)}$$

The relationship,

$$\text{Osmotic pressure } (\Pi) \propto \text{Absolute temperature}$$

for a given concentration of solution, is analogous to Charles's law,

$$\text{Gas pressure} \propto \text{Absolute temperature}$$

for a given volume of gas.

Combining the effects of volume and temperature change on the pressure of a gas leads to an expression (p. 37),

$$pV = kT$$

where p is the pressure, V the volume, and T the absolute temperature of a given mass of gas; k is a constant. If 1 mol of gas (22.4 dm^3 at s.t.p.) is considered k is written as R and the gas equation results:

$$pV = RT$$

As osmotic pressure varies with volume of solution and absolute temperature in the same way a similar equation must hold. For a given mass of solute, therefore,

$$\Pi V = kT$$

where V is the volume of solution, T the absolute temperature, and k a constant. If 1 mol is considered the relationship becomes

$$\Pi V = RT$$

and the numerical value of R in this expression is the same as that in the gas equation.

The experimental value for the osmotic pressure of a 1 per cent solution of cane sugar (10 g dm^{-3}) is 66.37 kN m^{-2} at 273 K. The volume containing 1 mol of sucrose (342 g) is 34.2 dm^3, so that

$$(66.37 \times 10^3)(34.2 \times 10^{-3}) = R \times 273$$

This gives a value of R equal to 8.314 J K^{-1} mol^{-1}, which compares with the gas constant of 8.3143 (p. 37).

The same laws, then, govern osmotic and gas pressure, and there is a remarkable similarity between the behaviour of molecules distributed throughout a gas and solute molecules distributed throughout a solution. The osmotic pressure of a dilute solution is, in fact, equal to the pressure which the solute would exert if it existed as a gas at the same temperature occupying a volume equal to that of the solution.

As 1 mol of a gas occupies 22.4 dm^3 at 0°C and exerts a pressure of 101.325 kN m^{-2} it follows that 1 *mol of a solute dissolved in 22.4 dm³ of solution at 0°C will exert an osmotic pressure of* 101.325 kN m^{-2}. It does not matter what the solute is so long as it is in dilute solution and is not associated or dissociated. Like vapour pressure, freezing point depression and boiling point elevation, the osmotic pressure of a solution is a colligative property depending simply on the total number of solute particles dissolved in a solution.

7. Determination of relative molecular masses from osmotic pressure measurement. The fundamental statement that 1 mol of any solute in 22.4 dm³ of solution has an osmotic pressure of 101.325 kN m⁻² at 0°C can be written, and modified by unitary method, as follows:

Mass of solute/g	Volume of solution/dm³	Temperature/K	Osmotic pressure/kN m⁻²
1 mol	22.4	273	101.325
M_r	22.4	T	$101.325 \times \dfrac{T}{273}$
M_r	V	T	$101.325 \times \dfrac{T}{273} \times \dfrac{22\cdot4}{V}$
x	V	T	$101.325 \times \dfrac{T}{273} \times \dfrac{22\cdot4}{V} \times \dfrac{x}{M_r}$

A solution of x gramme of a solute of relative molecular mass, M_r, dissolved in V dm³ of solution at T K, has, therefore, an osmotic pressure (in kN m⁻²), Π given by

$$\Pi = \frac{T}{273} \times \frac{22.4}{V} \times \frac{x}{M_r} \times 101.325$$

Alternatively,

$$M_r = \frac{T \times 22.4 \times x \times 101.325}{\Pi \times 273 \times V}$$

By measuring the relevant quantities a value of M_r can be obtained, though the measurement of osmotic pressure is too difficult for the method to be widely used. It has an application, however, in the measurement of very high relative molecular masses of substances such as proteins. A 1 per cent solution of a protein of relative molecular mass 50000 would have an osmotic pressure of about 500 N m⁻² which can be measured. The freezing point depression of the same aqueous solution would be only 0.00037 K, which is too small for accurate measurement.

8. Relationship between vapour pressure and osmotic pressure The proportionality between the lowering of vapour pressure of a solution and the corresponding boiling point elevation and freezing point depression has already been shown (pp. 267 and 272). Osmotic pressure is another colligative property of solutions similarly proportional to lowering of vapour pressure. This proportionality is shown by the following argument.

Consider the arrangement as shown in Fig. 118. A tube, closed at its lower end by a semi-permeable membrane and containing a dilute solution, is immersed in pure solvent, the whole arrangement being enclosed under a bell-jar which was, originally, evacuated, and which, therefore, contains only solvent vapour.

When osmosis is complete, the osmotic pressure, Π, of the dilute solution will be given by,

$$\Pi = h \times \text{density of solution in tube}$$

The pressure at points on a level with A will be that of the vapour pressure of the solvent, p_{solv}, whilst that at points on a level with B will be that of the vapour pressure of the solution, p_{soln}. The pressure at A will be greater than that at B because of the mass of a column of solvent vapour of height h. The lowering of vapour pressure is, therefore, given by

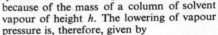

$$p_{\text{solv}} - p_{\text{soln}} = h \times \text{density of solvent vapour}$$

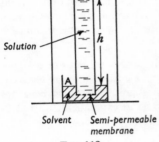

Solution

Solvent   Semi-permeable
               membrane

FIG. 118.

If the density of the solution in the tube and the density of the solvent vapour between A and B did not change as h was changed it would be clear that the osmotic pressure was proportional to the lowering of the vapour pressure $(p_{\text{solv}} - p_{\text{soln}})$.

Strictly speaking, both the density of the solution in the tube and the density of the vapour do change as h does. If h increases, the solution in the tube gets more dilute and the vapour in the bell-jar gets rarified just as the air does on ascending into the atmosphere. For small values of h, however, i.e. for dilute solutions, both these changes will be very slight, and it follows that, for dilute solutions, the osmotic pressure is proportional to the lowering of vapour pressure.

9. Effect of ionisation and association on osmotic pressure. If a solute ionises in solution, the measured osmotic pressure of the solution will be greater than if there was no ionisation. Association of the solute will give lowered osmotic pressure readings.

The effect of ionisation or association on osmotic pressure is just the same as it is on the lowering of the vapour pressure, the elevation of the boiling point or the lowering of the freezing point, as described on p. 276. Thus, for a binary electrolyte, AB, ionising to give A^+ and B^- ions, to a degree, α,

$$\frac{\text{Expected } \Pi \text{ if no ionisation}}{\text{Actual } \Pi \text{ caused by ionisation}} = \frac{1}{1 + \alpha}$$

Similarly (p. 277), if the actual osmotic pressure value as measured is used for calculating the relative molecular mass of the solute which has ionised,

$$\frac{\text{Real relative molecular mass}}{\text{Apparent relative molecular mass}} = \frac{1 + \alpha}{1}$$

10. The functioning of a semi-permeable membrane. The mechanism by which a semi-permeable membrane allows the passage of solvent molecules but prevents that of solute molecules is not well understood. It is probable that it depends on the particular membrane being used, and on the solvent and solute concerned. Three suggestions are worthy of consideration.

a. When the solute molecules are larger than those of the solvent it may be that a semi-permeable membrane can act as a molecular sieve. Some membranes, however, are still semi-permeable when the solute molecules are smaller than those of the solvent.

b. A semi-permeable membrane may provide a large number of fine capillaries. With pure solvent at one end, and solution at the other, there may be an air gap through which solvent vapour can diffuse from the solvent into the solution, which has a lower vapour pressure. If such a mechanism is correct, it is really air which is acting as the semi-permeable membrane as in f. on p. 282.

c. If a semi-permeable membrane can dissolve a solvent, but not a solute, solvent may pass through the membrane by a mechanism involving solution. Such, for example, is probably the mechanism by which benzenol (phenol) acts as the semi-permeable membrane in experiment h., p. 282. Water is soluble in benzenol, but calcium nitrate(v) is not.

It will be realised that all these suggestions are vague, and the precise functioning of a semi-permeable membrane is not known. Nor is the true nature of the forces responsible for osmosis covered by any very adequate theory.

11. Reverse osmosis. A semi-permeable membrane allows the passage of a solvent into a solution and, in so doing, a pressure can be built up. If a solution is separated from a solvent and a pressure, higher than the osmotic pressure, is applied to the solution, solvent will pass through the semi-permeable membrane from the solution. The phenomenon is known as reverse osmosis.

It can be used, theoretically, for obtaining drinking water from sea water but, as yet, only uneconomically.

QUESTIONS ON CHAPTER 26

1. Define osmosis and the osmotic pressure of a solution. Either show how you can demonstrate the existence of the former and how the latter can be measured for an aqueous solution, or show how the osmotic pressure of a solution is related to the lowering of the vapour pressure of the solvent. (O. & C.)

2. Describe an experiment to demonstrate the phenomenon of osmosis. A solution of 1 g of a substance X in 100 cm³ of water has an osmotic pressure of 67.33 kN m^{-2} (505 mm of mercury) at 7°C. What is the relative molecular mass of X? (O. Prelims.)

3. Calculate the osmotic pressure of a 1 per cent solution of glucose in water at 15°C, and the relative molecular mass of a sugar, a solution of which

had an osmotic pressure of 255 kN m^{-2} (2.522 atmospheres) at 20°C when the concentration was 34.2 g dm^{-3} of water.

4. A 3.42 per cent solution of sucrose ($M_r = 342$) was found to be isotonic with a 5.96 per cent solution of raffinose. Calculate the relative molecular mass of raffinose.

5. A saturated aqueous solution of a non-electrolyte of relative molecular mass 180 exerted an osmotic pressure of 52.82 kN m^{-2} (0.516 atmosphere) at 22°C. What is the solubility of the substance in g dm^{-3} of water at this temperature?

6. Calculate the osmotic pressure of a saturated solution of sucrose in water. Why is it, do you think, that the laws of osmotic pressure do not hold accurately for concentrated solutions?

7. Compare and contrast the phenomenon of osmosis with the behaviour of gases.

8. Explain what is meant by the osmotic pressure of a solution. How would you demonstrate its existence? How can it be measured? A solution of potassium chloride containing 5 g dm^{-3} was found to have an osmotic pressure of 304 kN m^{-2} at 18°C. What conclusions may be drawn from this observation?

9. (a) Why is osmotic pressure measurement sometimes advantageous for measuring very high relative molecular masses? (b) The osmotic pressure, at 25°C, of a solution containing 1.346 g of a protein in 100 cm^3 of solution is 971.8 N m^{-2}. Calculate the relative molecular mass of the protein.

10. A solution of 0.608 g of haemoglobin in 100 cm^3 of water gave an osmotic pressure of 202.6 N m^{-2} at 0°C. Assuming the solution to be ideal, calculate the relative molecular mass of haemoglobin.

11. What is (i) osmotic pressure, (ii) the van't Hoff factor? Indicate their importance in chemistry. Describe one method by which the osmotic pressure of a solution may be determined. The osmotic pressure of a solution containing 40 g of a non-electrolyte, Z, per dm^3 was 80 kN m^{-2} at 15°C. What was the relative molecular mass of Z? (S.)

12. The osmotic pressure of a solution of a sample of synthetic polyisobutylene in benzene at 25°C was 20.66 N m^{-2}. If the solution had a concentration of 2 g dm^{-3} what is the relative molecular mass of the polyisobutylene?

13. A solution of a colloid in water containing 12.3 g dm^{-3} has an osmotic pressure of 980.7 N m^{-2} (10 cm of water) at 25°C. What is the relative molecular mass if R has a value of 8.314 J K^{-1} mol^{-1}? (O. Schol.)

14. The apparent degree of ionisation of a 0.1M solution of sodium chloride is 85 per cent. What would be the concentration in g dm^{-3} of a solution of a non-electrolyte having a relative molecular mass of 180 which would be isotonic with sodium chloride solution? (Army and Navy.)

15. Explain what is understood by describing solutions as being isotonic. A solution of a non-electrolyte having a relative molecular mass of 180 and containing 15 g dm^{-3} was found to be isotonic with a solution of sodium chloride containing 2.64 g of the salt per dm^3. What is the apparent degree of ionisation of the sodium chloride? (Camb. 1st M. B.)

16. Distinguish between the meaning of the terms diffusion, osmosis, dialysis and sedimentation.

17. What is meant by the statements (a) osmosis is a colligative property, (b) isotonic solutions are not necessarily isosmotic, (c) osmotic pressure is not a pressure at all?

18. Describe the part played by osmosis in biological processes.

Chapter 27
Thermochemistry

CHEMICAL processes, such as reaction, combustion, solution, dilution, neutralisation, atomisation and hydrogenation are invariably associated with a corresponding energy change. A study of this energy change can help to throw light on the mechanism of the process involved.

In its simplest form, a study of the energy changes associated with chemical processes is concerned mainly with the evolution or absorption of heat, and is generally known as thermochemistry. A broader, and more complete, detailed study of all the energy changes constitutes what is known as thermodynamics (Chapter 48).

1. Heats of reaction. Chemical reactions are accompanied by some energy change, which, most obviously, involves the evolution or absorption of heat. The energy change may show up, however, in some other form. In an accumulator, chemical reactions release electrical energy; in chemiluminescent reactions, light energy is emitted; and in reactions involving an increase in volume, work may be done in expansion.

A reaction giving out heat to the surroundings is said to be *exothermic*; one absorbing heat from the surroundings is *endothermic*. The heat change is generally measured in joules or kilojoules,* and it is constant for a given reaction carried out under a fixed set of conditions.

The heat change associated with a reaction is conveniently written as part of, or alongside, the ordinary chemical equation for the reaction, the states of the reactants and products, and the temperature, being indicated if necessary. Thus

$$C(\text{graphite}) + O_2(g) = CO_2(g) + 393.4 \text{ kJ at } 25°C$$
$$C(\text{graphite}) + O_2(g) = CO_2(g); \Delta H(298 \text{ K}) = -393.4 \text{ kJ}$$

These equations mean that 393.4 kJ are evolved when 12 g of graphite react with 32 g of oxygen to form 44 g of carbon dioxide, the initial temperature of the reactants, and the final temperature of the products, being 25°C. When a gas is involved in such an equation it is assumed to be at a pressure of 101.325 kN m^{-2} unless otherwise indicated.

An evolution of heat to the surroundings means that the system as a whole loses heat and, therefore, finishes with a lower energy content. That is why ΔH *is written as negative for an exothermic reaction* and *positive for an endothermic one.*

The heat of a reaction is defined as the heat change when the reaction takes place between the masses of the reagents indicated by the equation for the reaction.

2. Definitions. The heat of reaction is a very wide term for there are so many different types of reaction. For some of the more important types,

* Heats of reaction are sometimes expressed in calories or kilocalories. 1 calorie = 4.1840 joule, or 1 kilocalorie = 4.1840 kilojoule.

the heat of reaction is re-named and, to some extent, re-defined, as follows:

a. *Heat of combustion* (p. 297). *The heat of combustion of a substance is the heat change when* **1 mol** *of it is completely burnt in oxygen.* The heat of combustion of methane, for example, is 890.2 kJ mol^{-1},

$$CH_4(g) + 2O_2(g) = CO_2(g) + 2H_2O(l); \quad \Delta H(298 \text{ K}) = -890.2 \text{ kJ}$$

b. *Heat of neutralisation* (p. 299). *The heat of neutralisation of an acid by a base is the heat change when* **1 gramme-equivalent** *of the acid is neutralised by a base, the reaction being carried out in dilute aqueous solution.* The heat of neutralisation of hydrochloric acid by sodium hydroxide, for example, is 57.33 kJ per gramme-equivalent,

$$HCl + NaOH = NaCl + H_2O; \quad \Delta H(298 \text{ K}) = -57.33 \text{ kJ}$$

c. *Heat of formation* (p. 300). *The heat of formation of a compound is the heat change when* **1 mol** *of the compound is formed from its elements under stated conditions.* Standard heats of formation are quoted at 25°C and 101.325 kN m^{-2}. The heat of formation of water, for example, is 285.9 kJ mol^{-1},

$$H_2(g) + \tfrac{1}{2}O_2(g) = H_2O(l); \quad \Delta H(298 \text{ K}) = -285.9 \text{ kJ}$$

d. *Heat of atomisation* (p. 301). *The heat required to convert* **1 gramme-atom or 1 mol** *of an element from its normal state at 25°C and 101.325 kN m^{-2} into free atoms.* The heat of atomisation of hydrogen, for example, is 432.7 kJ mol^{-1},

$$H_2(g) = 2H; \quad \Delta H = 432.7 \text{ kJ}$$

e. *Heat of hydrogenation* (p. 302). *The heat change when 1 mol of an unsaturated compound is completely converted into the corresponding saturated compound by reaction with gaseous hydrogen.* The heat of hydrogenation of ethene, for example, is 137.5 kJ mol^{-1},

$$C_2H_4(g) + H_2(g) = C_2H_6(g); \quad \Delta H = -137.5 \text{ kJ}$$

f. *Miscellaneous.* Other specific types of heats of reaction may be used. *Heats of dissociation, heats of crystallisation and heats of ionisation,* may, for example, be met. The heat changes associated with chemical changes other than reactions, e.g. *heats of dilution* (p. 295), *heats of solution* (p. 295) and *heats of sublimation* are also used.

3. Hess's law of constant heat summation. *Hess's law* (1840) *states that the heat evolved or absorbed in a chemical change is the same whether the change is brought about in one stage or through intermediate stages.*

Thus the heat change in a reaction A to C is the same whether the reaction takes place in one stage as

$$A \longrightarrow C$$

or in two stages as

$$A \longrightarrow B \longrightarrow C$$

This is a particular application of the law of conservation of energy. If Hess's law did not hold it would be possible to obtain energy without doing any work. If a change from A to B evolved 100 J when carried out in one way and only 90 J in another way it would be possible, by going from A to B by the first path and then back to A by the second, to obtain 10 J of heat from nowhere. This is contrary to all scientific experience.

The validity of Hess's law can be demonstrated experimentally by measuring the heat change when a reaction is brought about in two or more different ways. The result of making a solution of ammonium chloride from 36.5 g of hydrogen chloride, 17 g of ammonia and water in two ways are summarised below, the heats of reaction being given in kilojoules. The figures illustrate the truth of Hess's law.

	Method 1		Method 2
i	Mix the two gases $\Delta H = -176.1$	i	Dissolve the NH_3 in water $\Delta H = -35.2$
ii	Dissolve the product in water $\Delta H = +16.3$	ii	Dissolve the HCl in water $\Delta H = -72.4$
		iii	Mix the two solutions $\Delta H = -52.3$
	Overall reaction $\Delta H = -159.8$		Overall reaction $\Delta H = -159.9$

4. Thermochemical calculations. Hess's law enables thermochemical equations to be added and subtracted, and this is the procedure to adopt in all simple thermochemical calculations.

Given the heats of reactions,

i $C(s) + O_2(g) = CO_2(g)$; $\Delta H = -393.4 \text{ kJ}$
ii $2CO(g) + O_2(g) = 2CO_2(g)$; $\Delta H = -565.9 \text{ kJ}$

it is very simple to find the heat of the reaction

iii $2C(s) + O_2(g) = 2CO(g)$

This is done by taking 2 × i and subtracting ii. The heat of the reaction iii is, therefore, $2 \times (-393.4) - (-565.9)$, i.e. -220.9 kJ. And for half the quantities, so that the heat of reaction can be expressed in terms of 1 mol of carbon monoxide,

$$C(s) + \tfrac{1}{2}O_2(g) = CO(g); \quad \Delta H = -\frac{220.9}{2} = -110.45 \text{ kJ}$$

The heat of formation of carbon monoxide is, therefore, $110.45 \text{ kJ mol}^{-1}$.

Not all thermochemical calculations are as easy as this but it is generally a matter of adding and/or subtracting the given equations, multiplied or divided throughout wherever necessary (pp. 300 and 302).

5. Factors affecting heats of reactions. The heat of a reaction depends on the conditions under which the reaction is carried out as indicated below.

a. *State of reagents and products.* Latent heat is evolved or absorbed in changes of state, so that the state of reagents and products in a reaction affects the heat of the reaction.

The heat of reaction between hydrogen gas and solid iodine to form hydrogen iodide gas, for example, is 51.9 kJ.

$$H_2(g) + I_2(s) = 2HI(g); \quad \Delta H = 51.9 \text{ kJ}$$

Using iodine vapour, the value is -9.2 kJ, and the reaction is, now, exothermic,

$$H_2(g) + I_2(g) = 2HI(g); \quad \Delta H = -9.2 \text{ kJ}$$

The difference is due to the heat change in vaporising solid iodine,

$$I_2(s) = I_2(g); \quad \Delta H = 61.1 \text{ kJ}$$

b. *Allotropic modifications.* A heat change is involved in the conversion of one allotrope into another, so that the particular allotrope used in a reaction affects the heat of the reaction.

For example

$$S \text{ (rhombic)} + O_2(g) \quad = SO_2(g); \quad \Delta H = -296.97 \text{ kJ}$$
$$S \text{ (monoclinic)} + O_2(g) = SO_2(g); \quad \Delta H = -296.63 \text{ kJ}$$

The difference in the heats of reaction is due to the heat change in the conversion of rhombic into monoclinic sulphur,

$$S \text{ (rhombic)} = S \text{ (monoclinic)}; \quad \Delta H = -0.34 \text{ kJ}$$

Similarly,

$$C \text{ (graphite)} = C \text{ (diamond)}; \quad \Delta H = +2.1 \text{ kJ}$$

c. *Reactions involving solutions.* There is a heat change, known as the heat of solution, when a solute dissolves in a solvent to form a solution. The value of the heat change depends on the concentration of the solution formed, for dilution of a solution also causes a heat change known as the *heat of dilution.*

The heat of solution most commonly used is that *at infinite dilution.* This *is defined as the heat change when* 1 *mol of a substance dissolves in such a large volume of solvent that addition of more solvent produces no further heat of dilution.*

The effect of heat of solution on the heat of a reaction is illustrated in the following example:

$$\tfrac{1}{2}H_2(g) + \tfrac{1}{2}Cl_2(g) \longrightarrow HCl(g); \quad \Delta H = -92.21 \text{ kJ}$$
$$HCl(g) + aq \longrightarrow HCl(aq) \quad \Delta H = -72.38 \text{ kJ}$$
$$\therefore \tfrac{1}{2}H_2(g) + \tfrac{1}{2}Cl_2(g) + aq \longrightarrow HCl(aq); \quad \Delta H = -164.7 \text{ kJ}$$

The term (aq), in a thermochemical equation, indicates a solution so dilute that it can be regarded as infinitely dilute, i.e. any heat of dilution can be neglected.

6. Reactions at constant pressure or constant volume. A reaction carried out in an open vessel takes place at a constant pressure, but one carried out in a sealed container takes place at constant volume. The heat of reaction may be affected by the conditions chosen.

Heats of reaction measured at constant volume are symbolised by ΔU, and

$$\Delta U = -q_v$$

where q_v is the heat evolved in the reaction. For an exothermic reaction, q_v will be positive, giving a negative value for ΔU; for an endothermic reaction, q_v will be negative, and ΔU will be positive. ΔU is known as the change in intrinsic or internal energy (p. 297).

Heats of reaction measured at constant pressure are symbolised by ΔH, and

$$\Delta H = -q_p$$

ΔH is referred to as the *enthalpy* (or heat content) change (p. 297).

a. *Reaction producing an increase in gas volume.* Such an increase will necessitate expansion, if the reaction is carried out at constant pressure, and, consequently, work must be done against the pressure. Some of the energy which, at constant volume, would be evolved as heat will, at constant pressure, be used up in doing this work. Therefore,

$$q_v = q_p + \text{Work done in expansion}$$
$$\text{or} \qquad \Delta H = \Delta U + \text{Work done in expansion}$$

For an increase in volume of ΔV at a constant pressure of p, the work done in expansion will be equal to $p\Delta V$. If the volume increase is due to the formation of 1 mol of gas, $p\Delta V$ will be equal to RT or $8.3T$ joule (p. 38). For an increase in gas volume of n mol, the work done in expansion will be $8.3nT$ joule, so that

$$\Delta H = \Delta U + 8.3nT$$

The reaction of 1 mol of zinc with excess dilute hydrochloric acid at 18°C, for example, produces 151.5 kJ if the reaction is carried out in an open vessel at constant (atmospheric) pressure. There is an increase in gas volume of 1 mol, for volume changes other than that due to the formation of hydrogen can be neglected, so that the heat of reaction at constant volume would be given by

$$\Delta U = \Delta H - 8.3T = -151\,500 - (8.3 \times 291) = -153\,915 \text{ J}$$

which is in good agreement with experimental measurement.

b. *Reaction producing a decreased gas volume.* Work will have to be done on a system such as this, at constant pressure, to bring about the contraction in volume. For a volume contraction of n mol of gas, the work done by the constant pressure will be $8.3nT$ joule, so that

$$\Delta H = \Delta U - 8.3nT$$

For the reaction

$$2CO + O_2 = 2CO_2$$

the heat of reaction at constant pressure, and 17°C, is $\Delta H = -569\,000$ J. There is a decrease in gas volume of 1 mol so that the heat of reaction at constant volume is given by

$$\Delta U = \Delta H + 8.3T = -569\,000 + (8.3 \times 290) = -566\,593 \text{ J}$$

There may, therefore, be significant differences between the ΔH and ΔU values for a reaction involving gases. For reactions involving only solids and/or liquids, volume changes are never very large and ΔH and ΔU values are approximately equal.

7. Enthalpy change and enthalpy diagrams. The heat of reaction measured at constant pressure, ΔH, is known as the enthalpy (or heat content) change for the reaction. The corresponding value at constant volume, ΔU, is known as the change in intrinsic or internal energy. The two are related (pp. 296 and 536) as follows:

$$\Delta H = \Delta U + \text{Work done in expansion}$$

As most chemical reactions are carried out in open containers at constant, atmospheric pressure, ΔH values are used more commonly than ΔU values. It is, too, convenient to summarise the ΔH changes in an enthalpy diagram, as is done below for the information given at the bottom of page 295.

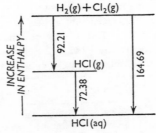

8. Heat of combustion. *The heat of combustion of a substance is the heat change when 1 mol of it is completely burnt in oxygen.*

Measurements are made in a bomb-calorimeter (Fig. 119). A known mass of the substance under test is placed (in a thin glass tube if volatile) in a platinum crucible at the centre of a steel-alloy, cylindrical bomb. The bomb is filled through a complicated valve system, with oxygen, at a pressure of about 2.5×10^3 kN m^{-2} (25 atmospheres), to ensure rapid and complete combustion, and immersed in a calorimeter, usually containing water.

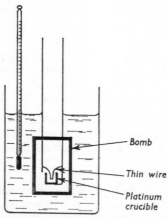

The combustion of the substance is started by passing a current through a thin wire of iron or platinum in contact with the substance. The heat evolved on burning is determined by measuring the rise in temperature of the liquid in the calorimeter. The amount of heat evolved can be found by seeing what amount of electrical energy is required, through a heating coil, to give the same rise of temperature, or, alternatively, the thermal capacity of the apparatus can be measured electrically or by using a substance whose heat of combustion is known.

FIG. 119. Outline arrangement of a bomb calorimeter. The complicated engineering details necessary to maintain pressure inside the bomb are not shown.

For a slow combustion, or when only a small amount of heat is liberated, it is more accurate to make readings adiabatically, the whole set-up being surrounded by a water jacket which is maintained at the same temperature as the inner group. This allows heat losses by radiation or convection to be neglected. The outer jacket can be maintained at the correct temperature by having one junction of a thermocouple in the liquid of the outer jacket and another in the liquid of the inner calorimeter. If any temperature differences arise, the thermocouples are so arranged, through an electrical circuit, to heat or cool the outer jacket.

Measurements made in a bomb calorimeter, at constant volume, are ΔU values, but they can be converted into values at constant pressure, ΔH, using the equation on p. 296. The accuracy of the measurements with a modern bomb calorimeter is estimated at 2 parts in 10000.

Some typical heats of combustion, in kJ mol^{-1}, are given below:

Substance	ΔH(298 K)	Substance	ΔH(298 K)
Hydrogen gas . .	−286	Ethyne gas . . .	−1560
Graphite . . .	−393	Liquid benzene . .	−3278
Methane gas . .	−890	Liquid methylbenzene .	−3917
Carbon monoxide gas .	−283	Methanol liquid . .	−715

Calorific values for fuels, expressed in calories per gramme, B.Th.U. per gramme or B.Th.U. per cubic foot are also measured in a bomb calorimeter.

9. Heat of neutralisation. *The heat of neutralisation of an acid by a base is the heat change when* 1 *gramme-equivalent of the acid is neutralised by a base, the reaction being carried out in dilute aqueous solution.*

Heats of neutralisation may be measured by mixing solutions of acid and base in a calorimeter. Accurate results can only be obtained by paying very detailed attention to temperature measurement and heat losses, but the general principle of the methods used can be illustrated, with fair results, by using a vacuum flask surrounded by a felt-packed box as a calorimeter (Fig. 120).

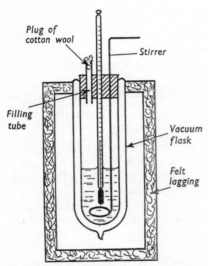

Fig. 120. Use of a vacuum flask calorimeter for measurement of heat of neutralisation.

The thermal capacity of the calorimeter must first be measured. This is done by measuring the temperature of the empty vacuum flask, adding 100 g of water at about 40°C, and measuring the fall in temperature of the water. Then

$$\text{Thermal capacity in joules } (c) \times \frac{\text{Rise in temperature of flask}}{} = 100 \times \frac{\text{Fall in temperature of water}}{} \times 4.184$$

100 cm^3 of a N solution of an alkali is then placed in the emptied flask, and its temperature is recorded (t_1°C). 100 cm^3 of a N solution of an acid, at a measured temperature (t_2°C), is then added. The highest temperature (t_3°C) reached by the mixture is recorded. Thorough stirring and a thermometer graduated in tenths of a degree are essential.

Taking the relative density and the specific heat capacity of the alkaline, acidic and salt solutions concerned as unity, the heat evolved by the neutralisation process is given by

$$[(100 + c)(t_3 - t_1) + 100(t_3 - t_2)] \times 4.184 \text{ joule}$$

The heat of neutralisation for the acid and alkali chosen will be ten times this value.

For a strong acid and a strong base (p. 408) the heat of neutralisation is effectively constant at 57.33 kJ. This is because strong acids and strong bases, and the salts they form, are all completely ionised in dilute solution (p. 366). Thus the reaction between any strong acid and any strong base, e.g.

$$HCl + KOH \longrightarrow KCl + H_2O$$
$$H^+ + Cl^- + K^+ + OH^- \longrightarrow K^+ + Cl^- + H_2O$$

is simply the formation of unionised water from H^+ and OH^- ions,

$$H^+(aq) + OH^-(aq) \longrightarrow H_2O(l); \qquad \Delta H = -57.33 \text{ kJ}$$

The constancy of the heat of neutralisation of any strong acid and any strong base provides simple, but convincing, evidence that strong acids and bases are, in fact, completely ionised.

With a weak acid and/or a weak base, neutralisation also produces a heat change due to the reaction

$$H^+(aq) + OH^-(aq) \longrightarrow H_2O(l)$$

but, during the reaction, previously unionised acid and/or base is converted into ions and this involves a further heat change known as the *heat of ionisation*, which may be either positive or negative. The heat of neutralisation of an acid–base reaction involving either weak acids or weak bases may, therefore, be either greater or smaller than 57.33 kJ. Effects other than the heat of ionisation, such as association or hydration may also affect the issue.

10. Heat of formation. *The heat of formation of a compound is the heat change when 1 mol of the compound is formed from its elements, under stated conditions.*

To achieve some standardisation it is customary to choose 25°C and 101.325 kN m^{-2} pressure as the conditions under which heats of formation are quoted, and such figures are known as *standard heats of formation*.

They cannot, generally, be measured directly because most compounds cannot be made directly from their component elements. Use has therefore to be made of Hess's law so that heats of formation can be calculated from other experimentally determined heats of reaction.

Methane, for instance, cannot be made directly from carbon and hydrogen but its heat of formation can be calculated from the heats of combustion of methane, graphite and hydrogen, as follows:

i $\quad CH_4(g) + 2O_2(g) = CO_2(g) + 2H_2O(l); \quad \Delta H(298 \text{ K}) = -890.2 \text{ kJ}$
ii $\quad C(graphite) + O_2(g) = CO_2(g); \qquad\qquad \Delta H(298 \text{ K}) = -393.4 \text{ kJ}$
iii $\quad 2H_2(g) + O_2(g) = 2H_2O(l); \qquad\qquad\quad \Delta H(298 \text{ K}) = -571.5 \text{ kJ}$

Adding ii and iii, and subtracting i, gives the result

$$C(graphite) + 2H_2(g) = CH_4(g); \qquad \Delta H^\ominus(298 \text{ K}) = -74.7 \text{ kJ}$$

so that the standard heat of formation of methane is -74.7 kJ mol^{-1}.

Other standard heats of formations in kJ mol^{-1} for some common compounds are summarised in the table below.

The superscript $^\ominus$ written in $\Delta H^\ominus(298 \text{ K})$ is used to indicate a standard heat of formation.

Such figures are useful because the heat of reaction at 25°C of any reaction can be readily calculated by subtracting the standard heats of formation of the products from those of the reagents.

Substance				$\Delta H^\ominus$	Substance				$\Delta H^\ominus$
$H_2O(l)$	.	.	.	-285.8	$CH_4(g)$	.	.	.	-74.7
$HCl(g)$	.	.	.	-92.3	$C_2H_6(g)$	.	.	.	-84.7
$HBr(g)$	.	.	.	-36.23	$C_3H_8(g)$	.	.	.	-103.9
$HI(g)$	.	.	.	25.94	$C_2H_4(g)$	.	.	.	52.3
$SO_2(g)$	.	.	.	-297	$C_2H_2(g)$	.	.	.	226.7
$SO_3(g)$	.	.	.	-395.2	$C_6H_6(l)$	.	.	.	49.05
$NO(g)$	.	.	.	90.37	$C_6H_6(g)$	.	.	.	82.92
$CO(g)$	.	.	.	-110.5	$CH_3OH(l)$	.	.	.	-238.6
$CO_2(g)$	.	.	.	-393.4	$C_2H_5OH(l)$	.	.	.	-277.6

In the reaction

$$\begin{array}{ccccc} C_2H_6(g) & + & 3\tfrac{1}{2}O_2(g) & = & 2CO_2(g) & + & 3H_2O(l) \\ -84.7 & & 0 & & 2(-393.4) & & 3(-285.8) \\ & & & & -786.8 & & -857.4 \end{array}$$

the heats of formation of the reagents and products are as shown so that the heat of the reaction is $-786.8 - 857.4 + 84.7$, i.e. -1559.5 kJ.

11. Heats of atomisation. The standard heat of formation of a compound is the heat change when 1 mol of the compound is formed from its elements at 25°C and 101.325 kN m^{-2} pressure (p. 300). Under these conditions the elements are conventionally allotted zero heat contents or enthalpies. This, however, is a purely conventional arrangement, convenient in calculations where only energy *differences* between compounds before and after reaction are concerned.

The heat of formation of a compound from its isolated atoms is a more important quantity when the binding forces or bond energies in a compound are under consideration. Such a heat of formation can only be obtained by taking into account heats of atomisation or dissociation, i.e. *the heat necessary to convert 1 mol of an element from its standard state at 25°C and 101.325 kN m^{-2} pressure into free atoms.* To form free atoms from an element in its normal state involves the breaking down of a crystal lattice for a solid element, or the rupture of interatomic bonds in a molecule for a gas.

Values for heats of atomisation are not easy to measure and are mainly obtained by spectroscopic methods beyond the scope of this book. Some typical results are summarised below

$H_2(g)$	432.2 kJ	$Cl_2(g)$	239.3 kJ
C(graphite)	713.6 kJ	$Br_2(g)$	190.4 kJ
$O_2(g)$	493.3 kJ	$I_2(g)$	149.0 kJ

To obtain the heat of formation of methane from its atoms, it is necessary to calculate as follows.

The standard heat of formation of methane is -74.7 kJ mol^{-1} (p. 301), i.e.

i $C(graphite) + 2H_2(g) = CH_4(g);$ $\Delta H = -74.7$ kJ

The heats of atomisation of graphite and hydrogen may be written in the following form

ii $C(graphite) = C(free\ atoms);$ $\Delta H = 713.6$ kJ
iii $2H_2(g) = 4H(free\ atoms);$ $\Delta H = 864.4$ kJ

Subtraction of ii plus iii from i gives the result,

$C(free\ atoms) + 4H(free\ atoms) = CH_4(g);$ $\Delta H = -1652.7$ kJ

and -1652.7 kJ mol^{-1} is the heat of formation of methane from its atoms, a very different figure from the standard heat of formation of methane (-74.7 kJ mol^{-1}).

There are four C—H bonds in a methane molecule, and it follows that the average bond energy of the C—H bond in methane is a quarter of 1652.7, i.e. 413.2 kJ. This may also be called the heat of formation of the C—H bond, if the sign is reversed, i.e.

$C(free\ atom) + H(free\ atom) = C—H(bond);$ $\Delta H = -413.2$ kJ

The main use of heats of formation from free atoms is in obtaining such bond energies (p. 180).

12. Heats of hydrogenation. *The heat of hydrogenation is the heat change when 1 mol of an unsaturated compound is completely converted into the corresponding saturated compound by reaction with gaseous hydrogen.*

The values of heats of hydrogenation are of importance in showing the peculiarities of conjugated systems (p. 193). The heat of hydrogenation of 1-butene, for example, is 126.8 kJ mol^{-1},

and this really corresponds to the change,

It might, therefore, be expected that the heat of hydrogenation of buta-1,3-diene

buta-1, 3-diene

Cyclohexene

would be 2×126.8 kJ mol^{-1}, whereas the actual value is found to be 238.9, i.e. 14.7 kJ smaller.

An even greater discrepancy is found between the heats of hydrogenation of cyclohexene (119.7 kJ mol^{-1}) and benzene (208.4 kJ mol^{-1}). If benzene had three C=C bonds in its molecule, the expected heat of hydrogenation would be 3×119.7, i.e. 359.1 kJ mol^{-1}. The actual value is 150.7 kJ lower than this.

Such discrepancies are found with every conjugated system and are accounted for by resonance (p. 193) or by the delocalisation of π-electrons (p. 194).

QUESTIONS ON CHAPTER 27

1. Give an account of the experimental determination of the heat of combustion of benzene. Given the thermochemical equations (both at 25°C),

$$2C_6H_6(l) + 15O_2 = 12CO_2 + 6H_2O; \quad \Delta H = -6689 \text{ kJ}$$
$$2C_2H_2(g) + 5O_2 = 4CO_2 + 2H_2O; \quad \Delta H = -2595 \text{ kJ}$$

calculate the heat of reaction for the formation of benzene from ethyne at 25°C according to the equation,

$$3C_2H_2(g) = C_6H_6(l) \tag{C. Schol.}$$

2. Define heat of combustion and heat of formation. State Hess's law. (a) For the reaction,

$$3C_2H_2(g) = C_6H_6(g)$$

calculate the heat of reaction, stating whether heat is absorbed or evolved. (b) Calculate the heat of combustion of ethyne gas.

You are given the following heats of formation in kJ mol^{-1}: $C_6H_6(g)$, 82.84 absorbed; $CO_2(g)$, 393.3 evolved; $C_2H_2(g)$, 226.8 absorbed; $H_2O(g)$, 241.8 evolved. (O. & C.)

3. Explain the terms: heat of neutralisation, heat of formation, heat of reaction and heat of combustion. Why is the heat of neutralisation of a strong acid and a strong base constant?

Given that the heat of combustion of benzene is 3278 kJ mol^{-1} (evolved) and that the heats of formation of carbon dioxide and water are 393.4 and 285.8 kJ mol^{-1} (evolved) respectively, calculate the heat of formation of benzene, stating whether heat is absorbed or evolved. (O. & C.)

4. The heats of combustion of hydrogen, carbon and methane are 292.9, 376.6 and 753.1 kJ mol^{-1} respectively. Calculate the heat of formation of methane.

5. If the heat of combustion of hydrogen is 285.8 kJ mol^{-1}, what mass of ice at 0°C could be converted into steam at 100°C by the combustion of 1000 dm^3 of hydrogen measured at s.t.p.? (L.H. of fusion of ice = 330.5 J g^{-1}. L.H. of vaporisation of water = 2255 J g^{-1}.)

6. State Hess's law. The heat of combustion of ethanol is -1430 kJ and the heats of formation of carbon dioxide and water are -393.4 and -285.8 respectively, all measurements referring to similar conditions. What is the heat of formation of ethanol under these conditions?

7. Do the following figures agree with Hess's law?

$$\begin{array}{ll} H_2(g) + Cl_2(g) = 2HCl(g) & \Delta H = -184.1 \text{ kJ} \\ H_2(g) + I_2(s) = 2HI(g) & \Delta H = +506 \text{ kJ} \\ 2HI(g) + Cl_2(g) = I_2(s) + 2HCl(g) & \Delta H = -234.3 \text{ kJ} \end{array}$$

8. The heats of neutralisation of 0.2M hydrochloric acid and 0.1M sulphuric acid by 0.2M sodium hydroxide, in 0.2N solution, are 573.2 and 656.9 kJ respectively. The heat of the reaction

$$NaCl + \tfrac{1}{2}H_2SO_4 = \tfrac{1}{2}Na_2SO_4 + HCl$$

using 0.2M sodium chloride and 0.1M sulphuric acid using 0.2N solutions is $\triangle H = -292.9$ kJ. What conclusions can you draw about the relative strengths of the two acids?

9. What temperature rise would be expected if 100 cm^3 of 0.5M hydrochloric acid and 100 cm^3 of 0.5M sodium hydroxide solution were mixed in a calorimeter of thermal capacity 150.6 J K^{-1}? Take all solutions as having a relative density and a specific heat capacity of unity, and neglect heat losses.

10. The heat of formation of AgCl(s) is -127 kJ and the heat liberated when 1 mol of silver chloride is precipitated from silver nitrate(v) and sodium chloride solutions is 65.48 kJ. Calculate the heat of the reaction,

$$Ag(s) + \tfrac{1}{2}Cl_2(g) + H_2O = Ag^+(aq) + Cl^-(aq)$$

11. The thermal capacity of a bomb calorimeter is 2.259 kJ K^{-1}. When 1 g of benzenecarboxylic acid is completely burnt in the bomb, the rise in temperature of 1 200 g of water in the calorimeter is 3.65 K. Calculate the approximate value of the heat of combustion of benzenecarboxylic acid per mole at constant volume.

12. Using the standard heats of formation for ethane and ethanol, given on p. 301, calculate the heat of the reaction,

$$C_2H_6(g) + \tfrac{1}{2}O_2 = C_2H_5OH(l)$$

13. Given that

$$CO(g) + H_2(g) = H_2O(g) + C(s); \quad \Delta H_{298} = -131.3 \text{ kJ}$$

calculate the heat of reaction at constant pressure and 25°C when 1 g of carbon reacts completely with water vapour to form carbon monoxide and hydrogen. What would the value be at constant volume?

14. Calculate the heat of the reaction

$$2Al + Cr_2O_3 = Al_2O_3 + 2Cr$$

given that the heats of formation of aluminium oxide and chromium(III) oxide are $-1 590$ and $-1 130$ kJ respectively. Comment on points of interest associated with the high heat of formation of aluminium oxide.

15. State Hess's law and define heat of formation. The heats of combustion of carbon and carbon monoxide are, respectively, $+393.3$ and $+284.5$ kJ. (The $+$ sign denotes heat liberated.) Calculate the heat of the reaction,

$$2C + O_2 = 2CO \qquad \text{(Army and Navy.)}$$

16. State and account for Hess's law of constant heat summation. Explain its value in the determination of the heats of formation of organic compounds. The heats of formation of water and carbon dioxide are 285.8 and 405.4 kJ mol^{-1} respectively, at 15°C and constant pressure. The heats of combustion of methane and ethene at 15°C and constant pressure are 886.6 and 1 423 kJ mol^{-1} respectively. Calculate the heats of formation of methane and ethene at 15°C (a) at constant pressure and (b) at constant volume. (Gas constant, R, is 8.3 J K^{-1} mol^{-1}.) (W.)

17. The heat of formation of hydrogen selenide is -811.7 kJ if amorphous selenium is used and $-1 050$ kJ if metallic selenium is used. Calculate the heat

change in converting amorphous into metallic selenium.

18. The heat of formation of the C—C link (as in ethane) is -341.4 kJ, and that of the C=C link (as in ethene) is -610.9 kJ, and that of the C≡C link (as in ethyne) is -803.7 kJ. How would you expect the reactivity of these three hydrocarbons to differ in the light of these data? (O. & C. S.)

19. Illustrate the importance of calorimetric measurements in chemistry. (O. Schol.)

20. Define (a) the heat of combustion of a compound, and (b) the heat of reaction. Outline an experiment to find (a).

$$Au(OH)_3 + 4HCl = HAuCl_4 + 3H_2O + 96.23 \text{ kJ}$$
$$Au(OH)_3 + 4HBr = HAuBr_4 + 3H_2O + 154.0 \text{ kJ}$$

On mixing 1 mol of tetra bromauric(III) acid with 4 mol of hydrochloric acid 2.092 kJ are absorbed. Calculate the percentage of tetra bromauric(III) acid which has been changed into tetra chlorauric(iii)acid.

21. How can it be decided (a) which is the stabler of two allotropes, (b) whether a molecule is likely to have a conjugated system of C=C bonds?

Chapter 28
Rate and equilibrium constants

1. Introduction. A chemical reaction is simply a rearrangement of atoms. Before reaction, various atoms are linked together in different patterns; after reaction, a rearrangement is found to have taken place.

Why do some chemicals react together, whilst others don't? Can the products of a reaction be predicted theoretically? Why are some reactions very rapid, possibly explosive, when others are very slow? Can the rate of any reaction be predicted? What is the precise mechanism by which atoms rearrange themselves in the course of a chemical reaction? What is the effect on a chemical reaction of changing such factors as the concentrations of the reagents, the pressure and the temperature? Why do catalysts sometimes speed up a chemical reaction, and how do they do it?

A student sometimes thinks that all the information required to be known about a reaction is contained in the simple, balanced equation for the reaction, but this is a very limited and false outlook. The questions posed above are not answered by a simple equation. They involve, in fact, a study of *thermodynamics*, of *chemical kinetics* and of *reaction mechanisms*.

Whether or not a particular reaction is likely to take place, and some indication of the most favourable conditions under which it will take place, can be ascertained by a theoretical consideration of the heat, and other energy, changes associated with the reaction. This is the province of chemical thermodynamics, a brief account of which is given in Chapter 48.

Thermodynamic arguments give little or no information, however, about the possible *rates* of chemical reactions. This is the domain of chemical kinetics, an outline of which is given in Chapters 28–32.

Finally, the atomic rearrangements involved in a chemical reaction are studied under the general heading of reaction mechanisms (Chapter 31).

2. Rate or velocity constants. The rate of a reaction can be obtained by measuring either the amount of reagents used up, or the amount of products formed, in unit time. The rate is found to be dependent on the concentration of the reagents and on temperature, and, sometimes, on pressure and catalysis. The relationship between reaction rate and concentration is discussed in this chapter; the effect of temperature, pressure and catalysis is dealt with in Chapter 29.

Chemical reactions vary very greatly in complexity. Some involve only one step whilst others may pass through a number of intermediate steps each having a different rate. There is, as a result, no single comprehensive theory of reaction kinetics.

a. *Simple, one-step reactions.* Most simple, one-step reactions are governed by the *law of mass action* first put forward by Guldberg and Waage in 1864. It states that *the rate of reaction, at a constant temperature, is proportional to the product of the active masses of the reacting substances.*

In the following discussion active masses * will be taken as being equal to molecular concentrations, i.e. the activity coefficients will be regarded as unity. The discussion will also be limited to homogeneous reactions taking place in one phase. Heterogeneous reactions are discussed on p. 318.

For a simple, one-step reaction

$$A + B \longrightarrow \text{products of reaction}$$

the law of mass action states that

$$\text{Rate of reaction} \propto [A] \times [B]$$

where [A] is a conventional way of representing the concentration of A in mol dm^{-3}. It follows that

$$\text{Rate of reaction} = k \times [A] \times [B]$$

and k is known as the rate or velocity constant for the reaction.

If A and B are gases and partial pressures are used to express their concentrations, then

$$\text{Rate of reaction} = k' \times p_A \times p_B$$

For a more general reaction,

$$xA + yB + zC \longrightarrow \text{products of reaction}$$

the rate is given by

$$\text{Rate of reaction} = k \times [A]^x \times [B]^y \times [C]^z = k' \times p_A{}^x \times p_B{}^y \times p_C{}^z$$

The rate constant for a given reaction is a constant under given conditions. The actual rate of reaction varies, for as the reactants are used up so the rate decreases until, eventually, the reaction ceases.

b. *Complex reactions.* For a complex reaction taking place in steps it is still, generally, possible to find an expression relating the reaction rate and concentration, but it may not be a simple expression and more than one rate constant may be involved. The rate of formation of hydrogen bromide from hydrogen and bromine (p. 342), for example, is related to the concentrations by the following expression

$$\text{Rate} = \frac{k \times [H_2] \times [Br_2]^{\frac{1}{2}}}{m + [HBr]/[Br_2]}$$

where k and m are rate constants.

* In many reactions, the active masses of the reactants may be taken as their molecular concentrations. These are most conveniently measured in moles per dm^3, or partial pressure, for gases; in moles per dm^3 for a solution; or as mole fractions for a liquid mixture (p. 227). The active mass of a substance is only equal to its molecular concentration, however, if there is no interaction or interference between the molecules concerned. Many chemical systems do not approach this ideal state and there is some interference, particularly in reactions involving ions. In such cases the molecular concentration has to be multiplied by an activity coefficient (p. 378) to obtain the effective active mass,

Active mass = Molecular concentration × Activity coefficient

This reaction provides an example of a simple, overall reaction equation,

$$H_2 + Br_2 = 2HBr$$

being related to a much more complex kinetic expression.

Alternatively, the kinetic expression can be simpler than the overall equation would suggest. Thus the oxidation of iodide ions by peroxo-disulphate(VI) ions is represented by the equation

$$2I^-(aq) + S_2O_8{}^{2-}(aq) = I_2(s) + 2SO_4{}^{2-}(aq)$$

but the rate of the reaction is given by

$$\text{Rate} = k \times [I^-] \times [S_2O_8{}^{2-}]$$

3. Reversible reactions. Equilibrium constants. Many reactions do not go to completion and then stop, because the products of the reaction themselves react to reform the original reactants.

Phosphorus pentachloride, for example, decomposes, on heating in a closed vessel, to form chlorine and phosphorus trichloride. But chlorine and phosphorus trichloride will also react together to form phosphorus pentachloride. Eventually the sealed vessel contains all three chemicals, and the composition of the mixture, at a given temperature, will be found to be the same whether phosphorus pentachloride, or an equivalent amount of a mixture of chlorine and phosphorus trichloride, is taken as the starting material. The reaction is said to be reversible,

$$PCl_5 \rightleftharpoons PCl_3 + Cl_2$$

and the mixture finally obtained is called the *equilibrium mixture.*

When the equilibrium mixture has been obtained in a reversible reaction, the system is not static. The equilibrium is a dynamic one, the rate of the forward reaction being just equalled by the rate of the back reaction.

For a reversible reaction

$$A + B \rightleftharpoons C + D$$

it can be shown, experimentally or by a thermodynamic argument, that

$$\frac{[C] \times [D]}{[A] \times [B]} = K$$

where K is a constant known as the equilibrium constant. It is conventional to put the products of the reaction in the numerator in an equilibrium constant expression, and it must be emphasised that the concentrations in such an expression are the concentrations *in the equilibrium mixture.*

For a more general reaction,

$$xA + yB \rightleftharpoons rC + sD$$

the equilibrium constant is given by

$$K = \frac{[C]^r \times [D]^s}{[A]^x \times [B]^y}$$

and this is the mathematical expression of what is sometimes called *the law of chemical equilibrium*.

If the forward and back reactions involved in a reversible reaction are both simple, one-step reactions then their rate constants will be related to the equilibrium constant. For a reversible reaction

$$A + B \rightleftharpoons C + D$$

the rates of the forward and back reactions will be

$$\text{Forward rate} = k' \times [A] \times [B] \qquad \text{Back rate} = k'' \times [C] \times [D]$$

where k' and k'' are the rate constants concerned. At equilibrium

$$\text{Forward rate} = \text{Back rate}$$
$$\therefore k' \times [A] \times [B] = k'' \times [C] \times [D]$$
$$\therefore \frac{[C] \times [D]}{[A] \times [B]} = \frac{k'}{k''} = K$$

In this expression the concentration terms are expressed in mol dm^{-3}, and K is sometimes written as K_c when these units are used. For reactions involving gases, the concentrations may be expressed in terms of partial pressures; the corresponding value of the equilibrium constant in these units is written as K_p. K_c is referred to as the equilibrium constant on a concentration basis; K_p as the constant on a pressure basis.

MEASUREMENT OF RATES OF REACTIONS AND RATE CONSTANTS

The following examples will illustrate the methods adopted for measuring reaction rates and rate constants. The first two examples cannot give accurate numerical results and are for demonstration purposes only.

4. Reaction of sodium thiosulphate(VI) solution and an acid. A solution of sodium thiosulphate(VI) reacts with acids to form a sulphur precipitate,

$$S_2O_3^{2-}(aq) + 2H^+(aq) \longrightarrow H_2O(l) + SO_2(g) + S(s)$$

and the time of the first appearance of the precipitate is a measure of the rate of the reaction, for the more rapidly the reaction is taking place the smaller will be the time required to build up a visible precipitate of sulphur. The rate of reaction is, therefore, inversely proportional to the time taken for the sulphur first to appear.

If 5 cm³ portions of 0.5M hydrochloric acid are added to separate 5 cm³ portions of 1.5M, M, 0.5M and 0.25M solutions of sodium thiosulphate(VI) and the times when the sulphur first appears are recorded, it will be found that the reciprocal of the times are proportional to the original concentration of sodium thiosulphate, i.e.

$$\text{Rate of reaction} \propto \frac{1}{\text{Time}} \propto \text{Concentration of thiosulphate(VI)}$$

In this experiment a constant concentration of acid is used.

By mixing the same acid and hypo solutions at different temperatures it can be shown that rise of temperature increases the rate of reaction.

5. Reaction of potassium bromate(V) and potassium iodide in acid solution. Potassium bromate(V), in acid solution, reacts as bromic(V) acid, and potassium iodide, in acid solution, reacts as hydriodic acid,

$$BrO_3^-(aq) + 6I^-(aq) + 6H^+(aq) \longrightarrow 3H_2O(l) + 3I_2(s) + Br^-(aq)$$

If a little starch is added to the mixture the iodine is detected by the appearance of a blue colour, and the development of the blue colour to match a standard colour can be used to measure the rate of the reaction.

It can readily be shown that the rate is proportional to the concentrations of the potassium bromate(V) and potassium iodide solutions used (p. 341).

6. Decomposition of hydrogen peroxide. The decomposition of hydrogen peroxide is very slow at room temperature,

$$2H_2O_2(aq) \longrightarrow 2H_2O(l) + O_2(g)$$

but it is speeded up by adding a catalyst such as manganese(IV) oxide.

The rate of the reaction can be measured by passing the oxygen evolved

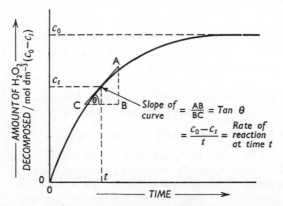

FIG. 121. The decomposition of hydrogen peroxide.

into a gas burette. The rate is then expressed in terms of the volume of oxygen collected per second. Initially, when the concentration of hydrogen peroxide is high, the rate of evolution of oxygen will be high. As more and more hydrogen peroxide is decomposed, the rate will decrease.

Alternatively, the rate of the reaction can be measured by withdrawing samples of the reaction mixture, in a pipette, at definite time intervals. These samples are immediately added to a measured excess of standard potassium manganate(VI) (permanganate) solution which prevents any further decomposition of hydrogen peroxide. By back titration, the concentration of hydrogen peroxide in the sample can be calculated. This will be equal to the concentration of hydrogen peroxide in the reaction mixture at the time the sample was withdrawn.

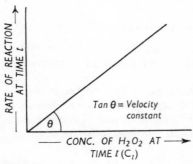

FIG. 122.

If the original concentration of hydrogen peroxide at time zero, i.e. before adding the catalyst, is known the change in concentration with time can be found.

Let the original concentration of hydrogen peroxide at time zero be c_o mol dm^{-3}, and the concentration after time t min be c_t mol dm^{-3}. Then the amount of hydrogen peroxide decomposed in t min will be $(c_o - c_t)$ mol dm^{-3}. The average rate of the reaction over the period of t min will be $\left(\dfrac{c_o - c_t}{t}\right)$ mol dm^{-3} min^{-1}. This is only an average rate, and to get values for more definite rates at definite times a graphical method can be used.

Plotting $(c_o - c_t)$ against t will give a curve as in Fig. 121. At time zero, no hydrogen peroxide will have been decomposed. Eventually, when it has all decomposed, the curve will flatten out indicating the end of the reaction. This will happen when c_t becomes zero, i.e. when there is no concentration of hydrogen peroxide, or when $(c_o - c_t)$ becomes equal to c_o.

The slope of the curve in Fig. 121, at any time, t, will give the rate of the reaction, $\dfrac{c_o - c_t}{t}$, at time t. The corresponding concentration of hydrogen peroxide will be c_t.

Plotting the rates obtained by taking the slopes of the curve in Fig. 121 against the corresponding c_t values will give a straight line as in Fig. 122. This proves that, for this reaction,

$$\text{Rate of reaction} = k[\text{H}_2\text{O}_2]$$

The slope of the line in Fig. 122 gives the value of the rate constant, k.

It may well have been that the rate had to be plotted against c_t^2 to give a straight line. This would have indicated that

$$\text{Rate of reaction} = k'[H_2O_2]^2$$

The numerical values obtained in this typical method of following the way in which the rate of a reaction changes with time can also be treated more mathematically, as explained on p. 337. Such a treatment does not necessitate the relatively inaccurate and tedious procedure of measuring the slope of a curve.

7. Hydrolysis of an ester in dilute acid solution. An ester, such as methyl methanoate (formate), is hydrolysed by water to give an acid and an alcohol,

$$H \cdot COOCH_3(l) + H_2O(l) \longrightarrow H \cdot COOH(l) + CH_3OH(l)$$

Under some conditions this reaction is reversible (see p. 315), but in the presence of excess dilute acid the reaction goes practically to completion, at a reasonable rate.

The rate of decomposition of hydrogen peroxide, described in (6), was followed by measuring the amount of hydrogen peroxide used up in definite time intervals. The rate of hydrolysis of methyl methanoate is followed by measuring how much methanoic acid is formed in definite time intervals.

This is done by withdrawing samples of the reaction mixture at definite time intervals. These samples are immediately diluted, and titrated against standard alkali. The dilution is necessary to prevent any further reaction in the sample withdrawn for measurement. The titration results enable the concentration of methanoic acid formed to be calculated, though in doing this the amount of alkali necessary to react with the dilute acid originally present must be taken into account. This latter amount will be constant throughout the experiment.

As time passes, more and more methanoic acid is formed so that the titration values for successive samples rise. Eventually they reach a maximum value which indicates that the reaction has gone completely and that no further methanoic acid is being formed.

From this final titration value, the total number of moles of methanoic acid formed per dm^3 can be calculated. This must be equal to the original number of moles of methyl methanoate present per dm^3, as decomposition of 1 mol of methyl methanoate produces 1 mol of methanoic acid.

Let this *final* concentration of methanoic acid, i.e. the *initial* concentration of methyl methanoate, be c_o mol dm^{-3}. Then, if c_t is the concentration of methanoic acid measured after time t, the concentration of methyl methanoate remaining undecomposed after time t must be $(c_o - c_t)$, i.e. the original amount minus the amount converted into acid.

Plotting c_t against time gives a curve as in Fig. 123, and the slope of the curve at various times gives the rate of the reaction at those times. These rates plotted against the corresponding $(c_o - c_t)$ values will give a straight line, showing that

$$\text{Rate of reaction} = k[\text{H·COOCH}_3]$$

The slope of the line will give the value of k, the rate constant.

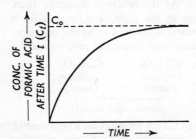

FIG. 123. Formation of methanoic acid by hydrolysis of methyl methanoate.

The concentration of water in this reaction is essentially constant as it is present in such a large excess (see p. 340), and the action of the acid added at the start is catalytic (see p. 354).

As in **6**, the experimental results can be interpreted more mathematically (see p. 337).

8. Reaction between hydrogen peroxide and potassium iodide in acid solution. The Harcourt–Esson reaction.

This reaction can be used to provide another experimental method of measuring reaction rates. The reaction takes place according to the equation,

$$2\text{HI} + \text{H}_2\text{O}_2 \longrightarrow 2\text{H}_2\text{O} + \text{I}_2$$

If it is carried out in the presence of sodium thiosulphate(VI), the iodine formed reacts to re-form hydrogen iodide,

$$\text{I}_2 + \text{Na}_2\text{S}_2\text{O}_3 \longrightarrow 2\text{NaI} + \text{Na}_2\text{S}_4\text{O}_6$$
$$2\text{NaI} + \text{acid present} \longrightarrow 2\text{HI}$$

so that the concentration of hydrogen iodide remains constant throughout.

Hydrogen peroxide, of known concentration, is added to some dilute sulphuric(VI) acid containing some starch solution and 1 cm³ of standard sodium thiosulphate(VI) solution. At time zero, some potassium iodide is added.

As iodine is formed, it reacts with the thiosulphate(VI), but when all the thiosulphate(VI) has been used up the mixture turns blue. The time, t_1, is recorded, and a further 1 cm³ portion of thiosulphate(VI) is added. When the mixture once more turns blue, the time, t_2, is again recorded, and a further 1 cm³ portion of thiosulphate(VI) is added. This is continued for ten or twelve added portions of thiosulphate(VI). The rate of formation of iodine gets slower and slower as time passes so that the time for the reappearance of the blue colour lengthens.

In time t_1, iodine equivalent to 1 cm³ of the standard thiosulphate(VI) is formed. In time t_2, iodine equivalent to 2 cm³ of the thiosulphate(VI) is formed. The amount of iodine formed can, therefore, be plotted against time, and the rate of the reaction at any time can be obtained by taking the slope of this curve. At any time, the concentration of hydrogen peroxide is equal to the original concentration minus the hydrogen peroxide equivalent to the iodine formed. The rate of the reaction can, therefore, be plotted against the concentration of hydrogen peroxide, when it will be found that

$$\text{Rate} \propto \text{Concentration of H}_2\text{O}_2$$

9. The inversion of cane sugar. It is very convenient if a reaction rate can be followed by a purely physical method and this can be done for the inversion of cane sugar.

Cane sugar, or sucrose, is dextrorotatory, but it is hydrolysed, in acid solution, to fructose, which is laevorotatory and glucose, which is dextrorotatory. As the optical activity of fructose is greater than that of glucose, the mixture is, in fact, laevorotatory. As the reaction proceeds, then, polarised light is rotated less and less to the right, and, eventually is rotated to the left. That is why the process is referred to as the inversion of cane sugar, and the mixture of fructose and glucose as invert sugar,

$$C_{12}H_{22}O_{11} \ + \ H_2O \ \longrightarrow \ C_6H_{12}O_6 \ + \ C_6H_{12}O_6$$

Sucrose		Fructose	Glucose
(dextro)		(laevo)	(dextro)

Invert sugar (laevo)

The rate of the reaction is measured by taking readings on a polarimeter. If p_o, p_t and p_∞ are the polarimeter readings at the start, after time t, and at the end of the reaction respectively, then $(p_o - p_\infty)$ is proportional to the original concentration of cane sugar, $(p_o - p_t)$ is proportional to the amount of cane sugar used up after time t, and $(p_t - p_\infty)$ is proportional to the concentration of cane sugar after time t. The + or − values of the polarimeter readings must, of course, be taken into account.

The amount of cane sugar used up, $(p_o - p_t)$, can be plotted against time, and the slope of the curve at various times will give the rate of the reaction at those times. If these rates are plotted against the corresponding $(p_t - p_\infty)$ values it will be found that a straight line results showing that

Rate $\propto (p_t - p_\infty) \propto$ Concentration of cane sugar

10. Use of other physical methods. Physical methods, other than the measurement of optical rotation, can also be used.

Reactions in which there is a change in gas volume can be followed by measuring the change in pressure with time. For reactions with a smaller volume change, perhaps reactions in solution, it may be possible to follow the rate of the reaction by carrying it out in a dilatometer (p. 222).

If coloured substances are involved, colorimetric methods can be employed, and the study of absorption spectra is very similar. Changes in refractive index, dielectric constant, and other physical properties can also be useful.

EQUILIBRIUM MIXTURES. MEASUREMENT OF EQUILIBRIUM CONSTANTS

The state of affairs in a reversible reaction at equilibrium is illustrated in section **11** below, whilst sections **12–16** show how equilibrium constants can be measured.

11. Hydrolysis of bismuth(III) chloride. A dilute solution of bismuth(III)

chloride in a little hydrochloric acid is clear and colourless, but addition of water produces a white suspension. This is because bismuth(III) chloride is hydrolysed by water to form the white, insoluble bismuth(III) chloride oxide (bismuth oxychloride),

$$BiCl_3(aq) + H_2O(l) \rightleftharpoons BiOCl(s) + 2HCl(aq)$$

The reaction is reversible and

$$\frac{[BiOCl] \times [HCl]^2}{[BiCl_3] \times [H_2O]} = K_c$$

The shifting of the equilibrium to the left or the right, in this reaction, can be shown by adding more hydrochloric acid or more water. Addition of water produces more bismuth(III) chloride oxide and the mixture becomes opalescent; the equilibrium shifts from left to right. Addition of acid makes the mixture clear again as the equilibrium shifts from right to left.

12. Reaction between an acid and an alcohol. Acids and alcohols react to form esters and water, e.g.

$$CH_3 \cdot COOH + C_2H_5OH \rightleftharpoons CH_3 \cdot COOC_2H_5 + H_2O$$

To find the equilibrium constant for such a reaction,

$$K_c = \frac{[CH_3 \cdot COOC_2H_5] \times [H_2O]}{[CH_3 \cdot COOH] \times [C_2H_5OH]}$$

it is necessary to measure the concentrations of the substances present *in the equilibrium mixture* at a particular temperature.

This is done by sealing known masses of ethanoic (acetic) acid (a mol) and ethanol (ethyl alcohol) (b mol) in a glass tube at a definite temperature, say 50°C. After some hours at this temperature, the acid and the alcohol will have reacted to form the equilibrium mixture. The tube is then cooled, and the amount of acid it contains is found by titration with standard alkali. The purpose of the cooling is to 'freeze' the equilibrium. It is essential that the equilibrium should not shift whilst the equilibrium concentrations are being measured, and 'freezing' the equilibrium means cooling to a sufficiently low temperature that the equilibrium will not shift appreciably during the measurement of the concentrations.

If there are x mol of acid present at equilibrium, it follows that $(a - x)$ mol of acid have been converted into ester and water. Since 1 mol of ester and 1 mol of water are formed from 1 mol of acid, there must be $(a - x)$ mol of ester and of water present in the equilibrium mixture. Moreover $(a - x)$ mol of alcohol must have been used up, so that the amount of alcohol present at equilibrium will be $\{b - (a - x)\}$ mol.

For a sealed tube of volume V dm³ the equilibrium concentrations, in mol dm⁻³, would be as follows,

Acid $\dfrac{x}{V}$ Ester $\dfrac{a-x}{V}$

Alcohol $\dfrac{b-(a-x)}{V}$ Water $\dfrac{a-x}{V}$

and the equilibrium constant will be given by

$$K_c = \frac{\left(\dfrac{a-x}{V}\right) \times \left(\dfrac{a-x}{V}\right)}{\left(\dfrac{x}{V}\right) \times \left(\dfrac{b-(a-x)}{V}\right)} = \frac{(a-x)^2}{x \times (b-(a-x))}$$

In this reaction, with an equal number of molecules on each side of the equation, the volume of the container cancels out in the equilibrium expression and it need not be known. For a reaction involving a change in the number of molecules involved, the value of V must be known (see p. 321).

13. Reaction between hydrogen and iodine.
Hydrogen and iodine react, reversibly, to form hydrogen iodide,

$$H_2(g) + I_2(g) \rightleftharpoons 2HI(g)$$

To find the equilibrium constant, known masses of hydrogen and iodine are kept in a sealed tube at, say, 450°C, until the equilibrium is established. The equilibrium mixture is then 'frozen' by rapid cooling; rapid so that the equilibrium has not time to shift and adjust itself to the equilibrium which would, given time, exist at the lower temperature.

The amount of iodine in the equilibrium mixture at 450°C is then measured by titration with standard sodium thiosulphate(VI) solution. As a check, too, the amounts of hydrogen and hydrogen iodide can be measured.

If the original amounts of hydrogen and iodine were a mol and b mol respectively, and if x mol of iodine are converted into hydrogen iodide, then $2x$ mol of hydrogen iodide will be formed, and the equilibrium concentrations in mol dm⁻³, in a vessel of volume V dm³, will be

Hydrogen, $\dfrac{a-x}{V}$ Iodine, $\dfrac{b-x}{V}$ Hydrogen iodide, $\dfrac{2x}{V}$

The equilibrium constant will be given by,

$$K_c = \frac{[HI]^2}{[H_2] \times [I_2]} = \frac{\left(\dfrac{2x}{V}\right)^2}{\left(\dfrac{a-x}{V}\right) \times \left(\dfrac{b-x}{V}\right)} = \frac{4x^2}{(a-x) \times (b-x)}$$

V, once again, cancelling out. The fact that V does not appear in the final expression means that the composition of the equilibrium mixture is independent of V, and, consequently, independent of pressure.

14. Use of flow method. The equilibrium mixture resulting from gas reactions can be obtained by passing the reactant gases through a tube, kept at a steady temperature, at such a rate that equilibrium is established. The equilibrium gas mixture coming out of the tube can then be 'frozen', and analysed. If necessary the attainment of equilibrium within the hot region of the tube can be expedited by including a catalyst.

Such a flow method was used, for example, by Haber in his investigation of the formation of ammonia from nitrogen and hydrogen.

15. The reaction $I_2 + I^- \rightleftharpoons I_3^-$. This is typical of a reversible reaction whose equilibrium constant is measured by a partition method. The equilibrium constant is given by

$$K_c = \frac{[I_3^-]}{[I_2] \times [I^-]}$$

and it is necessary to measure, or calculate, the various concentrations in an equilibrium mixture.

This can be done by dissolving iodine and a known mass of potassium iodide, i.e. I^- ions, in a known volume of water and adding some tri- or tetra-chloromethane or benzene. Free iodine molecules distribute themselves between the aqueous and non-aqueous solvents, but I^- and I_3^- do not do this as the ions are insoluble in the non-aqueous solvent.

The concentration of iodine in the non-aqueous layer can be measured by titrating a portion of it and $[I_2]$ in the aqueous layer can be calculated from this value if the partition coefficient of iodine between water and the non-aqueous solvent is known, or is separately measured.

Titration of the aqueous layer gives the total iodine concentration in that layer. This is equal to $[I_2]$ plus $[I_3^-]$. As $[I_2]$ is known, $[I_3^-]$ can be found.

The total iodide ion concentration in the aqueous layer can be calculated from the original mass of potassium iodide added. This total iodide ion concentration is equal to $[I^-]$ plus $[I_3^-]$. As $[I_3^-]$ is known, $[I^-]$ can be found.

The three equilibrium concentrations, $[I_2]$, $[I_3^-]$ and $[I^-]$ give the value for K_c, and the same value will be obtained starting with different amounts of iodine, potassium iodide and water.

16. Measurement of gas pressure. For equilibrium mixtures in which the number of gaseous molecules changes as the equilibrium shifts it is possible to measure equilibrium constants by measuring the pressure exerted by the equilibrium mixture. This is particularly convenient when it is required to measure equilibrium constants at different temperatures. The equilibrium constant for the dissociation of iodine molecules into iodine atoms provides a simple example.

If 1 mol of iodine is taken, in a container of volume, V, the pressure, p, exerted at a temperature, T, at which the iodine is in the vapour state, will be given by RT/V if the iodine did not dissociate. As the iodine does, in fact, dissociate, the measured pressure, P, at T, when the equilibrium has been established, will be equal to $p(1 + \alpha)$, where α is the degree of dissociation. This is because the dissociation gives an increased number of particles,

$$I_2 \rightleftharpoons 2I$$
$$(1 - \alpha) \quad 2\alpha$$

The total number of particles becomes $(1 + \alpha)$, so that the partial pressure of I_2 molecules will be $P(1 - \alpha)/(1 + \alpha)$ whilst the partial pressure of I atoms will be $P2\alpha/(1 + \alpha)$. The equilibrium constant is, therefore,

$$K_p = \frac{(p_I)^2}{(p_{I_2})} = \frac{(P2\alpha)^2(1 + \alpha)}{(1 + \alpha)^2(1 - \alpha)P} = \frac{4\alpha^2 P}{1 - \alpha^2} = \frac{4\alpha^2 p}{1 - \alpha}$$

HETEROGENEOUS REACTIONS

The application of the law of equilibrium to heterogeneous reactions, such as those between gases and solids, is difficult because of the problem of interpreting the meaning of the active mass of a solid. In some heterogeneous reactions, the law cannot be applied, as the surface area and nature of the solid seem to be the predominant factors. There are, however, some selected heterogeneous reactions for which the experimental results can be related to the law of equilibrium by assuming that the active masses of the solids involved remain constant throughout the reaction.

17. Thermal dissociation of calcium carbonate. Solid calcium carbonate undergoes thermal dissociation on heating,

$$CaCO_3(s) \rightleftharpoons CaO(s) + CO_2(g)$$

and the pressure of carbon dioxide in equilibrium with solid calcium carbonate and solid calcium oxide, at any one temperature, is found to be constant. This pressure is sometimes referred to as the *dissociation pressure* of calcium carbonate. Its variation with temperature is shown graphically in Fig. 173 on p. 529.

The law of equilibrium shows that

$$K = \frac{[CaO] \times [CO_2]}{[CaCO_3]}$$

and, if [CaO] and $[CaCO_3]$ be taken as constant, then K' is equal to $[CO_2]$. Such a constant concentration of carbon dioxide would give a carbon dioxide pressure dependent only on temperature, for the value of K will change as temperature does. In this way, then, the law of equilibrium can account for the experimental facts.

The system can also be considered from the point of view of the phase rule. This is done on p. 529, where more details are given.

18. Reaction of steam and iron. The equilibrium constant for the reaction between steam and iron, is given by

$$3Fe(s) + 4H_2O(g) \rightleftharpoons Fe_3O_4(s) + 4H_2(g)$$

$$K_c = \frac{[Fe_3O_4] \times [H_2]^4}{[Fe]^3 \times [H_2O]^4}$$

If the active masses of the solids concerned are taken as constant,

$$K'_c = \frac{[H_2]^4}{[H_2O]^4} \quad \text{or} \quad K''_c = \frac{[H_2]}{[H_2O]} \quad \text{or} \quad K_r = \frac{p_{H_2}}{p_{H_2O}}$$

Such relationships can be tested by measuring the partial pressures of hydrogen and steam in equilibrium with different mixtures of iron and iron oxide. It is found that the ratio p_{H_2}/p_{H_2O} is constant at any one temperature and independent of the amounts of the iron and iron oxide present.

(Questions on this Chapter will be found on pp. 327–9.)

Chapter 29
Effects of catalysis, pressure and temperature on reactions

1. Effect of catalysts on reactions.

a. *Effect on rate and rate constant.* A catalyst alters the rate of a chemical reaction, and, therefore, alters the rate constant of the reaction. In fact, the effectiveness of two catalysts in speeding up a reaction is compared, quantitatively, by measuring the rate constants of the reaction in the presence of each catalyst. The better the catalyst, the higher the rate constant. The functioning of catalysts is considered on p. 357.

b. *Effect on equilibrium mixture and equilibrium constant.* A catalyst does not alter the composition of the equilibrium mixture in a reversible reaction, and, therefore, does not alter the value of the equilibrium constant. A catalyst does, however, speed up both forward and back reactions in a reversible reaction, to the same extent, so that the same equilibrium mixture is obtained more rapidly in the presence of a catalyst.

If a catalyst did affect an equilibrium constant it would defy the law of conservation of energy. A catalyst, for example, which speeded up the change from A to B more than the reversed change from B to A could be used to obtain energy from nowhere. For, in a reversible reaction,

$$A \rightleftharpoons B + x \text{ joule}$$

addition of the catalyst would cause a shift in the equilibrium from left to right. An equilibrium mixture of A and B, without catalyst, could, then, be made to evolve heat simply by adding the catalyst.

2. Effect of pressure on reactions.

a. *Effect on rate and rate constant.* Pressure has very little effect on solids and liquids and does not greatly affect reactions involving only solids and/or liquids.

Reactions involving gases, however, are speeded up by increase of pressure because increasing the pressure of a gas increases its effective concentration. Such increased concentration causes an increased rate of reaction, but the rate constant for the reaction remains unchanged and is not affected by pressure.

b. *Effect on equilibrium mixture and equilibrium constant.* A change of pressure does not change equilibrium constants but it does change the composition of the equilibrium mixture in any reaction in which there is a volume change. This is because change of pressure affects the magnitude of the volume change.

Consider a reaction in which there is an increase in volume, e.g.

$$\underset{a \text{ mol}}{PCl_5} \rightleftharpoons \underset{b \text{ mol}}{PCl_3} + \underset{c \text{ mol}}{Cl_2}$$

and let the amounts of each substance present at equilibrium be as shown.

If, then, the volume containing the equilibrium mixture is V dm^3 the equilibrium constant will be given by

$$K_c = \frac{[PCl_3][Cl_2]}{[PCl_5]} = \frac{\left(\dfrac{b}{V}\right)\left(\dfrac{c}{V}\right)}{\left(\dfrac{a}{V}\right)} = \frac{bc}{aV}$$

As K must remain constant, an increase in pressure, which will lower the volume, V, must lead to a decrease in b and c and an increase in a. This means that the equilibrium will shift from right to left, i.e. the equilibrium mixture will contain more phosphorus pentachloride and less chlorine and trichloride the higher the pressure. Conversely, the equilibrium mixture will contain more chlorine and trichloride and less pentachloride the lower the pressure.

For a reaction in which there is a decrease in volume, e.g.

$$N_2 + 3H_2 \rightleftharpoons 2NH_3$$

the effect of pressure is reversed. The higher the pressure the more ammonia there will be in the equilibrium mixture and vice versa.

In a reaction where the volume of the products is the same as that of the reactants, pressure has no effect on the composition of the equilibrium mixture, though increase in pressure will enable the equilibrium mixture to be attained more quickly in a gaseous reaction.

c. *Le Chatelier's principle.* The conclusions reached in b. can also be deduced from the widely applicable principle of Le Chatelier. This states that *any system in equilibrium shifts the equilibrium, when subjected to any constraint, in the direction which tends to nullify the effect of the constraint.*

Changing the pressure on an equilibrium mixture is applying a constraint to a system in equilibrium. *Increase of external pressure will*, therefore, *cause the equilibrium to shift in the direction which will bring about a lowering of pressure.* In a chemical reaction, this means that the equilibrium will shift in the direction which produces the smaller number of gas molecules.

Increase in pressure therefore causes an equilibrium shift from left to right in the reactions

$$N_2 + 3H_2 \rightleftharpoons 2NH_3$$
$$2SO_2 + O_2 \rightleftharpoons 2SO_3$$
$$NH_3 + HCl \rightleftharpoons NH_4Cl$$

or from right to left in the reactions

$$PCl_5 \rightleftharpoons PCl_3 + Cl_2$$
$$N_2O_4 \rightleftharpoons 2NO_2$$

Decreasing the pressure has the opposite effect, and pressure has no effect on the composition of the equilibrium mixture in a reaction in which no volume change takes place.

3. Effect of temperature on reactions.

a. *Effect on rate and rate constant.* The rate of a chemical reaction is increased by raising the temperature, and the rate constant for a reaction is only a constant at a fixed temperature.

This would be expected from kinetic theory considerations, for increase in temperature will increase molecular motion and raise the rate of inter-molecular collisions.

For the majority of reactions the rate constant, and hence the rate of reaction, is about doubled for a rise in temperature of 10 K. This means an increase of about 10 per cent for each 1 K rise.

Kinetic theory calculations show, however, that the increase in the total number of inter-molecular collisions is only about 2 per cent for each 1 K rise in temperature. The increase in rate of a reaction, as temperature is increased, is, therefore, greater than the corresponding increase in the total number of collisions. Moreover, calculation shows that the total number of collisions at any temperature is greater than the number of molecules which actually react. Only a certain proportion of the total number of molecules react on collision, and this proportion rises more rapidly with increase in temperature than does the total number of collisions.

Arrhenius first suggested, in 1889, that a molecule would only react on collision if it had higher than the average energy, i.e. if it was activated, the necessary energy for reaction to occur being known as the *activation energy*. This has proved to be a very useful idea, for it can be shown that activation energy, E, is the factor which links rate constant with temperature, the relationship being (p. 330),

$$\frac{d \ln k}{dT} = \frac{E}{RT^2}$$

The matter is considered in more detail in Chapter 30.

b. *Effect on equilibrium mixture and equilibrium constants.* In a reversible reaction, change of temperature will affect the rates of both forward and back reactions, but not necessarily to the same extent. Change of temperature, therefore, causes a shift of equilibrium and a change in equilibrium constant, but the change depends on whether exothermic or endothermic reactions are concerned.

The way in which equilibrium constants change with temperature is found, both theoretically and experimentally (p. 331), to be governed by the relationship

$$\frac{d \ln K}{dT} = \frac{\Delta H}{RT^2}$$

where ΔH is the heat of reaction.

The expression gives, on integrating,

$$\ln K = -\frac{\Delta H}{RT} + c$$

where c is a constant, if it is assumed that ΔH does not change with temperature. This relationship can be used to show how the equilibrium mixture obtained in a reversible reaction will change with temperature.

For a reversible reaction which is exothermic in the forward direction, e.g.

$$N_2 + 3H_2 \rightleftharpoons 2NH_3; \quad \Delta H = -92.37 \text{ kJ}$$

the value of ΔH is negative. The value of $-\Delta H/RT$ will, therefore, be positive, and the greater T is the smaller will be its value. K will, therefore, be smaller the greater T is and as

$$K = \frac{[NH_3]^2}{[H_2]^3 \times [N_2]}$$

less ammonia will be found in the equilibrium mixture the higher the temperature is. In other words, increase in temperature shifts the equilibrium from right to left. Conversely, a lowering of temperature will shift the equilibrium from left to right to give an equilibrium mixture containing more ammonia.

For a reversible reaction which is endothermic in the forward direction, e.g.

$$N_2 + O_2 \rightleftharpoons 2NO; \quad \Delta H = +179.9 \text{ kJ}$$

the value of ΔH will be positive, and the value of $-\Delta H/RT$ will be negative. Increase in T will make $-\Delta H/RT$ smaller, so that K will be bigger. Increase in T therefore causes a shift of equilibrium from left to right. Decrease in T causes a shift from right to left.

c. *Le Chatelier's principle.* The shifting of an equilibrium by change of temperature can also be predicted by applying Le Chatelier's principle. *If the temperature of a system in equilibrium is increased, the equilibrium will shift in the direction which absorbs heat.* Lowering of the temperature will shift the equilibrium in the direction evolving heat.

Thus in the reversible reaction

$$N_2 + 3H_2 \rightleftharpoons 2NH_3; \quad \Delta H = -92.37 \text{ kJ}$$

increase in temperature shifts the equilibrium from right to left, and decrease in temperature will shift it from left to right. The equilibrium mixture will contain less ammonia at high temperatures, and more at low temperatures.

In the reaction,

$$N_2 + O_2 \rightleftharpoons 2NO; \quad \Delta H = +179.9 \text{ kJ}$$

increase in temperature will shift the equilibrium from left to right, and decrease in temperature will shift it from right to left.

4. Summary. The table given below summarises the effects of a catalyst, temperature and pressure on the rate, rate constant, composition of equilibrium mixture and equilibrium constant of a reaction.

	Rate of reaction	*Rate constant*	*Composition of equilibrium mixture*	*Equilibrium constant*
Addition of catalyst	Changed	Changed	Unchanged	Unchanged
Increase of pressure	Increased for gas reactions	Unchanged	Changed if there is a change in volume	Unchanged
Decrease of pressure	Decreased for gas reactions	,,	,,	,,
Increase of temperature	Increased	Increased	Changed	Changed
Decrease of temperature	Decreased	Decreased	Changed	Changed
Mathematical relationship		$\dfrac{d \ln k}{dT} = \dfrac{E}{RT^2}$		$\dfrac{d \ln K}{dT} = \dfrac{\Delta H}{RT^2}$

The change in the position of equilibrium between A and A′ as the pressure ór temperature is changed may be summarised as follows:

$$A \rightleftharpoons A' + \text{Heat evolved}$$
(Smaller   (Larger
volume)   volume)

⟵ Increase in pressure ——
—— Decrease in pressure ⟶
⟵ Increase in temperature ——
—— Decrease in temperature ⟶

APPLICATION TO SOME INDUSTRIAL REACTIONS

If a reversible reaction is to be carried out industrially the conditions must be chosen which will give the most economic yield of the required product. The main factors are (a) the proportion of the required product in the equilibrium mixture, and (b) the time taken for the equilibrium mixture to be formed. These factors can be partially controlled by changes of pressure and/or temperature, and the use of catalysts.

5. The Haber process. The exothermic formation of ammonia from hydrogen and nitrogen,

$$N_2 + 3H_2 \rightleftharpoons 2NH_3; \qquad \Delta H = -92.37 \text{ kJ}$$

was thoroughly investigated by Haber between the years 1905 and 1915.

Theory shows that increase of pressure will increase the proportion of ammonia in the equilibrium mixture (p. 321), and that decrease in temperature will have the same effect (p. 323). These conclusions were borne out by experimental studies, the results of which are summarised in Fig. 124.

High pressure clearly gives a high yield of ammonia, but the higher the pressure the greater the cost of equipment to produce and maintain the pressure. Most ammonia plants are, therefore, operated at a pressure of

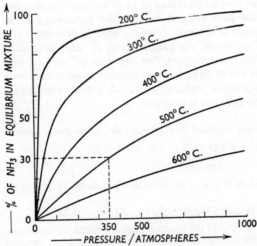

FIG. 124. The percentage of ammonia in the equilibrium mixture obtained from a 3:1 mixture of hydrogen and nitrogen at different temperatures and pressures.

about 350×10^2 kN m^{-2} (350 atmospheres), but pressures as low as 120×10^2 kN m^{-2} (120 atmospheres) or as high as $1\,000 \times 10^2$ kN m^2 (1 000 atmospheres) are used.

Low temperatures give the highest yield of ammonia, but, at a low temperature, the rate of reaction is slow so that the equilibrium mixture is only obtained slowly. A compromise between a reasonable rate of reaction and a reasonable yield of ammonia is obtained by operating at temperatures of about 500°C.

To expedite the attainment of equilibrium, a catalyst, consisting of iron oxides mixed with promoters such as aluminium(III) oxide and potassium hydroxide, is used. In operation, the catalyst is reduced to iron.

An initial gas mixture of hydrogen and nitrogen in the ratio 3:1 by volume is used, and the gas mixture must be freed from carbon monoxide, carbon dioxide, oxygen, water vapour and sulphur compounds to avoid poisoning (p. 353) the catalyst. As only about 30 per cent of ammonia is formed at 350×10^2 kN m^{-2} (350 atmospheres) and 500°C, even if time is allowed for the full establishment of equilibrium, the gas mixture is recirculated, the ammonia formed being removed by solution in water.

6. The contact process. In the contact process, sulphur dioxide and oxygen are converted into sulphur trioxide,

$$2SO_2 + O_2 \rightleftharpoons 2SO_3; \quad \Delta H = -188.3 \text{ kJ}$$

Increase in pressure will give an increased yield of sulphur trioxide, but the effect is small and the yield is good even at low pressures such as 200 kN m^{-2} (2 atmospheres) which are, therefore, used in practice.

As in the synthesis of ammonia, the greatest yield of sulphur trioxide would be obtained at low temperatures, but the rate of reaction and attainment of equilibrium would be very slow. A compromise temperature of about 450°C is used in practice, together with a catalyst of vanadium(v) oxide. A yield of about 98 per cent is obtained.

7. The Bosch process. If steam is passed through hot coke, water gas is formed,

$$\underset{\text{(Steam)}}{H_2O} + C \longrightarrow \underset{\underbrace{}_{\text{Water gas}}}{CO + H_2}; \quad \Delta H = +131.4 \text{ kJ}$$

Carbon monoxide obtained in this way reacts with more steam to g.ve more hydrogen,

$$CO(g) + H_2O(g) \rightleftharpoons CO_2(g) + H_2(g); \quad \Delta H = -41 \text{ kJ}$$

This reversible reaction is the basis of the Bosch process for preparing hydrogen.

As there is no change in the number of molecules during the reaction, pressure has no effect on the equilibrium mixture formed from carbon monoxide and steam. The process is, therefore, operated at a low pressure, just sufficient to maintain a flow of gases.

Lowering the temperature will give a greater yield of hydrogen, but, as in (5) and (6), a compromise has to be struck. A temperature of about 500°C is used, together with a catalyst mixture of iron and chromium oxides.

The yield of hydrogen is also improved by using an excess of steam which pushes the equilibrium from left to right.

8. Synthesis of nitrogen oxide. Nitrogen oxide can be synthesised from nitrogen and oxygen,

$$N_2(g) + O_2(g) \rightleftharpoons 2NO(g); \quad \Delta H = 179.9 \text{ kJ}$$

Pressure has no effect on the composition of the equilibrium mixture. As the reaction is endothermic, raising the temperature increases the yield of nitrogen oxide, and to obtain anything like a reasonable yield a very high temperature has to be used. Even at 2000°C, the yield of nitrogen oxide is only about 2 per cent.

The process has been operated industrially by passing air through an electric arc. The resulting gas mixture, containing some nitrogen oxide, is cooled whereupon the nitrogen oxide is converted into nitrogen dioxide, which can then be made into nitric(v) acid. Some nitrogen and oxygen react together, too, in the atmosphere, when there are lightning flashes.

QUESTIONS ON CHAPTERS 28 AND 29

1. What is meant by the rate of a chemical reaction? Describe a method of measurement of the rate of one chemical reaction, and explain to what use the result may be put. (O. Schol.)

2. When 60 g of ethanoic acid were heated with 46 g of ethanol until equilibrium was reached, 12 g of water and 58.7 g of ethyl ethanoate were formed. What is the equilibrium constant for the reaction between acid and alcohol? What mass of ester would be formed under the same conditions starting from 90 g of ethanoic acid and 92 g of ethanol?

3. Describe the experimental method of measuring the rate of reaction for any one chosen reaction.

4. When 1 mol of ethanol reacts with 1 mol of ethanoic acid until an equilibrium is obtained, there is then present in the mixture 0.333 mol of alcohol and acid, and 0.666 mol of ester and water. Calculate the amount of ester present at equilibrium when (a) 3 mol of ethanol react with 1 mol of acid, (b) 92 g of ethanol react with 60 g of ethanoic acid, (c) 1 mol of ethanol reacts with 1 mol of acid in the presence of 1 mol of water.

5. State the law of mass action and show how you would use it to derive the equilibrium constant of a reaction

$$2A + B = C + 2D + Q \text{ joule}$$

in which the reactants and resultants are all gases. What steps could be taken (a) to obtain as large a yield as possible of C from a given amount of B, and (b) to obtain the yield as quickly as possible? (O. & C.)

6. Describe how the equilibrium between hydrogen, iodine and hydrogen iodide could be measured at, say, 300°C. How would you expect the equilibrium to vary with (a) increasing temperature, (b) increasing pressure? Give your reasons. The reaction at 300°C is slightly exothermic. (O. & C. S.)

7. Starting with equimolecular quantities of ethanol and ethanoic acid, the position at equilibrium consists (in molecular quantities) of $\frac{1}{3}$ ethanol, $\frac{1}{3}$ acid, $\frac{2}{3}$ ester and $\frac{2}{3}$ water. In what molecular proportions must the acid and alcohol be mixed in order to obtain a 90 per cent yield of ester from the quantity of ethanol used?

8. The reaction

$$2NH_3 \rightleftharpoons N_2 + 3H_2$$

is said to be a 'balanced' reaction. What is meant by this? Describe some experiments that might be carried out to show that the reaction is a balanced one. (O. Schol.)

9. What do you understand by the terms reversible reaction, rate constant, equilibrium constant? The equilibrium constant of the reaction

$$C_2H_5OH + CH_3COOH \rightleftharpoons CH_3COOC_2H_5 + H_2O$$

at 25°C is 4.0. Calculate the mass of ethyl ethanoate that can be formed at 25°C from 100 g of ethanol and 100 g of ethanoic acid.

10. What do you understand by chemical equilibrium? Suggest methods for investigating it in two of the systems: (a) $CaCO_3 \rightleftharpoons CaO + CO_2$, (b) $I_2 + I^- \rightleftharpoons I_3^-$, (c) $2NH_3 \rightleftharpoons N_2 + 3H_2$. What generalisations would you expect to be able to draw from these results?

11. Derive expressions for the equilibrium constant for the reaction represented by the following equation:

$$2A + B = C + 2D$$

if the original concentrations of A and B are a and b mol dm^{-3} respectively and if (i) x mol dm^{-3} of C are found in the equilibrium mixture, (ii) y mol dm^{-3} of A are converted into C and D, or (iii) the equilibrium concentration of B is $a/2$.

12. Describe the effect on the equilibrium mixture arising from the following reaction,

$$CO(g) + 2H_2(g \rightleftharpoons CH_3OH(g); \quad \Delta H = -92.05 \text{ kJ}.$$

of (a) increased pressure, (b) increased temperature, (c) increased partial pressure of hydrogen, (d) the presence of a catalyst.

13. Derive an expression for the equilibrium constant for the reaction

$$3H_2 + N_2 \rightleftharpoons 2NH_3$$

in terms of a, b, x and P, where a and b are the original numbers of moles of hydrogen and nitrogen, respectively, and x is the number of moles of nitrogen which react at a total pressure of P N m^{-2}.

14. Give an account of some uses of Le Chatelier's principle.

15. Use Le Chatelier's principle to predict the effect of (i) increased pressure, (ii) decreased temperature on (a) the melting of ice, (b) the solution of sodium chloride in water, (c) the solution of ammonia in water, (d) the formation of nitrogen oxide from nitrogen and oxygen, (e) the formation of sulphur trioxide from sulphur dioxide and oxygen.

16. 4 g of potassium iodide was dissolved in 500 cm^3 of water, and about 1 g of iodine was dissolved in 100 cm^3 of benzene. The two solutions were then mixed and allowed to stand. Subsequent titrations showed that 10 cm^3 of the benzene layer was equivalent to 5.1 cm^3 of M/10 sodium thiosulphate(VI) solution, whilst 50 cm^3 of the aqueous layer was equivalent to 2.9 cm^3 of M/10 thiosulphate. The distribution coefficient of iodine between benzene and water is 130. Calculate the value of the equilibrium constant for the equilibrium,

$$KI + I_2 \rightleftharpoons KI_3$$

17. Calculate the percentage by volume of oxygen converted into nitrogen oxide when air (79 per cent N_2 and 21 per cent O_2 by volume) is heated to 2000°C, given that at this temperature the equilibrium constant, K, for the reaction

$$\tfrac{1}{2}N_2(g) + \tfrac{1}{2}O_2(g) = NO(g)$$

is 2.9×10^{-2}. (W. S.)

18. Distinguish between thermal dissociation and thermal decomposition. Formulate the equilibrium constant in terms of the degree of dissociation and the total pressure for the gaseous system:

$$2NO_2 \rightleftharpoons 2NO + O_2$$

Nitrogen dioxide is dissociated to the extent of 56.5 per cent at 494°C and 99 kN m^{-2} pressure. At what pressure will the dissociation be 80 per cent at 494°C? (W. S.)

19. Discuss the effects of temperature and pressure on the composition of the equilibrium mixture in a reversible reaction. The total gas pressure in a system originally consisting of a sample of solid ammonium hydrogensulphide (NH_4HS) at 25°C is 0.66 atmospheres. Calculate the value of the equilibrium constant, K, in terms of partial pressures, explaining the reasoning underlying any assumptions you may make. What would be the partial pressure of ammonia in the mixture if 0.1 atmospheres of this gas were added to the system? (O. & C.)

20. A solution containing 12.7 g of iodine and 166.1 g of potassium iodide in 1 dm^3 of water is shaken up with 1 dm^3 of benzene. Given that the partition

coefficient for iodine between benzene and water is 400, and that the equilibrium constant for the reaction $I_2 + I^- = I_3^-$ is 730 (concentrations expressed in mol dm^{-3}), calculate (a) the concentration of iodine molecules in the benzene, and (b) the concentration of I_3^- ion water. (Neglect the solubilities of water and potassium iodide in benzene and of benzene in water. Be careful to state the units in which concentrations are expressed.) (O. & C. S.)

21. 1 g of hydrogen and 127 g of iodine are allowed to attain equilibrium in an evacuated container of volume 10 dm^3 at a temperature of 450°C. At this temperature K_c is equal to 50. What is (a) the value of K_p, (b) the total pressure in the container, and (c) the partial pressure of hydrogen in the container?

22. The densities of diamond and graphite are 3.5 and 2.3 g cm^{-3} respectively, and the change from graphite to diamond is represented by the following equation,

$$C \text{ (graphite)} \rightleftharpoons C \text{ (diamond)}; \quad \Delta H = +1.883 \text{ kJ}$$

Is the formation of diamond from graphite favoured by (a) high or low temperature, (b) high or low pressure? Explain your answers.

23. Explain what is meant by the law of mass action. Illustrate its application by reference to (a) the solubility of a sparingly soluble salt, (b) the distribution of a solid between two mutually immiscible liquids, (c) the vaporization of a liquid. (C. Schol.)

24. Two solid compounds, A and B, dissociate at 15°C into gaseous products as follows:

$$A \rightleftharpoons A' + H_2S$$

$$B \rightleftharpoons B' + H_2S$$

The pressure at 15°C over excess solid A was 50 mm and over excess solid B was 68 mm. Find (a) the dissociation constant of A and of B (units, mm^2); (b) the relative number of molecules of A' and of B' in the vapour over a mixture of the solids; (c) the total pressure of gas over the mixture. (S.

Chapter 30
Activation energy

1. Activation energy. Kinetic theory considerations show that the kinetic energy is distributed throughout a gas in such a way that some molecules have high, and some low, energy. Furthermore, Maxwell's distribution law (p. 71) states that the number of molecules, n, with energy greater than a value, E, is related to the total number of molecules, n_0, by the expression

$$n = n_0 e^{-E/RT}$$

where R is the gas constant and T the absolute temperature.

If reaction only occurs between activated molecules, as first suggested by Arrhenius in 1889 (p. 322), then the rate of a reaction would be expected to be proportional to the number of activated molecules. For an activation energy of E, the number of molecules with energy greater than E, i.e. the number of molecules which can take part in reaction, will be n. Both the rate of a reaction and its rate constant, for given concentrations of reagents, will, therefore, be proportional to n, so that

$$\text{Rate of reaction} \propto k \propto Ae^{-E/RT}$$

where A is a constant, n_0 for given concentrations being fixed.

On taking logarithms, this relationship becomes

$$\ln k = C - \frac{E}{RT}$$

where C is a constant. This expression, linking rate constant with temperature, can also be written in a different way, for, if it assumed that C is independent of temperature, differentiation gives the result

$$\frac{\mathrm{d} \ln k}{\mathrm{d}T} = \frac{E}{RT^2}$$

Such relationships between rate constant and temperature, derived from the kinetic theory, can be shown to hold good for many reactions by plotting $\ln k$ against $1/T$. Straight lines result, and values of E, activation energies, can be obtained from the slopes of the lines.

In the thermal decomposition of dinitrogen oxide at high temperatures, for example,

$$2N_2O \longrightarrow 2N_2 + O_2$$

the values of $\log_{10} (k \times 1000)$ were found to be 4.064 at 1125 K and 3.575 at 1085 K. As

$$\log_{10} k_1 - \log_{10} k_2 = -\frac{E}{R \times 2.303}\left[\frac{1}{T_1} - \frac{1}{T_2}\right]$$

it can be calculated that the energy of activation is 285.6 kJ.

2. Effect of temperature on number of activated molecules. Two important points arise from the idea of activation energy.

a. Rise in temperature increases the number of activated molecules much more than the number of molecules with average energy. For a gas at an absolute temperature of T, the average kinetic energy per mole is $3/2 \times RT$ (p. 69). At 1000 K, the average energy will be approximately 12500 J mol^{-1}; at 2000 K, it is 25000 J mol^{-1}. Doubling the absolute temperature doubles the *average* energy.

The fraction of molecules having an energy greater than, say, 83140 J at 1000 K is given by

$$e^{-83\,140/8\,314} = 0.0000457 \quad \text{or} \quad 0.00457 \text{ per cent}$$

but the fraction with energy greater than 83140 J at 2000 K is

$$e^{-83\,140/16\,628} = 0.00677 \quad \text{or} \quad 0.677 \text{ per cent}$$

The temperature change which doubles the number of molecules with average energy will, therefore, increase the fraction of activated molecules by more than a hundred. That is why increase in temperature has such a marked effect on reaction rates.

The fraction of molecules having energy greater than 83140 J at 300 K, i.e. at slightly above room temperature, will be

$$e^{-83\,140/2\,494} = 3.6 \times 10^{-15}$$

b. The smaller the activation energy for a reaction, the more rapidly will it take place. For a reaction between two molecules with activation energy of 83140 Jmol^{-1} the velocity will be reasonably high at room temperature, but if the activation energy is 160000, a temperature of about 400°C will be necessary to achieve a reasonable rate.

3. Activation energy and heat of reaction. In the relationship between rate constant and temperature,

$$\frac{\mathrm{d}\ln k}{\mathrm{d}T} = \frac{E}{RT^2}$$

the activation energy is the important factor. The relationship between equilibrium constant and temperature is very similar in form,

$$\frac{\mathrm{d}\ln K}{\mathrm{d}T} = \frac{\Delta H}{RT^2}$$

but, here, the heat of reaction is the important factor.

This similarity between these two expressions is not unexpected, for an equilibrium constant is a ratio of two rate constants (p. 308), and the heat of reaction for a reversible reaction is related to the activation energies of the forward and back reactions.

An exothermic reaction may be pictured as following a path such as that shown in Fig. 125. Although the products of the reaction have lower energy than the reactants there is an energy hump or barrier which must be over-come before the products can be formed. The corresponding reaction path for an endothermic reaction is shown in Fig. 126.

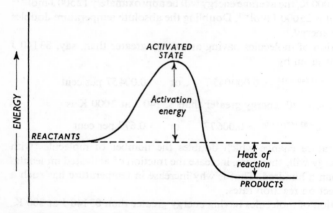

FIG. 125. Reaction path for an exothermic reaction.

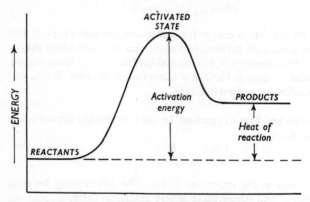

FIG. 126. Reaction path for an endothermic reaction.

For a generalised reversible reaction, represented as

$$A \rightleftharpoons B + Q \text{ joule}; \quad \Delta H = -Q \text{ joule}$$

the reaction path can be written as in Fig. 127. The activation energy for the forward reaction will be E_A; for the back reaction, it will be E_B. The heat of reaction is the difference between the two activation energies.

The rate of the forward reaction will be equal to $k_1[A]$ and that of the back reaction to $k_2[B]$, the equilibrium constant, K, being given by k_1/k_2.

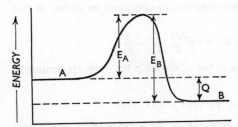

FIG. 127. Reaction paths for a reversible reaction, A $\rightleftharpoons$ B + Q joule.

For the forward reaction,

$$\frac{d \ln k_1}{dT} = \frac{E_A}{RT^2}$$

and for the back reaction,

$$\frac{d \ln k_2}{dT} = \frac{E_B}{RT^2}$$

As K is equal to k_1/k_2, it follows that

$$\frac{d \ln K}{dT} = \frac{d \ln k_1}{dT} - \frac{d \ln k_2}{dT}$$

$$= \frac{E_A}{RT^2} - \frac{E_B}{RT^2} = \frac{E_A - E_B}{RT^2}$$

$$= \frac{-Q}{RT^2}$$

and, using ΔH instead of $-Q$,

$$\frac{d \ln K}{dT} = \frac{\Delta H}{RT^2}$$

which, on integration, gives

$$\ln K = C - \frac{\Delta H}{RT}$$

where C is a constant, if it is assumed that ΔH does not change with temperature.

One further, qualitative conclusion can be drawn from the nature of the reaction path for an endothermic reaction. The activation energy must be greater than the heat of reaction, so that the activation energy of any endothermic reaction with a high heat of reaction is bound to be high. Such reactions are unlikely, therefore, to take place very readily at low temperatures.

4. Equilibrium constant and heat of reaction. The relationships between equilibrium constant and temperature mean that values of heats of

reactions can be obtained from equilibrium constants measured at two different temperatures, for

$$\ln K_2 - \ln K_1 = -\frac{\Delta H}{R}\left\{\frac{1}{T_2} - \frac{1}{T_1}\right\}$$

A particularly impressive use of the relationship lies in its application to the ionisation of water,

$$H_2O(l) \rightleftharpoons H^+(aq) + OH^-(aq)$$

By measuring the equilibrium constant, or the ionic product, at different temperatures (p. 391), the heat of the reaction can be shown to be $\Delta H = 57.33$ kJ, which is in agreement with the measured heat of neutralisation of any strong acid and any strong alkali (p. 300). For example,

$$HCl + NaOH \longrightarrow NaCl + H_2O; \qquad \Delta H = -57.33 \text{ kJ}$$

and this reaction, if full ionisation is assumed, is really,

$$H^+(aq) + OH^-(aq) \longrightarrow H_2O(l); \qquad \Delta H = -57.33 \text{ kJ}$$

Alternatively, if an equilibrium constant can be measured at one temperature the value at some other temperature can be calculated if the heat of reaction is known.

QUESTIONS ON CHAPTER 30

1. The values of $\log_{10} (k \times 1000)$ for the reaction

$$2N_2O \longrightarrow 2N_2 + O_2$$

are 4.064 and 3.575 at 1125 and 1085 K respectively. Calculate the energy of activation for the reaction.

2. The fraction of molecules with a higher energy than E is given by $e^{-E/RT}$. Use this relationship to plot a graph of the fraction of molecules having greater energy than E against E.

3. If the activation energy for a reaction is 83.14 kJ what will be the approximate ratio of the rate constant of the reaction at 27°C to that at 37°C? What will the ratio be for a reaction with an activation energy of 53.59 kJ?

4. Calculate the activation energy for a reaction which doubles in rate between 27°C and 37°C.

5. A very large number of reactions have activation energies between about 60 and 250 kJ mol^{-1}. Why is this?

6. Can you suggest any device, using ball-bearings, which would be helpful in the demonstration of the general meaning of activation?

7. If the degree of dissociation of water is 1.93×10^{-7} at 42°C and 0.39×10^{-7} at 2°C, calculate the heat of reaction for the reaction

$$H_2O(l) \rightleftharpoons H^+(aq) + OH^-(aq)$$

8. If the equilibrium constant for a reaction is 2.9×10^{-5} at 947°C and 10.4×10^{-5} at 1047°C calculate the approximate heat of the reaction.

9. In the thermal dissocation of solid ammonium hydrogensulphide,

$$NH_4HS(g) \rightleftharpoons H_2S(g) + NH_3(g)$$

the total pressure at equilibrium was found to be 23.33 kN m^{-2} at 9.5°C and 66.79 kN m^{-2} at 25.1°C. Calculate the heat of the reaction.

10. The change of equilibrium constant with absolute temperature for the contact process is represented by the equation, $\ln K_p = (94560/RT) - (89.37/R)$. Calculate (a) the value of K_p at 527°C, and (b) the heat of the reaction.

11. The equilibrium constants for the dissociation of oxygen,

$$O_2 = 2O$$

are 9.2×10^{-27} at 527°C and 8.0×10^{-16} at 927°C. Calculate the heat of dissociation.

12. The rate constants for the decomposition of dinitrogen pentoxide at different temperatures are given below:

Temp/°C	25	35	45	55	65
k/min^{-1}	0.00203	0.00808	0.0299	0.0900	0.292

Plot log k against the reciprocal of the Absolute temperature, and use your graph to find the average activation energy for the decomposition.

13. Distinguish carefully between the meaning of activation energy and free energy. Would you expect there to be any relationship between the numerical values of these quantities for a given reaction?

Chapter 31
Order, molecularity and mechanism of reactions

1. Meaning of Terms.

a. *Order*. It has been explained in the preceding pages that experimental measurements on the rate of a simple reaction,

$$x\text{A} + y\text{B} \longrightarrow \text{Products of reaction}$$

may show that

$$\text{Rate of reaction} \propto [\text{A}]^x[\text{B}]^y$$

If this be so for a reaction, then *the total order of the reaction is said to be* (x + y), *the order with respect to A being* x, *and with respect to B,* y.

The order of a reaction need not, however, be a whole number and is not necessarily related, numerically, to the chemical equation usually written for the reaction. It is determined solely by the orders of the concentration terms in the expression which best fits the rate–concentration relationship as measured experimentally.

Thus the order of the reaction usually represented by the equation,

$$2\text{N}_2\text{O}_5 \longrightarrow 4\text{NO}_2 + \text{O}_2$$

is 1, for experimental measurements show that the rate of reaction is proportional to $[\text{N}_2\text{O}_5]$.

At 450°C, the order of the reaction represented by the equation

$$\text{CH}_3\cdot\text{CHO} \longrightarrow \text{CH}_4 + \text{CO}$$

is 1.5, for experimental results show that the rate of reaction is proportional to $[\text{CH}_3\cdot\text{CHO}]^{1.5}$.

The kinetic order of a reaction depends entirely upon experimental kinetic measurements but, when the measurements are complicated, as for a chain reaction, the order cannot be given as a simple number (p. 342).

b. *Molecularity*. The molecularity of a reaction is more concerned with the theoretical reaction mechanism. A reaction may take place in stages, and, if so, the overall rate of the reaction, as measured, will be determined by the rate of the slowest stage which is known as the *rate-determining step. The molecularity of a reaction is defined as the number of atoms or molecules taking part in the rate-determining step. The molecularity must be a whole number.*

Very often the molecularity of a reaction is the same as its order. The decomposition of hydrogen iodide into hydrogen and iodine, for example,

$$2\text{HI} \longrightarrow \text{H}_2 + \text{I}_2$$

is a second order reaction because the rate is proportional to $[\text{HI}]^2$. It is also a bimolecular reaction (molecularity = 2) because two molecules of hydrogen iodide must come together before the reaction can take place.

On the other hand the reaction between hydrogen and deuterium,

$$H_2 + D_2 \longrightarrow 2HD$$

is a bimolecular reaction, but has an order of 1.5.

2. First-order reactions. In a first-order reaction such as

$$A \longrightarrow B + C$$

the rate is proportional to the concentration of A.

If the initial concentration of A, at zero time ($t = 0$) is a mol dm^{-3}, and if x mol dm^{-3} of A decompose in time t, then $(a - x)$ mol dm^{-3} of A will remain undecomposed after time t. At time t, the rate of the reaction is given by dx/dt and

$$\frac{dx}{dt} = k(a - x)$$

where k is the rate constant.

It follows that

$$dt = \frac{dx}{k(a - x)}$$

which, on integration, gives

$$t = -\frac{1}{k} \ln (a - x) + C \text{ (a constant)}$$

When $t = 0$, $x = 0$, i.e. no A has decomposed at zero time, so that

$$C = \frac{1}{k} \ln a$$

Therefore

$$t = -\frac{1}{k} \ln (a - x) + \frac{1}{k} \ln a = \frac{1}{k} \ln \frac{a}{a - x}$$

or $\qquad k = \frac{1}{t} \ln \frac{a}{a - x} = \frac{2.3}{t} \log_{10} \frac{a}{a - x}$ (as $\ln 10 = 2.3$)

For a given reaction under a given set of conditions, the validity of the above relationship can be proved by plotting t against $\ln \dfrac{a}{a - x}$, i.e. $\ln \dfrac{\text{initial concentration}}{\text{concentration at time } t}$. A straight line will result, if the reaction is of the first order, and the rate constant, k, for the reaction can be measured from the slope of the line.

For a first order reaction, the time for any definite fraction to be completed is independent of the original concentration. Thus the time for half the reaction to be completed, i.e. for a to become $a/2$, will be given by

$$t = \frac{1}{k} \ln \frac{a}{a - a/2} = \frac{1}{k} \ln 2$$

This time, known as the *half-life time*, is independent of a, and a similar result is obtained by considering fractions other than a half.

3. Second-order reactions. For a second-order reaction, e.g.

$$A + B \longrightarrow C + D$$

the rate will be proportional to [A][B].

The initial concentrations of A and B may be equal or different. If the initial concentrations of A and B are both a mol dm^{-3}, and if, after time t, x mol dm^{-3} of A and of B have reacted, then

$$\frac{dx}{dt} = k(a - x)^2$$

On integration this gives

$$t = \frac{1}{k} \times \frac{x}{a(a - x)} \qquad \text{or} \qquad k = \frac{1}{t} \times \frac{x}{a(a - x)}$$

and the validity of this relationship can be proved by plotting t against $\frac{x}{a(a - x)}$. Such a plot will give a straight line for a second-order reaction, and the value of the rate constant can be obtained from the slope of the line.

For a second-order reaction, the time for a definite fraction of the reaction to take place is not independent of the initial concentration. *For equal initial concentrations of both reagents, the time for a definite fraction to take place is inversely proportional to the initial concentration.*

Thus, for a to become $a/2$ in the above example, i.e. for half the reaction to take place,

$$t = \frac{1}{k} \times \frac{a/2}{a(a - a/2)} = \frac{1}{ka}$$

For a second-order reaction with different initial concentrations of a mol dm^{-3} and b mol dm^{-3}, the expression relating time and concentration is obtained by integrating

$$\frac{dx}{dt} = k(a - x)(b - x)$$

and is

$$t = \frac{1}{k(a - b)} \times \ln \frac{b(a - x)}{a(b - x)}$$

4. Determination of the order of a reaction.

a. *Graphical method.* Plotting t against $\ln \frac{a}{a - x}$ will give a straight line for a first order reaction, whereas plotting t against $\frac{x}{a(a - x)}$ will give a straight line for a second-order reaction, with equal initial concentrations.

The order of a reaction can, therefore, be obtained by finding out which plot of the experimental data gives a straight line.

b. *Half-life method.* The half-life for a first-order reaction is independent of the initial concentration. For a second-order reaction it is inversely proportional to the initial concentration, and for a third-order reaction it is inversely proportional to the square of the initial concentration.

In general, the half-life times, t_1 and t_2, for a reaction with initial concentrations of c_1 and c_2, are related by

$$\frac{t_1}{t_2} = \left(\frac{c_2}{c_1}\right)^{n-1}$$

where n is the order of the reaction. The value for n which fits the relationship can be found from measured values of t_1, t_2, c_1 and c_2.

The time taken for fractions of the reaction, other than a half, to be completed may also be used.

c. *Ostwald's isolation method.* If all the reactants in a reaction except one are present in very large excess the order with respect to the one reactant can be measured as in a. or b. above. By taking each reactant in turn, in this way, the total order can be obtained.

5. Some examples of reaction mechanisms. Chemical reactions are not usually as simple as chemical equations indicate. There are many complicating factors. The products of a reaction may react to reform the reactants, as in reversible reactions. The products may partially react to form some other products, thus giving rise to a series of consecutive reactions. The reactants may react together in more than one way, so that side reactions exist. Chain reactions may build up.

The elucidation of reaction mechanisms is, therefore, very difficult, and detailed knowledge is not very great. Some of the details to be fitted into a general reaction scheme may have to be, in the first instance, inspired guesses, so that some of the information on reaction mechanisms is, sometimes, unreliable.

Some of the problems which arise, and the sort of answers that can be obtained are mentioned in the following survey of some typical reactions.

a. *The thermal decomposition of dinitrogen pentoxide.* This reaction is one of the very few reactions which are both first order and unimolecular. The stoichiometric equation for the reaction is

$$2N_2O_5 \longrightarrow 4NO_2 + O_2$$

but experimental measurements show that the rate is proportional to $[N_2O_5]$.

The reaction must, therefore, take place in stages, probably as follows:

$$N_2O_5 \xrightarrow{\text{Slow}} NO_2 + NO_3$$
$$NO_2 + NO_3 \xrightarrow{\text{Fast}} NO_2 + NO + O_2$$
$$NO + NO_3 \xrightarrow{\text{Fast}} 2NO_2$$

b. *Radioactive decay*. The rate of disintegration of a radioactive element is proportional to the number of atoms of the element present, as described on p. 122. Radioactive decay is, therefore, a unimolecular process exhibiting first-order kinetics.

c. *Pseudo-first-order reactions*. The reactions in a. and b. are, in all probability, both first-order and truly unimolecular. Many reactions can, however, be made to exhibit first-order kinetics by carrying them out under special conditions.

Simple examples are provided by the hydrolysis of esters (p. 312) or of sucrose (p. 314). These reactions are carried out in the presence of a large excess of water so that the concentration of water remains virtually constant. As a result, the rates of the reactions are found to be proportional to the concentrations of ester and sucrose respectively. The reactions are, however, probably bimolecular.

Similarly, the reaction of hydriodic acid and hydrogen peroxide in acid solution is really bimolecular (p. 313), but if it is carried out in the presence of sodium thiosulphate(VI) the iodide ion concentration is made to remain constant, and the reaction shows first-order kinetics.

Reactions which are not truly first order, but which show first-order kinetics under special conditions are called pseudo-first-order reactions.

d. *The bromination of propanone (acetone)*. The bromination of propanone in dilute aqueous solution is represented by the equation,

$$CH_3 \cdot CO \cdot CH_3 + Br_2 \longrightarrow CH_3 \cdot CO \cdot CH_2Br + HBr$$

The rate of the reaction is found to be proportional to the concentration of propanone, but independent of the bromine concentration. It is clear that the reaction must take place in stages, and that the rate-determining step does not involve bromine. The suggested stages are

$$CH_3 \cdot CO \cdot CH_3 \xrightarrow{\text{Slow}} CH_3 \cdot \underset{\underset{OH}{|}}{C}{=}CH_2$$

$$CH_3{-}\underset{\underset{OH}{|}}{C}{=}CH_2 + Br_2 \xrightarrow{\text{Fast}} CH_3 \cdot CO \cdot CH_2Br + HBr$$

so that the reaction rate is really controlled by the slow change of the ketonic form of propanone into an enolic form.

e. *The dissociation of hydrogen iodide*. This reaction,

$$2HI \longrightarrow H_2 + I_2$$

has been very thoroughly studied. It is bimolecular and has an order of 2.

It is thought to be an example of a simple collision reaction between activated molecules, and there is a fairly good agreement between the rate constant as measured experimentally and as calculated theoretically.

The activation energy is found to be 184.2 kJ mol^{-1} from the variation of rate constant with temperature (p. 330).

The total number of collisions between hydrogen iodide can be calculated from the kinetic theory, and the fraction of molecules having energy greater than the activation energy is equal to $e^{-E/RT}$. The number of collisions between activated molecules can, therefore, be calculated; it works out to be 3.25×10^{17} molecules per cubic decimetre per second for a concentration of hydrogen iodide of 1 mol dm^{-3} at 101.325 kN m^{-2} pressure and 556 K.

This is the number of molecules which should react so that the reaction rate is 3.25×10^{17} molecules dm^{-3} s^{-1}, or $3.25 \times 10^{17}/6.02 \times 10^{23}$, i.e. 5.3×10^{-7}, mol dm^{-3} s^{-1}. This is the theoretical value for the velocity constant. The experimental value, under the same conditions, is 3.5×10^{-7} mol dm^{-3} s^{-1}. The agreement is remarkably good when the number of factors involved in the theoretical derivation is considered.

f. *The oxidation of hydriodic acid by hydrogen peroxide in acid solution.* The overall equation for this reaction is

$$2HI + H_2O_2 \longrightarrow I_2 + 2H_2O$$

It is found to have an order of 2, and the rate-determining step is probably

$$I^- + H_2O_2 \xrightarrow{\text{Slow}} H_2O + IO^-$$

followed by the fast reactions,

$$H^+ + IO^- \xrightarrow{\text{Fast}} HIO$$

$$HIO + H^+ + I^- \xrightarrow{\text{Fast}} H_2O + I_2$$

g. *The saponification of an ester.* The reaction between an ester. e.g. ethyl ethanoate and a solution of an alkali such as sodium hydroxide,

$$CH_3 \cdot COOC_2H_5 + OH^- \longrightarrow CH_3 \cdot COO^- + C_2H_5OH$$

is found, as the equation would suggest, to have an order of 2.

h. *The reaction of hydrobromic and bromic(v) acids.* This reaction provides an example of an order very different from that which the stoichiometric equation would indicate. The overall equation is,

$$5HBr + HBrO_3 \longrightarrow 3Br_2 + 3H_2O$$

The expression which fits the measured rate of the reaction is

$$\text{Rate} = k[H^+]^2[Br^-][BrO_3^-]$$

which shows that the reaction has an order of 4.

The rate-determining step may be

$$2H^+ + Br^- + BrO_3^- \longrightarrow HBrO + HBrO_2$$

but the subsequent fast reactions can only be guessed.

i. *The hydrolysis of alkyl halides.* The hydrolysis of bromomethane by a dilute solution of potassium hydroxide in a solvent of 80 per cent ethanol and 20 per cent water is found to be of second order, and of first order with respect to both the bromomethane and OH^- ion concentrations.

$$\text{Rate} = k[CH_3 \cdot Br][OH^-]$$

It is suggested that the reaction proceeds via the formation of a transition state or activated complex,

$$HO^- + CH_3Br \longrightarrow \{HO....CH_3....Br\} \longrightarrow HO-CH_3 + Br^-$$

Such a process is bimolecular, and a similar mechanism accounts for the hydrolysis of bromoethane under similar conditions.

But for higher halides such as tertiary butyl bromide, the kinetics of the hydrolysis are first order. The rate is independent of the OH^- ion concentration, the mechanism postulated being,

$$Bu \cdot Br \xrightarrow{\text{Slow}} Bu^+ + Br^-$$

$$Bu^+ + OH^- \xrightarrow{\text{Fast}} Bu \cdot OH$$

This is a unimolecular process.

The bimolecular process is referred to as an S_N2 reaction: S for substitution, $_N$ for nucleophilic (the OH^- ion is a nucleophilic reagent, p. 411, seeking a nucleus for its electrons), and 2 for bimolecular. The unimolecular process is an S_N1 reaction.

The way in which a halide will undergo hydrolysis depends mainly on the nature of the halide and how readily it can ionise, and on the ionising power of the solvent. Thus the S_N2 hydrolysis of bromomethane in ethanol–water solution is replaced by an essentially S_N1 hydrolysis in methanoic acid solution, i.e. in a better ionising medium. In ethanol–water solution, prop-2-yl bromide (isopropyl bromide) undergoes both S_N1 and S_N2 hydrolysis side by side.

The mechanism by which a halide undergoes hydrolysis under different conditions can be accounted for by electronic theories of organic chemistry.

6. Chain reactions. The reaction between hydrogen and iodine, like the dissociation of hydrogen iodide (p. 341), is a bimolecular reaction taking place by collision between activated molecules, but the reactions of hydrogen with bromine and chlorine are very different in mechanism.

The experimental measurements of the rate of formation of hydrogen bromide from hydrogen and bromine are found to fit the complicated expression,

$$\text{Rate} = \frac{k[H_2][Br_2]^{\frac{1}{2}}}{m + [HBr]/[Br_2]}$$

where k and m are constants. Clearly the idea of a numerical value for the order of such a reaction is of no value.

The theoretical interpretation of these experimental results involves a chain reaction. The first stage is thought to be the slight dissociation of bromine molecules into atoms,

$$Br_2 \longrightarrow 2Br$$

which is followed by the two reactions,

ii $\qquad$ $Br + H_2 \longrightarrow HBr + H$

iii $\qquad$ $H + Br_2 \longrightarrow HBr + Br$

As a result, two molecules of hydrogen bromide are formed, but an atom of bromine has also been formed and this can react with more hydrogen, as in ii, and thus set off a series of successive reactions, known as a chain reaction. If this was all that happened the reaction would go on until all the hydrogen or bromine present was used up.

The uninhibited propagation of the chain reaction is, however, limited by reactions such as

iv $\qquad$ $H + HBr \longrightarrow H_2 + Br$

v $\qquad$ $Br + Br \longrightarrow Br_2$

Reaction i is called a chain-initiating reaction; reactions ii and iii are chain propagating reactions; reactions iv and v are chain breaking, chain terminating or chain limiting reactions.

Such reactions can be interpreted theoretically by assuming that a *stationary state* is set up soon after the reaction is started. The rate of formation of any atom will be equal to its rate of removal by further reaction. By equating the necessary rate reaction expressions it is possible to obtain an overall rate expression in agreement with that found experimentally.

In the formation of hydrogen bromide by the suggested series of reactions, the reaction ii is endothermic and is relatively slow. With iodine, instead of bromine, the analogous reaction is still more endothermic and so slow that the chain reaction taking place in the formation of hydrogen bromide does not take place in the formation of hydrogen iodide. The direct collision mechanism is more advantageous for hydrogen iodide.

With chlorine, however, the reaction

$$Cl + H_2 \longrightarrow HCl + H$$

is rapid, so that the chain reaction mechanism takes place very easily, and often explosively, in the formation of hydrogen chloride from hydrogen and chlorine.

These chain reactions are propagated by atoms; they are sometimes known as *atomic chains*. Others occur in which the propagation is due to free radicals of short life.

7. Free Radicals. Radicals formed by the rupture of a covalent bond in such a way that each atom joined by the bond retains one of the electrons from the shared pair, are known as free radicals. Bond rupture of this type is said to be *homolytic*. Thus the homolytic fission of a covalent bond between a group A and a group B gives free radicals of A and B,

$$A \overset{\times}{\underset{\bullet}{}} B \longrightarrow A \times + \bullet B$$

Free radicals are most commonly obtained by thermal or photochemical decomposition.

a. *Thermal decomposition*. Paneth first used this method in 1929 to show the existence of free radicals. A stream of pure nitrogen or helium at about 250 N m^{-2} (2 mm) pressure was saturated with tetramethyl lead vapour by passing it over liquid tetramethyl lead, $Pb(CH_3)_4$. The mixture then passed into a quartz tube heated at point A to 500–600°C. The thermal decomposition of the tetramethyl lead was shown by the formation of a deposit of lead at A.

The gaseous products passing along the tube were found to be capable of reacting with cold metallic deposits of lead, zinc or antimony placed in the tube at B, beyond A. The products of these reactions were identified as metallic methyl compounds, indicating that the initial tetramethyl lead had decomposed, at 500–600°C, into lead and free methyl radicals.

These free radicals had only a very short life, for they would not react with metals at B if the distance from A to B was too great. By varying the distance, AB, and using gas flows of different speeds it was possible to measure the half-life period of the methyl radical; the value is about 10^{-2} s.

These experiments show that free radicals can be obtained by thermal decomposition, and the method of reaction with a metal also enables alkyl free radicals to be detected.

b. *Photochemical decomposition* (p. 347). Absorption of a quantum of light of wavelength 0.3 μm (3000 Å) by a molecule is equivalent to the absorption of about 420 kJ mol^{-1}, and this might be quite sufficient, if absorbed advantageously, to split a chemical bond, as bond energies are of the order 200–400 kJ mol^{-1} (p. 180).

Propanone (acetone), and other ketones, can, for instance, be split into free radicals by ultra-violet light,

$$CH_3 \cdot CO \cdot CH_3 \longrightarrow CH_3 \cdot CO + CH_3$$

and tetramethyl lead and alkyl halides can also be split into free radicals photochemically. The existence of the alkyl free radicals can be shown by their reaction with metallic deposits as in a. The splitting of chlorine molecules into atoms by light is a similar process.

In *flash photolysis*, a very powerful flash of light with energy of about 10^5 joule and a duration of about 10^{-4} s is passed through a gas mixture, and the free atoms and radicals formed can be investigated by taking a continuous photographic record of the absorption spectrum from the gas mixture.

Free radicals, like free atoms, are highly reactive and they play an important part as transient intermediates in many reactions.

8. Reactions involving free radicals. Free radicals may participate in chemical reactions either by undergoing a simple recombination after the formation or by taking part in a chain reaction.

Simple recombination of ethyl radicals probably takes place in the Kolbe reaction, the electrolysis of a solution of the sodium or potassium salt of a carboxylic acid. The main product of the reaction is an alkane, but by-products are always formed. Sodium propanoate, for example, yields mainly butane, but also ethane, ethene and ethyl propanoate. The reaction mechanism suggested to account for these experimental facts involves the formation of free propanoate radicals, which then split up to form free ethyl radicals,

$$C_2H_5 \cdot COO^- \longrightarrow C_2H_5 \cdot COO + 1e$$
$$C_2H_5 \cdot COO \longrightarrow C_2H_5 + CO_2$$

The free ethyl radicals may form butane by simple recombination

$$2C_2H_5 \longrightarrow C_4H_{10}$$

and ethane, ethene and ethyl propanoate may also be formed by the reactions,

$$2C_2H_5 \longrightarrow C_2H_6 + C_2H_4$$
$$C_2H_5 + C_2H_5 \cdot COO \longrightarrow C_2H_5 \cdot COOC_2H_5$$

It seems very likely that a great many reactions, particularly those of organic compounds, involve chain reactions in which free radicals are concerned, but not many reactions have been elucidated beyond doubt.

Chlorine and methane will not react in the dark, but the reaction can be initiated by heating or by the action of ultra-violet light. Once started, the reaction proceeds rapidly. The chain reaction thought to be taking place is made up of the following stages,

$$Cl_2 \longrightarrow 2Cl \qquad \text{Initiating}$$
$$\left. \begin{array}{l} Cl + CH_4 \longrightarrow HCl + CH_3 \\ CH_3 + Cl_2 \longrightarrow CH_3Cl + Cl \end{array} \right\} \text{Propagating}$$
$$\left. \begin{array}{l} 2Cl \longrightarrow Cl_2 \\ 2CH_3 \longrightarrow C_2H_6 \end{array} \right\} \text{Limiting}$$

Further chlorination of the methane can take place via the reactions,

$$Cl + CH_3 \cdot Cl \longrightarrow HCl + CH_2Cl$$
$$CH_2Cl + Cl_2 \longrightarrow CH_2Cl_2 + Cl$$

The thermal decomposition of ethanal (acetaldehyde), at high temperatures,

$$CH_3 \cdot CHO \longrightarrow CH_4 + CO$$

is another reaction which might take place by a chain reaction. The measured order of the reaction is $1\frac{1}{2}$, and the possible reaction mechanism involves the stages

$$CH_3 \cdot CHO \longrightarrow CH_3 + CHO \qquad \text{Initiating}$$

$$\left.\begin{array}{l} CH_3 + CH_3 \cdot CHO \longrightarrow CH_4 + CH_3 \cdot CO \\ CH_3 \cdot CO \longrightarrow CH_3 + CO \\ CHO \longrightarrow CO + H \\ H + CH_3 \cdot CHO \longrightarrow H_2 + CH_3 \cdot CO \end{array}\right\}\text{Propagating}$$

$$\left.\begin{array}{l} CH_3 + CH_3 \longrightarrow C_2H_6 \\ CH_3 + CH_3 \cdot CO \longrightarrow CH_3 \cdot CO \cdot CH_3 \\ 2CH_3 \cdot CO \longrightarrow CH_3 \cdot CO \cdot CO \cdot CH_3 \end{array}\right\}\text{Limiting}$$

By applying the stationary state method (p. 343) to these separate reactions, the theoretical order for the overall reaction is found to be $1\frac{1}{2}$.

9. Characteristics of chain reactions. The majority of chain reactions have two characteristics which often enable the existence of a chain reaction to be recognised.

a. *Catalysis by free radicals.* If a chain reaction depends on free radicals then the introduction of those radicals into the reacting mixture ought to expedite the reaction. There are many examples where this is found to be so.

Ethane and propane can be chlorinated almost completely in the dark at 150°C if a trace of tetraethyllead vapour is present, to provide ethyl radicals. Similarly, the decomposition of butane at 500°C can be brought about very easily by the addition of dimethylmercury vapour. The use of tetraethyllead as an anti-knock in petrol is almost certainly connected with its ability to facilitate the decomposition of hydrocarbons, and free radicals play a very important part in the cracking of fuel oils.

Ethanal (acetaldehyde) does not undergo thermal decomposition at 300°C but addition of a little azomethane, to provide methyl radicals, causes instant decomposition. The chain reaction between hydrogen and chlorine can be brought about not only by heat or light, but also by a small amount of sodium vapour, which forms chlorine atoms by the reaction,

$$Na + Cl_2 \longrightarrow NaCl + Cl$$

b. *Inhibition.* Just as chain reactions are catalysed by the free radicals which propagate the chains, so they are inhibited if anything is done to remove the vital free radicals. Introduction of substances which readily react with free radicals slows down or stops chain reactions by terminating the chains.

Nitrogen oxide, which is itself a molecule with an odd electron (p. 192), is particularly effective in inhibiting many chain reactions, presumably by a process such as,

$$CH_3 + NO \longrightarrow CH_3NO \longrightarrow H_2O + HCN$$

Similarly, oxygen inhibits the reaction between chlorine and hydrogen or alkanes by reacting with the chlorine atoms,

$$O_2 + Cl \longrightarrow ClO_2$$

Because of these possibilities of chains being broken, chain reactions are generally very sensitive to the presence of impurities.

10. Photochemistry. The preceding discussion of free radicals shows that light can initiate chemical reactions, and the chemical effects of visible, infra-red and ultra-violet light are studied in photochemistry. Such chemical effects may be very useful as, for example, in the important process of photosynthesis, in photographic processes, and in other reproduction processes such as blueprint paper.

Only light which is absorbed by a system can give rise to chemical effects, and this is sometimes referred to as the first law of photochemistry. It was first suggested as early as 1818 by Grotthus and Draper. Einstein suggested, in 1908, that *the absorption of one quantum of light activates one atom or molecule of absorbent,* and this statement is sometimes known as the second law of photochemistry.

The energy of one quantum of light is equal to $h\nu$, where h is Planck's constant (p. 105) and ν the frequency of the light used. On a mole basis, the energy is $Nh\nu$ where N is the Avogadro's constant, and this amount of energy is known as *an einstein*. For visible light of wavelength 0.6 μm (6000 Å), its value is 188.3 kJ mol^{-1}; for ultra-violet light of wavelength 0.2 μm (2000 Å) it is 5648 kJ mol^{-1}.

11. Quantum yield. A molecule activated by the absorption of light energy might not undergo any chemical reaction. It may, however, be decomposed by the energy, and the resulting atoms or free radicals may initiate a chain reaction. If the absorption of one quantum of light energy simply causes the decomposition of one molecule, without any subsequent reaction, the *quantum yield* of the change, defined as the *number of molecules decomposed or formed per quantum absorbed,* will be 1. Such a state of affairs, however, is rare. More commonly, the initial absorption of one quantum leads to a series of changes corresponding to quantum yields with very varied numerical values.

The effect of ultra-violet light on a mixture of hydrogen and chlorine, for example, is to decompose the chlorine molecules. The chlorine atoms so formed initiate a chain reaction,

$$Cl_2 + h\nu \longrightarrow 2Cl$$
$$Cl + H_2 \longrightarrow HCl + H$$
$$H + Cl_2 \longrightarrow HCl + Cl$$

giving an overall quantum yield for the reaction of between 10^4 and 10^6. By comparison, the quantum yield for the decomposition of ethanedioic (oxalic) acid in the presence of uranyl salts,

$$UO_2^{2+} + h\nu \longrightarrow {}^*UO_2^{2+}$$
$$\text{(uranyl ion)} \qquad \text{(activated)}$$
$${}^*UO_2^{2+} + H_2C_2O_4 \longrightarrow CO + CO_2 + H_2O + UO_2^{2+}$$

by light of wavelengths between 0.25 and 0.45 μm is 0.5.

The quantum yield of this latter reaction is so constant that the reaction can be used to measure the amount of light being absorbed by measuring the amount of ethanedioic acid decomposed by titration with standard potassium manganate(VII) (permanganate). For every two molecules of ethanedioic acid decomposed, one quantum is absorbed.

12. Radiation chemistry. Atoms and molecules can be activated and/or decomposed by radiation other than that of visible, ultra-violet and infra-red light. Still more marked effects can be produced by subjecting substances to the effect of high energy radiation such as X-rays and γ-rays or beams of protons or neutrons or α- or β-particles.

Such radiations have sufficiently high energy (that of γ-radiation may be as high as 2000 MeV per quantum) to cause ionisation in the matter by which they are absorbed; they are called *ionising radiations* whether they are electromagnetic, such as X- or γ-rays, or particulate, such as α- and β-rays. The electrons released in the original ionisation also have high energy and bring about still further ionisation.

The resulting chemical changes are widespread and complex and cannot yet be fully interpreted. A typical example is provided by the radiation decomposition (radiolysis) of water by α-particles. The main products are hydrogen and hydrogen peroxide, and possible processes include

$$H_2O \xrightarrow{\alpha\text{-particles}} H_2O^+ + e$$
$$H_2O^+ + e \longrightarrow H + OH$$
$$H_2O^+ + H_2O \longrightarrow H_3O^+ + OH$$
$$H + H \longrightarrow H_2$$
$$OH + OH \longrightarrow H_2O_2$$

Radiation chemistry is still in its infancy. It is of great interest, partially because of the known biological effects of radiation and the connection with such diseases as cancer.

QUESTIONS ON CHAPTER 31

1. 5 cm³ portions of M/15 sodium thiosulphate(VI) solution and M/10 hydrochloric acid were mixed at different temperatures, the time for the appearance of a sulphur precipitate being recorded. The results obtained were,

Temp/°C	25	32	35	44	49	61	68
Time/s	60	43	25	16	11	6	4

Plot a graph of log 1/time against 1/absolute temperature and comment on its significance.

2. The polarimeter readings in an experiment to measure the rate of inversion of cane sugar were as follows:

Time/min	0	10	20	30	40	80	∞
Angle	32.4	28.8	25.5	22.4	19.6	10.3	−14.1

What is the order of the reaction?

3. The rate of inversion of cane sugar is of the first order. If 25 per cent of a sample of cane sugar is hydrolysed in 60 seconds, how long will it take for 50 per cent to be hydrolysed?

4. A solution of hydrogen peroxide was titrated against an M/50 solution of potassium manganate(VII), and 25 cm³ of the peroxide solution were found to be equivalent to 46.1 cm³ of the manganate(VII) solution. On adding colloidal platinum to a sample of the peroxide solution decomposition began. At intervals,

25 cm³ portions were withdrawn and titrated, rapidly, against the same manganate(VII) solution. The results obtained were as follows:

Time/min	5	10	20	30	50
cm³ of KMnO₄	37.1	29.8	19.6	12.3	5.0

Find, either graphically or mathematically, the value of the velocity constant for the decomposition of the peroxide solution.

5. The rate constant of a unimolecular reaction $AB \longrightarrow A + B$ is 1.4×10^{-4} s^{-1}. Explain what this statement means. In what time is half the compound AB changed into A and B? (O. Schol.)

6. Methoxymethane decomposes, under certain conditions, into methane, carbon monoxide and hydrogen:

$$(CH_3)_2O \longrightarrow CH_4 + CO + H_2$$

A sample of the ether was found to exert an initial pressure of 40 kN m^{-2}. After 10 seconds, the pressure had risen by 1.08 kN m^{-2}. How long will it take for the pressure to reach 80 kN m^{-2}?

7. In the reaction between nitrogen oxide and hydrogen to form nitrogen and water, the times for half change are 140 and 102 second at initial gas pressures of 38.4 and 45.39 kN m^{-2} respectively. What is the order of the reaction? Suggest a possible mechanism.

8. In the thermal decomposition of dinitrogen oxide the half-life in seconds rises from 255 to 470 to 860 as the initial pressure is changed from 39.47 to 18.53 to 7 kN m^{-2}. What is the order of the reaction?

9. Explain how you would follow the decomposition of diazoacetic ester which decomposes in aqueous solution with the evolution of nitrogen according to the following equation,

$$N_2CH \cdot COOEt + H_2O \longrightarrow HOCH_2 \cdot COOEt + N_2$$

Describe the apparatus that you would use and outline the sort of measurements you would make. What form of rate equation would you expect the decomposition to follow? (O. Schol.)

10. The hydrolysis of methyl methanoate in dilute aqueous solution is catalysed by hydrogen ions. Describe the experiments you would make to verify this statement, and to discover how the rate of hydrolysis depends on the concentration of the ester and of the hydrogen ion.

The reaction between propanone and iodine (to give iodopropanone) is catalysed by hydrogen ions. The rate of the reaction is proportional to the product of the propanone concentration and of the hydrogen ion concentration, but is independent of the iodine concentration. What can you infer from this information? (O. Schol.)

11. What factors may influence the rate of a chemical reaction? Illustrate your answer with reference to (a) the saponification of ethyl ethanoate, (b) the reaction between sulphur dioxide and oxygen, and (c) the preparation of oxygen from potassium chlorate(V). For any one of these reactions describe how the rate could be measured experimentally. (O. Prelims.)

12. Some chemical reactions take place at great speed; others proceed very slowly. What are the reasons for these differences? Illustrate your answer with examples. (O. Schol.)

13. What is the order of a reaction in which half the reagents react in half an hour, three-quarters in one hour and seven-eighths in one and a half hours?

14. The thermal decomposition of phosphine into phosphorus and hydrogen is found to be a first-order reaction. Suggest a possible mechanism.

15. The rate of reaction between potassium iodate(v) and sulphuric(IV) acid solutions is found to be proportional to the product of the concentrations of potassium iodate(v) and sulphuric(IV) acid. The overall reaction is represented by the equation

$$KIO_3 + 3H_2SO_3 \longrightarrow KI + 3H_2SO_4$$

Suggest a likely mechanism for the reaction.

16. The rate of reaction between phenylamine (aniline) and iodine to form 3-iodophenylamine in the presence of potassium iodide and a dilute buffer solution is found to decrease with increasing iodide concentration and increase with increasing pH. Suggest a possible mechanism of the reaction.

17. Organic esters can be hydrolysed in both acid and alkaline solutions. Compare the two reaction mechanisms.

18. What is the value of an einstein, in joules, for red light of frequency 4×10^{14} Hz?

Chapter 32
Catalysis

1. Characteristics and examples of catalysis. Berzelius first used the word catalysis, in 1835, to summarise the effect on a variety of reactions of the addition of a substance which was able to speed up the reaction but itself remained chemically unchanged. Nowadays *a catalyst is defined as a substance which will alter the rate of a chemical reaction, itself remaining chemically unchanged at the end of the reaction.*

The detailed functioning of a catalyst is not, in every case, well understood, but there are well-established features characteristic of catalytic action, which are described below.

a. *A catalyst is chemically unchanged at the end of a reaction.* Qualitative and quantitative analysis show that a catalyst undergoes no chemical change in a catalytic reaction, but it may undergo a physical change.

Manganese(IV) oxide (manganese dioxide), used as a catalyst in the thermal decomposition of potassium chlorate(V), is powdered in the course of the reaction, and the surface of platinum gauze, used in the catalytic oxidation of ammonia to nitric(V) acid, is roughened.

b. *A small amount of catalyst often affects the rate of a reaction for a long time.* As a catalyst is not used up it will go on functioning for a considerable time; theoretically, but not practically, for ever. Moreover, in many reactions, only a minute amount of catalyst is needed. Copper(II) ions, at a concentration of about 1 g in 10^9 dm³, catalyse the oxidation of sodium sulphate(IV) solutions, and a concentration of 2 g of colloidal platinum in 10^6 dm³ catalyses the decomposition of hydrogen peroxide.

In some reactions, however, the rate of reaction is directly proportional to the concentration of catalyst present. Catalysis by H^+ or OH^- ions (p. 357) is usually of this type.

c. *The effect of a solid catalyst is dependent on its state of subdivision.* A solid lump of platinum, for example, will have less catalytic activity than colloidal or spongy platinum, or platinised asbestos. Finely divided nickel is a better catalyst than lumps of solid nickel.

d. *A catalyst will not affect the composition of an equilibrium mixture* (see p. 320).

e. *Catalysts are often specific in their action.* One catalyst will alter the rate of one reaction without necessarily having any affect at all on other reactions.

Different catalysts, moreover, can bring about completely different reactions. Ethanol vapour, for example, is dehydrated to ethene or ethoxyethane (diethyl ether) by passing over hot aluminium(III) oxide,

$$C_2H_5OH(g) \xrightarrow{Al_2O_3} C_2H_4(g) + H_2O(g)$$
$$2C_2H_5OH(g) \xrightarrow{Al_2O_3} C_2H_5 \cdot O \cdot C_2H_5(g) + H_2O(g)$$

whereas it is dehydrogenated to ethanal (acetaldehyde) or ethyl ethanoate by passing over hot copper,

$$C_2H_5OH(g) \xrightarrow{Cu} CH_3 \cdot CHO(g) + H_2(g)$$
$$2C_2H_5OH(g) \xrightarrow{Cu} CH_3 \cdot COOC_2H_5(g) + 2H_2(g)$$

Similarly, methanoic (formic) acid vapour is dehydrated by passing over hot aluminium(III) oxide,

$$H \cdot COOH(g) \xrightarrow{Al_2O_3} CO(g) + H_2O(g)$$

but dehydrogenated by passing over hot zinc oxide,

$$H \cdot COOH(g) \xrightarrow{ZnO} CO_2(g) + H_2(g)$$

Enzymes, which catalyse so many and such important processes, are particularly specific in their action. Urease, which can be extracted from soya beams, will catalyse the hydrolysis of carbamide (urea) even when present only to the extent of 1 part in 10 millions, but it has no effect on the hydrolysis of methyl carbamide.

Other remarkable chemical changes catalysed by enzymes include the conversion of starch into maltose by diastase, maltose into glucose by maltase, glucose into ethanol and carbon dioxide by zymase, and starch into sugars by enzymes, such as pepsin, present in saliva juices.

'Natural' catalysts can, in fact facilitate many complex reactions which simply cannot be carried out in a laboratory.

f. *Negative catalysts*. A catalyst alters the rate of a chemical reaction, and in industry the catalysts which speed up a reaction are generally of most importance. Some catalysts, however, retard a reaction; they are known as negative catalysts. Typical examples are

Reaction	Negative catalyst
$2H_2O_2 \longrightarrow 2H_2O + O_2$	Dilute acids. Propane–1,2,3–triol (Glycerol)
$2C_6H_5 \cdot CHO + O_2 \longrightarrow 2C_6H_5 \cdot COOH$	Benzene-1,4–diol
$2H_2 + O_2 \longrightarrow 2H_2O$	Iodine. Carbon monoxide
$H_2SO_3 + Air \longrightarrow H_2SO_4$	Benzenol. Tin(II) chloride

Negative catalysts may function by poisoning a catalyst (p. 353) already present in the reaction mixture, by being preferentially adsorbed on the walls of the reaction vessel, or by breaking a chain of reactions. Tetraethyl lead (anti-knock), for example, is added to petrol to stop chain reactions which cause pre-ignition or knocking.

Other everyday examples of the use of negative catalysts are provided by antioxidants in rubber and plastics to retard atmospheric oxidation which causes perishing, stabilisers in rocket fuels, and rust inhibitors in anti-freeze mixtures for use in internal combustion engines.

g. *The effect of a catalyst is often enhanced by adding promoters.* The activity of a catalyst is sometimes very greatly increased by the addition of a small amount of other substances, which may have no catalytic effect on their own. Such substances are known as *promoters*.

A remarkable example is provided by the reaction between iodide and peroxodisulphate ions,

$$2I^-(aq) + S_2O_8^{2-}(aq) \longrightarrow I_2(s) + 2SO_4^{2-}(aq)$$

This reaction is catalysed by an iron(II) sulphate(VI) solution as dilute as M/32000. But further addition of M/2500000 copper(II) sulphate(VI) solution increases the rate of the reaction still further even though the copper(II) sulphate(VI) solution will not act catalytically by itself.

In the Haber process, the efficiency of an iron catalyst is promoted, by adding oxides such as iron(III) oxide, potassium oxide and aluminium oxide.

Other reactions where mixtures are used to obtain the maximum catalytic effect are

Reaction	Catalyst mixture
$2CO + O_2 \longrightarrow 2CO_2$	Hopcalite ($MnO_2 + CuO$)
$CO + 2H_2 \longrightarrow CH_3OH$	Zinc and chromium(III) oxides
Water gas $+ H_2 \longrightarrow$ Hydrocarbons (Fischer–Tropsch process)	Iron or cobalt oxide and thorium oxide

h. *Catalysts can be poisoned.* Just as small amounts of some substances promote catalysis, so other substances poison a catalyst and render it in-effective. Examples are given below

Reaction	Catalyst	Catalyst poison
$2H_2 + O_2 \longrightarrow 2H_2O$	Platinum	CO, H_2S or CS_2
$2SO_2 + O_2 \longrightarrow 2SO_3$	Platinum	Arsenic compounds
$C_2H_4 + H_2 \longrightarrow C_2H_6$	Copper	CO or mercury
$2H_2O_2 \longrightarrow 2H_2O + O_2$	Platinum	HCN, HS_2 or $HgCl_2$

Less of the catalyst poison than is necessary to cover the surface of the catalyst with a monomolecular layer will, nevertheless, still poison the catalyst.

The action of some catalyst poisons is, too, specific, so that a poison can be used to retard an undesirable reaction. Ethanal vapour passed over copper at 300°C is dehydrogenated to ethanal (acetaldehyde) (p. 352), but the ethanal formed then tends to decompose into methane and carbon monoxide. This decomposition can be retarded by poisoning the copper catalyst by including a little water vapour with the alcohol vapour. The rate of change of ethanol into ethanal is not affected by this treatment.

Water, in this example, is the catalyst poison, but the commonest poisons, listed above, are also dangerous physiological poisons, which possibly prevent important body reactions taking place by retarding the catalytic effect of enzymes.

i. *Transitional elements and many of their compounds have marked catalytic activity.* This is illustrated by the following examples of reactions in which transitional elements or their compounds are used as catalysts. It seems probable that the variable valency of the transitional elements is, in some way, related to their catalytic activity.

Reaction	Catalyst
$2SO_2 + O_2 \longrightarrow 2SO_3$	Pt or V_2O_5
$C_2H_4 + H_2 \longrightarrow C_2H_6$	Finely divided
(Catalytic hydrogenation)	Ni
$CO + H_2O \longrightarrow CO_2 + H_2$	Fe_2O_3
$2KClO_3 \longrightarrow 2KCl + 3O_2$	MnO_2
$2CH_3 \cdot CHO + O_2 \longrightarrow 2CH_3 \cdot COOH$	Manganese(II)ethanoate
$C_6H_6 + Cl_2 \longrightarrow C_6H_5 \cdot Cl + HCl$	Fe filings
$C_6H_5 \cdot Cl + NH_3 \longrightarrow C_6H_5 \cdot NH_2 + HCl$	Cu_2O
$2C_6H_6 + 2HCl + O_2 \longrightarrow 2C_6H_5 \cdot Cl + 2H_2O$	$CuCl_2$
(Raschig's reaction)	
$C_6H_6 + HCl + CO \longrightarrow C_6H_5 \cdot CHO + HCl$	$Cu_2Cl_2 + AlCl_3$
(Gattermann's reaction)	
Bleaching powder suspension $\xrightarrow{\text{Warm}} O_2$	Co or Ni salts

2. Autocatalysis. One of the products of a reaction may act as a catalyst for the reaction. The phenomenon is known as autocatalysis, and the initial reaction rate rises, as the catalytic product is formed, instead of decreasing steadily. The plot of reaction rate against time shows a maximum.

Examples of autocatalytic reactions, with the product acting as the catalyst underlined, are represented by the following equations:

$$CH_3 \cdot COOCH_3 + H_2O \longrightarrow \underline{CH_3 \cdot COOH} + CH_3OH$$

$$2KMnO_4 + 5H_2C_2O_4 + 3H_2SO_4 \longrightarrow 2\underline{MnSO_4} + K_2SO_4 + 8H_2O + 10CO_2$$

$$2AsH_3 \longrightarrow 2\underline{As} + 3H_2$$

The oxidising action of concentrated nitric(v) acid is also catalysed by nitrogen dioxide, so that the acid is a poor oxidising agent when it contains no dissolved nitrogen dioxide but a very strong agent when it is saturated with the dioxide, as in fuming acid. Moreover, nitrogen dioxide is formed as a product when the acid does oxidise,

$$2HNO_3 \longrightarrow H_2O + 2NO_2 + [O]$$

so that the oxidising action is autocatalytic. This accounts for the marked increase in reaction rate when copper is dipped into concentrated nitric(v) acid. By comparison, there is no reaction if carbamide (urea) is present, for carbamide reacts with any nitrogen dioxide present or formed.

The atmospheric oxidation of linseed oil, an important process in the drying of paints and varnishes, is also autocatalytic, the catalyst being a peroxide. Catalysts added to linseed oil mixtures to speed up the drying process are known as siccatives.

3. Theories of catalytic action. There are so many different catalysts that classification is not easy, and there is certainly no all-embracing theory of catalytic action. Hinshelwood has written that 'there is no sense or profit in talking about theories of catalytic reactions in general' because 'the methods of operation of catalysts are as diverse as the modes of chemical change'.

In the widest sense, catalysts can be regarded as homogeneous or heterogeneous. In *homogeneous catalysis*, the reaction takes place in one phase and the catalyst is uniformly distributed and is itself in the same phase. It is generally to be found in reactions taking place in the liquid phases, or in solution. In *heterogeneous* or *surface* or *contact* catalysis, the catalyst is not in the same phase as the reacting mixture. This type of catalysis is found in gas reactions catalysed by a solid.

The two most likely mechanisms of catalytic activity involve the formation of intermediate compounds and the adsorption of one or more of the reactants. Intermediate compound formation is to be found most commonly in homogeneous catalysis; adsorption, in heterogeneous catalysis.

a. *Intermediate compound formation.* It is suggested that a reaction represented by the equation,

$$A + B \longrightarrow D$$

might take place, in the presence of a catalyst, C, via the stages

$$A + C \longrightarrow AC$$
$$AC + B \longrightarrow D + C$$

each stage being faster than the direct reaction.

The catalyst does play a part in the reaction, being converted into an intermediate compound, but is eventually reformed. If the formation and reaction of the intermediate compound are reactions with lower energies of activation than the direct reaction, then the effect of the catalyst in speeding up the reaction is explained. In simple language, a big hill is surmounted by taking an alternative, two-stage route involving two smaller hills.

The suggestion is certainly feasible, but the actual isolation of intermediate compounds, which would prove their existence, is very difficult, perhaps because the intermediate compounds, by their very nature, are unstable. In general, then, the intermediate compounds suggested as being formed in a catalytic reaction are usually conjectural.

A mechanism commonly suggested for the thermal decomposition of potassium chlorate(v) in the presence of manganese(IV) oxide, for example, is represented by the equations,

$$2KClO_3 + 6MnO_2 \longrightarrow 6MnO_3 + 2KCl$$
$$6MnO_3 \longrightarrow 6MnO_2 + 3O_2$$

but the suggested intermediate compound, MnO_3, has never been isolated, and the detailed mechanism for this reaction is undoubtedly very much more complicated. Indeed, potassium chloride and oxygen are not the only products.

In some reactions, where the 'catalyst' must be present in fairly large quantities, the formation of intermediate compounds can be proved. In the Friedel–Crafts' reaction, for example, using aluminium chloride as the 'catalyst', the reaction mechanism is represented by the following equations,

$$CH_3 \cdot Cl + AlCl_3 \longrightarrow [CH_3]^+[AlCl_4]^-$$
$$C_6H_6 + [CH_3]^+[AlCl_4]^- \longrightarrow C_6H_5 \cdot CH_3 + AlCl_3 + HCl$$

with $[CH_3]^+[AlCl_4]^-$ as the intermediate compound. Similarly crystals of nitrosylsulphuric(VI) acid, $NO \cdot HSO_4$, can be isolated in the lead chamber process for making sulphuric(VI) acid, and the equations

$$2NO_2 + 2H_2O + 2SO_2 \longrightarrow 2SO_5NH_2 \text{ (sulphonitronic acid)}$$
$$2SO_5NH_2 + NO_2 \longrightarrow 2NO \cdot HSO_4 + H_2O + NO$$
$$2NO \cdot HSO_4 + SO_2 + 2H_2O \longrightarrow 3H_2SO_4 + 2NO$$
$$3NO + 1\tfrac{1}{2}O_2 \longrightarrow 3NO_2$$

give a better indication of the reaction mechanism than the simpler equations

$$2NO_2 + 2H_2O + 2SO_2 \longrightarrow 2H_2SO_4 + 2NO$$
$$2NO + O_2 \longrightarrow 2NO_2$$

In both the Friedel–Crafts' and the lead chamber reactions, however, it can be argued that the aluminium chloride and the nitrogen dioxide are not genuine catalysts because they must be present in reasonably high concentration.

The suggested formation of intermediate compounds is probably most usefully applied in the field of acid-base catalysis (p. 357).

b. *Adsorption by catalysts.* In a typical example of heterogeneous catalysis, such as a reaction between two gases catalysed by a solid, it is suggested that at least one of the reacting gases is adsorbed on the solid surface, with a resultant weakening of the bonds in the molecules concerned rendering them more reactive. If the adsorption is chemisorption (p. 496), something not unlike an intermediate compound is being formed.

Subsequent reaction takes place between adsorbed molecules, or between one adsorbed molecule and a molecule in the gaseous phase. Finally, the reaction products are desorbed from the surface of the catalyst. In some reactions, too, the rates of diffusion to and from the catalyst surface are of importance.

The very real importance of surface effects in some chemical reactions can very readily be shown. The bromination of ethene (ethylene), for example, takes place quite readily at 200°C in a glass flask,

$$C_2H_4(g) + Br_2(g) \longrightarrow C_2H_4Br_2(g)$$

Coating the inside of the glass flask with paraffin wax stops the reaction almost completely, whilst increasing the glass surface area by the addition of glass beads to the flask increases the reaction rate considerably.

Yet too little is known about the mechanism of adsorption at solid surfaces for this adsorption theory of catalytic action to be widely applicable in any detail.

4. Acid–base catalysis. Catalysis by acids and/or bases provides one of the commonest types of homogeneous catalysis.

Many reactions are well known to be catalysed by H^+ and/or OH^- ions. Examples are provided by the hydrolysis of esters (p. 312), the inversion of cane sugar (p. 314) and the decomposition of hydrogen peroxide (p. 311).

The rate of inversion of cane sugar, in particular, is proportional to the H^+ ion concentration, and measurement of the reaction rate can be used as a method of measuring unknown H^+ ion concentrations (p. 404). In this reaction, the anions present are not of primary importance.

For some reactions, however, both H^+ and OH^- ions have a catalytic action. The rate of mutarotation of glucose, for example, is approximately proportional to the H^+ ion concentration at pH less than three and approximately proportional to the OH^- ion concentration at pH greater than six. For pH between 3 and 6, the rate of the change is independent of pH. Similarly, the rate of iodination of propanone reaches a minimum at pH of about 4. As the pH is changed (either lowered or increased) the reaction rate increases.

The functioning of H^+ ions as catalysts is thought to be due to the formation of an intermediate compound between the ions and one of the reactants. Similarly, the functioning of OH^- ions is thought to be due to their ability to accept a proton from one of the reagents. In keto–enol tautomerism, for example, the acid catalysed change is represented by the equations

whilst the base catalysed change is represented by

If H^+ and OH^- ions act as catalysts in this way it would be expected that other proton donors and proton acceptors would catalyse reactions in the same sort of way, and this is, in fact, so.

Nitramide, for example, decomposes in aqueous solution into water and dinitrogen oxide,

$$HN_2NO_2(aq) \longrightarrow N_2O(g) + H_2O(l)$$

The reaction rate can be measured by measuring the rate of evolution of gas. In alkaline solution, the reaction is catalysed by OH^- ions,

$$NH_2NO_2 + OH^- \longrightarrow H_2O + NHNO_2^-$$
$$NHNO_2^- \longrightarrow N_2O + OH^-$$

In a buffer solution of ethanoic acid and ethanoate ions, the OH^- ion concentration is too low to catalyse the reaction, but ethanoate ions, acting as a base, do catalyse the reaction,

$$NH_2NO_2 + Ac^- \longrightarrow HAc + NHNO_2^-$$
$$NHNO_2^- \longrightarrow N_2O + OH^-$$
$$OH^- + HAc \longrightarrow H_2O + Ac^-$$

The reaction is said to display *general basic catalysis*. Other reactions display *general acid catalysis*, and the keto–enol transformation is subject to both general acid and general base catalysis. The transformation will not take place in an aprotic solvent (p. 409) such as a pure hydrocarbon, but addition of an acid, e.g. ethanoic acid, or a base, e.g. ethylamine, enables the reaction to take place.

Since a solvent such as water can act as either an acid or a base it may itself act as a catalyst. It has, in fact, been reported that many common reactions do require a trace of moisture. Ammonia and hydrogen chloride, for example, will not react if they have been dried for long periods over phosphorus(v) oxide.

It is possible, then, that many reactions in solution may be catalysed by the solvent. Addition of salts to such solvents will affect the ionic concentrations, leading to so-called *salt effects*.

QUESTIONS ON CHAPTER 32

1. Describe and illustrate, with examples, the essential features of a catalyst. (O. & C.)

2. What do you understand by the term catalyst? Give three examples of catalysis, and state how you would show experimentally that manganese(iv) oxide catalyses the decomposition of heated potassium chlorate(v). (Camb. 1st M. B.)

3. What are the essential characteristics of a catalyst? What general explanations can you give of catalytic activity? How would you find out whether copper(ii) oxide catalysed the decomposition of potassium chlorate(v)? (O. & C.)

4. State the characteristic features of catalysis. Define the following terms, quoting one example of each: (*a*) homogeneous catalysis, (*b*) heterogeneous catalysis, (*c*) catalyst poison, (*d*) autocatalysis. What influence has a catalyst on the composition of the final equilibrium mixture of a reaction? Mentioning essential experimental details, describe concisely how you would find out whether a given mineral acid acted as a catalyst for the hydrolysis of ethyl ethanoate by water. (N.)

5. State two important characteristics of a catalyst. For each of the two main

processes for manufacturing sulphuric(VI) acid outline one mechanism which has been suggested as accounting for the action of the catalyst. What is meant by (a) negative catalysis, (b) autocatalysis, (c) poisoning of catalysts? Quote one example of negative catalysis and one example of autocatalysis. Describe one industrial consequence of the poisoning of catalysts. (N.)

6. What effect can a catalyst have on (a) the velocity of a reaction, (b) the equilibrium position in a reversible reaction? What types of catalyst are recognised and what theories have been advanced to explain their action? Name three industrial processes in which catalysts are used, stating the catalyst in each case. (O. & C.)

7. A catalyst is something which speeds up a reaction but does not start it. Criticise this definition of a catalyst.

8. How would you demonstrate, experimentally, that the reaction between ethanedioic (oxalic) acid and potassium manganate(VII) was autocatalytic and that Mn^{2+} ions were responsible for this?

9. Draw the general shape of the rate of reaction-time graph for an autocatalytic reaction.

10. Explain why copper dipped into concentrated nitric(V) acid reacts slowly at first and then with increasing speed. Explain, also, why the copper will not react if some carbamide is added to the nitric(V) acid. What is the effect of (a) hydrogen peroxide, (b) sodium nitrate(III) on the reaction between copper and nitric(V) acid?

11. Give examples, with equations where possible, of chemical reactions which are catalysed by (a) a metallic surface, (b) a colloidal metal, (c) hydrogen ions.

12. Comment on the following: (a) The reaction between solutions of sodium chloride and silver nitrate(V) is said to be instantaneous. (b) Hydrogen and oxygen normally need heating before they will combine, but on a palladium surface they will combine at ordinary temperatures. (c) Although a catalyst affects the speed of a reaction it does not alter the position of equilibrium. (d) Several reactions of chlorine are accelerated by light. (O. & C. S.)

13. Write an essay on 'Enzymes'.

14. The rate constants for the decomposition of nitramide at 15°C are summarised, as follows, together with the corresponding ionic concentrations:

Rate const. $\times 10^5$	212	246	383	526	726	800
$[H^+] \times 10^5$	8.2	7.8	4.0	3.5	0.98	1.7
$[CH_3 \cdot COOH] \times 10^4$	146	162	136	169	67	126
$[CH_3 \cdot COO^-] \times 10^4$	35·5	41.4	68.3	96.1	136	158

What conclusions can you draw?

15. The following equations have been suggested to account for the catalytic action of manganese(IV) oxide on the thermal decomposition of potassium chlorate(V),

$$2MnO_2 + 2KClO_3 \longrightarrow 2KMnO_4 + Cl_2 + O_2$$
$$2KMnO_4 \longrightarrow K_2MnO_4 + MnO_2 + O_2$$
$$K_2MnO_4 + Cl_2 \longrightarrow 2KCl + MnO_2 + O_2$$

Describe carefully what experiments you would carry out to attempt to prove the validity or falseness of these equations.

16. What is meant by the statement that negative autocatalysis occurs in the esterification of an organic acid?

Chapter 33
Ionic theory and conductivity

Faraday's laws (p. 89) were the first outcome of quantitative measurements on the passage of electricity through solutions, and these laws, together with many other observations and measurements to be described in this group of chapters, are accounted for by the ionic theory first suggested by Arrhenius in 1887.

The ionic theory, in its broadest outline, is concerned with the formation, the number and the speed of ions in a liquid or a solution.

1. Formation of ions. *Any compound, which, in solution (most commonly in water) or in the fused (molten) state, will conduct an electric current, and be decomposed by it, is called an electrolyte.* Other substances are known as non-electrolytes.

The ionic theory supposes that a molten electrolyte, or a solution of an electrolyte, contains free, electrically charged atoms or radicals known as ions. These ions are responsible for the electrical conductivity of the liquid or solution, and for many of its physical and chemical properties. That is why the number of ions present, and their speeds, are matters of importance.

The following points regarding the formation of ions should be noted carefully.

a. Any electrolyte gives both positively and negatively charged ions, but the total charge on the positively charged ions is exactly balanced by that on the negatively charged ions, e.g.

$$MgCl_2 \longrightarrow Mg^{2+} + 2Cl^-$$
$$FeCl_3 \longrightarrow Fe^{3+} + 3Cl^-$$
$$(NH_4)_2SO_4 \longrightarrow 2NH_4^+ + SO_4^{2-}$$

Solutions of electrolytes are, therefore, electrically neutral though they contain free positive and negative ions.

b. Positive ions are called *cations*, as they migrate towards the cathode during electrolysis; they are formed from atoms or radicals by the loss of electrons. Negative ions, known as *anions*, are formed by the gain of electrons. For example

$$Na \longrightarrow Na^+ + 1e \qquad Cl + 1e \longrightarrow Cl^-$$
$$\text{A cation} \qquad\qquad\qquad \text{An anion}$$

The ions of metals, and the ammonium and hydrogen ions, are, generally, positively charged; those of non-metals and acid radicals are, generally, negatively charged.

c. The loss or gain of electrons by an atom or a radical affects its properties, so that an ion and an atom of the same element are quite distinct. The one is electrically charged; the other is neutral.

d. Ions in solution are, generally, solvated. This means that the simple ion is combined with some solvent molecules. The H^+ ion, for example, has a molecule of water combined with it in aqueous solution; it exists as the oxonium or hydronium, H_3O^+, ion. Such hydration or solvation of an ion has often to be taken into account but when it is not of particular significance the simple ion is usually written in chemical equations.

2. The degree of ionisation. Weak and strong electrolytes. Experimental results show, as will be seen, that there are two general types of electrolytes known as weak and strong electrolytes.

a. *Weak electrolytes.* Weak electrolytes, e.g. most organic acids and bases, water and ammonia solution, are comparatively poor conductors in solution. This is due to the fact that only a fraction of their molecules split up into ions in solution. The solution formed contains some ions, but these are in equilibrium with unionised molecules, e.g.

$$H_2O \rightleftharpoons H^+ + OH^-$$
$$HEt \rightleftharpoons H^+ + Et^-$$
Ethanoic acid   Ethanoate ion

The lack of complete ionisation accounts for the low conductivity.

The fraction, or percentage, of molecules which are ionised is known as the degree of ionisation. In M/10 ethanoic acid solution, for example, the degree of ionisation is 0.0134 or 1.34 per cent. Such a solution contains mainly ethanoic acid molecules, together with relatively few hydrogen and ethanoate ions. In pure water there are mainly H_2O molecules and only very, very few H^+ and OH^- ions (p. 390).

One of the most important contributions of Arrhenius was his suggestion (p. 366) that *the degree of ionisation of a weak electrolyte increases as the solution containing the ions is diluted*, until at infinite dilution (zero concentration) the degree of ionisation rises to 1 or 100 per cent. In M/1000 ethanoic acid solution, for example, the degree of ionisation is 0.134 or 13.4 per cent. This rather unexpected proposition is of major importance in accounting for the behaviour of weak electrolytes.

The actual number of ions in a solution of a weak electrolyte depends, therefore, on (i) the total concentration of the solution, i.e. on the number of molecules which might provide ions, and (ii) the degree of ionisation, i.e. the extent to which any molecules present are split up into ions.

b. *Strong electrolytes.* Strong electrolytes, e.g. mineral acids, alkalis and most salts, give solutions with much higher conductivities than weak electrolytes. This is because strong electrolytes are more or less completely ionised when in the molten state or in solution. In other words, their degree of ionisation is 1 or 100 per cent, or nearly so. Strong electrolytes, in fact, often exist as ions in the solid form, e.g. sodium chloride (p. 169).

If it is assumed that a strong electrolyte is fully ionised in a solution of any dilution, the number of ions present in the solution will depend only on the concentration of the solution.

3. The speed of ions. Ionic interference. The ionic theory assumes that the passage of an electric current through a solution of an electrolyte depends both on the number of ions present in the solution and on their speed.

For a strong electrolyte, fully ionised, the number of ions present in a solution will be proportional to the concentration of the solution, but the electrical conductivity of such solutions is *not* proportional to their concentrations.

The experimental effect of concentration on the conductivity of a solution of a strong electrolyte is now regarded as being due to a variation in ionic speed. In a solution containing a high concentration of ions, a high degree of ionic interference is envisaged, which effectively decreases ionic speed. As a solution is diluted, such ionic interference gets smaller and smaller, and it is this change in the extent of ionic interference which is used to account for the way in which the conductivity of a solution of a strong electrolyte changes with concentration (p. 366).

This idea of ionic interference is a valuable addition to the original ideas of the ionic theory.

4. Summary of ionic theory. The important fundamental ideas of the ionic theory may be summarised as follows:

a. A solution of an electrolyte contains free ions.

b. Passage of a current through a solution of an electrolyte depends on (i) the number, (ii) the speed of the ions present.

c. In a solution of a weak electrolyte the degree of ionisation increases as the dilution increases, until, at infinite dilution (zero concentration), there is complete ionisation.

d. In a solution of a strong electrolyte there is always complete ionisation, i.e. the degree of ionisation is approximately 1 or 100 per cent, but ionic interference limits the movement of the ions. This ionic interference becomes less as the dilution increases.

CONDUCTIVITY OF SOLUTIONS OF ELECTROLYTES

5. Conductivity.* Solutions of electrolytes, like metallic conductors, obey Ohm's law except under abnormal conditions. The current flowing through a given solution under given conditions is, therefore, proportional to the

* The term electrolytic conductivity is used when the context demands it.

reciprocal of the resistance of the solution, and this quantity is known as the conductance of the solution; it is expressed in ohm^{-1} (Ω^{-1}). A solution with a resistance of 10 Ω has a conductance of 0.1 Ω^{-1}. Remember that high resistance means low conductance, and vice versa.

To compare the resistances of different substances the idea of resistivity is used. Similarly, for conductance, conductivity is used. The resistivity of a conductor is defined as the resistance between opposite faces of a unit cube of the conducting material. As resistance is proportional to length and inversely proportional to cross-sectional area, it follows that

$$R = \frac{\rho l}{a} \quad \text{or} \quad \rho = \frac{Ra}{l}$$

where R is the resistance, l is the length, a is the area, and ρ is the resistivity of the material concerned.

If R is measured in ohms, l in metres, and a in square metres, ρ will be in ohm metre (Ω m) units. These are the basic SI units but Ω cm and $\mu\Omega$ m units are also commonly used.

The reciprocal of the resistivity ($1/\rho$) is known as the conductivity (κ). The basic SI units of conductivity are ohm^{-1} metre^{-1} (Ω^{-1} m^{-1}) but Ω^{-1} cm^{-1} units are very common. 1 Ω^{-1} cm^{-1} = 100 Ω^{-1} m^{-1}.

The importance of the measurement of the conductivities of solutions of electrolytes is concerned with the way in which they vary with the concentration of the solution as described on p. 365.

6. Measurement of conductivity of solutions. The conductivity of a solution is obtained by measuring its resistance using a modified Wheatstone

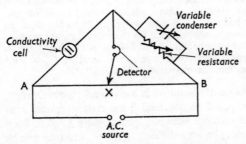

FIG. 128. Wheatstone bridge circuit for measuring conductivity.

bridge circuit. Because direct current causes a back e.m.f. (p. 439) due to polarisation, a rapidly alternating current must be used, and a telephone receiver or an oscilloscope is used in the Wheatstone bridge circuit to detect the balance point.

A typical arrangement is shown in Fig. 128. The variable resistance and

the position of X along a wire, AB, are changed until no current passes
through the detector. At this balance point,

$$\frac{\text{Resistance of Conductivity Cell}}{\text{Resistance of Variable Resistance}} = \frac{\text{AX}}{\text{XB}}$$

and the value of the resistance of the conductivity cell is obtained.

The a.c. source can be an induction coil, but in modern work a vacuum tube
oscillator is used; it is quieter and gives a more symmetrical current. For accurate
work the capacitance of the conductivity cell is balanced by having a variable
condenser in parallel with the variable resistance. The conductivity cell containing
the solution under test is made of insoluble glass or fused silica and must be
thoroughly washed and steamed before use. The shape of the cell varies as shown
in Fig. 129. The electrodes are generally of thick platinum foil, firmly fixed in
position and coated with a layer of platinum black to decrease polarisation effects.

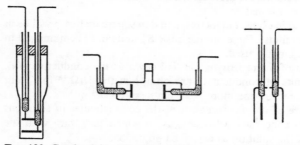

FIG. 129. Conductivity cells. The right-hand one is a dipping cell.

Specially purified water, known as conductivity water, must be used for accurate
work in making the solution. Such water is obtained by distilling good distilled
water, containing a little potassium permanganate(VII), in a hard glass distillation
flask using a tin or resistance-glass condenser. Alternatively, high-grade ion-
exchange resins (p. 500) can be used. Ultra-pure water, made by the repeated
distillation under reduced pressure, must be used for the most accurate work.

Ordinary distilled water has a conductivity of $5–10 \times 10^{-6}$ ohm^{-1} cm^{-1}.
Conductivity water of quite good quality will have a value of about 0.5×10^{-6}
ohm^{-1} cm^{-1}. Ultra-pure water may have a value of 0.05×10^{-6}.

From the measured resistance of a solution in a given cell, the resistivity,
and hence the conductivity, could be obtained if the cross-sectional area
of the electrodes, a, and the distance between them, l, were known. Such
measurements would, however, be very difficult to make, and they can be
avoided by making use of what is known as the *cell constant*. For a given
cell, both l and a are constant, and the ratio, l/a, is called the cell constant.

As R is equal to $l/\kappa a$ it follows that

$$\text{Conductivity} = \frac{1}{\text{Measured resistance}} \times \text{Cell constant}$$

A cell constant can be obtained from the cell dimensions, but it is more
commonly measured by using a solution, such as M/10 potassium chloride,

whose conductivity is known. Once measured, the cell constant for a cell is fixed so long as the physical dimensions of the cell are not altered in any way. To ensure this the electrodes in a conductivity cell must be rigidly fixed in their relative positions.

7. Variation of conductivity with concentration. The conductivity of a solution varies with its concentration as shown in Fig. 130. The full curve, with the maximum, is only obtained when a wide range of concentrations is possible. In many cases a saturated solution is obtained before the maximum is reached so that only the left hand part of the curve is obtainable.

Some substances, such as potassium chloride, give solutions with much higher conductivities than others, such as ethanoic acid. Some typical figures are given below, concentration being expressed in mol dm^{-3} and conductivity in ohm^{-1} cm^{-1} $\times$ 10^3.

Concentration		0.0001	0.001	0.01	0.1	1	2	3
Conduc-tivity	KCl	0.013	0.12	1.2	11.2	98.2	185.2	264.9
	Etha-noic acid	0.0107	0.041	0.143	0.46	1.32	1.60	1.62

Using mol m^{-3} units the given concentration figures would have to be multiplied by 10^3; using Ω^{-1} m^{-1} units the given conductivity figures would have to be multiplied by 10^2.

For equal concentrations the differences between the two sets of figures is so marked that plotting the figures for each type of substance on conductivity–concentration graphs of the same scale is well nigh impossible.

Those substances with high conductivities are known as *strong electrolytes*; *they include most mineral acids, alkalis and most salts.* Substances with comparatively small conductivities are known as *weak electrolytes*; *they include most organic bases and acids.* There is no absolutely sharp line of demarcation between strong and weak electrolytes, for a few substances exhibit intermediate behaviour, but the distinction does play a very large part in considerations of ionic theory, as will be seen.

As a solution gets more and more concen-

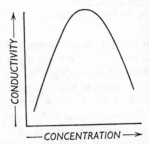

FIG. 130. The general shape of a conductivity–concentration curve for both weak and strong electrolytes. The curves for both types of electrolyte show a maximum, but it is well nigh impossible to plot both types of curve on the same scale. If judged on the same scale, the curve for a strong electrolyte lies well above that for a weak electrolyte.

trated there are more and more solute particles in a given volume of the solution. Because it is the solute that causes electrical conductivity (pure water and other solvents are very poor conductors) it might, on this account, be expected that conductivity would rise with increasing concentration, and, at first, for dilute solutions, it does so. At higher concentrations, however, the conductivity reaches a maximum and then begins to fall, and some factor other than the increase in the number of solute particles on increasing concentration must enter in.

For a weak electrolyte, this factor is the degree of ionisation which, according to the ionic theory, decreases as the concentration increases. The maximum in the conductivity–concentration curves is due to the composite

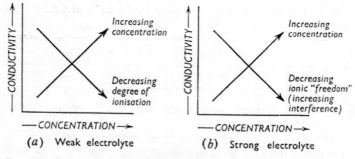

Fig. 131. The composite factors which give rise to the maxima in conductivity–concentration curves for (a) weak, and (b) strong electrolytes.

effect of an increase in the total number of solute particles and a decrease in the degree of ionisation. As the concentration increases, there are more molecules but a lower proportion of them split up into ions (Fig. 131).

For a strong electrolyte the other factor is the increase in ionic interference (decrease in ionic freedom) as the concentration increases. A strong electrolyte is fully, or almost fully, ionised at all concentrations. At high concentrations, however, the ions will interfere with each other so that their freedom of movement and their speed are restricted. As a solution gets more concentrated, the ionic interference increases and the ionic speed decreases. For a strong electrolyte, then, increasing concentration gives more ions, in a fixed volume of solution, but the ions that are present interfere with each other, which gives lower ionic speeds (Fig. 131).

Both the change in the degree of ionisation for a weak electrolyte and the change in ionic interference for a strong electrolyte can be investigated by making use of the conception of molar conductivity.

8. Molar and equivalent conductivity. Comparison of the conductivities of, say, 1M and 0.1M acid solutions is not a fair comparison for M acid contains 1 mol of acid per cubic decimetre whereas 0.1M acid only contains 0.1 mol. To compare them on equal terms, volumes of solutions containing equal amounts of solute ought to be considered, and this is done by comparing molar conductivities instead of conductivities.

The molar conductivity of a solution is the electrolytic conductivity (κ) of the solution divided by its concentration, i.e.

$$\text{Molar conductivity } (\Lambda) = \frac{\text{Electrolytic conductivity } (\kappa)}{\text{Concentration}}$$

The symbol Λ_c is used to denote the molar conductivity of a solution at a concentration c.

In basic SI units concentration is expressed in mol m^{-3} and electrolytic conductivity in Ω^{-1} m^{-1} giving Ω^{-1} m^2 mol^{-1} units for molar conductivity. If concentration is expressed in mol cm^{-3} units and conductivity in Ω^{-1} cm^{-1} units then the units of molar conductivity are Ω^{-1} cm^2 mol^{-1}.

The use of equivalent conductivities has some advantages over the use of molar conductivities, but the decline in the usage of the concept of equivalents has led to a preference for molar conductivity over equivalent conductivity.

The use of the word 'molar' is, however, not strictly correct. In the SI system of units, 'molar' means 'divided by amount of substance', but in the term molar conductivity it means 'divided by concentration'.

Equivalent conductivity is very similar to molar conductivity, concentration being expressed in gramme-equivalents per volume instead of in moles per volume. When the relative molecular mass of a solute is equal to its equivalent mass there is no numerical difference between the molar and equivalent conductivities.

9. Variation of molar conductivity with concentration. The following figures illustrate the general way in which molar conductivity values vary with concentration; the molar conductivities are expressed in Ω^{-1} cm^2 mol^{-1} and the concentrations in mol dm^{-3}.

Concentration		0.0001	0.001	0.01	0.1	1	2	3
Λ_c	KCl	129.1	127.3	122.4	112.0	98.3	92.6	88.3
	Ethanoic acid	107.0	41.0	14.3	4.6	1.32	0.80	0.54

Using mol m^{-3} units the given concentrations have to be multiplied by 10^3; using mol cm^{-3} units the multiplication factor is 10^{-3}. To convert the Λ_c values from Ω^{-1} cm^2 mol^{-1} units into Ω^{-1} m^2 mol^{-1} units it is necessary to multiply by 10^{-4}.

The figures are best summarised, graphically, by plotting the molar conductivity against the reciprocal of the concentration (sometimes referred to as the dilution of the solution) or against the square root of the concentration, and this is done in Figs. 132 and 133.

It will be seen that the effect of increasing concentration on the molar conductivity of a solution of a strong electrolyte is rather different than for a weak electrolyte. For a strong electrolyte, Λ values approach a maximum

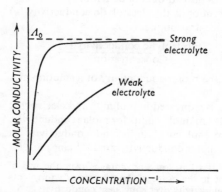

FIG. 132. The general shape of molar conductivity–(concentration)$^{-1}$ curves for strong and weak electrolytes. That for a strong electrolyte reaches a maximum at low concentrations, and the Λ_0 value can be obtained by extrapolation from the experimental curve. That for a weak electrolyte would, theoretically, reach a maximum at a low concentration, but readings cannot be obtained, experimentally, at such low concentrations.

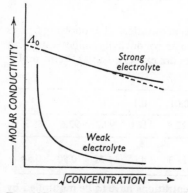

FIG. 133. The general shape of molar conductivity–(concentration)$^{\frac{1}{2}}$ curves for strong and weak electrolytes. Λ_0 values for strong electrolytes can readily be obtained from such curves by extrapolation, but this cannot be done for weak electrolytes.

value at low concentrations. For a weak electrolyte, Λ values also reach a maximum at low enough concentrations, but the concentration may be so low that it is impossible to measure the resulting low conductivity.

The maximum value for the molar conductivity of an electrolyte, reached at low concentrations, is known as the molar conductivity at zero concentration; it is symbolised by Λ_0. It may also be known as the molar conductivity at infinite dilution and be symbolised by Λ_∞. Its numerical value, for a strong electrolyte, is best obtained by extrapolation from the molar conductivity—(concentration)$^{\frac{1}{2}}$ graph, which, at low concentrations is almost a straight line. Numerical values of Λ_0 for a weak electrolytes cannot be obtained in the same way. They must be obtained indirectly, as described in the following section.

10. Kohlrausch's law. Measurement of Λ_0 for a weak electrolyte. Kohlrausch, who did much of the early experimental work on measuring the conductivities of solutions, noticed that the difference between the Λ_0 values for two salts, which were strong electrolytes, having the same anion or the same cation, was always constant. For example, using Λ_0 values in Ω^{-1} cm^2 mol^{-1}

$$\Lambda_0(NaCl) - \Lambda_0(NaNO_3) = 3.7 \qquad \Lambda_0(KCl) - \Lambda_0(NaCl) = 21.1$$
$$(108.9) \qquad (105.2) \qquad\qquad (130.0) \qquad (108.9)$$
$$\Lambda_0(KCl) - \Lambda_0(KNO_3) = 3.7 \qquad \Lambda_0(KNO_3) - \Lambda_0(NaNO_3) = 21.1$$
$$(130.0) \qquad (126.3) \qquad\qquad (126.3) \qquad (105.2)$$

Such results can be accounted for only by assuming that the Λ_0 value of an electrolyte is the sum of two terms, one for the anion and another for the cation. These terms are known as the molar conductivities of the ions concerned (they were called ionic mobilities by Kohlrausch), and Kohlrausch's law (1876) states that *the molar conductivity at zero concentration of an electrolyte is equal to the sum of the molar conductivities of the ions produced by the electrolyte.*

Thus,

$$\Lambda_0(NaCl) = \Lambda(Na^+) + \Lambda(Cl^-) \qquad \Lambda_0(KCl) = \Lambda(K^+) + \Lambda(Cl^-)$$

This is why $\Lambda_0(NaX) - \Lambda_0(KX)$, which is equal to $\Lambda(Na^+) - \Lambda(K^+)$ is independent of the nature of X.

Once the numerical value of one molar conductivity can be obtained, others follow from Λ_0 values. Molar conductivities of ions can be obtained from transport number measurements, described on p. 458, but our present purpose does not require any single value of a molar conductivity of an ion.

The immediate usefulness of Kohlrausch's law is that it provides a method for finding the Λ_0 value for weak electrolytes from Λ_0 measurements on strong electrolytes. The Λ_0 value for ethanoic acid (a weak electrolyte), for example, can be obtained from the Λ_0 values of hydrochloric acid, sodium chloride and sodium ethanoate (all strong electrolytes). For

$$\Lambda_0(HCl) + \Lambda_0(NaEt) - \Lambda_0(NaCl) = \Lambda(H^+) + \Lambda(Et^-)$$
$$[\Lambda(H^+) + \Lambda(Cl^-)] + [\Lambda(Na^+) + \Lambda Et^-] - [\Lambda(Na^+) + \Lambda(Cl^-)]$$

and
$$\Lambda(H^+) + \Lambda(Et^-) = \Lambda_0(HEt)$$
Therefore
$$\Lambda_0(HEt) = \Lambda_0(HCl) + \Lambda_0(NaEt) - \Lambda_0(NaCl)$$

To summarise, Λ_0 values for strong electrolytes can be obtained from Λ_c—(concentration)$^{\frac{1}{2}}$ graphs. Λ_0 values for weak electrolytes can be obtained by measurement of the Λ_0 values for selected strong electrolytes.

11. Reasons for variation of Λ_c with concentration. Any effect on the conductivity of a solution due simply to dilution lowering the amount of solute present is ruled out in considering molar conductivities. At any concentration, the figures for Λ_c refer to the same amount of solute between the electrodes in a conductivity cell. If the amount of solute is not, then, changed, why is it that concentration has any effect at all on equivalent conductivity? The explanation is different for weak and strong electrolytes.

a. *Weak electrolytes.* The increase in Λ_c with decreasing concentration for a weak electrolyte is due, as first suggested by Arrhenius, to the increase in degree of ionisation with decreasing concentration. This produces an increasing number of ions as the solution gets more and more dilute until, eventually, at zero concentration, even a weak electrolyte is fully ionised, and Λ_c reaches the limiting value of Λ_0.

Arrhenius took the argument a stage further and suggested that the degree of ionisation, in a solution of concentration c, might be equal to the ratio Λ_c/Λ_0. Values for α obtained in this way from conductivity measurements were found to be in good agreement with the α values obtainable from vapour pressure, freezing point, boiling point and osmotic pressure measurements (p. 276), and this provided very firm evidence in favour of the ionic theory.

α is a ratio of two molar conductivity values and has no units. The same value for α will be obtained irrespective of the units used for molar conductivity (Ω^{-1} cm^2 mol^{-1} or Ω^{-1} m^2 mol^{-1}) so long as the same units are used for both the Λ_c and the Λ_0 values.

b. *Strong electrolytes.* Arrhenius assumed that the explanation given for weak electrolytes also applied for strong electrolytes, but this cannot be so as it is now known that strong electrolytes are fully ionised, or almost so, at any concentration and, often, in the solid state.

In a solution of a weak electrolyte, conductivity depends almost entirely on the total number of ions present, for there is little ionic interference and the velocities of the ions are reasonably constant at all concentrations.

This is not so in a solution of a strong electrolyte. Here, the ionic interference between the large number of ions present is considerable and has a large effect on the velocities of the ions. Thus, although there may be a large number of ions present, they are not entirely free to play their full part in the conduction of electricity, or in other chemical effects.

As a solution is diluted, the ionic interference diminishes and the change in Λ_c with concentration for a strong electrolyte is due to this decrease in ionic interference. The Λ_c-(concentration)$^{-1}$ curve is, essentially, an ionic interference–(concentration)$^{-1}$ curve. At zero concentration, ionic interference is negligible, and Λ_c reaches its limiting value of Λ_0.

The ratio Λ_c/Λ_0 for a strong electrolyte is almost equal to the α values obtained for strong electrolytes from vapour pressure, freezing point, boiling point and osmotic pressure measurements. But the agreement, such as it is, is of little significance, for the whole conception of the degree of ionisation for a strong electrolyte has little meaning, for the degree is always approximately 1 or 100 per cent.

The Λ_c/Λ_0 value for a strong electrolyte is, therefore, called the apparent degree of ionisation or, better, the *conductance ratio*.

Attempts have been made, notably by Debye and Hückel, to treat ionic interference quantitatively, and some success has been achieved. Further details are given on p. 377.

QUESTIONS ON CHAPTER 33

1. What are the essential differences between a sodium atom and a sodium ion?
2. What is meant by the term cell constant, and for what purpose is a cell constant used? (a) The resistivity of M/50 potassium chloride solution is 361 Ω cm, and a conductivity cell containing such a solution was found to have a resistance of 550 Ω. What is the cell constant? (b) The same cell filled with M/10 zinc sulphate(VI) solution had a resistance of 72 Ω. What is the conductivity of M/10 zinc sulphate(VI) solution?
3. A conductivity cell with electrodes 2 cm^2 in area and 1 cm apart has a resistance of 7.25 ohm when filled with 5 per cent potassium chloride solution. What is (a) the cell constant, (b) the conductivity of the potassium chloride solution? If the cell was filled with M/50 potassium chloride, with a resistivity of 361 Ω cm, what would the cell resistance be?
4. What would be the resistance of a conductivity cell with electrodes of cross-sectional area 4 cm^2 and 2 cm apart when the cell is filled with pure water of conductivity 0.8×10^{-6} ohm^{-1} cm^{-1}? What current would flow through the cell under an applied potential difference of 10 volt?
5. What is meant by the term conductivity of a solution? How can conductivity be measured? How does it vary with the concentration of a solution, and how is the variation accounted for in terms of the ionic theory?
6. Using the data given on p. 365, plot the conductivity of (a) potassium chloride, (b) ethanoic acid against concentration.
7. If the conductivity of sulphuric(VI) acid in M/2000 solution is x what is (a) the resistivity, (b) the equivalent conductivity, and (c) the molar conductivity?
8. Why was the conception of molar conductivity introduced when that of conductivity was already established? How can the molar conductivity of M/100 ethanoic acid be measured?
9. Using the figures given on p. 367, plot the molar conductivity of (a) potas-

sium chloride, (b) ethanoic acid against (i) the reciprocal of concentration and (ii) the square root of concentration. From your graphs read off the Λ_0 value for potassium chloride and the conductance ratio in M/50 potassium chloride solution.

10. If the current carrying capacity of ions in a solution depends on their total number and their speed, how would you expect the conductivity of (a) ethanoic acid, (b) sodium chloride to vary with (i) the dilution of the solution, (ii) temperature?

11. How did Arrhenius account for the change in the molar conductivity of an electrolyte with dilution, and how have his ideas had to be modified?

12. Explain, briefly, how the degree of ionisation of a weak electrolyte might be obtained by measurements of (a) osmotic pressure, (b) freezing point, (c) boiling point, (d) vapour pressure, and (e) conductivity.

13. Calculate the Λ_0 values for (a) methanoic acid, (b) ammonia solution from the Λ_0 values given below:

$$NaCl = 113 \qquad NH_4Cl = 134.1$$
$$NaOH = 225.2 \qquad Na\ methanoate = 101.2 \qquad HCl = 397.8$$

14. What would you write to convince a fifth-form boy that the degree of ionisation of a weak electrolyte increased as the concentration of its solution got less and less?

15. The conductivity of dichloroethanoic acid at a dilution of 8 dm³ is 0.023 8 Ω^{-1} cm^{-1}. The Λ_0 value for this acid is 385. Calculate the degree of ionisation at a dilution of 8 dm³.

16. The molar conductivities of Na^+, K^+ and Cl^- ions are 43.4, 64.6 and 65.5 respectively. What approximate raio of sodium chloride and potassium chloride is required to give a solution with a Λ_0 value of (i) 119.5, (ii) 124.8?

17. If the conductivity of an M solution of sodium chloride is x Ω^{-1} cm^{-1} what is (a) the conductivity of the solution in Ω^{-1} m^{-1} (b) the molar conductivity of the solution in Ω^{-1} cm^2 mol^{-1} and (c) the molar conductivity of the solution in Ω^{-1} m^2 mol^{-1}?

18. What are the advantages and disadvantages of expressing molar conductivities of solutions in (a) Ω^{-1} cm^2 mol^{-1} units, and (b) Ω^{-1} m^2 mol^{-1} units?

19. Express the data given in the tables on pages 375 and 377 in concentration units of mol m^{-3}, conductivity units of Ω^{-1} m^{-1} and molar conductivity units of Ω^{-1} m^2 mol^{-1}.

20. What is the relationship between the molar and equivalent conductivities of solutions of (a) sodium chloride, (b) sulphuric acid, (c) potassium hydroxide, and (d) calcium hydroxide?

Chapter 34
Ostwald's dilution law

1. Derivation of law. The partial ionisation of molecules into ions produces an equilibrium mixture of ions and molecules, which, for a binary electrolyte, AB, can be represented as

$$AB \rightleftharpoons A^+ + B^-$$

Ostwald applied the idea of the law of equilibrium (p. 309) to this equilibrium, in 1888, and arrived at an important relationship known as Ostwald's Dilution Law.

If the original concentration of AB in a solution is c mol dm^{-3} and if the degree of ionisation is α, the concentrations of AB, A^+ and B^- (in mol dm^{-3}) at equilibrium will be

$$\begin{array}{ccc} AB & \rightleftharpoons A^+ & + B^- \\ c(1-\alpha) & c\alpha & c\alpha \end{array}$$

According to the equilibrium law

$$\frac{[A^+][B^-]}{[AB]} = K$$

and, therefore,

$$\frac{(c\alpha)(c\alpha)}{c(1-\alpha)} = K \quad \text{or} \quad \frac{c\alpha^2}{(1-\alpha)} = K$$

If α is small as compared with 1, i.e. for a weak electrolyte in a solution where the degree of ionisation is low, $(1-\alpha)$ will be approximately equal to 1 so that

$$c\alpha^2 \simeq K \quad \text{or} \quad \alpha \simeq \sqrt{\frac{K}{c}}$$

These two relationships, the first accurate and the second an approximation, are expressions of Ostwald's dilution law. *For a weak binary electrolyte, with a small degree of ionisation, the degree of ionisation is proportional to the square root of the reciprocal of the concentration.*

K, which is the equilibrium constant for the reaction

$$AB \rightleftharpoons A^+ + B^-$$

is known as the *dissociation or ionisation constant* of AB. If c is expressed in mol dm^{-3} units K is also in these units.

2. Validity of Ostwald's dilution law. The validity of Ostwald's dilution law can be tested by measuring α values at various concentrations, and by seeing whether they fit the theoretically obtained expression. If the α values are obtained from conductivity measurements, i.e. from Λ_c/Λ_0 values, the accurate expression of Ostwald's dilution law becomes

$$\frac{c(\Lambda_c/\Lambda_0)^2}{(1 - \Lambda_c/\Lambda_0)} = \frac{c\Lambda_c^2}{\Lambda_0(\Lambda_0 - \Lambda_c)} = K$$

In testing this expression for ethanoic acid, a weak electrolyte, the following results are obtained:

$c/\text{mol dm}^{-3}$	1.011	0.253	0.0316	0.00198	0.00494
α	0.00372	0.00838	0.0239	0.092	0.176
$K \times 10^5/\text{mol dm}^{-3}$	1.405	1.759	1.846	1.841	1.853

It will be seen that the value of K is reasonably constant when c is small, i.e. in dilute solutions. Variations in the value of K in more concentrated solutions is attributed to some ionic interference of the type found more noticeably with strong electrolytes. As a wide approximation, *Ostwald's dilution law holds good for weak electrolytes in dilute solutions.*

This is far from so, however, for strong electrolytes. Taking potassium chloride as a typical example, and using Λ_c/Λ_0 values as being equal to α, the results obtained are

$c/\text{mol dm}^{-3}$	1	0.2	0.02	0.001	0.0001
α	0.7565	0.8310	0.9234	0.9802	0.9936
$K/\text{mol dm}^{-3}$	2.350	0.8154	0.2221	0.0485	0.0154

There is no constancy at all about the values of K, and *Ostwald's dilution law does not hold for strong electrolytes, because the ionic interference is so high.*

This failure of Ostwald's dilution law, as applied to strong electrolytes, was a severe stumbling block in the historical development of the ionic theory. Originally it could not be explained, and it became known as the *anomaly of strong electrolytes.* Pending any theoretical treatment, many empirical formulae were devised to try to fit the experimental variation of Λ_c with concentration for a strong electrolyte. Kohlrausch suggested,

$$\Lambda_0 - \Lambda_c = \text{a constant} \times \sqrt{c}$$

and this holds reasonably well at low concentrations (see Fig. 133).

Eventually the distinction between weak and strong electrolytes became clearer and the idea of ionic interference was introduced to account for the behaviour of strong electrolytes (see section **6**).

3. Dissociation constants for weak monobasic acids. Ostwald's dilution law only holds for weak electrolytes in dilute solution, but, within these limitations, it is extremely valuable. Its importance lies in the fact that it relates the degree of ionisation of a weak electrolyte, *which is variable*, with the dissociation constant of the electrolyte, which, at a definite temperature, *is fixed*. The law relates three quantities, α, c and K, and if any two of these are known the third is easily calculated.

Such calculations are particularly common in relation to aqueous solutions of weak acids. The degree of ionisation of a weak acid in a solution of known concentration can be measured by conductivity and other methods (p. 276), and, from such measurements, the dissociation constant of the acid (or the acidity constant) can be obtained. Some typical values for weak, monobasic acids are given below in mol dm^{-3}:

Methanoic acid 2.1×10^{-4} Benzenol (phenol) 1×10^{-10}
Ethanoic acid 1.8×10^{-5} Chloric(I) acid 9.6×10^{-7}
Chloroethanoic acid 1.55×10^{-3} Hydrocyanic acid 7.2×10^{-10}
Benzenecarboxylic acid 6.7×10^{-5} Nitric(III) acid 5×10^{-4} .

Acids with high dissociation constants are stronger than those with low ones. At the same concentration, an acid with a higher dissociation constant will have a higher degree of ionisation, i.e. it will ionise to give a higher concentration of H^+ ions. The calculation of the degree of ionisation of a weak acid, and the resulting H^+ ion concentration, in a solution of the weak acid is very common, as in the following examples:

a. M/10 *ethanoic acid.* The dissociation constant of ethanoic acid is 1.8×10^{-5} mol dm^{-3}. In M/10 solution, the concentration is 0.1 mol dm^{-3}. Using the approximate form of Ostwald's dilution law,

$$\alpha^2 = 10 \times 1.8 \times 10^{-5}$$
$$\therefore \alpha = \sqrt{1.8 \times 10^{-4}} = 1.34 \times 10^{-2}$$

The degree of dissociation of ethanoic acid in M/10 solution is, therefore, 1.34×10^{-2} or 1.34 per cent.

Complete ionisation of ethanoic acid, at a concentration of 0.1 mol dm^{-3} would give 0.1 mol dm^{-3} of H^+ ion. Ionisation of a fraction, 0.0134, of the ethanoic acid would give a H^+ ion concentration of 0.1×0.0134, i.e. 0.00134 mol dm^{-3}. This is, therefore, the concentration of H^+ ion in M/10 ethanoic acid.

Notice that the approximate form of Ostwald's dilution law has been used in this calculation because the α value is small as compared with 1.

b. *M/1000 ethanoic acid.* Here, c is 0.001 so that

$$\alpha^2 \simeq 1000 \times 1.8 \times 10^{-5}$$
$$\therefore \alpha \simeq \sqrt{1.8 \times 10^{-2}} = 0.134$$

The degree of ionisation in M/1000 ethanoic acid is, therefore, approximately 0.134 or 13.4 per cent, and the resulting H^+ ion concentration is 0.001×0.134 or 0.000134 mol dm^{-3}.

The α value in this example is not very small and the use of the approximate form of Ostwald's dilution law is inaccurate. Using the accurate form

$$\alpha^2 = 1000 \times 1.8 \times 10^{-5}(1 - \alpha)$$

or $$\alpha^2 + (1.8 \times 10^{-2})\alpha - (1.8 \times 10^{-2}) = 0$$

which gives a value of 0.125 for α.

4. Dissociation constants for weak polybasic acids. Polybasic acids ionise in stages and have more than one dissociation constant. Carbonic acid, which is dibasic, has two dissociation constants, represented as follows

$$H_2CO_3 \rightleftharpoons HCO_3^- + H^+ \qquad\qquad HCO_3^- \rightleftharpoons CO_3^{2-} + H^+$$

$$K_1 = \frac{[H^+] \cdot [HCO_3^-]}{[H_2CO_3]} = 3 \times 10^{-7} \qquad K_2 = \frac{[H^+] \cdot [CO_3^{2-}]}{[HCO_3^-]} = 6 \times 10^{-11}$$

Some other typical values for dissociation constants of polybasic acids are given below in mol dm^{-3}:

Phosphoric(v) acid, $K_1 = 1.1 \times 10^{-2}$ Sulphuric(iv) acid, $K_1 = 1.2 \times 10^{-2}$
$\qquad\qquad\qquad\quad K_2 = 2 \ \ \times 10^{-7}$ $\qquad\qquad\qquad\qquad K_2 = 1 \ \ \times 10^{-7}$
$\qquad\qquad\qquad\quad K_3 = 3.6 \times 10^{-12}$ Hydrogen sulphide, $K_1 = 9 \ \ \times 10^{-8}$
$\qquad\qquad\qquad\qquad\qquad\qquad\qquad\qquad\qquad\qquad\quad K_2 = 1 \ \ \times 10^{-15}$

For many inorganic oxy-acids it is a fairly general rule that the successive dissociation constants, K_1, K_2, K_3 . . . are in the approximate ratio of $1:10^{-5}$: 10^{-10} . . .

5. Dissociation constants for weak bases. Just as there are weak and strong acids, so there are weak and strong bases. Strong bases, such as potassium and sodium hydroxide, are fully ionised in solution, and Ostwald's law does not apply to them. It does, however, apply to solutions of weak bases.

The degree of ionisation of a weak base in a solution of known concentration can be measured by conductivity or other methods, and the dissociation constant of the base can be obtained from the results. Some typical values are given below. The higher the value, the stronger the base.

Ammonia 1.8×10^{-5} $\qquad\qquad$ Carbamide 1.5×10^{-14}
Phenylamine 4.6×10^{-10} $\qquad$ Hydrazine $\ \ \ 1 \times 10^{-6}$

From the dissociation constant of a weak base it is a simple matter to calculate the degree of ionisation and the corresponding OH$^-$ ion concentration in a solution of any concentration.

The dissociation constant for ammonia in aqueous solution, for example, is 1.8×10^{-5} mol dm^{-3}. In M/10 solution, therefore,

$$\alpha^2 = 10 \times 1.8 \times 10^{-5}$$

$$\therefore \alpha = \sqrt{1.8 \times 10^{-4}} = 1.342 \times 10^{-2}$$

The degree of ionisation in M/10 ammonia solution is, therefore, 1.342×10^{-2} or 1.342 per cent.

Complete ionisation into OH$^-$ ions at a concentration of 0.1 mol dm^{-3} would give 0.1 mol dm^{-3} of OH$^-$ ion,

$$NH_3 + H_2O \longrightarrow NH_4^+ + OH^-$$

Ionisation of a fraction 0.01342 would give an OH$^-$ ion concentration of 0.1×0.01342, i.e. 0.001342 mol dm^{-3}. This is, therefore, the concentration of OH$^-$ ions in M/10 aqueous ammonia.

6. The anomaly of strong electrolytes. Ionic interference. The theoretical treatment of ionic interference, undertaken originally by Debye and Hückel, is based on the idea that the electrical attraction between positive and negative ions in a solution prevents the ions acting as entirely isolated, single particles, except in very dilute solutions.

In an ionic crystal, a positive ion is surrounded by negative ions, and vice versa (p. 170). Debye and Hückel suggested that, in a solution, a positive ion is surrounded by an ionic atmosphere of negative ions, whilst a negative ion is surrounded by an ionic atmosphere of positive ions.

Ionisation is complete, in so far as there are no individual molecules, or only very few, of a strong electrolyte in a solution, and in so far as the attractive forces between ions in a solution are not so strong as they are in an ionic crystal. But random distribution of ions throughout a solution is not complete.

The chemical functioning, and the freedom of movement, of any ion surrounded by an ionic atmosphere with opposite charge, will clearly be different from that of a free ion. So far as motion under an applied e.m.f., i.e. conductivity, is concerned, two electrical effects will be important.

First, a positive ion moving towards the cathode will tend to drag its ionic atmosphere with it. This will result in an asymmetric ionic atmosphere with fewer negative ions in front of the positive ion, and more behind. Such an asymmetric ionic atmosphere will exert a retarding force on the positive ion and reduce its velocity.

Secondly, a negative ionic atmosphere will, as a whole, move towards the anode, so that the central positive ion will, in effect, be 'moving against the stream'.

Debye, Hückel and Onsager treated such considerations mathematically and derived what is generally known as the Onsager equation. As applied to an electrolyte giving two univalent ions, the equation is

$$(\Lambda_0 - \Lambda_c) = \left\{ \frac{82.48}{\eta\sqrt{DT}} + \frac{8.20 \times 10^5 \times \Lambda_0}{(DT)^{3/2}} \right\} \sqrt{c}$$

where η is the viscosity of the medium, D is its dielectric constant, T the absolute temperature, and c the concentration.

Such an equation gives some idea of the complexity of the matter. The Onsager equation, nevertheless, provides good agreement with experimental data, so long as the solutions considered are dilute enough. The equation, in fact, is of the same general type as the empirical equation put forward by Kohlrausch (p. 374).

Further support for the Debye–Hückel–Onsager treatment is proved by the fact that the conductivity of a solution increases if very high frequency a.c. of about 3×10^6 Hz, or very high field strengths of about 10^5 volt cm^{-1}, are used. The ionic atmosphere cannot 'follow' the rapidly changing a.c. field, and the very high field strength gives ionic velocities which leave the ionic atmosphere completely behind.

The theory does not, however, account at all quantitatively for the experimental results from concentrated solutions of strong electrolytes. An alternative approach, due mainly to Bjerrum, replaces the idea of ionic atmosphere by one in which two oppositely charged ions associate in solution to form an ion-pair.

Whilst, therefore, the general idea of ionic interference accounts qualitatively, and to some extent quantitatively, for the experimental results, the anomaly of strong electrolytes is very far from fully resolved.

7. Activity coefficients. Ostwald's dilution law is derived by applying the law of equilibrium to an equilibrium involving ionisation, e.g.,

$$AB \rightleftharpoons A^+ + B^-$$

In the simple derivation, the concentrations of AB, A^+ and B^- in mol dm^{-3} are used, but, as has been seen, such a treatment gives an expression which does not hold for strong electrolytes.

The statement of the law of mass action says that it is the *active masses* of the chemicals concerned which should be considered, and active mass is not equal to concentration when ionic interference enters in. Instead

Active mass = Concentration × Activity coefficient

The more accurate expression for the dissociation constant of AB is, therefore, given by

$$K = \frac{([A^+] \times f_A^+) \times ([B^-] \times f_B^-)}{([AB] \times f_{AB})}$$

where f_A^+, f_B^- and f_{AB} are the activity coefficients of A^+, B^- and AB respectively.

In very dilute solutions, where ionic interference is negligible, activity coefficients are approximately equal to 1 so that the simple derivation of the Ostwald dilution law (p. 373) holds.

It is not possible to measure experimentally the activity coefficient of an individual ion, but the mean coefficient of an electrolyte, taken, for a binary electrolyte, as the square root of the product of the individual ionic activities, can can be measured, but the details are beyond the scope of this book.

Experimental measurements show, however, that the mean activity coefficient for all strong electrolytes containing only univalent ions is approximately 0.80 in M/10 solution, 0.90 in M/100 solution and 0.96 in M/1000 solution. In more dilute solution, the mean activity coefficient approaches 1.

QUESTIONS ON CHAPTER 34

1. The molar conductivity of an M/32 solution of a weak acid is 9.2 Ω^{-1} cm^2 mol^{-1}. If the molar conductivity at zero concentration is 389 Ω^{-1} cm^2 mol^{-1}, what is the dissociation constant of the acid?

2. The dissociation constant of ethanoic acid is 1.8×10^{-5} mol dm^{-3}. Calculate the values of the degree of ionisation at dilutions of 10, 100, 1000, 10000 and 100000 dm^3, and plot the degree of ionisation against the dilution.

3. The dissociation constant of benzenecarboxylic acid is 6.7×10^{-5} mol dm^{-3}. What are the concentrations of the solutions, in mol dm^{-3}, in which benzenecarboxylic acid is (a) 10 per cent, (b) 25 per cent, (c) 50 per cent, and (d) 90 per cent ionised?

4. The molar conductivities of ethanoic acid in various solutions (V is the volume containing 1 mol of solute) are given below:

V/dm^3	13.57	54.28	434.2	1737	6948	∞
Λ_0/Ω^{-1} cm^2 mol^{-1}	6.09	12.09	33.22	63.60	116.8	387.9

Do these figures agree with Ostwald's dilution law? What conclusion would you draw from the figures as to the value of the dissociation constant of ethanoic acid?

5. On what basic principles is the derivation of Ostwald's law, and its testing by the use of Λ_c/Λ_0 values, based? Which principle is it that is false so far a strong electrolytes are concerned?

6. The dissociation constant for ammonia in aqueous solution is 1.99×10^{-5} mol dm^{-3}. What is the concentration of OH$^-$ ions in an 0.01M solution?

7. The dissociation constant of methanoic acid is 2.1×10^{-4} mol dm^{-3}. Using the relationship $c\alpha^2 = K$, calculate the percentage degree of ionisation of methanoic acid in 0.00001M solution. Why is the answer you get absurd? What is the correct answer?

8. A dibasic acid has two dissociation constants, usually written as K_1 and K_2 and referred to as the primary and secondary dissociation constants. For carbonic acid, for instance, K_1 is 3×10^{-7} and K_2 is 6×10^{-11} mol dm^{-3}. Why is K_1 usually bigger than K_2?

9. Compare the OH$^-$ ion concentration in M/100, M/1000, M/10000 and M/100000 solutions of (i) ammonia solution, (ii) potassium hydroxide. The dissociation constant of ammonia is 1.99×10^{-5} mol dm^{-3}; potassium hydroxide may be regarded as fully ionised.

10. Applying the law of mass action to the ionic equilibrium show how the (equivalent) electrical conductivity of dilute ethanoic acid varies with its concentration. State, very briefly, how the conductivity of dilute hydrochloric acid depends on concentration, and how it is affected, qualitatively, by temperature. Although dilute ethanoic acid smells of ethanoic acid, dilute hydrochloric, in contrast to the concentrated acid, has no smell whatever. Why is this so? (B.)

11. Define molar conductivity of a solution of an electrolyte. The molar conductance of a solution of ethanoic acid containing 0.03 mol dm^{-3} is 8.50 units at 18°C, and 14.7 units at 100°C. The molar conductivity of ethanoic acid at zero concentration at 18°C is 347 units and at 100°C is 773 units. Calculate (a) the degree of dissociation of ethanoic acid at 18°C and 100°C when the concentration is 0.03 mol dm^{-3}, and (b) the dissociation constants of ethanoic acid at these two temperatures. From your results deduce whether the dissociation of ethanoic acid is exothermic or endothermic. (C. Schol.)

12. Describe the experiments you would make in order to determine the relation between the degree of dilution of a solution of ethanoic acid and its molar conductivity. How are the results of such experiments explained? The molar conductivity of ethanoic acid at zero concentration is 388 Ω^{-1} cm^2 mol^{-1}. The molar conductivity of a solution containing 0.3 g of ethanoic acid in 50 cm^3 of water is 4.6 Ω^{-1} cm^2 mol^{-1}. Calculate the freezing point of this solution. (A solution containing 34.2 g of cane sugar, $C_{12}H_{22}O_{11}$, in 100 g of water freezes at -1.85°C.) (C. Schol.)

13. What do you understand by (a) the law of mass action, (b) an equilibrium constant? If the dissociation constant for ethanoic acid at 18°C is 1.8×10^{-5} mol dm^{-3}, calculate the concentration of hydrogen ions in decinormal solutions of ethanoic acid at 18°C.

14. Derive the mathematical expression for Ostwald's dilution law for an electrolyte, one molecule of which ionises to give n cations and m anions.

15. The conductivity of a 0.05M solution of ethanoic acid is 4.4×10^{-4} Ω^{-1} cm^{-1}. The molar conductivities of H$^+$ and Et$^-$ ions are 310 and 77 Ω^{-1} cm^{-1} respectively. Calculate the ionisation constant of ethanoic acid.

16. What is meant by (a) the conductivity, (b) the molar conductivity, and (c) the equivalent conductivity of a solution of an electrolyte.

The molar conductivity of a 0.031 25M solution of a weak monobasic acid is 9.2. At zero concentration the value is 389. Calculate the ionisation constant of the acid.

17. The degree of ionisation of propanoic acid in 0.1M solution is 0.011 33. What would the degree of ionisation be, under the same conditions, in 0.01M solution?

18. Why is the value of the ionisation constant of hydrofluoric acid low (6×10^{-4}) whilst that of hydrochloric acid is very high?

19. What is meant by saying that the K_α value for the $Al(H_2O)_6^{3+}$ ion is 1×10^{-5}?

20. A glass tube closed at each end by a platinum black electrode contains hydrochloric acid. Explain, giving your reasons, how the electrical resistance of this cell changes if (a) the concentration of hydrochloric acid is doubled, (b) the distance between the electrodes is halved, (c) the temperature is changed.

How is the molar conductivity of a solution obtained from conductivity measurements, and how does it vary with concentration? What is the significance of the value of the molar conductivity at zero concentration and how is this value obtained for ethanoic acid?

21. Look up the values of the ionisation constants of methanoic, ethanoic, propanoic, and butanoic acids. Comment on the values. Do likewise for the halogenated ethanoic acids.

Chapter 35
Solubility product

1. Meaning of solubility product. In a *saturated* solution of a sparingly soluble electrolyte, such as silver chloride, simultaneous equilibria will be set up between undissolved solid, dissolved (but unionised) silver chloride, and free silver and chloride ions,

$$
\underset{\text{(Undissolved solid)}}{\text{AgCl}} \rightleftharpoons \underset{\substack{\text{(Dissolved, but} \\ \text{unionised)}}}{\text{AgCl}} \rightleftharpoons \underset{\text{(Ions in solution)}}{\text{Ag}^+(\text{aq}) + \text{Cl}^-(\text{aq})}
$$

Applying the law of equilibrium (p. 309), it follows that

$$
\frac{[\text{Ag}^+][\text{Cl}^-]}{[\text{AgCl}]} = K
$$

but, in a saturated solution, [AgCl] will be constant, at a fixed temperature, so that

$$
[\text{Ag}^+][\text{Cl}^-] = \text{a constant, } K_s(\text{AgCl})
$$

The constant, $K_s(\text{AgCl})$, is known as the solubility product of silver chloride. It is equal to 1×10^{-10} (p. 386) and the units are $\text{mol}^2 \text{ dm}^{-6}$ if the ionic concentrations are expressed in mol dm^{-3}.

In general, for an electrolyte, A_xB_y, ionising as follows,

$$
A_xB_y \rightleftharpoons x\text{A}^+ + y\text{B}^-
$$

the solubility product is given by

$$
K_s(A_xB_y) = [\text{A}^+]^x[\text{B}^-]^y
$$

where [A$^+$] and [B$^-$] are the ionic concentrations *in a saturated solution*.

It must be emphasised that the conception of solubility product is only accurately valid if the active masses (p. 307) of the ions concerned are used, but for dilute solutions, i.e. solutions of sparingly soluble electrolytes, accurate results can be obtained by using ionic concentrations expressed in mol dm^{-3}. Moreover, the general qualitative idea is applicable to more concentrated solutions.

The main use of solubility products is concerned with the conditions under which an electrolyte will dissolve, or will come out of solution as a solid, i.e. form a precipitate. It is important to remember that, for an electrolyte, AB, with $K_s(\text{AB})$ equal to $[\text{A}^+][\text{B}^-]$,

  i a solution in which $[\text{A}^+][\text{B}^-]$ is *less than* $K_s(\text{AB})$ is not saturated, and more AB can be dissolved in it;

  ii a solution in which $[\text{A}^+][\text{B}^-]$ is *equal to* $K_s(\text{AB})$ is saturated, and

  iii if anything be done to a solution which tends to make $[\text{A}^+][\text{B}^-]$ *greater than* $K_s(\text{AB})$ then solid AB will be precipitated. Evaporation of a saturated solution, for instance, tends to increase the ionic concentrations of the ions present. As a result, solid is deposited.

2. The common ion effect. The equilibrium existing in a solution of an electrolyte, AB,

$$AB \rightleftharpoons A^+ + B^-$$

will be affected by addition of either more A^+ or more B^- ions, and changes brought about in such a way are known as common ion effects.

a. *Precipitation of silver ethanoate.* The equilibrium in a saturated solution of silver ethanoate is

$$AgEt \rightleftharpoons Ag^+ + Et^-$$

Addition of concentrated solutions of either silver nitrate (Ag^+ ions) or sodium ethanoate (Et^- ions) precipitates silver ethanoate.

b. *Purification of common salt.* Sodium chloride, as obtained from natural sources, is usually contaminated with small amounts of calcium and magnesium chlorides. Both these impurities are deliquescent, and they must be removed in the manufacture of free-running table salt.

This can be done by passing hydrogen chloride gas into a saturated solution of the impure sodium chloride. This addition of chloride ions precipitates sodium chloride, but calcium and magnesium chloride impurities remain in solution as the solution is not saturated so far as they are concerned.

Because sodium chloride is not sparingly soluble the argument cannot be applied quantitatively. Moreover, the use of water in hydration of the H^+ ions on adding the hydrogen chloride will lead to an increase in ionic concentrations of Na^+ and Cl^- ions.

c. *Salting out of soap.* In the manufacture of soap, taken, for simplicity, as being pure sodium octodecanoate (stearate), a solution of common salt is added to the solution of sodium octodecanoate to obtain the solid soap.

Addition of Na^+ ions, from the common salt, increases the value of $[Na^+][Oc^-]$ in the solution until it eventually reaches $K_s(NaOc)$. Further addition of Na^+ ions precipitates sodium stearate.

The process is known as the salting out of soap. Again, it is doubtful whether the process can be explained solely in terms of solubility product. Some of the sodium octodecanoate is in colloidal solution (p. 507) and addition of common salt will coagulate it, and hydration of Na^+ ions enters in as in b.

3. Solubility product and analysis. There are some important analytical examples of the application of the solubility product principle.

a. *Precipitation of hydroxides in Group* 3. In Group 3 of the routine analytical scheme widely used, iron(III), aluminium or chromium(III) hydroxides may be precipitated by using a mixture of solid ammonium chloride and ammonia solution as the precipitating reagent. Use of ammonia solution by itself, or sodium hydroxide solution, could cause precipitation of the hydroxides of manganese(II), nickel(II), cobalt(II), zinc, magnesium and calcium.

In order to limit the precipitation to iron(III), aluminium and chromium(III) hydroxides, sufficient OH^- ion must be added to exceed their solubility products, which are relatively low, but the amount of OH^- ion added must not be big enough to exceed the higher solubility products of the other hydroxides listed below.

	$Zn(OH)_2$	1×10^{-17}
	$Co(OH)_2$	2×10^{-16}
$Fe(OH)_3$ $\quad 1 \times 10^{-38}$	$Fe(OH)_2$	1×10^{-15}
$Al(OH)_3$ $\quad 1 \times 10^{-33}$	$Mn(OH)_2$	1×10^{-14}
$Cr(OH)_3$ $\quad 1 \times 10^{-30}$	$Ni(OH)_2$	1×10^{-14}
	$Mg(OH)_2$	6×10^{-12}
	$Ca(OH)_2$	8×10^{-6}

Low solubility products. Precipitated in presence of NH_4Cl	High solubility products. Not precipitated in presence of NH_4Cl

Use of ammonia solution alone would give too high a concentration of OH^- ions,

$$NH_3(g) + H_2O(l) \rightleftharpoons NH_4^+(aq) + OH^-(aq)$$

but the presence of solid ammonium chloride, i.e. of a high concentration of NH_4^+ ions, suppresses the ionisation of the ammonia solution and, in this way, limits the OH^- ion concentration.

It is interesting to note that iron(II) hydroxide is not precipitated in the presence of ammonium chloride. That is why any iron(II) compound which might be present is oxidised, after Group 2, to iron(III), by boiling with concentrated nitric(V) acid. The resulting iron(III) compound is precipitated as a hydroxide in Group 3.

Manganese(II) hydroxide is the most likely hydroxide to come down, unwanted, as a precipitate in Group 3. It has not got the lowest solubility product of the hydroxides which might, wrongly, be precipitated, but the hydroxides of zinc, cobalt(II) and nickel(II), even if precipitated initially, are soluble in the excess ammonium hydroxide present (p. 483). A solution containing a high concentration of Mn^{2+} ions will give a precipitate in Group 3 if insufficient ammonium chloride is present, i.e. if the ionisation of the ammonia solution is not suppressed sufficiently.

b. *Precipitation of sulphides in Groups 2 and 4.* The sulphides precipitated in Group 2 have lower solubility products than those precipitated in Group 4. The values are given below.

HgS	3×10^{-54}		CoS	3×10^{-26}
CuS	3×10^{-42}		ZnS	1×10^{-23}
CdS	4×10^{-29}		NiS	4×10^{-21}
PbS	4×10^{-28}		MnS	1.5×10^{-15}

Lower solubility products.   Higher solubility products.
Precipitated in Group 2   Precipitated in Group 4

A high concentration of S^{2-} ions would precipitate all these sulphides from solutions containing the necessary metallic ions, and to limit the precipitation to those in the left-hand column the S^{2-} ion concentration must be limited.

This is achieved by passing hydrogen sulphide into an *acidic* solution, in Group 2. Hydrogen sulphide ionises,

$$H_2S \rightleftharpoons H^+ + HS^- \qquad\qquad HS^- \rightleftharpoons H^+ + S^{2-}$$

but in the presence of acid, i.e. of H^+ ions, the ionisation is suppressed so that only a limited concentration of S^{2-} ions is produced.

To get the concentration of S^{2-} ions high enough, yet not too high, requires reasonably careful control of the H^+ ion concentration before the hydrogen sulphide is passed in. That is why analytical schemes suggest that the acid concentration after Group 1 should be adjusted to approximately 0.5M, or give some alternative method for controlling the concentration of acid.

The use of a too concentrated acid solution, i.e. too small a concentration of S^{2-} ions, together with a weak solution of a metallic salt might prevent complete precipitation of all the Pb^{2+} or Cd^{2+} ions. With a weak acid solution, i.e. too high a concentration of S^{2-} ions, and a strong solution of salt, precipitation of cobalt(II) sulphide may occur when it is not wanted.

In Group 4, the solution is no longer acidic, but alkaline. Ionisation of hydrogen sulphide is, therefore, not suppressed by the presence of H^+ ions, but encouraged by removal of H^+ ions through reaction with OH^- ions. A concentration of S^{2-} ions high enough to precipitate the sulphides of nickel(II), cobalt(II), zinc and manganese(II) is, therefore, attained.

c. *Use of potassium chromate(VI) as an indicator.* In a silver nitrate(V)–chloride titration, the silver nitrate(V) is run into the chloride solution to which some potassium chromate(VI) solution is added to act as an indicator.

The first addition of silver nitrate(V) precipitates silver chloride, but not silver chromate(VI). This is because silver chloride has a lower solubility (0.00146 g dm^{-3}) than silver chromate(VI) (0.0284 g dm^{-3}). When sufficient silver nitrate(V) has been added to precipitate all the silver chloride, further addition precipitates brick-red silver chromate(VI), and this serves as the end point.

A more detailed consideration will show why silver chloride is precipitated before the silver chromate(VI). The solubility products concerned are

$$K_s(AgCl) = [Ag^+][Cl^-] = 1 \times 10^{-10}$$
$$K_s(Ag_2CrO_4) = [Ag^+]^2[CrO_4{}^{2-}] = 2.5 \times 10^{-12}$$

On the addition of 0.1M silver nitrate(V) solution to 0.1M sodium chloride solution containing potassium chromate(VI) at a concentration of 0.01M the following changes take place.

The initial $[Cl^-]$ will be 10^{-1} and the initial $[CrO_4{}^{2-}]$ is 10^{-2}. If so little silver nitrate(V) is added to give $[Ag^+]$ of, say, 10^{-10} then

$$[Ag^+][Cl^-] = 10^{-11} \quad \text{and} \quad [Ag^+]^2[CrO_4{}^{2-}] = 10^{-22}$$

Neither the solubility product of silver chloride nor that of silver chromate(VI) is exceeded so that there will be no precipitation.

Addition of more silver nitrate(V) to give $[Ag^+]$ of 10^{-8}, will give

$$[Ag^+][Cl^-] = 10^{-9}$$

so that the silver chloride will be precipitated, but silver chromate(VI) will not, as

$$[Ag^+]^2[CrO_4{}^{2-}] = 10^{-18}$$

When precipitation of silver chloride is completed, the mixture will be saturated with silver chloride so that the $[Ag^+]$ will be 10^{-5}, and $[Ag^+]^2[CrO_4{}^{2-}]$ will be 10^{-12}. A very small further addition of silver nitrate(V) will raise $[Ag^+]^2 [CrO_4{}^{2-}]$ above the solubility of silver chromate(VI) (2.5×10^{-12}) which will be precipitated.

Potassium chromate(VI) can only be used as an indicator in neutral solution as it is soluble in acids. Its action as an indicator is not like that of the dye-stuffs used in acid–alkali titrations (p. 396) where only one or two drops are required. Sufficient potassium chromate(VI) is required to produce enough silver chromate(VI) to saturate the solution so that silver chromate(VI) can be precipitated.

4. Solubility of sparingly soluble salts of weak acids in strong acids. A salt of a weak acid, sparingly soluble in water, is often soluble in a strong acid. Calcium ethanedioate (oxalate), for example, is only sparingly soluble in water, but soluble in dilute hydrochloric acid. This is sometimes regarded as the 'turning out' of a weak acid by a stronger one,

$$CaOx(s) + 2HCl(aq) \longrightarrow CaCl_2(aq) + H_2Ox(aq)$$

In a saturated solution of calcium ethanedioate in water,

$$[Ca^{2+}][Ox^{2-}] = K_s(CaOx)$$

but addition of a strong acid, i.e. of H^+ ions, reduces the ethanedioate ion concentration by the formation of ethanedioic acid,

$$Ox^{2-} + 2H^+ \longrightarrow H_2Ox$$

This reduction in the ethanedioate ion concentration enables more calcium ethanedioate to go into solution.

Addition of Cl^- ions, along with H^+ ions, does not materially affect the position, for Ca^{2+} and Cl^- ions do not combine together for the product they would form, $CaCl_2$, is a salt which is soluble and fully ionised.

5. Calculation of solubility product from solubility. The solubility product of a sparingly soluble salt can be calculated from its solubility, measured as described in the following section.

The solubility of silver chloride, for example, at 18°C, is 0.001 46 g dm^{-3}. The relative molecular mass of silver chloride is 143.5, so that the solubility is approximately 10^{-5} mol dm^{-3}.

Silver chloride ionises according to the equation

$$AgCl \rightleftharpoons Ag^+ + Cl^-$$

and, as the solution is necessarily dilute because of the low solubility, it is reasonable to assume that ionisation is complete. As 1 mol of silver chloride gives 1 mol of Ag^+ and of Cl^-, the concentrations of these ions in a saturated solution are

$$[Ag^+] = 10^{-5} \text{ mol dm}^{-3} \qquad [Cl^-] = 10^{-5} \text{ mol dm}^{-3}$$

The solubility product of silver chloride,

$$K_s(AgCl) = [Ag^+][Cl^-]$$

is, therefore, approximately equal to 1×10^{-10} mol^2 dm^{-6}.

6. Measurement of the solubility of a sparingly soluble salt. This experimental measurement is not easy, but it can be carried out by conductivity measurements.

The measured conductivity of a saturated solution of silver chloride, at 18°C, for example, is 1.274×10^{-6} Ω^{-1} cm^{-1}. At the same temperature, the conductivity of the conductivity water used in making the saturated solution was 0.054×10^{-6}. The conductivity due to the silver chloride, is, therefore 1.220×10^{-6} Ω^{-1} cm^{-1}.

If the solubility of silver chloride at 18°C is x g dm^{-3}, it is also $x/143.5$ mol dm^{-3}. The concentration of a saturated solution of silver chloride is, therefore, $x/143.5$ mol dm^{-3} or 10^{-3} $x/143.5$ mol cm^{-3}.

The molar conductivity, Λ_c, for the saturated solution is given by

$$\Lambda_c = \frac{\text{Conductivity}}{\text{in } \Omega^{-1} \text{ cm}^{-1}} \div \frac{\text{Concentration}}{\text{in mol cm}^{-3}} = 1.22 \times 10^{-3} \times \frac{143.5}{x}$$

and, as the solution is very dilute, this can be taken as being equal to Λ_0, i.e.

$$\Lambda_0 = 1.22 \times 10^{-3} \times \frac{143.5}{x} \; \Omega^{-1} \text{ cm}^2 \text{ mol}^{-1}$$

The Λ_0 value for silver chloride can also be obtained from the molar conductivities of Ag^+ and Cl^- ions, or from the Λ_0 values for silver nitrate(v), sodium chloride and sodium nitrate(v) (p. 369). Thus.

$$\Lambda_0(AgCl) = \Lambda(Ag^+) + \Lambda(Cl^-)$$
$$\Lambda_0(AgCl) = \Lambda_0(AgNO_3) + \Lambda_0(NaCl) - \Lambda_0(NaNO_3)$$

Such measurements give a value for $\Lambda_0(AgCl)$ of $120\ \Omega^{-1}\ cm^2\ mol^{-1}$. Therefore,

$$1.22 \times 10^{-3} \times \frac{143.5}{x} = 120$$

or

$$x = \frac{1.22 \times 10^{-3} \times 143.5}{120} = 0.00146$$

The solubility of silver chloride is 0.00146 g dm^{-3}.

7. Calculation of solubility from solubility product. If the solubility product of a substance is known it is an easy matter to calculate the solubility.

Silver chromate(vi), for example, has a solubility product of 2.5×10^{-12} $mol^3\ dm^{-9}$. This means that, in a saturated solution of silver chromate(vi)

$$[Ag^+]^2[CrO_4^{2-}] = 2.5 \times 10^{-12}$$

Such a solution will be very dilute, so that the silver chromate(vi) can be regarded as fully ionised,

$$Ag_2CrO_4 \longrightarrow 2Ag^+ + CrO_4^{2-}$$

If the solubility of silver chromate(vi) is x mol dm^{-3} then

$$[Ag^+] = 2x \text{ mol } dm^{-3} \qquad [CrO_4^{2-}] = x \text{ mol } dm^{-3}$$

Therefore,

$$(2x)^2 x = 2.5 \times 10^{-12}$$
$$x = \sqrt[3]{0.625 \times 10^{-12}} = 0.855 \times 10^{-4}$$

The relative molecular mass of silver chromate(vi) is 333, giving a solubility of $0.855 \times 10^{-4} \times 333$, i.e. 0.0284 g dm^{-3}.

QUESTIONS ON CHAPTER 35

1. Account for the following:

(a) Zinc sulphide alone is precipitated when hydrogen sulphide is passed into an aqueous solution containing Zn^{2+} and Mn^{2+} ions and dilute acetic acid.

(b) Magnesium hydroxide is not precipitated from solutions of magnesium salts by ammonia solution in the presence of ammonium chloride.

(c) A concentrated solution of calcium chloride gives no precipitate with ammonia solution, but with sodium hydroxide solution a precipitate forms immediately.

(d) To ensure complete precipitation of sulphides in Group 2 of the analysis scheme it is important to control the acidity.

2. The solubility product of magnesium hydroxide is 3.4×10^{-11} mol^3 dm^{-9}. Calculate its solubility in grammes per litre.

3. If the solubility of silver chloride in water is 1.3×10^{-5} mol dm^{-3}, what will it be in 0.1M sodium chloride solution?

4. If one molecule of bismuth(III) sulphide ionises to give two Bi^{3+} ions and three S^{2-} ions, what is the numerical relationship between the solubility of bismuth(III) sulphide expressed as x g dm^{-3} and the solubility product, K, of bismuth(III) sulphide?

5. If the solubility of potassium chlorate(VII) is x mol dm^{-3} in water containing y mol dm^{-3} of potassium chloride, what will the solubility be in pure water?

6. (a) The dissociation constant for hydrogen sulphide into hydrogen and sulphide ions is 10^{-22} mol^2 dm^{-6}, and the concentration of a saturated solution of the gas in water is approximately M/10. What will be the S^{2-} ion concentration in a saturated solution of hydrogen sulphide in (i) M hydrochloric acid, (ii) M/10 hydrochloric acid?

(b) Cadmium sulphide is first precipitated by passing hydrogen sulphide into an acidic solution when the acid is 2M and the concentration of Cd^{2+} ions is 0.01 mol dm^{-3}. What is the solubility product of cadmium sulphide?

7. (a) The solubility product of silver chloride is 1.2×10^{-10} mol^2 dm^{-6}. What is the Ag^+ ion concentration when (i) 5 cm^3, (ii) 10 cm^3 of M/10 silver nitrate(V) solution is added to 10 cm^3 of M/10 sodium chloride solution?

(b) If potassium chromate(VI), with a solubility product of 1.6×10^{-12} mol^3 dm^{-9}, is present in sufficient quantity to give a CrO_4^{2-} ion concentration of 0.01 mol dm^{-3}, what Ag^+ ion concentration is necessary for silver chromate(VI) to be precipitated?

8. A saturated solution of calcium sulphate(VI) (solubility product = 2.4 $\times 10^{-5}$) will precipitate strontium sulphate(VI) (solubility product = 2.8×10^{-7} mol^2 dm^{-6}) from a solution of a strontium salt. Assuming ideal conditions, what is the minimum concentration of Sr^{2+} ions required to give a precipitate of strontium sulphate(VI) if equal volumes of saturated calcium sulphate(VI) solution and the solution containing the Sr^{2+} ions are mixed? What minimum Sr^{2+} ion concentration would give a precipitate with 0.05M sulphuric(VI) acid under the same conditions?

9. Explain what is meant by the term solubility product and why it can be given an approximately constant value for a given salt, solvent and temperature. The solubility products of silver chloride and iodide are approximately 10^{-10} and 10^{-16} mol^2 dm^{-6} respectively. What will be the effect of (a) adding sodium chloride solution to a saturated solution of silver chloride, (b) shaking up solid silver chloride with a solution of potassium iodide, and (c) shaking up solid silver iodide with a solution of hydrochloric acid? (O. Schol.)

10. What is meant by the term solubility product? Give two examples of the application of this concept to qualitative analysis. The solubility of lead sulphate(VI) in water at 17°C is 0.035 g dm^{-3}. Calculate (a) the solubility product of lead sulphate(VI), and (b) the solubility of lead sulphate(VI) (in g dm^{-3}) in a 0.01M solution of sodium sulphate(VI) at the same temperature. (Assume complete dissociation of both solutes.) (O. & C. S.)

11. What do you understand by the term solubility product? The solubility of silver chloride in water at 20°C is 1.507 mg dm^{-3}. What would be the solubility in 0.001M potassium chloride? Assume both salts to be completely dissociated. Explain why silver chloride dissolves in ammonia solution but not in dilute hydrochloric acid, and why barium phosphate(V) dissolves in hydrochloric acid but not in ammonia solution. (O. Prelims.)

12. The solubility of silver chloride in water at 20°C is 1.507 mg dm^{-3}. What will be the solubility in 0.001M potassium chloride solution?

13. If the solubility products of calcium sulphate(VI) and barium sulphate(VI) are 2.25 × 10^{-4} and 1.0 × 10^{-10} mol^2 dm^{-6} respectively, calculate the corresponding solubilities in g dm^{-3}.

· 14. Explain why calcium carbonate is insoluble in water but soluble in dilute hydrochloric acid, whilst lead carbonate is insoluble in both water and dilute hydrochloric acid.

15. Explain why (a) zinc sulphide is insoluble in water but soluble in dilute hydrochloric acid, (b) mercury(II) sulphide is insoluble in water and dilute hydrochloric acid, (c) copper(II) sulphide is insoluble in water and dilute hydrochloric acid, but soluble in aqua regia.

Chapter 36
pH values and indicators

1. Ionic product of water. It is sometimes said that pure water is a non-conductor. What is really meant is that pure water has such a low conductivity that, in some cases, it can be neglected. There are, however, many considerations in which the small conductivity of water cannot be neglected.

The fact that even the purest water ever obtained is a conductor (p. 364) shows that water is very feebly ionised,

$$H_2O(l) \rightleftharpoons H^+(aq) + OH^-(aq)$$

and application of the law of equilibrium gives the expression,

$$K = \frac{[H^+][OH^-]}{[H_2O]}$$

In pure water, the concentration of water molecules is so very large as compared with those of H^+ and OH^- ions that it can be regarded as constant. In a mixture of $1\,000\,000$ water molecules with one H^+ and one OH^- ion, a four-fold change in the ion concentrations to $4H^+$ and $4OH^-$ will still leave $999\,997$ water molecules. The concentration of water molecules has hardly altered, and it can be taken as constant.

This means that, in water,

$$[H^+][OH^-] = \text{a constant}, K_w$$

and the constant is known as the ionic product of water.

The numerical value of this ionic product can be obtained from experimental measurements. The conductivity of the purest conductivity water ever made is $0.054 \times 10^{-6}\ \Omega^{-1}\ cm^{-1}$. The concentration of water is 18 g per 18 cm³ or 1 mol per 18 cm³ or $(1/18)$ mol cm⁻³.

The molar conductivity of water is, therefore, given by

$$\Lambda_c = \frac{\text{Conductivity in } \Omega^{-1}\ cm^{-1}}{\text{Concentration in mol cm}^{-3}}$$
$$= 0.054 \times 10^{-6} \times 18 = 9.72 \times 10^{-7}\ \Omega^{-1}\ cm^2\ mol^{-1}$$

The Λ_0 value for water (545) can be obtained either from the molar conductivities of H^+ and OH^- ions or from Λ_0 values for hydrochloric acid, sodium hydroxide and sodium chloride,

$$\Lambda_0(H_2O) = \Lambda(H^+) + \Lambda(OH^-)$$
$$\Lambda_0(H_2O) = \Lambda_0(HCl) + \Lambda_0(NaOH) - \Lambda_0(NaCl)$$

The degree of ionisation of water, equal to Λ_c/Λ_0 is, therefore, $9.72 \times 10^{-7}/545$ or approximately 18×10^{-10}. This very low value is a measure of how very feebly water is ionised. It means that only $0.000\,000\,18$ per cent of water molecules are, in fact, ionised.

Water has a concentration of 18 g $(18\ cm^3)^{-1}$ or 1 mol $(18\ cm^3)^{-1}$ or

$(1\,000 \div 18)$ mol dm^{-3}. If it was fully ionised the resulting concentrations of H$^+$ and OH$^-$ would be $1\,000 \div 18$ mol dm^{-3}. But, as the degree of ionisation is only 18×10^{-10} it follows that

$$[\text{H}^+] = [\text{OH}^-] = \frac{1\,000}{18} \times 18 \times 10^{-10} = 10^{-7}$$

The concentrations of H$^+$ and of OH$^-$ in pure water are, therefore, 10^{-7} mol dm^{-3}, so that the ionic product is 10^{-14} mol^2 dm^{-6},

$$K_w = [\text{H}^+][\text{OH}^-] = 10^{-7} \times 10^{-7} = 10^{-14}$$

2. Ionic product and temperature. The figure of 10^{-14} for the ionic product of water is a convenient one for general use, but the accurate value depends on the temperature as is shown in the following table:

$T/^\circ\text{C}$	18	25	40	75
$K_w/\text{mol}^2\,\text{dm}^{-6}$	0.61×10^{-14}	1×10^{-14}	2.92×10^{-14}	16.9×10^{-14}

The considerable change in K_w with temperature is due to the change in the ionisation constant, K, of water,

$$K = \frac{[\text{H}^+][\text{OH}^-]}{[\text{H}_2\text{O}]}$$

The equilibrium constant for a reaction is related to temperature by the expression

$$\frac{\text{d}\ln K}{\text{d}T} = \frac{\Delta H}{RT^2}$$

By taking the value of K_w, which is a multiple of K, at two different temperatures it is possible to calculate ΔH for the reaction

$$\text{H}^+ + \text{OH}^- \longrightarrow \text{H}_2\text{O}$$

The calculated value is found to be in close agreement with the experimentally measured value (p. 334).

3. pH values. In any aqueous solution, the equilibrium

$$\text{H}_2\text{O(l)} \rightleftharpoons \text{H}^+\text{(aq)} + \text{OH}^-\text{(aq)}$$

will shift, according to the conditions, but the equilibrium constant will remain constant, at a fixed temperature, so that the ionic product of water

$$[\text{H}^+][\text{OH}^-] = 10^{-14}$$

will also remain constant.

In pure water [H$^+$] equals [OH$^-$], and this equality means that pure water is neutral. In an acidic solution [H$^+$] will be greater than [OH$^-$], and in an alkaline solution [OH$^-$] will be greater than [H$^+$]. Summarising,

Neutral solution $\quad [\text{H}^+] = [\text{OH}^-] = 10^{-7}$

Acidic solution $\quad [\text{H}^+] > [\text{OH}^-] \quad [\text{H}^+] > 10^{-7} \quad [\text{OH}^-] < 10^{-7}$

Alkaline solution $\quad [\text{OH}^-] > [\text{H}^+] \quad [\text{OH}^-] > 10^{-7} \quad [\text{H}^+] < 10^{-7}$

the ionic concentrations being expressed in mol dm^{-3}.

The very wide range, from a strongly acidic solution with H^+ concentration of about 10^{-1} mol dm^{-3} to a strongly alkaline solution with H^+ concentration about 10^{-13} mol dm^{-3}, made a more convenient scale for expressing H^+ concentration desirable. Such a scale was suggested by Sørensen, in 1909. On this scale, *the negative logarithm of the H^+ concentration in mol dm^{-3} in a solution is called the pH of the solution, p* for potenz, meaning strength. Thus

$$pH = -\log_{10}[H^+] = \log_{10}\frac{1}{[H^+]}$$

For a neutral solution, with $[H^+] = [OH^-] = 10^{-7}$ the pH is 7. For an acid solution with $[H^+] > 10^{-7}$, the pH is less than 7. For an alkaline solution, with $[H^+] < 10^{-7}$ the pH is greater than 7.

4. pH changes during acid–alkali titrations. As an alkaline solution is run into an acidic solution during a titration there is a change in the pH value and, at the end point, the change may be a sharp one which can be detected by using an indicator (p. 395).

The way in which the pH changes during the course of a titration depends upon the nature of the acid and alkali used.

a. *Titration of strong alkali against strong acid*, e.g. 0.1M sodium hydroxide against 25 cm^3 of 0.1M hydrochloric acid.

Before starting the titration the pH of the 0.1M acid, fully ionised to give $[H^+]$ equal to 10^{-1} mol dm^{-3}, will be 1.

After the addition of 5 cm^3 of 0.1M alkali, there will be present the equivalent of 20 cm^3 of 0.1M acid in 30 cm^3 of solution. 20 cm^3 of 0.1M acid contains $\frac{10^{-1} \times 20}{1000}$ mol of H^+, and the concentration of H^+ will, therefore, be $\frac{10^{-1} \times 20}{1000} \times \frac{1000}{30}$ mol dm^{-3}, i.e. 0.0667 mol dm^{-3}. The pH is calculated as follows:

$$pH = -\log_{10} 0.0667 = -(\bar{2}.8240) = 1.176$$

When 25 cm^3 of 0.1M alkali have been added to the acid, the mixture will be just neutral with a pH of 7.

With 26 cm^3 of 0.1M sodium hydroxide added, the equivalent of 1 cm^3 of 0.1M alkali will be present in 51 cm^3 of solution. 1 cm^3 of 0.1M alkali contains $\dfrac{10^{-1} \times 1}{1000}$ mol of OH^- ion, so that the concentration of OH^- ion

will be $\dfrac{10^{-1} \times 1}{1000} \times \dfrac{1000}{51}$ mol dm^{-3}, i.e. 1.96×10^{-3} mol dm^{-3}. The concentration of H^+ will, therefore, be $\dfrac{10^{-14}}{1.96 \times 10^{-3}}$ so that the pH is given by

$$pH = -\log_{10}\left(\dfrac{10^{-11}}{1.96}\right) = -(\overline{12}.7077) = 11.292$$

Further figures are included in the following table, and these are plotted in Fig. 134a:

cm³ of 0.1M NaOH added	0	5	10	20	24	24.9	25	25.1	26
pH	1	1.176	1.367	1.954	2.690	3.155	7	10.683	11.292

The graph shows that there is a quite sudden change in pH, from 4 to 10, around the end point, and when equivalent amounts of acid and alkali are present the solution is exactly neutral, with a pH of 7. In this type of titration the end point is also the neutral point and the equivalence point.

b. *Titration of a strong alkali against a weak acid*, e.g. 0.1M sodium hydroxide solution against 25 cm³ of 0.1M ethanoic acid.

The calculation of the pH changes during this type of titration is not so easy as during the titration between strong alkali and strong acid, because the weak acid cannot be taken as fully ionised and its actual degree of ionisation will change throughout the titration. The matter can, however, be treated qualitatively.

0.1M ethanoic acid is not fully ionised so its H^+ ion concentration will be less than that of 0.1M hydrochloric acid, i.e. its pH will be higher.

The effect of adding a solution of a strong alkali, which is fully ionised, will be more marked than the effect of the same alkali on a strong acid. The pH change in a titration of 0.1M sodium hydroxide against 25 cm³ of 0.1M ethanoic acid is, therefore, of the form as shown in Fig. 134b, and this is typical of any titration of a strong alkali against a weak acid.

The change in pH from 7.5 to 10.5 at the end point of the titration is not so great as that in a titration of a strong alkali against a strong acid. Moreover, the pH at the end point, when equivalent amounts of alkali and acid are present, is greater than 7.

In other words, a mixture of equivalent amounts of a strong alkali and a weak acid forms an alkaline solution. This is because the salt formed is, in fact, hydrolysed, i.e. interacts with water (p. 414). With sodium hydroxide and ethanoic acid, the salt formed is sodium ethanoate. This is fully

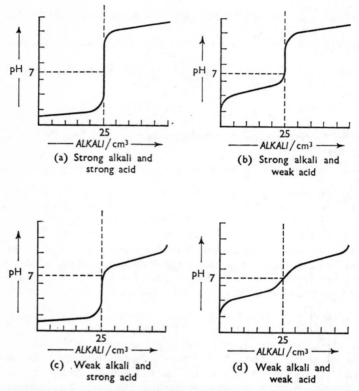

Fig. 134. Titration curves showing the pH changes as a 0.1M solution of a strong or weak alkali is added to 25 cm³ of a 0.1M solution of a strong or weak acid. The precise shape of the curves depends on the actual strengths of the alkalis and acids concerned. Slightly different curves are obtained, too, if solutions other than 0.1M are used.

ionised, but the Na^+ and Et^- ions interact with OH^- and H^+ ions from water,

$$NaEt \longrightarrow Na^+ + Et^-$$

The interaction of the ions is not equal. Et^- and H^+ ions combine appreciably to form ethanoic acid which, being a weak acid, is not highly ionised. Na^+ ions, on the other hand, do not combine at all appreciably with OH^-

ions because the sodium hydroxide which would be formed is a strong electrolyte and fully ionised.

The overall effect is a removal of H^+ ions from solution which is not counterbalanced by any similar removal of OH^- ions, and the resulting solution of sodium ethanoate is alkaline.

c. *Titration of a weak alkali against a strong acid*, e.g. 0.1M ammonia solution against 0.1M hydrochloric acid.

In the titration of a weak alkali against a strong acid, the sudden change in pH at the end point occurs between about 3.5 and 6.5 as shown in Fig. 134c. The pH at the end point, when equivalent amounts of alkali and acid are present, is less than 7, i.e. the mixture is acidic. This is accounted for by hydrolysis as follows:

$$
\begin{array}{ccc}
NH_4Cl \longrightarrow & NH_4^+ & + & Cl^- \\
& + & & + \\
H_2O \rightleftharpoons & OH^- & + & H^+ \\
& \updownarrow & & \updownarrow \\
& NH_3 + H_2O & & HCl
\end{array}
$$

Removal of OH^- ions from the solution in the formation of unionised ammonia is not balanced by an equal removal of H^+ ions, so that the concentration of H^+ ions is greater than that of OH^- ions.

d. *Titration of a weak alkali against a weak acid*, e.g. 0.1M ammonia solution against 0.1M ethanoic acid. There is only a gradual change in pH throughout this type of titration of the kind shown in Fig. 134d.

5. Neutralisation indicators. Indicators are substances which vary in colour according to the H^+ ion concentration of the solution or liquid to which they are added.

Such indicators can be used to measure H^+ ion concentrations (p. 403), or to detect changes in H^+ ion concentration or pH. Different indicators change colour over different ranges of pH, and the most useful are those having a distinct colour change over a narrow range of pH. Some of the commoner indicators are listed below:

Indicator	'Acid' colour	'Alkaline' colour	pH Range
Thymol blue . .	Red	Yellow	1.2–2.8
Methyl orange . .	Pink	Yellow	3.1–4.4
Methyl red . . .	Pink	Yellow	4.4–6.3
Bromothymol blue .	Yellow	Blue	6.0–7.6
Azolitmin (in litmus) .	Red	Blue	5.0–8.0
Cresol red . . .	Yellow	Red	7.2–8.8
Thymol blue . .	Yellow	Blue	8.0–9.6
Phenolphthalein .	Colourless	Red	8.3–10

It will be seen that thymol blue undergoes two colour changes at different pH values, and this is not unusual. Mixtures of selected indicators give a gradual change of colour over a wide range of pH; such mixtures are known as *universal indicators*.

In an acid–alkali titration, it is important in choosing an indicator which will give the correct end point to pick one which changes colour over the range in which there is a marked pH change during the titration (see Fig. 134). The ideal indicator would change colour over a range whose mid-point was the mid-point of the marked pH change occurring during the titration. This means that the choice of indicator is limited as follows:

Titration	Marked pH change	Indicator
Strong acid and strong alkali	4–10	Litmus (5–8) or almost any indicator
Weak acid and strong alkali	7.5–10.5	Phenolphthalein (8.3–10)
Strong acid and weak alkali	3.5–6.5	Methyl red (4.4–6.3)
Weak acid and weak alkali	No marked change	End point cannot be detected accurately by any indicator

6. The functioning of an indicator. In the simplest theory of indicator action, first suggested by Ostwald in 1891, it is assumed that all indicators are either weak acids or weak bases. The most commonly used indicators are, in fact, weak acids, the unionised molecule being one colour whilst the anion is a different colour.

The ionisation of such a weak acid indicator might be represented as

$$HIn \rightleftharpoons H^+ + In^-$$
$$\text{(One colour)} \qquad \text{(Different colour)}$$

In an acidic solution, where the concentration of H^+ ions is high, the indicator will be mainly present as HIn molecules. In alkaline solution, where H^+ ions are removed by combination with OH^- ions, the indicator will be mainly present as In^- ions.

Dilution by itself might be expected to change the colour of an indicator, for in dilute solutions, the greater dissociation of HIn molecules will give more In^- ions. It is, in fact, true that dilution can affect the colour of an indicator.

The equilibrium constant for the ionising indicator is given by

$$K_i = \frac{[H^+][In^-]}{[HIn]}$$

where K_i is the ionisation constant of the indicator.

When the indicator is at the mid-point of its colour change there will be as many HIn molecules present as In^- ions, i.e. $[HIn]$ will be equal to $[In^-]$, and, at this stage, K_i will be equal to $[H^+]$. Thus *the mid-point of an indicator's colour change occurs when the hydrogen ion concentration is equal to the ionisation constant of the indicator.* The corresponding pH value is known as the pK value for the indicator. Some typical numerical values are given below:

Thymol blue	1.51	Azolitmin	7.9
Methyl orange	3.6	Cresol red	8.3
Methyl red	5.0	Thymol blue	8.9
Bromothymol blue	7.1	Phenolphthalein	9.6

The pH at which an indicator is midway through its colour change, i.e. the pK of the indicator, is of value, but the range over which the colour change may be said to be completed is also useful. To decide this, theoretically, it is necessary to assume the conditions under which the colour change might be said to be complete. The assumption most generally made is that the colour change from the HIn colour to that of In^- ions will be complete when $[In^-]$ is equal to $10[HIn]$. Similarly, the colour change, in the other direction, will be complete when $[HIn]$ is equal to $10[In^-]$.

When $[In^-]$ is equal to $10[HIn]$,

$$K_i = 10[H^+] \qquad \text{or} \qquad [H^+] = \frac{K_i}{10}$$

and when $[HIn]$ is equal to $10[In^-]$

$$K_i = \frac{[H^+]}{10} \qquad \text{or} \qquad [H^+] = 10K_i$$

A $[H^+]$ ion concentration of K_i corresponds to a pH equal to pK; $[H^+]$ of $0.1K_i$ corresponds to a pH of pK $-$ 1; $[H^+]$ of $10K_i$ corresponds to a pH of pK $+$ 1.

The colour change of an indicator can, therefore, be summarised as:

	First change of colour	Mid-point of change	Colour change complete
$[H^+]$	$0.1K_i$	K_i	$10K_i$
pH	pK $-$ 1	pK	pK $+$ 1

The H^+ ion concentration has to change 100-fold to complete the colour change, or, in other words, the range over which an indicator changes colour completely is 1 unit of pH on each side of the pK value for the indicator.

These theoretical figures for the range of an indicator do not accurately coincide with the range as measured experimentally, as shown in the following table, but this is due to the arbitrary assumption made about the completion of the colour change in the theoretical considerations.

Indicator	pK value	'Theoretical' range	'Experimental' range
Methyl orange .	3.6	2.6–4.6	3.1–4.4
Methyl red . .	5.0	4.0–6.0	4.4–6.3
Bromothymol blue .	7.1	6.1–8.1	6.0–7.6
Phenolphthalein .	9.6	8.6–10.6	8.3–10.0

7. Indicators which are not acids or bases. Some indicators are neither acids nor bases, and their colour change is related to the existence of tautomeric forms. 4-nitrobenzenol, for example, acts as an indicator with a colour change from colourless to yellow and a pH range of 5–7. It exists in two tautomeric forms, I and II, I being colourless and stable in acid solution, whilst II is yellow and stable

$$
\begin{array}{ccccc}
\text{OH} & & \text{O} & & \text{O} \\
\bigcirc & \rightleftharpoons & \bigcirc & \rightleftharpoons & \bigcirc \quad + \ \text{H}^+ \\
\text{NO}_2 & & \text{NO·OH} & & \text{NO·O}^- \\
\text{I} & & \text{II} & & \text{III} \\
\text{(colourless)} & & \text{(yellow)} & &
\end{array}
$$

in alkaline solution. A small excess of acid or alkali upsets the equilibrium between I and II and produces a colour change. The change is probably assisted, too, by II functioning as a weak acid and ionising to give the anion, III.

The tautomeric forms of other indicators are more complicated but the same ideas apply.

8. Uses of indicators in special titrations. It is important to realise that the terms acidic, alkaline and neutral can be very misleading unless carefully understood in their particular context. A solution of a weak acid, e.g. carbonic acid, may have a pH of 5. Phenolphthalein would indicate that such a solution was acidic, but, so far as methyl orange is concerned, the solution would appear to be 'alkaline'. The solution would turn phenolphthalein colourless, but it would turn methyl orange yellow. Similarly, a solution with a pH of 8 would be alkaline (yellow) when tested with methyl orange, but 'acidic' (colourless) when phenolphthalein was used.

True neutrality occurs when the pH is 7, but, as has been seen (p. 395), the pH of mixtures formed when equivalent quantities of acids and alkalies are mixed may be less than or greater than 7 if the acid and alkali are not of equal strength. The end point of an acid–alkali titration is not always at a pH of 7.

a. *Titration of sodium carbonate.* Sodium carbonate is a salt of a strong alkali and a weak acid, and its solution in water, because of hydrolysis, is alkaline. It will react with solutions of strong acids, e.g.

$$\text{Na}_2\text{CO}_3 + 2\text{HCl} \longrightarrow 2\text{NaCl} + \text{H}_2\text{CO}_3$$

and the detection of the end point of such a reaction is affected by the fact that one of the products of the reaction is itself an acid.

It is, however, a very weak acid and the pH of the sort of carbonic acid solutions obtained in a typical carbonate–acid titration is about 5. Such a solution would be 'acid' to phenolphthalein (range 8.3–10) and turn it colourless, but it would not be acid to methyl orange (range 3.1–4.4), and would not turn it red.

The pH of 0.025M sodium carbonate solution is 11.5, so that methyl orange will be yellow in this solution. At the end point, in titration with 0.1M hydrochloric acid, the colour will change to red, but this colour change will not have been affected by the carbonic acid formed, as carbonic acid, at this concentration, will not affect methyl orange.

Carbonates can, therefore, be titrated against strong acids by using methyl orange as indicator.

b. *Titration of disodium tetraborate*(III), *borax* $Na_2B_4O_7$. This is a salt, like sodium carbonate, of a strong alkali and a weak acid, and its solution is alkaline. It will react quantitatively with a strong acid such as hydrochloric acid according to the equation,

$$Na_2B_4O_7 + 2HCl + 5H_2O \longrightarrow 2NaCl + 4H_3BO_3$$

and the end point of this reaction can be detected so long as methyl orange, which is not affected by the very weak boric(III) acid, H_3BO_3, is used as indicator. In fact, borax is a very good substance to use in standardising acid solutions.

Methyl orange added to 0.05M borax solution will be turned yellow and, at the end point in titration with 0.1M hydrochloric acid, it will be turned red. The boric(III) acid formed would not contribute to the colour change for the pH of 0.05M boric(III) acid is just over 5. The range of methyl orange is 3.1 to 4.4, so that it is yellow in 0.05M boric(III) acid solution.

Boric(III) acid is acid to phenolphthalein, with a range of 8.3 to 10, so that it can be titrated against solutions of strong alkalis using phenolphthalein as indicator. It is, however, better to carry out such a titration in the presence of propane-1,2,3-triol (glycerol), glucose or mannitol, all of which form stronger, complex monobasic acids with boric acid.

c. *Double indicator method.* A mixture of sodium carbonate and sodium hydroxide can be estimated by titration with standard acid, using two indicators.

Phenolphthalein is first added to the mixture to be titrated and the standard acid, say 0.1M hydrochloric acid, is run in. The colour of the indicator changes from pink to colourless when all the sodium hydroxide has been converted into sodium chloride and when the sodium carbonate has been converted into sodium hydrogencarbonate,

$$NaOH + HCl \longrightarrow NaCl + H_2O$$
$$Na_2CO_3 + HCl \longrightarrow NaHCO_3 + NaCl$$

The pH of an approximately 0.1M solution of sodium hydrogencarbonate is about 8 (p. 416).

The amount of acid required, say x cm^3, to decolorise the phenolphthalein is equivalent to all the hydroxide originally present plus half the total carbonate.

If methyl orange is now added to the mixture and the titration continued the sodium hydrogencarbonate will be converted into carbonic acid and sodium chloride,

$$NaHCO_3 + HCl \longrightarrow NaCl + H_2CO_3$$

The pH will change from about 8 to less than 3. The amount of acid required, say y cm³, will be equivalent to the hydrogencarbonate, i.e. to half the original carbonate.

The sodium carbonate present in the original mixture is, therefore, equivalent to $2y$ cm³ of acid, whilst the sodium hydroxide is equivalent to $(x - y)$ cm³.

A similar procedure can be adopted in estimating a mixture of sodium carbonate and sodium hydrogencarbonate. Here, the titration value (x cm³) using phenolphthalein is equivalent to half the sodium carbonate present,

$$Na_2CO_3 + HCl \longrightarrow NaHCO_3 + HCl$$

whilst the methyl orange titration value (y cm³) is equivalent to half the sodium carbonate plus all the sodium hydrogencarbonate.

The sodium carbonate present in the original mixtures is, therefore, equivalent to $2x$ cm³ of acid, whilst the sodium hydrogencarbonate is equivalent to $(y - x)$ cm³.

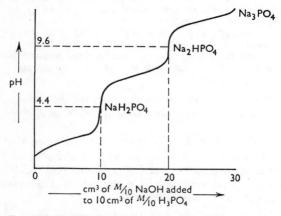

Fig. 135. The pH changes in a titration of 0.1M sodium hydroxide solution against 0.1M phosphoric(v) acid.

d. *Titration of phosphoric*(v) *acid.* Phosphoric(v) acid is tribasic so that it can react with an alkali such as sodium hydroxide in three stages,

i $H_3PO_4 + NaOH \longrightarrow NaH_2PO_4 + H_2O$
ii $NaH_2PO_4 + NaOH \longrightarrow Na_2HPO_4 + H_2O$
iii $Na_2HPO_4 + NaOH \longrightarrow Na_3PO_4 + H_2O$

to form two acid salts and one normal one.

The pH value of the various products, all in M/10 solution are

H_3PO_4	NaH_2PO_4	Na_2HPO_4	Na_3PO_4
1.5	4.4	9.6	$\simeq 12$

Addition of sodium hydroxide solution to phosphoric(v) acid solution containing methyl orange, until the colour changes completely to yellow, will give a solution of sodium dihydrogenphosphate(v). The amount of sodium hydroxide required will be equivalent to one-third of the phosphoric(v) acid present.

Using phenolphthalein, addition of sodium hydroxide will give a deep magenta colour when all the phosphoric(v) acid has been converted into disodium hydrogenphosphate(v) (pH = 9.6). But before then the phenolphthalein will become pale rose when the pH reaches 8.3, i.e. when about 93 per cent of disodium hydrogenphosphate(v) has been formed. It is, therefore, not easy to obtain an accurate end point using phenolphthalein as an indicator. The measured reading is, however, approximately equivalent to two-thirds of the total phosphoric(v) acid present (Fig. 135).

9. Buffer solutions.

A buffer solution, or a solution of reserve acidity or alkalinity, is one which maintains a fairly constant pH, even when small amounts of acid or alkali are added to it. Water, or simple aqueous solutions, do not maintain their pH value at all well because of the marked effect of impurities such as dissolved carbon dioxide absorbed from the air or silicates dissolved from a glass vessel.

Acid buffer solutions can be made by mixing a weak acid with a salt of the same weak acid; alkaline buffer solutions, by mixing a weak base with a salt of the weak base.

A simple acid buffer solution can be made from ethanoic acid and sodium ethanoate,

$$HEt \rightleftharpoons H^+ + Et^-$$
$$NaEt \longrightarrow Na^+ + Et^-$$

Sodium ethanoate is fully ionised, and the ethanoate ions it produces suppress the ionisation of the ethanoic acid so that the mixture contains more ethanoic acid molecules and more ethanoate ions than ethanoic acid alone.

The excess ethanoate ions react with any H^+ ions which might be added,

$$Et^- + H^+ \longrightarrow HEt$$

whilst the excess ethanoic acid molecules react with any added OH^- ions,

$$HEt + OH^- \longrightarrow H_2O + Et^-$$

Addition of small amounts of acid or alkali to the buffer solution of sodium ethanoate and ethanoic acid do not, therefore, greatly affect the pH of the mixture.

A numerical example, taking a buffer solution made of 0.1M ethanoic acid and 0.1M sodium ethanoate, illustrates the point quantitatively.

The Et^- ion concentration, assuming that it all comes from the fully ionised sodium ethanoate and neglecting the relatively small amount from the weakly ionised ethanoic acid, will be 0.1 mol dm^{-3}. The concentration of HEt molecules, neglecting the small degree of ionisation, will also be 0.1 mol dm^{-3}.

The dissociation constant of ethanoic acid is 1.8×10^{-5} mol dm^{-3}, i.e.

$$\frac{[H^+][Et^-]}{[HEt]} = 1.8 \times 10^{-5}$$

so that

$$[H^+] = \frac{1.8 \times 10^{-5} \times [HEt]}{[Et^-]}$$

In the mixture under consideration, the H$^+$ ion concentration will be 1.8×10^{-5} mol dm^{-3}, and the pH value will be 4.74.

If 10 cm^3 of M hydrochloric acid, i.e. 0.01 mol of H$^+$, is added to 1 dm^3 of the mixture, it will react with Et$^-$ ions to form HEt molecules. 0.01 mol of Et$^-$ will be used up, and 0.01 mol of HEt will be formed. The concentrations of Et$^-$ ions and HEt molecules, after the addition of hydrochloric acid, will be

$[Et^-] = 0.1 - 0.01 = 0.09$ mol dm^{-3} $[HEt] = 0.1 + 0.01 = 0.11$ mol dm^{-3}

The new H$^+$ ion concentration will be

$$[H^+] = \frac{1.8 \times 10^{-5} \times 0.11}{0.09} = 2.2 \times 10^{-5} \text{ mol dm}^{-3}$$

and the new pH, 4.66.

Addition of 10 cm^3 of M hydrochloric acid to 1 dm^3 of this buffer solution therefore lowers the pH from 4.74 to 4.66. Addition of 10 cm^3 of M hydrochloric acid to 1 dm^3 of water would produce 0.01M hydrochloric acid; the pH would be changed from 7 to 2.

An alkaline buffer solution contains a weak alkali together with one of its salts. A mixture of ammonia solution and ammonium chloride provides an example. The NH$_4$$^+$ ions from the fully ionised ammonium chloride suppress the ionisation of the ammonia solution so that the mixture contains a greater concentration of both ammonia molecules and ammonium ions than ammonia solution by itself.

Added OH$^-$ and H$^+$ ions are 'taken up' by the reactions,

$$NH_4^+ + OH^- \longrightarrow NH_4OH$$
$$NH_3 + H^+ \longrightarrow NH_4^+$$

It is so useful to be able to make, and keep, solutions with more-or-less constant pH that many buffer solution mixtures are marketed. Typical mixtures used include benzene–1,2–dicarboxylic (phthalic) acid and potassium hydrogenphthalate, sodium dihydrogenphosphate(v) (as the weak acid) and disodium hydrogenphosphate(v) (as the salt), and boric(III) acid and borax (p. 399). There are also many proprietary mixtures. The buffer mixtures are commonly provided in tablet form to be made up into aqueous solution as required.

10. Measurement of H$^+$ ion concentration or pH. The measurement of hydrogen ion concentrations in solutions is important in obtaining values for the dissociation constants of acids (p. 375), and many industrial and biological processes depend, too, on H^+ ion concentration and are controlled by its measurement. A variety of methods are available.

H^+ ions are invariably present as part of an equilibrium mixture, and measurement of their concentration requires a physical method which will not upset the equilibrium. It must also be pointed out that the following methods really measure the concentration of solvated H^+ ions, i.e. of H_3O^+ in aqueous solution (p. 406).

a. H^+ *ion concentration from conductivity measurements.* Conductivity measurements on a solution of an acid of known concentration will give the conductivity (p. 365), from which the molar conductivity, Λ_c, can be calculated. By measuring, also, the Λ_0 value for the acid, the degree of ionisation, Λ_c/Λ_0, can be obtained.

The H^+ ion concentration in a solution of any concentration can easily be calculated if the degree of ionisation at that concentration is known (p. 375), and the dissociation constant of the acid concerned can be obtained by applying Ostwald's dilution law (p. 373),

$$K = \frac{\alpha^2 c}{1 - \alpha}$$

b. H^+ *ion concentration from boiling point or freezing point measurements.* The degree of ionisation of a solute can be measured by comparing its effect on the boiling point or freezing point of a solvent with the calculated effect assuming no ionisation (p. 276).

If the solute used is an acid, the measured degree of ionisation can be used for obtaining the H^+ ion concentration or the dissociation constant of the acid, as in a.

c. H^+ *ion concentration using indicators.* If the same amount of the same indicator gives the same colour in the same amount of two different solutions, then the solutions must have the same pH, i.e. the same H^+ ion concentration.

H^+ ion concentrations in unknown solutions can, therefore, be measured if a set of buffer solutions of known pH is available. The approximate pH of an acid solution is first measured using universal indicator (p. 396). If its value is approximately 4, say, then methyl orange (range 3.1–4.4) is a suitable indicator to choose for further investigation. Two drops of methyl orange are added to 10 cm^3 of the acid solution and the colour obtained is matched against that given by 2 drops of methyl orange in 10 cm^3 portions of solutions of known pH varying from 3.1 to 4.4. By matching the colour of the unknown solution against that of the known solutions, the pH of the unknown solution can be found.

The colour matching can be done by eye or by using a colorimeter. The set of buffer solutions, each with a slightly different colour, can be replaced, too, by a coloured chart or by a set of coloured glasses. The use of a colorimeter with a set of coloured glasses provides a quick and simple method of measuring the pH values reasonably accurately.

d. H⁺ *ion concentration from measurements of reaction velocities.* Some reactions, e.g. the hydrolysis of esters or the inversion of cane sugar (p. 313) take place at speeds which are proportional to the H^+ ion concentration of the solution in which they are carried out. By comparing the measured rates of reaction in the presence of an acidic solution of known pH and of an unknown acidic solution it is possible to obtain the pH of the unknown solution. The method, however, is tedious.

e. H⁺ *ion by measurement of the e.m.f. of a cell.* This method is described on p. 341. It is the method incorporated into pH meters, which allow the pH of a solution to be read directly off the scale of the instrument when an electrode is immersed in the solution under test.

QUESTIONS ON CHAPTER 36

1. What is the approximate percentage decrease in H^+ ion concentration corresponding to an increase of 0.1 in pH?

2. What is the hydrogen ion concentration in solutions with pH of (*a*) 4, and (*b*) 3.6?

3. Calculate the pH of (*a*) 0.1M ethanoic acid solution if the dissociation constant of the acid is taken as 1.75×10^{-5} mol dm⁻³, (*b*) 0.01M hydrochloric acid, (*c*) 0.01M sodium hydroxide solution, (*d*) a solution with a hydrogen ion concentration of 2×10^{-6} mol dm⁻³.

4. The logarithmic nature of the pH scale means that the mean hydrogen ion concentration between 4 and 5 is not 4.5 but 4.3. Explain what this means.

5. Prove that the pH of an acid with dissociation constant, K, in a solution in which the degree of dissociation is α, is equal to $\log_{10} 1/K + \log_{10} \{\alpha/(1 - \alpha).\}$.

6. Calculate the approximate pH values of the solutions obtained by adding 20, 24, 24.9 and 25 cm³ of 0.1M sodium hydroxide to 25 cm³ of 0.1M hydrochloric acid. Why would the values be different if 0.1M ethanoic acid were used instead of hydrochloric acid, and what would be the effect in each case if ammonia solution were substituted for sodium hydroxide? (O. & C. S.)

7. Draw and explain the curves showing approximately the change of pH when 50 cm³ of 0.1M sodium hydroxide solution is run slowly with stirring into (*a*) 25 cm³ of 0.1M hydrochloric acid, (*b*) 25 cm³ of 0.1M ethanoic acid. What light do these curves throw on the indicators which may be used for the titration of sodium hydroxide with these acids? Why is it not possible to titrate ethanoic acid with ammonia solution, using ordinary indicators? (O. & C.)

8. Show how the formula expressing Ostwald's dilution law is derived. To what extent is it generally applicable? The dissociation constant of ethanoic acid in aqueous solution at 25°C is 1.75×10^{-5} mol dm⁻³. Calculate its degree of ionisation in 0.1M solution at 25°C, and the pH of the solution.

9. What is the pH of a 0.01M potassium hydroxide solution at 25°C if the ionisation constant of water at that temperature is 1×10^{-14}? What is the hydrogen ion concentration (in mol dm⁻³) of a solution which has a pH of 5.5?

10. Explain the meaning of the term dissociation constant as applied to acids and bases in solution. Calculate (*a*) the mass of hydrogen chloride required per dm³ to give a pH of 3, (*b*) the pH of a 0.01M solution of propanoic acid ($K = 1.45 \times 10^{-5}$ mol dm⁻²), and (*c*) the concentration in mol dm⁻³ of an aqueous ammonia solution ($K = 1.74 \times 10^{-5}$ mol dm⁻³) whose pH is 10.3.

11. A 0.025M solution in water of a monobasic acid was found to have a freezing point of $-0.06°C$. What is the dissociation constant and pK value of the acid?

12. What methods are used for determining the 'end points' of titrations? Outline the principles involved. (O. Schol.)

13. What is the evidence that when ammonia dissolves in water ammonium ions are formed? Will the pH of a solution of ammonium chloride in water be greater or less than 7? Give reasons (O. Schol.)

14. What is a buffer solution, and why are such solutions important? Given that the dissociation constant of ethanoic acid is 1.8×10^{-5} mol dm^{-3}, describe how you would prepare a buffer solution of pH 5. (O. Schol.)

15. What is a buffer solution? Explain its mode of action. Give two instances of the use of ammonia solution in the presence of ammonium chloride. (O. & C.)

16. What is the pH of a buffer solution made by dissolving 0.005 mol of methanoic acid ($K = 1.8 \times 10^{-4}$ mol dm^{-2}) and 0.007 mol of sodium methanoate in 1 dm^3 of aqueous solution? What is the effect on the pH of a tenfold dilution?

17. Calculate the pH of a mixture of 50 cm^3 of M ethanoic acid with 20 cm^3 of M sodium hydroxide. Point out, clearly, what assumptions you make in the calculation.

18. The K_i values for methyl red and methyl orange are 1×10^{-5} and 4×10^{-4} respectively. They both change colour from red in acid solution, through orange, to yellow in alkaline solution. Draw up a table to show their probable colours in solutions of pH 2, 3, 4, 5 and 6.

19. The acid dissociation constant of methyl orange is 4×10^{-4} mol dm^{-3}. The solubility product of magnesium hydroxide is 2×10^{-11} mol^3 dm^{-9}. Describe what will happen when molar sodium hydroxide solution is added dropwise to 25 cm^3 of 0.5M sulphuric acid in which 0.1 g of magnesium has been dissolved and containing a few drops of methyl orange. ($K_w = 10^{-14}$; Mg = 25.) (O. Schol.)

20. Explain the distinction between the terms 'the molarity of an acid' and 'the strength of an acid'. Discuss the way in which knowledge of the strength of an acid would affect your choice of the indicator to be used for determining its normality. (O. Schol.)

Chapter 37
Acids and bases

1. Oxonium or hydronium ion. An acid is often defined as a substance which, on dissolving in water, dissociates to produce hydrogen ions as the only positively charged ions. On this view, which was essentially that of Arrhenius, acidic properties are due to hydrogen ions, and the dissociation of an acid is conventionally represented by showing the formation of such ions, e.g.

$$H_2SO_4 \longrightarrow H^+ + HSO_4^-$$
$$H_2SO_4 \longrightarrow 2H^+ + SO_4^{2-}$$

Such a representation is now known to be a simplification, for H^+ ions do not exist in the presence of water. They are, in fact, hydrated to form H_3O^+ ions, known as oxonium or hydronium ions,

$$H^+ + H_2O \longrightarrow H_3O^+$$

It is possible, too, that the hydration may, under some conditions, be carried further to give $H(H_2O)_x^+$ or $H^+(aq)$ ions.

That H^+ ions are hydrated in the presence of water is to be expected on theoretical grounds. An isolated H^+ ion, or proton, is very minute, with a radius 50000 times less than that of a lithium ion which is itself a small positively charged ion. According to Fajans's rules (p. 165), an H^+ ion would have a very strong tendency to pass into the covalent state, and this it can do by combining with an H_2O molecule to form H_3O^+.

Experimental evidence for the hydration of H^+ ions comes from the measurement of transport numbers. The movement of water along with H^+ ions during electrolysis can be demonstrated (p. 455).

If it is the H_3O^+ ion which exists in the presence of water, and not the H^+ ion, the ionisation of water and of acids must, more accurately, be written as

$$H_2O(l) + H_2O(l) \rightleftharpoons H_3O^+(aq) + OH^-(aq)$$
$$HCl(g) + H_2O(l) \longrightarrow H_3O^+(aq) + Cl^-(aq)$$
$$H_2SO_4(l) + 2H_2O(l) \longrightarrow 2H_3O^+(aq) + SO_4^{2-}(aq)$$

It is, however, conventional to write H^+ instead of H_3O^+ in such equations and similar ones. Moreover, the numerical value for a hydrogen ion concentration is the same no matter whether H^+ or H_3O^+ is considered, so long as the concentration is expressed in mol dm^{-3}, as is generally done.

2. Bases and alkalis. The term base was introduced by Rouelle, in 1754, to denote a substance which reacted with an acid to form a salt and water only. A metallic oxide or hydroxide was regarded as the base from which salts could be made.

The term alkali is much more ancient, and was originally associated with the ashes from burnt wood. As chemistry developed, a distinction was made between mild alkalis, such as sodium and potassium carbonate, and caustic alkalis, such as sodium and potassium hydroxide.

Eventually the word alkali was most commonly used to denote a soluble base, and as sodium and potassium hydroxides are by far the commonest soluble bases the term is sometimes limited to these two substances.

From the viewpoint of early ionic theory a base was regarded as a substance which reacted with H^+ ions to form water, e.g.

$$PbO(s) + 2H^+(aq) \longrightarrow Pb^{2+}(aq) + H_2O(l)$$
$$NaOH(s) + H^+(aq) \longrightarrow Na^+(aq) + H_2O(l)$$

whilst a soluble base (alkali) was also defined as a substance which, on dissolving in water, dissociated to give OH^- ions.

3. Brønsted–Lowry theory. The definitions given in the preceding sections for acids, bases and alkalis are not entirely satisfactory, and apply only to aqueous solutions.

They were replaced, in 1922, by wider definitions put forward by Brønsted and Lowry;

An acid is a substance that will give up protons to a base.

A base is a substance that will accept protons from an acid.

An acid is a proton-donor and a base is a proton-acceptor.

The relationship between an acid and a base is then summarised by

$$\begin{array}{ccc} A & \rightleftharpoons & H^+ + B \\ \text{Acid} & & \text{Base} \\ \text{(Proton-donor)} & & \text{(Proton-acceptor)} \end{array}$$

A and B which may each be molecules, radicals or ions are said to be *conjugate or a conjugate pair*. B is called the conjugate base of the acid A, and A the conjugate acid of the base B. Typical conjugate pairs are underlined in the following examples:

$$\underline{HEt} \rightleftharpoons H^+ + \underline{Et^-} \qquad\qquad \underline{NH_4^+} \rightleftharpoons H^+ + \underline{NH_3}$$
$$\underline{H_2SO_4} \rightleftharpoons H^+ + \underline{HSO_4^-} \qquad\qquad \underline{H_2O} \rightleftharpoons H^+ + \underline{OH^-}$$

The recognition of ethanoate and hydrogensulphate(VI) ions as bases, along with ammonia and hydroxide ions, and that of ammonium ions and water as acids along with ethanoic acid and sulphuric(VI) acid shows how the conception of acids and bases is widened by the Brønsted–Lowry theory.

A is a strong acid if the equilibrium

$$A \rightleftharpoons H^+ + B$$

is well over to the right; B is a strong base if the equilibrium is well over to the left. Thus the conjugate base of a strong acid is a weak base and vice versa. Similarly, the conjugate acid of a strong base is a weak acid, and vice versa.

If the solvation of H^+ by a solvent is also taken into account the relationship between conjugate pairs in an aqueous solution is

$$A + H_2O \rightleftharpoons H_3O^+ + B$$

In such a change, water is accepting a proton from acid A, i.e. is acting as a base, whilst H_3O^+ ions are giving up a proton to base B, i.e. are acting as an acid. The change can be generalised as

$$\text{Acid (1)} + \text{Base (2)} \rightleftharpoons \text{Acid (2)} + \text{Base (1)}$$

with acid (1) conjugate with base (1), and acid (2) conjugate with base (2). Particular examples are

$$HEt + H_2O \rightleftharpoons H_3O^+ + Et^-$$
$$H_2SO_4 + H_2O \rightleftharpoons H_3O^+ + HSO_4^-$$
$$NH_4^+ + H_2O \rightleftharpoons H_3O^+ + NH_3$$
$$H_2O + H_2O \rightleftharpoons H_3O^+ + OH^-$$
$$\text{Acid (1)} + \text{Base (2)} \rightleftharpoons \text{Acid (2)} + \text{Base (1)}$$

It will be seen that water can be said to act as either an acid or a base.

4. Effect of solvent on acids and bases. As free H^+ ions cannot exist and are always solvated, the solvent plays a large part in acid–base considerations, and the Brønsted–Lowry theory is of particular value in elucidating the behaviour of acids and bases in different solvents.

Water, the most commonly used solvent, is a proton acceptor, i.e. basic or *protophilic*. In an aqueous solution of a strong acid, e.g. hydrochloric acid, the equilibrium

$$HCl(g) + H_2O(l) \rightleftharpoons H_3O^+(aq) + Cl^-(aq)$$

lies far over to the right, mainly because of the protophilic nature of the solvent water.

In a solvent of glacial ethanoic acid, which is not very ready to accept protons, the equilibrium

$$HCl(g) + CH_3 \cdot COOH(l) \rightleftharpoons CH_3 \cdot COOH_2^+(aq) + Cl^-(aq)$$

will not lie very far to the right, and hydrogen chloride is a weaker acid in glacial ethanoic acid solution than in aqueous solution. Differences in acid strength, not readily apparent in aqueous solution, become noticeable in glacial ethanoic acid solution, the order of decreasing strength for some common acids being $HClO_4 > H_2SO_4 > HCl > HNO_3$. In 0.005M solution in glacial ethanoic acid, for instance, the conductivity of chloric(vii) acid is more than fifty times that of nitric(v) acid.

In an *aprotic* solvent, which will not accept protons at all, e.g. a hydrocarbon such as methylbenzene (toluene), even the strongest acid exhibits no acidic properties. Hydrogen chloride, for example, is soluble in methylbenzene, but the solution is not acidic.

Acidic properties are also lacking in solutions of acids in *protogenic* solvents, e.g. liquid hydrogen fluoride or liquid hydrogen chloride, which are themselves proton donors or acidic. In such solvents, an 'acid' may, in fact, function as a 'base'. Nitric(v) acid, for example, acts as a base in liquid hydrogen fluoride solution. The equilibrium established is

$$\underset{\text{Acid}}{\text{HF(l)}} + \underset{\text{Base}}{\text{HNO}_3\text{(l)}} \rightleftharpoons \underset{\text{Acid}}{\text{H}_2\text{NO}_3{}^+} + \underset{\text{Base}}{\text{F}^-}$$

nitric(v) acid being forced to act as a proton acceptor, i.e. as a base.

Similarly, nitric(v) acid is forced to act as a base when mixed with concentrated sulphuric acid,

$$\text{HNO}_3 \longrightarrow \text{NO}_2{}^+ + \text{OH}^-$$

$$\text{HNO}_3\text{(l)} + 2\text{H}_2\text{SO}_4\text{(l)} \longrightarrow \text{NO}_2{}^+ + 2\text{HSO}_4{}^- + \text{H}_3\text{O}^+$$

The resulting $\text{NO}_2{}^+$, nitryl cation (nitronium ion), is of importance in nitration with mixtures of concentrated nitric(v) and sulphuric(vi) acids.

So far as the strength of a base is concerned, the solvent has the opposite effect to that which it has on acid strength. Basic properties are not evident if the solvent is aprotic, but a highly protogenic solvent leads to an increase in basic strength, whilst a protophilic solvent produces a decrease.

The effect of the solvent may be summarised as follows:

	Nature of solvent		
	Protophilic (Proton-accepting)	*Aprotic*	*Protogenic (Proton-donating)*
Strength of acid	Increased	Nil	Decreased
Strength of base	Decreased	Nil	Increased

5. Reactions in non-aqueous solvents. Chemical reactions in liquid ammonia solution have been extensively studied, following the original investigations of Franklin, and, more recently, many other non-aqueous solvents have been used.

Many of the reactions in such non-aqueous solvents are closely related to similar reactions in water. Liquid ammonia, for example, is slightly ionised, and can act as an acid or a base just as water can,

$$\text{NH}_3 + \text{NH}_3 \rightleftharpoons \text{NH}_4{}^+ + \text{NH}_2{}^-$$

$$\underset{\text{Acid}}{\text{H}_2\text{O}} + \underset{\text{Base}}{\text{H}_2\text{O}} \rightleftharpoons \underset{\text{Acid}}{\text{H}_3\text{O}^+} + \underset{\text{Base}}{\text{OH}^-}$$

The NH_4^+ and NH_2^- ions are related in just the same way as H_3O^+ and OH^- are.

A solution of ammonium chloride in liquid ammonia is acidic, just as a solution of hydrogen chloride in water is,

$$HCl \longrightarrow H^+ + Cl^-$$
$$NH_4Cl \longrightarrow NH_4^+ + Cl^-$$

Similarly, a solution of sodamide, $NaNH_2$, in liquid ammonia is basic, like a solution of sodium hydroxide in water,

$$NaOH \longrightarrow Na^+ + OH^-$$
$$NaNH_2 \longrightarrow Na^+ + NH_2^-$$

Moreover, solutions of ammonium chloride and sodamide, in liquid ammonia, can be titrated using phenolphthalein as an indicator, just as aqueous solutions of hydrogen chloride and sodium hydroxide can.

The small ionisation of liquid sulphur dioxide is probably represented by the equation,

$$2SO_2 \rightleftharpoons SO^{2+} + SO_3^{2-}$$

so that compounds, such as sulphur dichloride oxide (thionyl chloride), which dissolve in this solvent to give thionyl, SO^{2+}, ions are acidic,

$$SOCl_2 \rightleftharpoons SO^{2+} + 2Cl^-$$

whilst compounds, e.g. caesium sulphate(IV), which dissolve to give sulphate(IV), SO_3^{2-}, ions, are basic,

$$Cs_2SO_3 \rightleftharpoons 2Cs^+ + SO_3^{2-}$$

Analogous situations are found with other non-aqueous solvents which ionise slightly. Examples are provided by liquid hydrogen fluoride, liquid hydrogen sulphide, liquid hydrogen cyanide, and bromine trifluoride,

$$2HF \rightleftharpoons H_2F^+ + F^- \qquad 2HCN \rightleftharpoons H_2CN^+ + CN^-$$
$$2H_2S \rightleftharpoons H_3S^+ + SH^- \qquad 2BrF_3 \rightleftharpoons BrF_2^+ + BrF_4^-$$

6. The Lewis theory. The idea of acids and bases has been extended still further by Lewis, using the following definitions:

An acid is a substance that can accept a pair of electrons to form a covalent bond,

A base is a substance that can donate a pair of electrons to form a covalent bond.

For example,

$$H^+ + :\overset{\cdot\cdot}{\underset{\cdot\cdot}{O}}-H^- \rightleftharpoons H:\overset{\cdot\cdot}{\underset{\cdot\cdot}{O}}-H$$

Cl	H	Cl H
Cl ⁞B	:N× H ⟶	Cl ⁞B : N⁞ H
Cl	H	Cl H
Acid	Base	Covalent compound

Such ideas widen the conception of acids and bases, perhaps, too far, but the Lewis theory does bring out a close relationship between acid–base and oxidation–reduction reactions.

Acids, on the Lewis theory, are electron-pair acceptors, capable of sharing a pair of electrons with a base, which is an electron-pair donor. An oxidising agent also accepts electrons from a reducing agent, but, in this case, the transference of electrons from the reducing agent (electron donor) to the oxidising agent (electron acceptor) is complete and there is no *sharing* of electron pairs.

$$Fe^{3+} \quad + \quad 1e \quad \underset{\text{Oxidation}}{\overset{\text{Reduction}}{\rightleftharpoons}} \quad Fe^{2+}$$

Oxidising agent   Reducing agent
(Electron acceptor)   (Electron donor)

The terms *electrophilic* (electron-seeking) and *nucleophilic* (nucleus-seeking or electron-donating), widely used in organic chemistry, also apply, in the widest sense, to acids or oxidising agents (electrophilic), and bases or reducing agents (nucleophilic).

7. Amphoteric electrolytes. Some substances, known as amphoteric electrolytes or ampholytes, are capable of exhibiting both acidic and basic properties, combining with both bases and acids.

a. *Water*. Water can act as either an acid or a base (p. 408).

$$H_2O \quad + \quad H_2O \quad \rightleftharpoons \quad H_3O^+ \quad + \quad OH^-$$

Acid (1)   Base (2)   Acid (2)   Base (1)

and it can, therefore, be regarded as amphoteric.

b. *Metallic hydroxides*. Certain metallic hydroxides, e.g. zinc, aluminium, lead(II), chromium(III) and tin(II) hydroxides, are amphoteric. In terms of simple equations, they react with acids and with bases to form salts,

$$Zn(OH)_2 + 2HCl \longrightarrow ZnCl_2 + 2H_2O$$

$$Zn(OH)_2 + 2NaOH \longrightarrow Na_2Zn(OH)_4 \text{ (or } Na_2ZnO_2 + 2H_2O)$$

Disodium   Sodium
tetrahydroxozincate   zincate

Zinc hydroxide functions as a base by ionising to give OH^- ions which can accept a proton from an acid,

$$Zn(OH)_2 \longrightarrow Zn^{2+} + 2OH^-$$

It functions as an acid, by reacting with water to give H^+ ions,

$$Zn(OH)_2 + 2H_2O \longrightarrow Zn(OH)_4^{2-} + 2H^+$$

c. *Amino acids.* The most interesting ampholytes are those which contain quite separate acidic and basic groups within one molecule, and the commonest examples of this type are provided by amino acids, e.g. aminoethanoic acid, glycine, $CH_2(NH_2)\cdot COOH$.

Such acids will form salts with both acids and bases, e.g.

$$CH_2(NH_2)\cdot COOH + HCl \longrightarrow \{CH_2(NH_3)COOH\}^+ \ Cl^-$$
$$CH_2(NH_2)\cdot COOH + NaOH \longrightarrow \{CH_2(NH_2)\cdot COO\}^- \ Na^+ + H_2O$$

Aminoethanoic acid exists, both in the solid state and, to a large extent in solution, as $^+NH_3\cdot CH_2\cdot COO^-$ ions,

Such an ion, with a positive charge at one end and a negative charge at the other, is known as a *zwitterion* or a *dipolar ion.*

On this view, the acidic function of aminoethanoic acid is due to the NH_3^+ end,

whilst the basic function is due to the $-CO_2^-$ end,

The existence of dipolar ions in aminoethanoic acid is supported by X-ray analysis, and accounts for the surprisingly high melting point (above 230°C), the high dielectric constant in aqueous solution, and the ready solubility in water and sparing solubility in most organic (non-polar) solvents.

QUESTIONS ON CHAPTER 37

1. Explain the meaning of the following terms as applied to an acid or its aqueous solution; concentration; strength; basicity. What experiments would you make to show that (*a*) sulphuric(VI) acid is dibasic, (*b*) hydrochloric acid is stronger than sulphuric(VI) acid? (O. & C. S.)

2. The term 'strong acid' is very commonly used. What, precisely, does it mean?

3. Choose any two chemicals and explain how they can be regarded as acids on the various different theories of acid character.

4. An acid, in elementary chemistry, is sometimes defined as a compound which contains hydrogen replaceable by a metal, which will turn blue litmus red, and which will react with metallic carbonates to yield carbon dioxide. To what extent is this a satisfactory definition of an acid?

5. Comment on the definition of an acid as a substance giving rise, in solution, to a cation characteristic of the solvent.

6. Comment on the following statements: (a) The carbonate ion is a fairly strong base. (b) The ionisation constant of an acid, HA, in aqueous solution measures the competition for protons between water and A^- ions. (c) No acid, in aqueous solution, can be stronger than H_3O^+.

7. What is an acid? Illustrate, by means of examples, the factors influencing the strength of an acid. (O. Schol.)

8. Define what you mean by the strength of an acid, and describe methods of measuring it. Discuss the reaction between sulphuric(VI) acid and sodium chloride in terms of the relative strengths of sulphuric(VI) and hydrochloric acids. (O. Schol.)

9. What do you understand by the terms acid, base and salt? Into which of these classes would you place H_2O, $NaHSO_4$, NH_4^+, CuO, C_6H_5OH and CH_3COO^-? Give your reasons.

10. The most important acid is a proton. Discuss this statement.

11. What are the conjugate bases of the following when they act as monobasic acids: HCl, HNO_3, H_3PO_4, H_2O, H_3O^+, CH_3COOH, NH_4^+, HSO_4^- and HCO_3^-? Which of these conjugate bases can still be regarded as acids?

12. Illustrate the statement that salts which contain an ion which can act as a base are more soluble in dilute acids than in water.

13. Which of the following are to be regarded as acids or bases or both: HSO_4^-, HCO_3^-, $H_2PO_4^-$, NH_4^+, OH^-, H_2O, H_3O^+, CO_3^{2-}, NH_3? Give your reasoning.

14. The greater the number of oxygen atoms in a molecule of an oxy-acid of an element, the greater is the strength of the acid. Illustrate and discuss this statement.

Chapter 38
Hydrolysis of salts

1. Results of hydrolysis. A salt may be regarded as formed, together with water, by the reaction of an acid and a base, e.g.

$$HA \quad + \quad BOH \quad \longrightarrow \quad BA \quad + \quad H_2O$$

Acid   Base   Salt

where A is an anion and B a cation.

If such a reaction is reversible, a solution of a salt in water will contain some acid and some base, and if the acid and the base are not of equal strengths the solution will be either acidic or basic, the equilibrium between H^+ and OH^- ions in water being disturbed.

When this happens, salt hydrolysis is said to have taken place, and the nature and extent of the hydrolysis depends on the strengths of the acid and base from which the hydrolysed salt can be formed.

a. *Salt of strong acid and strong base.* Such salts, e.g. sodium chloride, sodium sulphate(VI) and potassium nitrate(V) are not hydrolysed, and their aqueous solutions are neutral, with a pH of 7.

In a solution of sodium chloride in water, for example, there are Na^+, Cl^-, H^+ and OH^- ions present, but, as sodium hydroxide and hydrochloric acid are both fully ionised there is no tendency for reaction between Na^+ and OH^- ions, or between H^+ and Cl^- ions. The H^+ and OH^- ion concentrations remain as they are in pure water.

b. *Salt of weak acid and strong base*, e.g. sodium ethanoate, potassium carbonate, potassium cyanide and sodium sulphide.

Taking sodium ethanoate (NaEt) as an example, the situation existing in its aqueous solution may be summarised as follows:

$$NaEt \longrightarrow Na^+(aq) + Et^-(aq)$$

$$H_2O \rightleftharpoons OH^-(aq) + H^+(aq)$$

NaOH   HEt

(Strong)   (Weak)

Formation of molecules of the weak ethanoic acid removes some H^+ ions from solution, but there is no compensating removal of OH^- ions, because sodium hydroxide is a strong electrolyte, fully ionised. The solution, therefore, contains excess OH^- ions and is alkaline, with a pH greater than 7.

The hydrolysis may be referred to as *anion hydrolysis*,

$$Et^-(aq) + H_2O(l) \rightleftharpoons HEt + OH^-(aq)$$

the Na^+ cation playing no significant part in the process.

c. *Salt of strong acid and weak base.* The situation in a solution of ammonium chloride may be summarised as follows:

$$NH_4Cl \longrightarrow NH_4^+(aq) \quad + \quad Cl^-(aq)$$

$$H_2O \; \underset{\longrightarrow}{\rightleftarrows} \; OH^-(aq) \quad + \quad H^+(aq)$$

$$NH_3 + H_2O \qquad\qquad HCl$$
$$\text{(Weak)} \qquad\qquad \text{(Strong)}$$

Here, the removal of OH^- ions from the solution leaves an excess of H^+ ions and the solution is acidic.

This is an example of *cation hydrolysis*,

$$NH_4^+(aq) + H_2O(l) \rightleftharpoons NH_4OH + H^+(aq)$$

As NH_4OH is itself unstable,

$$NH_4OH \longrightarrow NH_3 + H_2O$$

the hydrolysis may be written as

$$NH_4^+(aq) + H_2O(l) \rightleftharpoons NH_3(g) + H_2O(l) + H^+(aq)$$

or

$$NH_4^+(aq) \rightleftharpoons NH_3(g) + H^+(aq)$$

The hydrolysis is, therefore, partially due to the dissociation of NH_4^+ ions.

The acidity of a copper(II) sulphate(VI) solution may be explained in a similar way. An over-simplified summary shows

$$CuSO_4 \longrightarrow Cu^{2+}(aq) \quad + \quad SO_4^{2-}(aq)$$

$$2H_2O \; \underset{\longrightarrow}{\rightleftarrows} \; 2OH^-(aq) \quad + \quad 2H^+(aq)$$

$$Cu(OH)_2 \qquad\qquad H_2SO_4$$
$$\text{(Weak and} \qquad\qquad \text{(Strong)}$$
$$\text{insoluble)}$$

the acidity being attributed to the removal of OH^- ions in the formation of copper(II) hydroxide.

The hydrolysis of metallic cations is, however, more complicated, and the insoluble metallic hydroxide is rarely formed. Copper(II) sulphate(VI), for example, contains $Cu(H_2O)_4^{2+}$ ions rather than simple, unhydrated Cu^{2+} ions, and the hydrated ions dissociate in the same way as the NH_4^+ ion dissociates,

$$NH_4^+(aq) \longrightarrow NH_3(g) + H^+(aq)$$
$$Cu(H_2O)_4^{2+}(aq) \longrightarrow Cu(H_2O)_3OH^+(aq) + H^+(aq)$$
$$Cu(H_2O)_3OH^+(aq) \longrightarrow Cu(H_2O)_2(OH)_2 + H^+(aq)$$
$$Cu(H_2O)_2(OH)_2 \longrightarrow Cu(OH)_2 + 2H_2O$$

Formation of an insoluble metallic hydroxide is only the final stage, and is preceded by the formation of soluble hydroxyl complexes. Actual precipitation of insoluble hydroxides will only occur, generally, if the H^+ ion concentration is lowered by adding OH^- ions.

Hydrolysis of any AB salt, where B is the radical of a strong acid and A is a trivalent metallic ion or Zn^{2+}, Cd^{2+}, Hg^{2+} or Pb^{2+}, occurs by the formation of similar hydroxyl complexes.

d. *Salt of weak acid and weak base.* Both anion and cation hydrolysis will occur together, and the resulting solution may be acidic, neutral or alkaline depending on the relative extent to which they do occur.

A solution of ammonium ethanoate is approximately neutral (pH = 7), hydrolysis of Et^- ions being just about balanced by hydrolysis of NH_4^+ ions,

$$Et^-(aq) + H_2O(l) \longrightarrow HEt + OH^-(aq)$$
$$NH_4^+(aq) + H_2O(l) \longrightarrow NH_3(g) + H_2O(l) + H^+(aq)$$

Ammonium carbonate solution, in comparison, is alkaline.

2. Hydrolysis of acid salts. The pH of a solution of an acid salt is determined by the extent of the hydrolysis together with the extent of the dissociation of the hydrogen-containing anion.

For sodium hydrogencarbonate, $NaHCO_3$, anion hydrolysis leads to

$$HCO_3^-(aq) + H_2O(l) \longrightarrow H_2CO_3 + OH^-(aq)$$

whereas dissociation of the anion leads to

$$HCO_3^-(aq) \longrightarrow CO_3^{2-}(aq) + H^+(aq)$$

The hydrolysis predominates, as HCO_3^- is such a very, very weak acid ($K = 6 \times 10^{-11}$, p. 375), so that a solution of sodium hydrogencarbonate is alkaline, with a pH of 8.3 at a concentration of 0.1 mol dm^{-3}.

With sodium hydrogensulphate(IV),

$$HSO_3^-(aq) + H_2O(l) \longrightarrow H_2SO_3 + OH^-(aq)$$
$$HSO_3^-(aq) \longrightarrow SO_3^{2-}(aq) + H^+(aq)$$

The HSO_3^- ion is a stronger acid ($K = 1 \times 10^{-7}$) than HCO_3^- and it dissociates more than it is hydrolysed. A solution of sodium hydrogensulphate(IV) is, consequently, acidic, with a pH of about 5.

3. Hydrolysis constant. The general equation for the hydrolysis of a salt may be written as

$$BA + H_2O \rightleftharpoons HA + BOH$$

and the equilibrium constant for such a hydrolysis is given by

$$\frac{[HA][BOH]}{[BA][H_2O]} = K$$

As water is in very large excess in any aqueous solution, its concentration may be regarded as constant (p. 390) so that

$$\frac{[HA][BOH]}{[BA]} = K_h$$

and the resulting equilibrium constant is known as the hydrolysis constant of BA. In more general terms, it can be expressed as

$$\frac{[\text{Free Acid}][\text{Free base}]}{[\text{Unhydrolysed salt}]} = K_h$$

a. *Salt of weak acid and strong base. Anion hydrolysis.* The general equation becomes

$$B^+(aq) + A^-(aq) + H_2O(l) \rightleftharpoons HA + B^+(aq) + OH^-(aq)$$

or

$$A^-(aq) + H_2O(l) \rightleftharpoons HA + OH^-(aq)$$

and the hydrolysis constant is given by

$$\frac{[HA][OH^-]}{[A^-]} = K_h$$

In the solution,

$$[H^+][OH^-] = K_w \text{ (ionic product of water, p. 391) and}$$

$$\frac{[H^+][A^-]}{[HA]} = K_a \text{ (dissociation constant of HA, p. 375).}$$

It follows, by eliminating $[H^+]$, that

$$K_h = \frac{K_w}{K_a}$$

b. *Salt of strong acid and weak base. Cation hydrolysis.* The corresponding equations are

$$B^+(aq) + H_2O(l) \rightleftharpoons H^+(aq) + BOH$$

$$\frac{[H^+][BOH]}{[B^+]} = K_h \qquad [H^+][OH^-] = K_w \qquad \frac{[B^+][OH^-]}{[BOH]} = K_b$$

$$K_h = \frac{K_w}{K_b}$$

c. *Salt of weak acid and weak base.* This involves both anion and cation hydrolysis, and the equations concerned are

$$B^+(aq) + A^-(aq) + H_2O(l) \rightleftharpoons HA + BOH \qquad \frac{[HA][BOH]}{[B^+][A^-]} = K_h$$

$$[H^+][OH^-] = K_w \qquad \frac{[H^+][A^-]}{[HA]} = K_a \qquad \frac{[B^+][OH^-]}{[BOH]} = K_b$$

$$K_h = \frac{K_w}{K_a K_b}$$

d. *Summary.* The relationships between K_h, K_w, K_a and K_b are as follows:

Salt of weak acid and strong base	Salt of strong acid and weak base	Salt of weak acid and weak base
$K_h = \dfrac{K_w}{K_a}$	$K_h = \dfrac{K_w}{K_b}$	$K_h = \dfrac{K_w}{K_a K_b}$

4. Degree of hydrolysis. *The degree of hydrolysis of a salt is the fraction (or percentage) of the salt which is hydrolysed.*

a. *Salt of weak acid and strong base.* If 1 mol of a salt, BA, in V dm³ of solution has a degree of hydrolysis, h, the resulting concentrations at equilibrium will be given, in mol dm⁻³, by

$$A^-(aq) \;+\; H_2O(l) \;\rightleftharpoons\; HA \;+\; OH^-(aq)$$
$$\dfrac{1-h}{V} \qquad\qquad\qquad \dfrac{h}{V} \qquad \dfrac{h}{V}$$

It follows that

$$\frac{[HA][OH^-]}{[A^-]} \;=\; K_h \;=\; \frac{h/V \times h/V}{(1-h)/V} \;=\; \frac{h^2}{(1-h)V}$$

and this gives the relationship between degree of hydrolysis and hydrolysis constant. The degree of hydrolysis depends on the concentration of salt, i.e. on V.

If h is small as compared with 1,

$$K_h \simeq \frac{h^2}{V}$$

and the degree of hydrolysis bears the same relation to the hydrolysis constant as the degree of ionisation does to the ionisation constant (p. 373). As $K_h = K_w/K_a$,

$$\frac{K_w}{K_a} \simeq \frac{h^2}{V} \qquad \text{and} \qquad h \simeq \sqrt{\frac{K_w V}{K_a}}$$

The degree of hydrolysis of sodium ethanoate in 0.01M solution can easily be calculated as follows:

$$K_w = 10^{-14} \qquad\qquad\qquad K_{HEt} = 1.8 \times 10^{-5}$$

$$\therefore h = \sqrt{\frac{10^{-14} \times 100}{1.8 \times 10^{-5}}} = 2.54 \times 10^{-4} \quad \text{or} \quad 0.0254 \text{ per cent.}$$

In 0.1M solution, the degree of hydrolysis would be given by

$$h = \sqrt{\frac{10^{-14} \times 10}{1.8 \times 10^{-5}}} = 0.745 \times 10^{-4} \quad \text{or} \quad 0.00745 \text{ per cent.}$$

By comparison, the degree of hydrolysis of sodium cyanide is 3.77×10^{-2} or 3.77 per cent in 0.01M solution, and 1.18×10^{-2} or 1.18 per cent in 0.1M solution. These values are bigger than the corresponding values for sodium ethanoate, as the dissociation constant for hydrogen cyanide (7.2×10^{-10}) is smaller than that for ethanoic acid (1.8×10^{-5}).

For a salt of a weak acid and a strong base, the OH^- ion concentration is equal to h/V mol dm^{-3}. The H^+ ion concentration is, therefore, $K_w \times V/h$, and, as

$$h \simeq \sqrt{\frac{K_w V}{K_a}}$$

it follows that

$$[H^+] \simeq K_w V \sqrt{\frac{K_a}{K_w V}} = \sqrt{K_w K_a V}$$

The pH of the solution will be given by

$$pH = -\log [H^+] = -\tfrac{1}{2} \log K_w - \tfrac{1}{2} \log K_a - \tfrac{1}{2} \log V$$

The pH of 0.01M sodium ethanoate, for example, will be

$$-\tfrac{1}{2} \log 10^{-14} - \tfrac{1}{2} \log (1.8 \times 10^{-5}) - \tfrac{1}{2} \log 100$$

i.e. $7 + 2.37 - 1$,   or   8.37

b. *Salt of strong acid and weak base*,

$$B^+(aq) + H_2O(l) \longrightarrow H^+(aq) + BOH$$

As before,

$$K_h = \frac{h^2}{(1 - h)V} \simeq \frac{h^2}{V}$$

or, in this case,

$$\frac{K_w}{K_b} \simeq \frac{h^2}{V} \quad \text{and} \quad h \simeq \sqrt{\frac{K_w V}{K_b}}$$

The H^+ ion concentration is h/V, or

$$[H^+] = \frac{h}{V} = \sqrt{\frac{K_w V}{K_b}} \times \frac{1}{V} = \sqrt{\frac{K_w}{K_b V}}$$

so that

$$pH = -\log [H^+] = -\tfrac{1}{2} \log K_w + \tfrac{1}{2} \log K_b + \tfrac{1}{2} \log V$$

The dissociation constant for ammonia is 2.3×10^{-5}, so that, in 0.1M ammonium chloride solution,

$$h = \sqrt{\frac{10^{-14} \times 10}{2.3 \times 10^{-5}}} = 6.6 \times 10^{-5} \quad \text{or} \quad 0.0066 \text{ per cent}$$

and the pH is given by

$$pH = -\tfrac{1}{2} \log 10^{-14} + \tfrac{1}{2} \log (2.3 \times 10^{-5}) + \tfrac{1}{2} \log 10$$
$$= 7 - 2.32 + 0.5 = 5.18$$

c. *Salt of weak acid and weak base.*

$$B^+(aq) \quad + \quad A^-(aq) \quad + H_2O(l) \rightleftharpoons \quad HA \quad + \quad BOH$$

$$\frac{1-h}{V} \qquad\qquad \frac{1-h}{V} \qquad\qquad\qquad\qquad \frac{h}{V} \qquad\qquad \frac{h}{V}$$

The concentrations, at equilibrium, will be as shown, and

$$K_h = \frac{[HA][BOH]}{[B^+][A^-]} = \frac{h/V \times h/V}{(1-h)/V \times (1-h)/V} = \frac{h^2}{(1-h)^2}$$

Here, the relationship between K_h and h does not involve V, and the degree of hydrolysis is independent of the concentration of the solution.

The H^+ ion concentration is given by

$$[H^+] = \sqrt{\frac{K_w K_a}{K_b}}$$

and the pH by

$$pH = -\log [H^+] = -\tfrac{1}{2} \log K_w - \tfrac{1}{2} \log K_a + \tfrac{1}{2} \log K_b$$

5. Summary. The relationships may be summarised as follows:

a. *Salt of weak acid and strong base.*

$$K_h = \frac{K_w}{K_a} = \frac{h^2}{(1-h)V} \simeq \frac{h^2}{V} \qquad [H^+] \simeq \sqrt{K_w K_a V}$$

$$pH \simeq -\tfrac{1}{2} \log K_w - \tfrac{1}{2} \log K_a - \tfrac{1}{2} \log V$$

b. *Salt of strong acid and weak base.*

$$K_h = \frac{K_w}{K_b} = \frac{h^2}{(1-h)V} \simeq \frac{h^2}{V} \qquad [H^+] = \sqrt{\frac{K_w}{K_b V}}$$

$$pH = -\tfrac{1}{2} \log K_w + \tfrac{1}{2} \log K_b + \tfrac{1}{2} \log V$$

c. *Salt of weak acid and weak base.*

$$K_h = \frac{K_w}{K_a K_h} = \frac{h^2}{(1-h)^2} \qquad\qquad [H^+] = \sqrt{\frac{K_w K_a}{K_b}}$$

$$pH = -\tfrac{1}{2} \log K_w - \tfrac{1}{2} \log K_a + \tfrac{1}{2} \log K_b$$

6. Measurement of degree of hydrolysis. Degrees of hydrolysis can be calculated from K_h values, which, in turn, may be obtained from K_w, K_a and K_b values. The degree of hydrolysis of a salt may also be measured directly, and K_h values can be calculated from the measured value.

a. *Measurement of H^+ ion concentration.* When hydrolysis takes place, the H^+ ion concentration in the resulting solution is no longer 10^{-7} mol dm^{-3} as it is in pure water. Measurement of the actual H^+ ion concentration in a salt solution, using any of the methods described on pp. 402–4 can, therefore, be used to find the degree of hydrolysis of a salt.

The measured H^+ ion concentration in 0.1M sodium ethanoate, for example, is 1.34×10^{-9} mol dm^{-3}. The hydrolysis taking place is represented by

$$Et^- + H_2O \longrightarrow HEt + OH^-$$

The OH^- ion concentration must be $10^{-14}/(1.34 \times 10^{-9})$ mol dm^{-3} and this must be the number of mol dm^{-3} of Et^- ion which has been hydrolysed. The original concentration of Et^- ion was 10^{-1} mol dm^{-3}, so that the fraction hydrolysed must be

$$\frac{10^{-14}}{1.34 \times 10^{-9} \times 10^{-1}} = 7.46 \times 10^{-5} \quad \text{or} \quad 0.007467 \text{ per cent}$$

and this is the degree of hydrolysis (p. 418).

b. *Partition method.* Chemical analysis of the equilibrium mixture existing in a solution of a hydrolysed salt would upset the equilibrium, but the concentrations of the substances present can sometimes be found by a partition method.

Phenylamine (aniline) hydrochloride or phenyl ammonium chloride, for example, is a salt of the strong acid, hydrochloric acid, and the weak base, phenylamine. It is hydrolysed in the same sort of way as ammonium chloride (p. 415),

$$C_6H_5 \cdot NH_3{}^+ \rightleftharpoons C_6H_5 \cdot NH_2 + H^+$$

A known mass of phenyl ammonium chloride is shaken with a known volume of water, and a known volume of benzene is added. The free phenylamine formed by hydrolysis in the aqueous layer is soluble in the benzene layer. If the partition coefficient for phenylamine between water and benzene is known, the concentration of the free phenylamine in the aqueous layer can be obtained by measuring the amount of free phenylamine in the benzene layer. This can be done by precipitating the phenylamine in a withdrawn portion of the benzene layer by passing in hydrogen chloride and weighing the phenyl ammonium chloride formed. This will not upset the equilibrium in the aqueous layer.

The original amount of phenyl ammonium chloride is known, and the amount of free phenylamine formed from it (some in the aqueous, and some in the benzene, layer) is also known. The amount of H^+ ions formed, in moles, is equal to the total amount of phenylamine formed, in moles, and all the H^+ ion will be in the aqueous layer. The amount of unhydrolysed phenyl ammonium chloride will be equal to the original amount (in moles) minus the total amount of phenylamine formed (in moles), and all the unhydrolysed salt will be in the aqueous layer.

Thus the concentrations of phenyl ammonium chloride, phenylamine and H^+ ions in the aqueous layer can be calculated, and K_h obtained,

$$K_h = \frac{[C_6H_5 \cdot NH_2][H^+]}{[C_6H_5 \cdot NH_3{}^+]}$$

The degree of hydrolysis of the phenyl ammonium chloride can be calculated from the K_h value,

$$K_h = \frac{h^2}{(1-h)V}$$

QUESTIONS ON CHAPTER 38

1. What is meant by the term hydrolysis? Discuss briefly three examples from inorganic chemistry, and three from organic chemistry.

2. What do you understand by (a) the hydrolysis of a salt, (b) the hydrolysis of an ester, (c) the hydrolysis of an acid chloride? Give two examples of (a) and one each of (b) and (c). What tests would you apply in each case to show that hydrolysis has taken place? (O. & C.)

3. Give illustrative examples to show the difference between (a) anion hydrolysis, (b) cation hydrolysis, (c) hydration, (d) dehydration, (e) dehydrogenation.

4. Hydrolysis is a term used very widely in chemistry. Illustrate its various usages.

5. Explain why solutions of aluminium salts and of sodium hydrogensulphate(VI) are acidic whilst solutions of sodium sulphide, sodium phosphate(V) and sodium hydrogencarbonate are alkaline.

6. Explain why there is a large evolution of heat when anhydrous aluminium chloride is dissolved in water and why the resulting solution is acidic.

7. Explain why (a) a solution of sodium dihydrogenphosphate(V) is slightly acidic, (b) a solution of disodium hydrogenphosphate(V) is distinctly alkaline, and (c) a solution of sodium phosphate(V) is still more strongly alkaline.

8. Write equations to show the various stages in the hydrolysis of hydrated aluminium salts containing $Al(H_2O)_6{}^{3+}$ ions and of hydrated iron(III) salts containing $Fe(H_2O)_6{}^{3+}$ ions.

9. Solid, hydrated iron(III) alum is pale violet in colour. Solutions of iron(III) salts in water are generally yellow-brown, but become violet in colour if strongly acidified. Account for these facts in terms of the extent of hydrolysis into hydroxy-complexes.

10. What is the pH of a 0.01M solution of sodium ethanoate if the dissociation constant of ethanoic acid is 1.8×10^{-5} mol dm^{-3}?

11. Calculate the percentage degree of hydrolysis in an 0.1M solution of ammonium chloride if the ionisation constant of ammonia is 1.8×10^{-5} mol dm^{-3}.

12. What is the pH of the solution obtained when an 0.1M solution of an acid with dissociation constant 1.8×10^{-5} mol dm^{-3} is just neutralised by a strong base? What indicator would you suggest might be best used in a titration of this sort?

13. How many grammes of sodium ethanoate must be added to 500 cm^3 of water to give a solution with a pH of 8.52? The dissociation constant of ethanoic acid is 1.75×10^{-5} mol dm^{-3}.

14. If the dissociation constants of hydrocyanic acid and ammonia are 7.2×10^{-10} and 1.8×10^{-5} mol dm^{-3} respectively, what is the pH of an 0.05M solution of ammonium cyanide?

15. Would it be possible to estimate sodium hydroxide and ammonia in a solution of the two in water by direct titration with hydrochloric acid using two indicators, one of pK 11 and the other of pK 4?

16. Prove that, for a mixture of a weak acid and one of its salts,

$$pH = pK_a + \log [\text{salt}]/[\text{acid}]$$

Use the relationship to plot a graph showing how the pH of a mixture of a weak acid (ionisation constant $= 10^{-5}$) and one of its salts varies with the salt/acid concentration ratio.

17. The hydrolysis constant of ammonium chloride is 10^{-9}. The pH ranges of methyl orange and methyl red are 3.1–4.4 and 4.2–6.3 respectively. Which of these indicators would you prefer in the titration of ammonia with hydrochloric acid ending with ammonium chloride in (a) molar, (b) centimolar concentration?

18. The distribution coefficient, C_b/C_w, for phenylamine between benzene and water is 10 at 25°C. After 0.0273 mol of phenylamine hydrochloride had been shaken with 1 dm³ of water and 500 cm³ of benzene at 25°C until equilibrium had been established, 0.00165 mol of phenylamine were found in the benzene layer. Calculate the hydrolysis constant of phenylamine hydrochloride.

Chapter 39
Cells

1. The Daniell cell. Chemical reactions often release energy, most commonly as heat, but in electrical cells a chemical reaction takes place and releases some of its energy in the form of electricity.

The Daniell cell provides a simple example. Essentially it consists of a rod of zinc immersed in zinc sulphate(VI) solution and a rod of copper immersed in copper(II) sulphate(VI) solution, the two solutions being kept apart, though in electrical contact, by a porous pot (Fig. 136).

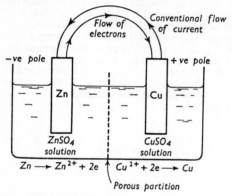

FIG. 136. A Daniell cell

If the zinc and copper rods are connected by a wire, a current will flow. The copper is referred to as the positive pole, and the zinc as the negative pole. The current is said to flow, conventionally, from the copper to the zinc in the external circuit. This corresponds, as will be seen, to a flow of electrons from zinc to copper in the external circuit.

How does the current originate? The zinc rod is in contact with $Zn^{2+}(aq)$ ions, and some zinc atoms from the rod form Zn^{2+} ions by the loss of two electrons,

$$Zn(s) \longrightarrow Zn^{2+}(aq) + 2e$$

The zinc ions pass into solution whilst the liberated electrons are free to flow away from the zinc rod. The copper rod is in contact with Cu^{2+} ions. Here, some Cu^{2+} ions deposit on the copper rod and are discharged (converted into atoms) by taking up electrons,

$$Cu^{2+}(aq) + 2e \longrightarrow Cu(s)$$

The flow of electrons from the zinc rod to the copper rod constitutes the current.

The two electrode processes, each comprising what is known as *a half-cell*, taken together make up the reaction,

$$Zn(s) + Cu^{2+}(aq) \longrightarrow Zn^{2+}(aq) + Cu(s)$$

and this is the chemical reaction taking place in the cell, liberating energy as electricity. As the reaction proceeds, the zinc rod decreases in mass whilst the copper rod increases. The concentration of Zn^{2+} ions increases, whilst that of Cu^{2+} ions decreases.

The direction of the current is controlled by the relative ease of ionisation of zinc and copper atoms. Zinc will form ions more readily than copper (p. 428) and this is why electrons flow from the zinc to the copper. If a rod of magnesium in magnesium sulphate(VI) solution replaced the copper rod in copper(II) sulphate(VI) solution, the flow of current would be reversed, i.e. the zinc rod would be the positive pole of such a cell. .

2. Electric potentials. The conversion of zinc atoms into zinc ions in one half of a Daniell cell means that the zinc rod becomes negatively charged

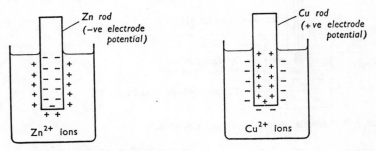

Zn rod
(−ve electrode
potential)

Cu rod
(+ve electrode
potential)

Zn^{2+} ions

Cu^{2+} ions

FIG. 137. Helmholtz double layers.

with respect to the solution surrounding it. If the cell is disconnected in the external circuit, this negative charge will build up making the loss of positively charged zinc ions more and more difficult, as the zinc rod will increasingly attract zinc ions rather than release them. Eventually, an equilibrium position will be established.

The situation is conveniently represented by what is known as a Helmholtz double layer (Fig. 137) and the zinc rod is said to have a negative electric potential. The tendency for a metal to lose ions to a surrounding solution is known as its *electrolytic solution pressure*, whilst the opposite tendency for ions to deposit on a metal from a solution is known as the *deposition pressure*. For a zinc rod in contact with an M solution of zinc ions, the electrolytic solution pressure is greater than the deposition pressure. This gives rise to the negative electric potential.

For a copper rod immersed in copper(II) sulphate(VI) solution of M concentration, the deposition pressure is greater than the electrolytic solution pressure. Cu^{2+} ions therefore deposit on the copper rod, giving it a positive charge in respect to the solution. As this positive charge builds up on the rod the deposition of further ions is hindered by repulsion, until, in the end, an equilibrium position is established. The copper rod is said to have a positive electric potential (Fig. 137).

The electric potential of a metal depends on the concentration of ions with which it is in contact. The electrolytic solution pressure of the metal remains constant but the deposition pressure decreases as the concentration of ions decreases. Electric potential values therefore fall as the concentration of the ions is lowered, and rise as the concentration is increased.

3. Measurement of standard electrode potentials. A single electric potential of a metal cannot be measured absolutely. To measure the potential difference between a metal and a solution necessitates the making of

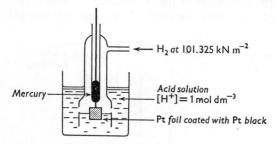

FIG. 138. A simple standard hydrogen electrode.

electrical connection between the metal and the solution. The connection to the solution could only be made by immersing some other metal in the solution, and this would introduce another electric potential.

The difficulty is overcome by using an arbitrary, relative scale of measurement based on a standard hydrogen electrode (Fig. 138).

Hydrogen gas, at $101.325 \text{ kN m}^{-2}$ pressure, is bubbled over a platinum electrode coated with platinum black and immersed in an acid solution with a H^+ ion concentration of 1 mol dm^{-3}. The hydrogen is adsorbed on the platinum black and an equilibrium is set up between the adsorbed layer of hydrogen and the hydrogen ions in the solution,

$$H(g) \rightleftharpoons H^+(aq) + 1e \qquad \text{or} \qquad \tfrac{1}{2}H_2(g) \rightleftharpoons H^+(aq) + 1e$$

The platinum black catalyses the setting up of this equilibrium.

There will, in fact, be an electric potential between the adsorbed hydrogen and the solution but this is arbitrarily taken as zero for the standard hydrogen electrode as described.

This, then, fixes a scale against which other relative electrode potentials, or electrode potentials, can be measured. The values depend on the metal used, on the concentration of metallic ions with which it is in contact, and on the temperature. When the ionic concentration is 1 mol dm^{-3} the electrode potential value is known as the standard electrode potential. Values are commonly quoted at 25°C and signified as $E^\ominus$.

The standard electrode potential of zinc, for example, is the e.m.f. of a cell consisting of a standard hydrogen electrode and a standard zinc

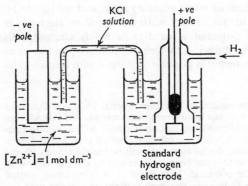

$[Zn^{2+}] = 1 \text{ mol dm}^{-3}$

Standard hydrogen electrode

FIG. 139. The arrangement for measuring the standard electrode potential of zinc against a standard hydrogen electrode. The e.m.f. of the cell is −0.76 volt, giving a standard electrode potential for zinc of −0.76 volt.

electrode. The two electrodes, or half-cells, are connected by a salt bridge of potassium chloride solution so that the overall e.m.f. can be measured as in the apparatus shown in Fig. 139.

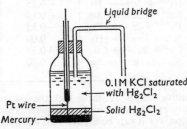

FIG. 140. A simple calomel electrode.

The e.m.f. of the cell is measured by using a potentiometer and the value at 25°C is measured as −0.76 volt. It is the electrode potential of the standard zinc electrode (−0.76) minus that of the standard hydrogen electrode (0).

Other standard electrode potentials can be measured in the same way. Experimentally, however, it is more convenient to use a reference electrode other than a standard hydrogen electrode, and this can be done so long as the electrode potential of the reference electrode used is known in relation to the standard hydrogen electrode.

A *calomel electrode* (Fig. 140) is often used as it is more permanent and more easily handled than a hydrogen electrode. The electrode potential of a decimolar calomel electrode relative to a standard hydrogen electrode is +0.334. It is, therefore, necessary to add 0.334 volt to the value of any electrode potentials measured against a calomel electrode to obtain the value relative to a standard hydrogen electrode. The standard electrode potential of zinc, for instance, referred to a decimolar calomel electrode is -1.094 volt. Other electrodes can be used instead of the calomel electrode.

4. Values of standard electrode potentials. The numerical values of standard electrode potentials are measured on an arbitrary scale, and the sign to be allotted to them is also a matter of choice. In this book an electrode is given a positive electrode potential when it is positively charged with respect to the solution with which it is in contact. Similarly, an electrode with a negative electrode potential is negatively charged with respect to its surrounding solution.

This sign convention is recommended by the International Union of Pure and Applied Chemistry, though it is not always used in America. The IUPAC sign convention has the advantage of providing that an electrode with a positive electrode potential forms the positive pole when combined with a standard hydrogen electrode. Moreover, in any cell, the positive pole will be the electrode with the highest positive electrode potential; the negative pole will be the electrode with the highest negative electrode potential. It is essential in interpreting quoted values of electrode potentials to be sure what sign convention is being used.

Values of some common standard electrode potentials at 25°C are given in the table below.

$Li^+ + 1e \longrightarrow Li$	-3.04		$Fe^{2+} + 2e \longrightarrow Fe$	-0.44
$K^+ + 1e \longrightarrow K$	-2.92		$Co^{2+} + 2e \longrightarrow Co$	-0.28
$Ba^{2+} + 2e \longrightarrow Ba$	-2.90		$Ni^{2+} + 2e \longrightarrow Ni$	-0.25
$Ca^{2+} + 2e \longrightarrow Ca$	-2.87		$Sn^{2+} + 2e \longrightarrow Sn$	-0.14
$Na^+ + 1e \longrightarrow Na$	-2.71		$Pb^{2+} + 2e \longrightarrow Pb$	-0.13
$Mg^{2+} + 2e \longrightarrow Mg$	-2.37		$H^+ + 1e \longrightarrow \frac{1}{2}H_2$	0.00
$Al^{3+} + 3e \longrightarrow Al$	-1.66		$Cu^2 + 2e \longrightarrow Cu$	$+0.34$
$Zn^{2+} + 2e \longrightarrow Zn$	-0.76		$Ag^+ + 1e \longrightarrow Ag$	$+0.80$

It will be seen that the most electropositive metals, i.e. those with the lowest electronegativity, have the highest negative standard electrode potentials. This is a disadvantage of the IUPAC convention, for it can be argued that a highly electropositive metal ought to be given a big positive electrode potential.

The electrode written in the table above as $Li^+ + 1e \longrightarrow Li$ may also be written as Li^+/Li, but it is conventional to write the more oxidised state first, i.e. the electrode process is taken as a reduction.

5. The e.m.f. of cells. An electrode with a negative standard electrode potential will be more negative than a standard hydrogen electrode, i.e. it will form the negative pole when combined with a standard hydrogen electrode.

The standard electrode potential of zinc is -0.76. When combined with a standard hydrogen electrode, as in Fig. 139, the cell is written as follows:

```
   ┌-----◄---- Flow of electrons in external circuit ----◄----┐
   ┆                                                          ┆
H₂/Pt │ H⁺(aq) (1 mol dm⁻³)        ┆      Zn²⁺(aq) (1 mol dm⁻³) │ Zn
   +ve pole                                              −ve pole
(H⁺ + 1e ⟶ ½H₂)                                    (Zn ⟶ Zn²⁺ + 2e)
```

It is conventional to write the hydrogen electrode on the left and the e.m.f. of a cell as written above is given a sign which indicates the polarity of the right-hand electrode. The e.m.f. of the cell is, then, -0.76 volt, this being made up of the standard electrode of the right-hand electrode minus that of the left-hand one. Writing the cell in this way is taken as implying the cell reaction

$$H_2(g) + Zn^{2+}(aq) \longrightarrow Zn(s) + 2H^+(aq)$$

and the negative e.m.f. value (corresponding to a positive ΔG value, (p. 542) indicates that it is the reverse of this reaction, i.e.

$$Zn(s) + 2H^+(aq) \longrightarrow Zn^{2+}(aq) + H_2(g)$$

which is spontaneous (with a positive e.m.f. value and negative ΔG). The liberation of hydrogen explains why metals with negative electrode potentials will, generally liberate hydrogen gas when placed in dilute acid.

A cell made up of a standard copper electrode and a standard hydrogen electrode is written as follows, and will have an e.m.f. of $+0.34 - 0.0$, i.e. $+ 0.34$ volt.

$$H_2/Pt \mid H^+(aq) \,(1 \text{ mol dm}^{-3}) \; \vdots \; Cu^{2+}(aq) \,(1 \text{ mol dm}^{-3}) \mid Cu$$
$$(\tfrac{1}{2}H_2 \longrightarrow H^+ + 1e) \qquad\qquad\qquad (Cu^{2+} + 2e \longrightarrow Cu)$$
$$+ \text{ ve pole}$$

In a Daniell cell, using standard electrodes, the arrangement is

$$Zn \mid Zn^{2+}(aq) \,(1 \text{ mol dm}^{-3}) \; \vdots \; Cu^{2+}(aq) \,(1 \text{ mol dm}^{-3}) \mid Cu$$
$$(Zn \longrightarrow Zn^{2+} + 2e) \qquad\qquad\qquad (Cu^{2+} + 2e \longrightarrow Cu)$$

and the e.m.f. is $+0.34 - (-0.76)$, i.e. $+1.10$ volt.

6. Concentration cells. The standard electrode potential is measured when an electrode is in contact with a solution of ions at a concentration of 1 mol dm⁻³. If the solution has a lower concentration of ions the electrode potential will have a lower numerical value. The standard electrode potential of copper, for example, is 0.344. The electrode potential of copper in contact with 0.01M copper (II) sulphate (VI) solution is 0.285.

This difference means that concentration cells can be set up, the e.m.f. originating solely from a concentration difference. A typical concentration cell is represented as follows:

$$\begin{array}{cc} \text{—ve Pole} & \text{+ve Pole} \\ \text{Cu} \mid \text{Cu}^{2+}(\text{aq})\ (0.01\text{M}) \vdots \text{Cu}^{2+}(\text{aq})\ (1\text{M}) \mid \text{Cu} \\ (\text{Cu} \longrightarrow \text{Cu}^{2+} + 2e) & (\text{Cu}^{2+} + 2e \longrightarrow \text{Cu}) \end{array}$$

Such a cell would give an initial e.m.f. of 0.344–0.285, i.e. 0.059 volt. Cu^{2+} ions will deposit on the right-hand electrode from the more concentrated solution more readily than on the left-hand electrode from the weaker solution. Copper atoms, therefore, form Cu^{2+} ions at the left-hand electrode, whereas Cu^{2+} ions form copper atoms at the right hand. Eventually, the two solutions will have equal Cu^{2+} ion concentrations, and no e.m.f. will be recorded.

A theoretical, thermodynamic treatment shows that the electrode potential, E_m, of a metal, M, in contact with its ions, M^{z+} at a concentration of $[M^{z+}]$ is related to its standard electrode potential, $E_m{}^{\ominus}$, by the relationship,

$$E_m = E_m{}^{\ominus} + \frac{RT}{zF} \log_e [M^{z+}]$$

where R is the gas constant, F the Faraday constant, T the absolute temperature and z the valency of the ion concerned.

Using numerical values of R and F, at a temperature of 25°C,

$$E_m = E_m{}^{\ominus} + \frac{0.059}{z} \log_{10} [M^{z+}]$$

It follows that the electrode potential of any metal is lowered by $0.059/z$ volt for each 10-fold lowering of the ion concentration, z being the valency of the ion.

The e.m.f. of a given concentration cell between two electrodes, 1 and 2, with ionic concentrations of c_1 and c_2 will be given by $E_2 - E_1$ or

$$\left\{ E_m{}^{\ominus} + \frac{0.059}{z} \log_{10} c_2 \right\} - \left\{ E_m{}^{\ominus} + \frac{0.059}{z} \log_{10} c_1 \right\} \quad \text{or} \quad \frac{0.059}{z} \log_{10} \frac{c_2}{c_1}$$

For the copper cell represented above, the e.m.f. will be 0.059 volt.

7. Measurement of H^+ ion concentration.

The H^+ ion concentration in a solution can be measured by measuring the electrode potential of a hydrogen electrode immersed in the solution. If the H^+ ion concentration in the solution is 1 mol dm^{-3}, the electrode potential will be equal to the standard hydrogen electrode potential, i.e. zero. For a H^+ ion concentration greater than 1 mol dm^{-3}, the electrode potential will be positive. For a concentration less than 1 mol dm^{-3}, it will be negative.

Measurement of the e.m.f. of the concentration cell

$$\begin{array}{cc} \text{H}_2/\text{Pt} \mid \text{H}^+(\text{aq})\ (1\ \text{mol dm}^{-3}) \vdots \text{H}^+(\text{aq})\ (\text{unknown}) \mid \text{H}_2/\text{Pt} \\ \text{standard} & \text{unknown} \\ \text{H electrode} & \text{H electrode} \end{array}$$

can, therefore, be used to measure unknown concentrations of H^+ ion. The e.m.f. will be equal to $0.059 \log_{10} c$, where c is the unknown H^+ ion concentration in mol dm^{-3}.

To get the value of the e.m.f. of the above cell it is more convenient to use some electrodes other than a standard hydrogen electrode. A calomel electrode (p. 427) can be used, for instance, the necessary numerical adjustment being made for the potential difference between a calomel electrode and the standard hydrogen electrode (p. 426).

A cell of the type described can be calibrated directly in units of pH and used to measure the pH of a solution directly. Such an instrument is known as a *pH meter*. A variety of such meters are available in a portable form. They often incorporate a *glass electrode*, which depends for its functioning on the fact that the potential drop across a thin glass membrane separating two solutions is related to the differences in pH of the solutions.

8. Potentiometric, electrometric or conductimetric titrations. The H^+ ion concentration, or pH, of a solution changes during an acid–alkali titration, there being a marked change at the end-point (p. 394). The end-point can be conveniently obtained by measuring the pH electrically, and continuously, as the titration is carried out.

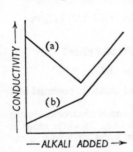

FIG. 141. The change in conductivity as a strong alkali is added to (a) a strong acid, (b) a weak acid.

This can be done using a pH meter immersed in the solution being titrated and recording the pH after each addition from the burette. In this way, pH curves, as shown on p. 394, can be obtained experimentally.

Alternatively, a hydrogen electrode can be immersed in the solution being titrated and connected to a reference electrode such as a calomel electrode. The change in pH during the titration can be followed by measuring the change in e.m.f. between the two electrodes.

Such a procedure is known as a potentiometric or electrometric titration. One of the advantages of the method is that it can be used with coloured solutions where indicators are not suitable.

In a similar way, the conductivity of a solution being titrated can be measured. At the end-point there is a sharp change of conductivity. Such a titration is known as a *conductimetric titration* (Fig. 141).

QUESTIONS ON CHAPTER 39

1. The theoretical treatment of the Daniell cell is simple and straightforward. What snags arise in practice?

2. What e.m.f. is obtainable from a cell made up of a standard zinc and a standard silver electrode? Which is the negative pole of the cell, and at which pole does oxidation take place?

3. What is meant by describing one metal as more electro-positive than

another? Briefly describe three experiments by which you could show that zinc is more electro-positive than copper.

4. The statement that zinc replaces copper from solution means that at equilibrium the ratio of copper(II) ion concentration to zinc ion concentration is very small. Explain, in more detail, what this statement means.

5. What are the advantages and the disadvantages of the sign convention chosen in this chapter for electrode potentials?

6. What will be the e.m.f. of a cell made up of a zinc rod immersed in 0.05M zinc sulphate solution and a decimolar calomel electrode?

7. The e.m.f. of a cell made up of a saturated calomel electrode and a standard hydrogen electrode is 0.248 volt. What will be the standard electrode potential of (a) lead, and (b) copper when referred to a saturated calomel electrode?

8. The standard electrode potential of zinc referred to the standard hydrogen electrode is -0.76. Referred to a decimolar calomel electrode it is -1.094. Will it be higher or lower than -0.76 when referred to a hydrogen electrode containing a concentration of hydrogen ions of 0.1 mol dm^{-3}, and will it be higher or lower than -1.094 when referred to a molar calomel electrode? Explain your reasoning.

9. What reactions take place (a) in each half-cell, and (b) overall, in the following cells?

i $H_2/Pt \mid HCl\,(0.1M) \mid Cl_2/Pt$ ii $Ag,AgCl(s) \mid HCl\,(1M) \mid H_2/Pt$
iii Cd amalgam $\mid CdSO_4(aq) \mid Hg_2SO_4(s) \mid Hg$

10. In what cells would the following overall reactions take place?

i $2AgBr + H_2 = 2Ag + 2HBr$ ii $H_2 + I_2 = 2HI$

11. How would you attempt to measure the standard electrode potential of a chlorine electrode?

12. What is the e.m.f. of a cell made up of two hydrogen electrodes, one in contact with a hydrogen ion concentration of 0.025 mol dm^{-3} and the other in contact with a hydrogen ion concentration of 10^{-8} mol dm^{-3}? Which is the negative pole of the cell?

Chapter 40
Electrolysis

1. The process of electrolysis. In an electrolytic process, a current is passed through the electrolyte to be decomposed, the electrolyte being in solution (usually in water) or in the molten state. The current is led in and out of the electrolyte through electrodes immersed in it, and electrodes which do not react with the electrolyte or any of the electrolysis products are generally chosen.

The electrode connected to the positive pole of the electricity supply is called the *anode*; that connected to the negative pole is called the *cathode*. The current is said to flow, conventionally, from the positive to the nega-

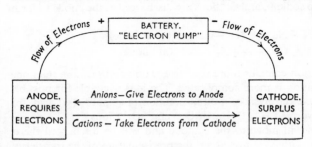

FIG. 142. The process of a simple electrolysis.

tive pole, but the actual flow of electrons, which is the current, is in the opposite direction, i.e. from negative to positive pole. *Electrons therefore pass into the cathode and out of the anode during electrolysis.* The source of the current is best regarded as an electron pump, a current of 1 ampere corresponding to a flow of approximately 6.25×10^{18} electrons per second.

To maintain the flow of electrons through the electrolyte, the ions present in the electrolyte move towards the electrodes, the negatively charged ions (anions) to the anode and the positively charged ions (cations) to the cathode. A more detailed consideration of the movement of ions is given on pp. 450–8 in the discussion of transport numbers.

The anions give up electrons to the anode and are discharged, whilst the cations are discharged at the cathode by taking electrons,

$$A^-(aq) \longrightarrow A + 1e \qquad\qquad C^+(aq) + 1e \longrightarrow C$$
(Anion) (Cation)

A and C will be the products of electrolysis, A being liberated at the anode and C at the cathode.

The giving up of electrons to the anode and the taking of electrons from the cathode maintains the flow of current, the whole process of a simple electrolysis being conveniently illustrated, in diagrammatic form, in Fig. 142.

The electrode process taking place at the anode (a giving of electrons) is an oxidation; that at the cathode (a taking of electrons) is a reduction (p. 467).

The quantitative relationship between the quantity of electricity and the amount of chemical decomposition are given by Faraday's laws, described on p. 89.

2. Electrode processes. The ions available for discharge in the electrolysis of molten sodium chloride are Na^+ and Cl^-. As there is only one cation and only one anion, there are no complications. Na^+ ions are discharged at the cathode to form sodium, and Cl^- ions are discharged at the anode to form chlorine,

$$Na^+ + 1e \longrightarrow Na \quad \text{(at cathode)}$$

$$Cl^- \longrightarrow \tfrac{1}{2}Cl_2 + 1e \quad \text{(at anode)}$$

When a solution contains more than one cation and/or more than one anion, the situation is more complicated. The ions available for possible discharge in the electrolysis of a dilute solution of sulphuric(VI) acid, for example, are hydrated H^+, OH^-, HSO_4^- and SO_4^{2-} ions.

The H^+ ions will move to the cathode and will be discharged there. Little is known, with certainty, about the precise nature of the transfer of electrons from the cathode, and the matter is complicated by the fact that it is really hydrated H^+ ions that are concerned. It is probable that hydrogen atoms are first formed, and that these then combine together into hydrogen molecules,

$$H^+(aq) + 1e \longrightarrow H \qquad 2H \longrightarrow H_2(g)$$

At the anode, three anions will be available for discharge, and there are various possible processes which could result in the transfer of electrons to the anode and the liberation of oxygen. It is, again, impossible to be precise about which of the various alternative processes actually takes place, but a likely mechanism is that OH^- ions are discharged more readily than the SO_4^{2-} or HSO_4^- ions,

$$OH^-(aq) \longrightarrow OH + 1e$$

and that the resulting hydroxide radicals then react together to form water and oxygen,

$$4OH \longrightarrow 2H_2O(l) + O_2(g)$$

If this mechanism is accepted, then the OH^- ion is said to be selectively discharged, and this idea of selective discharge is a helpful one.

3. Selective discharge. The idea of selective discharge means that the electrode process taking place at an anode or cathode is the one which will take place most easily.

Those ions which are most easily formed from their elements will be the least easily discharged. In terms of electrode potentials (p. 428), elements with a high negative electrode potential will be least readily discharged at a cathode, whilst elements with a high positive electrode potential will be least readily discharged at an anode.

Zn^{2+} ions, for example, will be selectively discharged if present, under comparable conditions, together with Na^+ ions, because the change

$$Zn^{2+}(aq) + 2e \longrightarrow Zn$$

will take place more readily than the change

$$Na^+(aq) + 1e \longrightarrow Na$$

Cu^{2+} ions will be discharged still more readily than Zn^{2+} ions.

Selective discharge from a mixture of electrolytes can, in fact, be used for the quantitative analysis of the metallic ions present, each metal being deposited in turn, and weighed.

The same arguments apply to anions. I^- ions will be selectively discharged in the presence of Cl^- ions because the change

$$I^- \longrightarrow I + 1e$$

takes place more easily than the change

$$Cl^- \longrightarrow Cl + 1e$$

On this basis, a list can be drawn up, following the electrode potential values, of the order in which common hydrated ions will, in general, be selectively discharged.

K^+	Ca^{2+}	Na^+	Mg^{2+}	Al^{3+}	Zn^{2+}	Fe^{2+}	Pb^{2+}	H^+	Cu^{2+}	Ag^+
-2.92	-2.87	-2.71	-2.37	-1.66	-0.76	-0.44	-0.44	-0.13 $\;$ 0	$+0.34$	$+0.80$

$\longrightarrow$ Increasing ease of discharge at cathode $\longrightarrow$

OH^-	I^-	Br^-	Cl^-	NO_3^-	SO_4^{2-}	F^-
$+0.40$	$+0.54$	$+1.07$	$+1.36$			$+2.85$

$\longleftarrow$ Increasing ease of discharge at anode $\longleftarrow$

4. Factors affecting selective discharge. The orders of selective discharge given at the end of the preceding section apply only for discharge from aqueous solutions containing ions at comparable concentrations which are approximately M, and for discharges involving zero, or low, overvoltages. High overvoltages, or the presence of an ion at a particularly high or low concentration, will affect the issue.

a. *Effect of overvoltage.* Depending on the nature of the electrode being used, it is found that the discharge of an ion is sometimes much more difficult than would be expected. This is particularly so, using certain electrodes, when gases are involved, e.g. in the discharge of H^+, OH^- and Cl^- ions.

So far as electrode potentials are concerned, H^+ ions should be discharged from a solution containing Zn^{2+} and H^+ ions of comparable concentration. If zinc electrodes be used, however, the Zn^{2+} ions will be selectively discharged before the H^+ ions. Hydrogen is said to have a high overvoltage at a zinc electrode. At a platinum electrode, with a low hydrogen overvoltage, H^+ ions would be discharged before Zn^{2+} ions.

Similarly, OH^- ions might be expected to be discharged at an anode in preference to any other anion, but the high oxygen overvoltage at most electrodes makes the discharge of OH^- ions more difficult.

Overvoltages at particular electrodes must, therefore, be taken into account in predicting the electrode process which will take place. Further details, and some numerical values, are given on page 440.

b. *Concentration effect.* An ion present at a low concentration is more difficult to discharge than the same ion at a higher concentration. By carefully balancing the concentrations of two metallic ions it is possible for them to be discharged together as an alloy. Brass, for instance, can be deposited in this way from suitable mixtures of zinc and copper sulphates(vi).

Moreover, the lowering of the concentration of an ion by adding a substance with which it will form a stable complex ion makes the discharge of the simple ion more difficult, and, perhaps, impossible. It is for this reason that cyanide solutions of metals are frequently used for electroplating purposes, the low concentration of metallic ions in the solutions being beneficial (p. 484).

The concentration of an ion must, therefore, be taken into account in predicting whether it is likely to be discharged or not.

5. Solution of anode. The liberation of electrons at an anode commonly originates from the discharge of an anion. It can, however, also originate from the ionisation of the anode.

In the electrolysis of copper(ii) sulphate(vi) solution, for example, using a copper anode, there are three possible anode processes, each liberating electrons,

$$\text{i} \quad SO_4^{2-}(aq) \longrightarrow SO_4 + 2e$$
$$\text{ii} \quad OH^-(aq) \longrightarrow OH + 1e$$
$$\text{iii} \quad Cu(s) \longrightarrow Cu^{2+}(aq) + 2e$$

Of these, iii takes place most easily, and the copper anode passes into solution as copper(ii) ions. It is generally referred to as a soluble anode (p. 442).

6. Summary of common electrolytic processes. The combination of overvoltage and concentration effects, and the possibility of the anode passing into solution, mean that the result of the electrolysis of a solution of any

Electrolyte	Electrode	At anode	At cathode	Notes
Molten NaCl (p. 445)	C	$Cl^- \longrightarrow Cl + 1e$ $2Cl \longrightarrow \underline{Cl_2}$	$Na^+ + 1e \longrightarrow \underline{Na}$	Only one anion and one cation present
Dil. H_2SO_4 (p. 438)	Pt	$OH^- \longrightarrow OH + 1e$ $4OH \longrightarrow 2H_2O + \underline{O_2}$	$H^+ + 1e \longrightarrow H$ $2H \longrightarrow \underline{H_2}$	(a) OH^- selectively discharged at anode (b) 2 vols. H_2 for 1 vol. O_2 (c) Acid concentration increases
NaOH soln. (p. 441)	Pt	$OH^- \longrightarrow OH + 1e$ $4OH \longrightarrow 2H_2O + \underline{O_2}$	$H^+ + 1e \longrightarrow H$ $2H \longrightarrow \underline{H_2}$	(a) H^+ selectively discharged at cathode (b) 2 vols. H_2 for 1 vol. O_2 (c) NaOH concentration increases
Conc. HCl (p. 442)	C	$Cl^- \longrightarrow Cl + 1e$ $2Cl \longrightarrow \underline{Cl_2}$	$H^+ + 1e \longrightarrow H$ $2H \longrightarrow \underline{H_2}$	(a) Cl^- discharged at anode because of concentration (b) One vol. H_2 for 1 vol. Cl_2 (c) Acid concentration decreases
CuSO$_4$ soln. (p. 442)	Pt	$OH^- \longrightarrow OH + 1e$ $4OH \longrightarrow 2H_2O + \underline{O_2}$	$Cu^{2+} + 2e \longrightarrow \underline{Cu}$	(a) OH^- selectively discharged at anode (b) Cu^{2+} selectively discharged at cathode (c) Solution converted into H_2SO_4
CuSO$_4$ soln. (p. 442)	Cu	$Cu \longrightarrow Cu^{2+} + 2e$	$Cu^{2+} + 2e \longrightarrow \underline{Cu}$	(a) Soluble anode (b) Cu^{2+} selectively discharged at cathode (c) No change in solution (d) Transfer of Cu from anode to cathode
Brine				

See section 11, p. 443.

one electrolyte can be varied by carrying out the electrolysis under different conditions. The concentration of the electrolyte can be changed, and different electrodes can be used. Temperature and current density can also be varied.

Typical results of straightforward electrolytic processes are summarised in the table on page 437, and more details are given later.

7. Mechanism of electrolysis.

The detailed mechanism of an electrolysis is complicated. There are many variable factors which affect the issue, and many unsolved problems of theoretical interpretation remain.

A consideration of the electrolysis of a normal solution of sulphuric(VI) acid using bright platinum electrodes introduces the main general principles associated with any electrolytic process.

The overall result is that hydrogen bubbles off from the cathode, and oxygen from the anode. Two volumes of hydrogen are produced for every one volume of oxygen, and the concentration of the acid increases. Water, in effect, is being decomposed. The changes can be summarised, in a simple way, as follows:

$$H_2SO_4$$
$$\downarrow$$

H$^+$ ions discharged,		OH$^-$ ions selectively discharged,
H$^+$ + 1e $\longrightarrow$ H	To 2H$^+$ SO$_4^{2-}$ Both ions	OH$^-$ $\longrightarrow$ OH + 1e
H atoms combine,	$\longleftarrow$ cathode H$^+$ OH$^-$ to anode $\longrightarrow$	OH radicals then combine,
2H $\longrightarrow$ H$_2$		4OH $\longrightarrow$ 2H$_2$O + O$_2$

$$\uparrow$$
$$H_2O$$

a. *Decomposition voltage.* If the voltage applied between the electrodes is gradually increased from zero, the current passing through the cell will rise only very slightly, and no significant electrolysis will take place, until a certain voltage is reached when electrolysis begins. Further increase in the applied voltage above this point will lead to more rapid electrolysis and a marked rise in current (Fig. 143a).

The voltage at which electrolysis first begins is known as the *experimental decomposition voltage.* Its value can be measured by using a circuit as shown in Fig. 144 to obtain current–voltage graphs (Fig. 143).

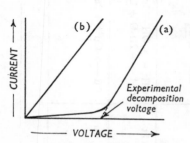

FIG. 143. Current–voltage curves for electrolysis of (*a*) 0.5M sulphuric(VI) acid solution using bright platinum electrodes, and (*b*) copper(II) sulphate(VI) solution using copper electrodes.

The experimental decomposition voltage for 0.5M sulphuric(VI) acid is approximately 1.70.

This minimum voltage is necessary to bring about the electrolysis of 0.5m sulphuric(vi) acid for two reasons. First, the formation of any hydrogen and oxygen at the electrodes changes their nature (they are said to be *polarised*), and the

$$\text{Pt} \mid \text{Dilute H}_2\text{SO}_4 \mid \text{Pt}$$

arrangement becomes a

$$\text{H}_2/\text{Pt} \mid \text{Dilute H}_2\text{SO}_4 \mid \text{O}_2/\text{Pt}$$

arrangement. This functions as a cell, which gives a back e.m.f. equal to the algebraic sum of the two electrode potentials under the conditions used. Such a back e.m.f. must be overcome before electrolysis can proceed.

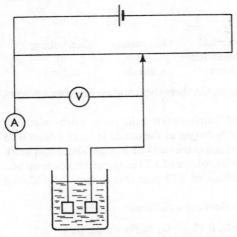

FIG. 144. Circuit for obtaining current–voltage graphs.

This back e.m.f. is equal to the *theoretical decomposition voltage*, for, once the back e.m.f. was overcome, it might be expected that electrolysis would take place. In some cases this is so, the theoretical decomposition voltage being equal to the experimental decomposition voltage.

$$\frac{\text{Theoretical decomposition}}{\text{voltage}} = \frac{\text{Electrode potential}}{\text{at anode}} + \frac{\text{Electrode potential}}{\text{at cathode}}$$

For 0.5M sulphuric(vi) acid, however, the experimental decomposition voltage (1.70) is greater than the theoretical value (1.23), and this is common, particularly when gases are being liberated at the electrodes. The necessity to apply a higher voltage than the theoretical decomposition voltage is due to overvoltage.

b. *Overvoltage.* The potential required to discharge a metallic ion at an electrode is generally equal, or very nearly equal, to the electrode potential of the metal concerned under the same conditions. To deposit zinc ions from a zinc sulphate(VI) solution with an ionic concentration of 1 mol dm^{-3}, for example, requires a cathode potential of approximately 0.76, that being the standard electrode potential for zinc.

But when gases are concerned in a discharge process, e.g. in the discharge of H$^+$, OH$^-$ and Cl$^-$ ions, the potential necessary may be considerably greater than the electrode potential concerned. The difference is known as the overvoltage. Its value depends on the particular ion, the particular electrode, and the particular way in which electrolysis is being carried out.

The origin of overvoltage is obscure, but one likely cause is a slow rate of change from atoms of gas into gas molecules at an electrode which does not catalyse such a change.

It follows that

$$\begin{matrix} \text{Experimental} \\ \text{decomposition} \\ \text{voltage} \end{matrix} = \begin{matrix} \text{Theoretical} \\ \text{decomposition} \\ \text{voltage} \end{matrix} + \begin{matrix} \text{Overvoltage} \\ \text{at} \\ \text{anode} \end{matrix} + \begin{matrix} \text{Overvoltage} \\ \text{at} \\ \text{cathode} \end{matrix}$$

and, when no overvoltage exists, the theoretical and experimental decomposition voltages are equal.

In the electrolysis of 0.5M sulphuric(VI) acid using bright platinum electrodes, the overvoltage of hydrogen at the anode is small (about 0.03 volt) but the overvoltage of oxygen at the cathode is large (about 0.44 volt). The experimental decomposition voltage of 1.70 is, therefore, made up of a theoretical decomposition voltage of 1.23 plus overvoltages of 0.03 and 0.44.

Other typical overvoltage values are as follows:

H$_2$ at Ag cathode, 0.15	O$_2$ at Pb anode, 0.31
H$_2$ at Zn cathode, 0.70	O$_2$ at Au anode, 0.53
H$_2$ at Hg cathode, 0.78	Cl$_2$ at polished Pt anode, 0.7

c. *Discharge potential.* The discharge potential of an ion at an electrode is equal to the electrode potential associated with the ion plus any overvoltage, i.e.

Discharge potential = Electrode potential + Overvoltage at the electrode

As overvoltages for metallic ions are always small, the discharge potential for a metallic ion is approximately equal to the electrode potential of the corresponding metal. The discharge potential for a non-metallic ion varies considerably with the nature of the electrode concerned, as the overvoltage varies so much.

d. *Resistance of electrolyte.* Application of the minimum experimental decomposition voltage will not produce electrolysis at a reasonable rate, as the current passing will be small (Fig. 143). To obtain a reasonable rate of electrolysis requires a reasonable current, and, to obtain this, the resistance of the various connections and the electrolyte have to be overcome. Any applied voltage, over and above the experimental decomposition voltage (which is necessary to overcome back e.m.f. and overvoltages) will be available for increasing the current according to Ohm's law. Thus

$$\text{Applied voltage} - \text{Experimental decomposition voltage} = \text{Current} \times \text{Resistance}$$

The higher the voltage available for overcoming cell resistance, the greater the current will be (Fig. 143), and this will give a greater rate of electrolysis.

8. Electrolysis of sodium hydroxide solution.

a. *Using platinum electrodes.* In this process, hydrogen is produced at the cathode, and oxygen at the anode. Two volumes of hydrogen are formed for each volume of oxygen. The concentration of the sodium hydroxide increases, water, in effect, being decomposed. The process may be summarised, simply, as follows:

$$NaOH$$
$$\downarrow$$

H^+ selectively discharged,
$H^+ + 1e \longrightarrow H.$ Both ions Na^+ OH^- To OH^- discharged,
Then, $\xleftarrow{\text{to cathode}}$ H^+ OH^- $\xrightarrow{\text{anode}}$ $OH^- \longrightarrow OH + 1e.$
$H + H \longrightarrow H_2$ Then,
                                                        $4OH \longrightarrow O_2 + 2H_2O$

$$\uparrow$$
$$H_2O$$

The experimental decomposition voltage for an M solution is approximately 1.7 volt, the same value as for 0.5 M sulphuric(VI) acid. This common value is explained by the fact that both processes are essentially the same, water being decomposed with liberation of hydrogen and oxygen at platinum electrodes.

At the anode, OH^- ions are discharged, as they are the only anions present; there will be an overvoltage of 0.44 volt. At the cathode there are two cations, Na^+ and H^+, and the H^+ ions are selectively discharged, even though they are present in only very low concentration.

The discharge potential for Na^+ ions from an M solution is approximately 2.71 volt. The concentration of H^+ ions in an M solution of sodium hydroxide will be 10^{-14} mol dm^{-3}, and the discharge potential for H^+ ions under such conditions will be about 0.83, neglecting any overvoltage. The overvoltage of hydrogen at a platinum electrode is, in fact, very low, so that the discharge of hydrogen ions will be easier than that of sodium ions.

b. *Using a mercury cathode.* If a mercury cathode is used, Na^+ ions will be discharged, as hydrogen has a high overvoltage at a mercury cathode (p. 440). Sodium amalgam will be formed, but if the concentration of sodium in the amalgam reaches a high enough value, the hydrogen overvoltage will be sufficiently lowered for hydrogen ions to be discharged, as at a platinum electrode.

9. Electrolysis of solutions of hydrogen halides. In the electrolysis of hydrochloric acid, hydrogen is liberated at the cathode, and chlorine and/or oxygen at the anode. Whether or not chlorine or oxygen or a mixture of the two is produced at the anode depends on a variety of factors, the discharge potentials of Cl^- and OH^- ions being fairly close under certain conditions.

Dilute hydrochloric acid, with a low concentration of chloride ions, will tend to produce mainly oxygen; concentrated hydrochloric acid will liberate mainly chlorine. At an electrode with a high overvoltage for oxygen and a lower one for chlorine, evolution of chlorine will be favoured.

Electrolysis of solutions of hydrogen iodide will invariably produce hydrogen and iodine, because discharge of I^- ions will always be easier than discharge of OH^- ions, because of oxygen overvoltage. On the other hand, electrolysis of fluoride solutions will always give oxygen at the anode, as discharge of F^- ions is so difficult.

10. Electrolysis of copper(II) sulphate(VI) solutions.

a. *Using platinum electrodes.* The cathode becomes plated with copper, and oxygen bubbles off from the anode. The electrolyte slowly loses its colour and is converted into sulphuric(VI) acid. The mechanism of the process can be simplified as follows:

$$CuSO_4$$

Cu²⁺ ions selectively discharged,   Both ions to cathode Cu^{2+} $SO_4{}^{2-}$ Both ions to anode OH^- ions selectively discharged,
$Cu^{2+} + 2e \longrightarrow Cu$ H^+ OH^- $OH^- \longrightarrow OH + 1e.$
                                                                                                                                    Then,
                                                                                                                                    $4OH \longrightarrow O_2 + 2H_2O$

$$H_2O$$

The experimental decomposition voltage is about 1.8 volt.

b. *Using copper electrodes.* In this process, copper is deposited on the cathode, and the anode goes into solution as Cu^{2+} ions. The electrolyte does not change, copper simply being transferred from the anode to the cathode.

Of the three possible anode processes,

i $SO_4{}^{2-}(aq) \longrightarrow SO_4 + 2e$
ii $OH^-(aq) \longrightarrow OH + 1e$
iii $Cu(s) \longrightarrow Cu^{2+}(aq) + 2e$

c takes place most easily. The copper anode is sometimes referred to as a soluble anode.

In this electrolysis, the electrodes do not become polarised as they remain Cu/Cu^{2+} electrodes throughout. As there are also no overvoltage effects, there is no decomposition voltage. Even the smallest voltage will bring about electrolysis, and increasing voltage will cause a steady rise in current, approximately following Ohm's law (Fig. 143b).

The theoretical decomposition voltage is also zero, the electrode potentials at the two electrodes being equal. In other words, the theoretical decomposition voltage is zero because there is really no decomposition taking place.

11. Electrolysis of brine. The electrolysis of a *dilute solution* of sodium chloride in water (weak brine) will generally yield oxygen at the anode and hydrogen at the cathode, the solution itself becoming more concentrated. As with solutions of sulphuric acid and sodium hydroxide, this is, again, an effective decomposition of water, summarised as follows:

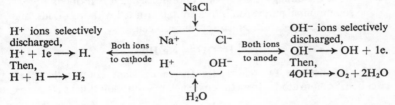

H$^+$ ions selectively discharged, H$^+$ + 1e $\longrightarrow$ H. Then, H + H $\longrightarrow$ H$_2$

Both ions to cathode

Both ions to anode

OH$^-$ ions selectively discharged, OH$^-$ $\longrightarrow$ OH + 1e. Then, 4OH $\longrightarrow$ O$_2$ + 2H$_2$O

The electrolysis of a *concentrated solution* of sodium chloride in water (brine) can, however, under suitable conditions, be used as an industrial

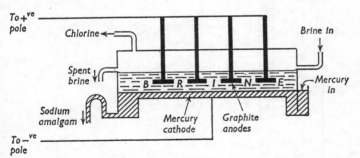

FIG. 145. Electrolysis of brine using a flowing mercury cathode.

source of sodium hydroxide, chlorine and hydrogen. The solution contains hydrated Na$^+$, H$^+$, Cl$^-$ and OH$^-$ ions, and special cells must be used to obtain the required products.

a. *Use of a flowing mercury cathode* (Fig. 145). In this type of cell, the anode is made of graphite, and the cathode consists of a stream of mercury flowing slowly across the bottom of the cell.

The theoretical decomposition voltage is about 3.2 volt, and overvoltages (mainly of chlorine at the graphite anode) of 0.2 volt give an experimental decomposition voltage of about 3.4 volt. In practice, a working voltage of about

4.4 volts is used, 1 volt being necessary to overcome the cell resistance and provide a reasonable current.

Cl^- ions are selectively discharged at the graphite anode, instead of OH^- ions. This is because the Cl^- ions are in much greater concentration than the OH^- ions, and because oxygen has a higher overvoltage than chlorine at a graphite electrode.

At the mercury cathode, Na^+ ions are discharged in preference to H^+ ions, because of the high hydrogen overvoltage at a mercury cathode (p. 440). The sodium content of the amalgam formed is not allowed to get very high as the amalgam is run off continuously as it is formed.

The sodium amalgam is run into a container fitted with iron grids, and water is added. This arrangement produces a cell,

$$Fe \mid H^+OH^- \mid Na/Hg$$

which, because the two electrodes are in constant contact, is short-circuited. Sodium atoms, from the amalgam, pass into solution as Na^+ ions,

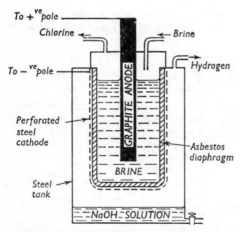

FIG. 146. Electrolysis of brine using a diaphragm cell.

and H^+ ions are discharged at the iron surfaces. Hydrogen gas bubbles off, and sodium hydroxide is obtained by evaporation.

b. *Use of a diaphragm cell.* In this type of cell (Fig. 146) the chlorine and sodium hydroxide are kept apart by a porous diaphragm, made of an asbestos composition, which separates the anode and cathode in adjacent compartments. The anode is made of graphite and the brine is contained in the anode compartment. The cathode is a steel network surrounding the asbestos diaphragm.

The solution in the anode compartment contains Na^+ and Cl^- ions, from the sodium chloride, together with H^+ and OH^- ions, from the water.

The Cl^- and OH^- ions move to the anode, where the Cl^- ions are selectively discharged because they are in much higher concentration than the OH^- ions. Chlorine bubbles off and is collected.

The Na^+ and H^+ ions move through the diaphragm to the cathode, where the H^+ ions are selectively discharged. The hydrogen formed bubbles off and is collected.

As Na^+ ions are not discharged at the cathode they increase in concentration around the cathode. OH^- ions also increase in concentration, for the discharge of H^+ ions causes more water molecules to ionise. This accumulation of both Na^+ and OH^- ions around the cathode means that the solution which slowly trickles through the diaphragm and the steel network cathode is, in effect, a solution of sodium hydroxide. This solution is collected, and most of the sodium chloride impurity is removed during evaporation to produce solid sodium hydroxide.

12. Electrolysis of fused sodium chloride. Electrolysis of fused sodium chloride is used, in the Downs' cell, for the manufacture of sodium. Magnesium and calcium are also made by similar methods. The same principles apply to the electrolysis of a fused salt as to that of an aqueous solution.

In the Downs' cell, a mixture of 40 per cent sodium chloride and 60 per cent calcium chloride is used to give a lower working temperature. Pure sodium chloride would require a temperature of about 800°C, whereas a mixture of sodium and calcium chlorides will melt at about 600°C. A graphite anode and a steel cathode are used, and the simple mechanism of the process is summarised as follows:

$$
\begin{array}{ccccccc}
 & & & \text{NaCl} & & & \text{Cl}^- \text{ discharged,} \\
 & & & \downarrow & & & \text{Cl}^- \longrightarrow \text{Cl} + 1\text{e.} \\
\text{Na}^+ \text{ discharged,} & \xleftarrow{\text{To}} & \text{Na}^+ & & \text{Cl}^- & \xrightarrow{\text{To}} & \text{Then,} \\
\text{Na}^+ + 1\text{e} \longrightarrow \text{Na} & \text{cathode} & & & & \text{anode} & \text{Cl} + \text{Cl} \longrightarrow \text{Cl}_2
\end{array}
$$

Although the mixed electrolyte used contains more Ca^{2+} ions than Na^+ ions, the Na^+ ions are selectively discharged at the temperature used. A similar process cannot, however, be used for making potassium, for the K^+ ions would not be discharged as readily as the Na^+ ions, and an alloy of potassium and calcium would result.

13. Electrolytic oxidation and reduction. The process taking place at the anode during electrolysis involves a giving of electrons and is, therefore, an oxidation. The process at the cathode involves a taking of electrons and is a reduction. The conditions in the vicinity of an anode can, therefore, be used for bringing about oxidation, whilst reduction can be brought about at a cathode.

a. *Anodic oxidation.* Some unusual oxidation reactions can be brought about by electrolytic methods, reactions which cannot, in some cases, be carried out in any other way. F^- ions, for instance, can be oxidised to fluorine only by anodic oxidation,

$$F^- \longrightarrow F + 1\text{e}$$

Other typical examples are provided by the preparation of peroxo-acids and the anodising of aluminium.

Electrolysis of concentrated sulphuric(VI) acid or concentrated solutions of potassium or ammonium sulphates(VI) using a platinum wire as an anode to give high current densities produces peroxodisulphuric(VI) acid, $H_2S_2O_8$. Its precise mode of formation is not known. It may be formed by a polymerisation of two hydrogensulphate radicals,

$$HSO_4^- \longrightarrow HSO_4 + 1e$$
$$2HSO_4 \longrightarrow H_2S_2O_8$$

or it may be that active oxygen at the anode oxidises SO_4^{2-} or HSO_4^- ions. It has also been suggested that hydroxyl radicals at the anode may bring about the oxidation, perhaps by polymerisation to hydrogen peroxide. By similar anodic oxidation, peroxocarbonates, e.g. $K_2C_2O_6$ can be obtained from carbonates, and peroxophosphates, e.g. $K_4P_2O_8$, from phosphates.

Aluminium is normally covered with a very thin film of oxide, and this film can be thickened by making aluminium the anode in an electrolysis of dilute sulphuric(VI) acid or chromic(VI) acid. The oxide film is increased by anodic oxidation of the aluminium, and the process is known as anodising. Anodised aluminium is resistant to corrosion, and the oxide film can also adsorb dyes so that anodised aluminium can be made in a variety of colours, with or without a pattern. It is widely used in making trays, restaurant table tops and ornaments.

b. *Cathodic reduction*. The discharge of any cation, e.g.

$$Cu^{2+}(aq) + 2e \longrightarrow Cu$$

is a cathodic reduction, but reduction processes involving molecules not necessarily directly involved in the electrolysis can also be achieved. Examples are the conversion of nitric(V) acid to hydroxylamine,

$$HNO_3 + 6H \longrightarrow NH_2OH + 2H_2O$$

and of a mixture of nitrobenzene and sulphuric(VI) acid to phenylamine (aniline). Lead cathodes are used.

Because hydrogen has a high overvoltage at a lead cathode the use of such a cathode will almost certainly produce hydrogen which is more active than ordinary hydrogen, but the mechanism of cathodic reduction is not fully understood, and many different products can be obtained under different conditions.

Nitrobenzene, for instance, can be reduced to phenyl azobenzene, azoxybenzene, phenylamine, phenylhydroxylamine and other products by cathodic reduction using different cathode materials, different current densities, different temperatures, and acidic or alkaline solutions.

14. Polarography. Polarography, introduced by Heyrovsky in 1922, provides a rapid and accurate method of analysis requiring only small quantities of material and simple equipment. It makes use of a dropping mercury cathode arranged, in its simplest form, as in Fig. 147. The dropping of the mercury continually renews the cathode surface, and this allows a series of cathode processes to take place at what is, essentially, the same cathode.

The e.m.f. applied between the cathode and anode can be steadily increased. If the mercury is dropping into a solution containing Cu^{2+} ions, there will be no appreciable current until the applied e.m.f. is high enough for the Cu^{2+} ions to be discharged at the cathode. When this discharge begins there will be a marked rise in current.

As Cu^{2+} ions are discharged other Cu^{2+} ions will diffuse towards the

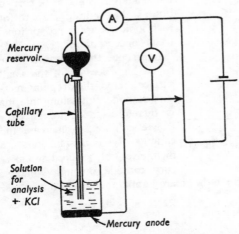

Mercury reservoir

Capillary tube

Solution for analysis + KCl

Mercury anode

FIG. 147. Arrangement of a dropping mercury cathode for obtaining a polarogram.

mercury cathode. Increase in the applied e.m.f. will cause increased rate of discharge, increased current and increased rate of diffusion. There is a limit to the rate of diffusion, however, depending on the bulk concentration of Cu^{2+} ions, and when this limit is reached further increase in e.m.f. will not cause any further increase in current as ions cannot diffuse to the cathode quickly enough. The current, at this stage, is called the *limiting current*, and the way in which the current changes as the e.m.f. is increased is shown in Fig. 148. The e.m.f. value when the current is half the limiting current is known as the *half-wave potential* and it is a characteristic of a particular ion.

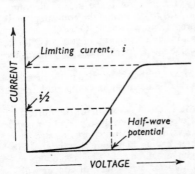

Limiting current, i

$\frac{1}{2}$

Half-wave potential

CURRENT

VOLTAGE

FIG. 148. Simplified change of current with voltage in polarography.

If other ions are present they, too, will cause a change in current, similar to that caused by Cu^{2+} ions, as they are discharged The half-wave potentials are different for different ions so that current–voltage graphs, called *polarograms* (Fig. 149), can be used to detect the presence of various ions in a mixture. The method can also be used for detecting molecules which are reduced at a mercury cathode under the conditions being used. Quantitative analysis can also be carried out as the limiting current for each ionic species is proportional to its concentration.

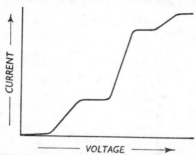

FIG. 149. A simplified polarogram showing the presence of other ions.

In practice, the resistance of the cell is high because of the small surface area of the dropping mercury cathode. Potassium chloride is, therefore, added to the electrolyte solution to increase the conductivity. Potassium ions are themselves only reduced at a mercury cathode at high voltages, and, therefore, do not interfere in the polarographic analysis for other ions.

QUESTIONS ON CHAPTER 40

1. If the electrical cost of producing 1 kg of magnesium by an electrolytic process is £x, what will it cost to produce y kg of aluminium by a similar process? Assume that the processes are 100 per cent efficient.

2. State Faraday's laws of electrolysis. What correspondence, if any, can you find between them and the laws of chemical combination, and how far do they indicate the electrical character of atomic structure? From what sources can evidence be obtained of the existence of ions in solution? (O. & C.)

3. State Faraday's laws of electrolysis. Describe, with the aid of diagrams, how you would use an electrolytic method (a) to determine the equivalent mass of a metal, (b) to purify a metal, (c) to prepare pure oxygen. (O. & C.)

4. Passage of a current of 1 000 ampere through a solution of sodium chloride in a diaphragm cell is found to produce 28.58 kg of chlorine per day. What is the efficiency of the process?

5. State Faraday's laws of electrolysis. How can they be applied to the measurement of (a) current, (b) the equivalent mass of an element or compound, (c) the charge on the electron? (O. & C.)

6. What substances are liberated in the electrolysis of dilute sulphuric acid using (a) platinum electrodes?, (b) lead electrodes? A current is passed through a voltameter containing copper(II) sulphate(VI) solution and copper electrodes in series with a sulphuric(VI) acid voltameter having platinum electrodes. If the current is stopped when 150 cm³ of hydrogen measured at 14°C and 98.65 kN m⁻² are liberated in the sulphuric(VI) acid voltameter, what mass of copper will be deposited on the cathode of the copper voltameter? (The pressure of water vapour at 14°C is 1.6 kN m⁻².) (Army and Navy.)

7. What do you understand by the term electrochemical equivalent of an element? How is the electrochemical equivalent related to the chemical equivalent of an element? Describe, with essential experimental details, how you would determine the electrochemical equivalent of an element such as silver. 0.406 g of a metal X was deposited from a solution by a current of 1 A flowing for 965

second. The metal formed a volatile chloride of vapour density 114. Calculate the relative atomic mass of the metal. (O. & C.)

8. Explain carefully what you mean by the terms 'electrolysis', 'electrolyte' and 'electrode'. If a current passing for 1200 s deposits 0.4818 g of metallic silver, what volume of (a) hydrogen, and (b) oxygen would you expect it to liberate in a voltameter (in the same circuit) where dilute sulphuric(VI) acid is being electrolysed?

9. State Faraday's law of electrolysis. A molar solution of sodium hydroxide is electrolysed at 0°C between platinum electrodes. What current would be required to produce from the cathode (a) 100 cm^3 (at s.t.p.) of dry hydrogen per minute, (b) 100 cm^3 of hydrogen at 0°C and 101.325 kN m^{-2} containing, as it actually would, water vapour? Take the vapour pressure of M sodium hydroxide as 101.4 N m^{-2} at 0°C.

Calculate the charge in coulombs on all the ions present in 1 cm^3 of M sodium hydroxide solution. (Avogadro constant = 6.02 × 10^{23}; Faraday constant = 96500 coulomb.) (B.)

10. (a) In the electrolysis of brine using a flowing mercury cathode, 1000 A liberate 30.16 kg of chlorine per day. If the working voltage of the cell is 4 volt, how many kilowatt hours of energy are required per 1016 kg of chlorine? (b) In a diaphragm cells, 2832 kilowatt hours are required per 1016 kg of chlorine. What advantages has the mercury cell over the diaphragm cell?

11. If a cylindrical rod of length 127 mm and diameter 19.05 mm is completely and evenly plated with nickel for 300 s, using a current of 2.5 A, what will be the thickness of the nickel plate. (The density of nickel is 8.9 g cm^{-3}.)

12. Describe any two industrial electrolytic processes and any two industrial electrothermal processes.

13. Write down the electrode processes which you would expect to take place in the electrolysis of M solutions of the following electrolytes, using platinum electrodes; (a) zinc sulphate(VI), (b) sodium sulphate(VI), (c) gold(III) chloride, (d) potassium nitrate(V), (e) cadmium iodide.

14. Explain the differences you would expect in the electrolysis of an approximately M solution of silver nitrate using (a) platinum and (b) silver electrodes.

15. What products would you predict from the electrolysis of hot brine, using carbon electrodes, and carrying out the electrolysis with constant and efficient stirring?

16. Discuss the electrolysis of potassium chloride (a) in the molten state, (b) in dilute aqueous solution, (c) in concentrated aqueous solution using a flowing mercury cathode. What do you think would happen if the mercury cathode in (c) was stationary?

17. Give an account of the electrolysis of fused carnallite.

18. Explain the meaning of the following terms: (a) back e.m.f., (b) depolariser, (c) discharge potential, (d) concentration polarisation.

Chapter 41
Transport numbers

1. Migration of ions. Hittorf discovered, in 1855, that the amount of an electrolyte in the vicinity of an electrode usually changes during electrolysis, and he attributed such changes to the fact that anions and cations in the solution of the electrolyte migrate to the electrodes with different speeds.

Both ions are responsible for carrying the current through the electrolyte, and Hittorf suggested that the fraction of the current which they each carried depended on their relative velocities,

$$\frac{\text{Current carried by anion}}{\text{Current carried by cation}} = \frac{\text{Velocity of anion } (v_A)}{\text{Velocity of cation } (v_C)}$$

The fraction of the total current carried by the anions was called the *transport number* of the anion (t_A); the fraction carried by the cations was the transport number of the cation (t_C). Clearly t_A must equal $(1 - t_C)$, and t_C must equal $(1 - t_A)$.

$$\frac{\text{Current carried by anion}}{\text{Total current}} \;=\; t_A \;=\; 1 - t_C$$

$$\frac{\text{Current carried by cation}}{\text{Total current}} \;=\; t_C \;=\; 1 - t_A$$

It follows that the following relationships hold,

$$\frac{t_A}{t_C} = \frac{v_A}{v_C} \qquad t_A = \frac{v_A}{v_A + v_C} \qquad t_C = \frac{v_C}{v_A + v_C}$$

2. Changes in electrolyte round electrodes. Transport numbers can be expressed, as in the preceding section, as ratios of current carrying capacity or as ratios of ionic velocities. They can also be expressed in terms of the changes in the amount of electrolyte around the anode and cathode during electrolysis, as will be seen in the following typical examples.

a. *Electrolysis with inert electrodes.* If a current of $96487n$ coulomb is passed through an electrolyte, CA, using inert electrodes, n mol of C^+ ion will be discharged at the cathode, and n mol of A^- ion will be discharged at the anode.

If the transport number of C^+ is t_C and of A^- is t_A, then the C^+ ions will carry $96487 n t_C$ coulomb towards the cathode whilst the A^- ions will carry $96487 n t_A$ coulomb towards the anode. In other words, $n t_C$ mol of C^+ ions will pass towards the cathode, whilst $n t_A$ mol of A^- ions will pass towards the anode.

After the passage of $96487n$ coulomb the situation, if the cell be divided into three compartments, may be summarised as follows:

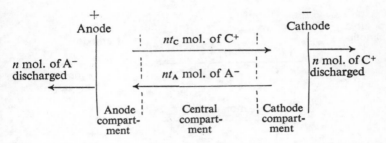

The amount of electrolyte in the central compartment will remain constant, for the ions passing out are replaced by an equal amount passing in. In the anode and cathode compartments, however, the changes in the amount of electrolyte may be summarised as follows, the amounts being given in moles:

Anode compartment	*Cathode compartment*
C^+. Loss of nt_C	C^+. nt_C in; n out
	Loss of $n - nt_C$
A^-. nt_A in; n out	or $n(1 - t_C)$
Loss of $n - nt_A$	or nt_A
or $n(1 - t_A)$	
or nt_C	A^-. Loss of nt_A

The total loss from the electrolyte is n mol of C^+ (discharged at cathode) and n mol of A^- (discharged at anode). nt_C mol of CA are lost from the anode compartment; nt_A are lost from the cathode compartment.

It follows that

$$\frac{\text{Loss round anode}}{\text{Loss round cathode}} = \frac{t_C}{t_A} = \frac{\text{Transport no. of cation}}{\text{Transport no. of anion}}$$

and

$$\frac{\text{Loss round anode}}{\text{Total loss}} = t_C \qquad \frac{\text{Loss round cathode}}{\text{Total loss}} = t_A$$

b. *Electrolysis with soluble anode.* With inert electrodes, ions discharged at the electrodes are removed from the solution of electrolyte. With a soluble anode, however, cations enter the anode compartment from the anode (p. 436).

The situation after $96487n$ coulomb have passed through a solution of an electrolyte CA, using a soluble anode of C, may be summarised as follows:

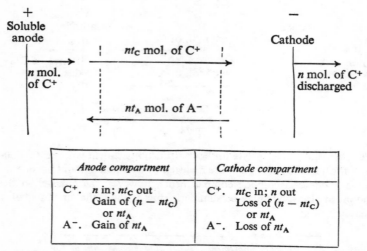

Anode compartment	Cathode compartment

C^+. n in; nt_C out
    Gain of $(n - nt_C)$
    or nt_A
A^-.   Gain of nt_A
	C^+. nt_C in; n out
	    Loss of $(n - nt_C)$
	    or nt_A
	A^-. Loss of nt_A

In this case there is no overall loss of electrolyte, n mol of C^+ being deposited at the cathode and liberated at the anode. It follows that

$$\frac{\text{Gain at anode (or loss at cathode)}}{\text{Deposit at cathode}} = \frac{nt_A}{n} = t_A$$

3. Measurement of transport numbers by Hittorf's method.

Hittorf made use of the relationships between transport numbers and the changes in an electrolyte in the vicinity of the electrodes to measure transport numbers. Care must be taken to use the correct relationship depending on whether inert electrodes are being used or not.

The most widely applicable form of Hittorf's apparatus is shown in Fig. 150. The apparatus is filled with a solution of an electrolyte of known concentration, and a current is passed through for some time. A low current is used to minimise thermal, and resulting diffusion, effects. After electrolysis has taken place, the solution in each

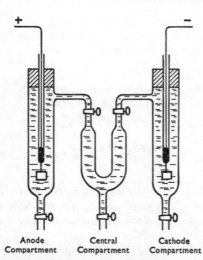

FIG. 150. Hittorf's apparatus for measuring transport numbers.

Anode Compartment   Central Compartment   Cathode Compartment of the three sections of the cell can be withdrawn separately and analysed. The change in the amount of electrolyte in the anode and cathode compartments can be obtained, and the lack of any change in the central compartment can be checked. The mass of the electrodes can also be measured, before and after electrolysis, and, as a further check, the total current

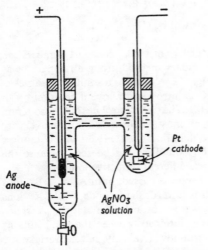

Fig. 151. Simple apparatus for measuring the transport numbers of Ag^+ and NO_3^- ions.

passed can be measured by connecting a copper or silver voltameter in series.

A rather simpler apparatus can often be used. The transport numbers of Ag^+ and NO_3^- ions, for example, can be measured, conveniently, in an apparatus as shown in Fig. 151. A solution of silver nitrate(v), whose concentration in grammes of silver nitrate(v) per kilogramme of water is known, is used. A low current is passed for some time, the number of coulombs being measured by connecting a copper or silver voltameter in series. Half the solution from the left-hand limb (anode compartment) is then withdrawn and analysed by measuring the mass and titrating with standard ammonium thiocyanate solution. It is important that the amount of solution withdrawn for analysis should include all the solution whose composition has changed, i.e. should cover what, in the theoretical treatment, is referred to as the anode compartment. This can be checked by withdrawing a further portion of solution from the left-hand limb, and finding whether it has the same composition as the original silver nitrate(v) solution.

The number of moles of silver nitrate(v) in the anode compartment before and after electrolysis can be calculated, and the gain in the anode compartment will then be known. The number of moles of silver deposited at the cathode will be equal to the number of coulombs passed, divided by 96487 so that (p. 452),

$$\frac{\text{Gain (in moles) of AgNO}_3 \text{ at anode}}{\text{No. of coulombs passed} \div 96487} = t_{\text{NO}_3}^{-}$$

The transport number for Ag^+ ions will be $(1 - t_{\text{NO}_3}^{-})$.

4. Transport numbers by moving boundary method. Transport numbers can also be measured by a moving boundary method, using an apparatus as shown in Fig. 152. A layer of a solution of an electrolyte, AX, is introduced above a solution of an electrolyte, BX. The two electrolytes have a common anion, X^-, but must be chosen so that the velocity of B^+ is less than that of A^+. Even with colourless solutions the boundary can be seen because the solutions have different refractive indices.

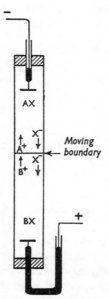

When a current is passed, both A^+ and B^+ ions move upwards towards the cathode, whilst X^- ions move downwards towards the anode. A sharp boundary is maintained between the two solutions as B^+ ions, with lower velocity, never overtake A^+ ions. On the other hand, B^+ ions never lag far behind, for if they do, the solution below the boundary gets diluted and its increasing resistance causes an increased potential drop which increases the velocity of B^+ ions.

The boundary, therefore, moves slowly upwards. If $96487n$ coulomb are passed, $96487nt_A^+$ will be carried by A^+ and nt_A^+ mol of A^+ will move upwards. If the concentration of A^+ ions was originally c mol cm^{-3}, and if the boundary moves x cm up a tube of cross-sectional area a cm^2, the number of moles of A^+ moving upwards will be axc. It follows that

$$nt_A^+ = axc$$

and the value of t_A^+ can be obtained by measuring a, x, c and n.

FIG. 152. The measurement of the transport number of A^+ by moving boundary method.

5. Transport number values. The value of the transport number for any one ion depends on the temperature, the nature of the other ions present, and the concentration of the ions concerned. Typical values, in 0.1M solution at 25°C are:

Solution of	Cation	Anion
AgNO$_3$ . . .	Ag$^+$, 0.47	NO$_3^-$, 0.53
KNO$_3$ . . .	K$^+$, 0.51	NO$_3^-$, 0.49
LiCl . . .	Li$^+$, 0.32	Cl$^-$, 0.68
NaCl . . .	Na$^+$, 0.38	Cl$^-$, 0.62
KCl . . .	K$^+$, 0.49	Cl$^-$, 0.51
HCl . . .	H$^+$, 0.83	Cl$^-$, 0.17

a. *Hydration of ions.* The transport number depends on the speed of migration of an ion during electrolysis, and this will be affected by the hydration or solvation of an ion. Thus, although Li$^+$ ions are smaller than Na$^+$ ions (p. 168) and might be expected to move more rapidly, they do, in fact, move more slowly because they are more hydrated than Na$^+$ ions. Li$^+$ ions, therefore, have a smaller transport number than Na$^+$ ions. Similarly, Na$^+$ ions have a smaller transport number than K$^+$ ions.

The hydration of ions in aqueous solution can be demonstrated by adding sugar or carbamide (urea) to a solution of an electrolyte undergoing electrolysis. The concentration of the sugar or carbamide is found to change in the anode and cathode compartments because more or less water moves into or out of the compartments along with the hydrated ions which are moving in or out.

b. *Migration and diffusion.* The transport number of Ag$^+$ ions, in 0.1M silver nitrate(v) solution, is 0.47, and that of NO$_3^-$ ions, 0.53. This means that, for every 100 Ag$^+$ ions discharged at the cathode in electrolysis of 0.1M silver nitrate(v) solution, only 47 are transported by the current. The other 53 must reach the cathode by diffusion.

As soon as electrolysis begins, Ag$^+$ ions close to the cathode will be discharged, and NO$_3^-$ ions will begin to migrate from the cathode to the anode. The concentration of silver nitrate(v) in the vicinity of the cathode will, therefore, decrease, and Ag$^+$ and NO$_3^-$ ions will diffuse into the region of lower concentration from the region of higher concentration farther away from the cathode.

Thus the movement of ions is due to electrolytic migration and to diffusion.

c. *Formation of complex ions* (p. 480). Transport number measurements can be used to show the formation of complex ions.

The measured transport number of Cd^{2+} ions in cadmium iodide solution, for example, falls from 0.444 in 0.005M solution to 0.003 in 0.025M solution. At still higher concentrations the transport number appears to be negative. The very large change with concentration is attributed to the formation of CdI$_4^{2-}$ ions,

$$CdI_2 \longrightarrow Cd^{2+}(aq) + 2I^-(aq)$$
$$Cd^{2+}(aq) + 4I^-(aq) \longrightarrow CdI_4^{2-}(aq)$$

In dilute solutions there are very few CdI$_4^{2-}$ ions so that Cd^{2+} ions exhibit their normal transport number. As the concentration is increased, however, more and more CdI_4^{2-} ions are formed. There comes a point when just as many CdI_4^{2-} ions pass into the anode compartment as Cd^{2+} ions pass out, and Cd^{2+} ions appear to have a zero transport number.

The existence of complex hexacyanoferrate(II) ions can also be shown. Transport number measurements on a solution of potassium hexacyanoferrate(II) show that all the iron passes to the anode, and that each atom of iron moving into the anode compartment has six CN radicals associated with it.

6. Electric mobility * of ions. The transport numbers of the ions from an electrolyte give only the *relative* speeds of the two ions,

$$\frac{t_A}{t_C} = \frac{\text{Speed of anion}}{\text{Speed of cation}}$$

The *electric mobility of an ion is defined as its velocity in metres per second* ($m\ s^{-1}$) *when moving under a potential gradient of* 1 *volt per metre.*

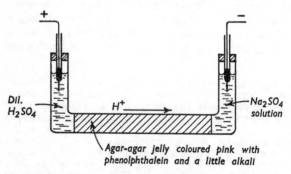

FIG. 153. Lodge's method for measuring the electric mobility of H^+ ions.

The same numerical value is obtained if expressed in cm s^{-1} for a potential gradient of 1 volt per cm.

Values for electric mobilities can, sometimes, be obtained by direct measurements using a method originated by Sir Oliver Lodge. In the apparatus shown in Fig. 153, for example, the rate of passage of H^+ ions towards the cathode can be obtained by measuring the rate at which the pink jelly is decolorised from left to right.

More accurate values for the electric mobilities of ions can be obtained, however, from measurements of transport numbers and Λ_0 values.

Consider a solution of a binary electrolyte, CA, containing x mol cm^{-3}, and having a degree of ionisation of α. Let such a solution be electrolysed in

---

* The older term was absolute velocity.

a tube of cross-sectional area 1 cm^2 between two electrodes 1 cm apart with a voltage drop of 1 volt between them.

There will be αx mol of C^+ and αx mol of A^- between the electrodes. C^+ ions will move with a velocity of v_C cm s^{-1} and A^- ions with a velocity of v_A cm s^{-1}. As 1 mol of an ion carries 96487 coulomb, C^+ ions will carry $96487\alpha x v_C$ coulomb per second, whilst A^- ions will carry $96487\alpha x v_A$ coulomb per second. The current passing through the electrolyte will, therefore, be $96487\alpha x(v_C + v_A)$ amp.

Because a centimetre cube is being considered and the potential difference between the electrodes is 1 volt, the current is equal to the conductivity of the electrolyte (k) in Ω^{-1} cm^{-1} (p. 363),

$$k = 96487\alpha x(v_C + v_A)$$

The molar conductivity of the electrolyte will be given by

$$\Lambda_x = 96487\alpha x(v_C + v_A) \times \frac{1}{x}$$
$$= 96487\alpha(v_C + v_A)$$

and, as $\Lambda_x/\Lambda_0 = \alpha$

$$\Lambda_0 = 96487(v_C + v_A)$$

If the Λ_0 value for an electrolyte is measured, the value of $(v_C + v_A)$ can be calculated. The corresponding value of v_C/v_A can be obtained from measured values of transport numbers, for v_C/v_A is equal to t_C/t_A. Individual values for v_C and v_A can be obtained in this way.

Taking potassium chloride as an example,

$$\Lambda_0 = 130 = 96487(v_{K^+} + v_{Cl^-})$$
$$\frac{t_{K^+}}{t_{Cl^-}} = \frac{0.49}{0.51} = \frac{v_{K^+}}{v_{Cl^-}}$$

from which it follows that

$$v_{K^+} = 6.6 \times 10^{-4} \text{ cm s}^{-1} \quad \text{and} \quad v_{Cl^-} = 6.9 \times 10^{-4} \text{ cm s}^{-1}$$

Other typical values for the electric mobilities of some simple ions, in cm s$^{-1} \times 10^4$, are given below.

Ion			Molar conductivity	Electric mobility (cm s$^{-1} \times 10^4$)
K^+	.	.	.	63.7	6.60
Cl^-	.	.	.	66.3	6.87
Na^+	.	.	.	42.6	4.42
NO^{3-}	.	.	.	62.6	6.49
H^+	.	.	.	330	34.20
OH^-	.	.	.	180	18.70

It will be seen that the electric mobilities of most ions are of the order of 5×10^{-4} cm s^{-1}, but that H^+ and OH^- ions have much higher values. Such high velocities for H^+ and OH^- ions are found only in hydroxylic solvents, such as water or ethanol, and it is suggested that the ions may, in such solvents, be passed on from one solvent molecule to another by a mechanism such as that depicted below:

7. Molar conductivity of ions. Kohlrausch discovered (p. 369) that the molar conductivity at infinite dilution of a solution was equal to the sum of the molar conductivities of the ions concerned, e.g.

$$\Lambda_0(NaCl) = \Lambda(Na^+) + \Lambda(Cl^-)$$

The similarity of this relationship to that between Λ_0 and the electric mobilities of the ions, e.g.

$$\Lambda_0(NaCl) = 96487 v_{Na^+} + 96487 v_{Cl^-}$$

suggests that the molar conductivity of an ion is equal to its electric mobility multiplied by 96487, and this, in fact, is the real significance of the term molar conductivity,

$$\text{Molar conductivity} = \text{Electric mobility} \times 96487$$

Values for molar conductivity, summarised in the table above, are, therefore, readily obtainable from transport number and Λ_0 values.

QUESTIONS ON CHAPTER 41

1. Distinguish between the meaning of the terms transport number of an ion, molar conductivity of an ion and electric mobility of an ion. Show how the terms are related.

2. Summarise the evidence which supports the suggestion that anions move towards the anode during electrolysis.

3. In a transport number measurement on a solution of silver nitrate(v) using platinum electrodes the mass of silver in the anode compartment fell from 10.075 g to 9.420 g, whilst that in the cathode compartment fell from 8.346 g to 7.517 g. Calculate the transport number of the NO_3^- ion.

4. In a measurement of the transport numbers of Ag^+ and NO_3^- ions using a solution of silver nitrate(v), a current which deposited 32.2 mg of silver in a silver voltameter was used. The loss at the cathode, and the gain at the anode, corresponded to 16.8 mg of silver. What is the ratio of the speed of the cation to that of the anion?

5. A solution of sodium chloride containing 0.6 g per 100 g of solution was electrolysed using a silver anode. 0.5 g of silver were deposited in a silver voltameter in series with the transport number cell during the electrolysis, and, after electrolysis, it was found that 47.51 g of solution from around the anode contained 0.1826 g of sodium chloride. Calculate the transport numbers of the Na^+ and Cl^- ions.

6. In an electrolysis of a 4 per cent solution of silver nitrate(v), 0.3208 g of silver were deposited on the cathode, whilst the amount of silver in the cathode compartment fell from 1.4751 to 1.3060 g. Calculate (a) the transport number of the NO_3^- ion, and (b) the loss of silver in the anode compartment.

7. In a measurement of transport number by the moving boundary method the boundary between a 0.02M solution of sodium chloride and a solution of cadmium chloride was found to move 60 mm in 2070 s in a tube of cross section 0.12 cm^2. The steady current flowing was 0.0016 amp. Calculate the transport number of the Na^+ ion.

8. The order in which the alkali metals become hydrated is

$$Li^+ > Na^+ > K^+ > Rb^+ > Cs^+$$

What experimental evidence can you put forward to support this statement, and what is the theoretical background to the statement?

Chapter 42
Corrosion

1. Metallic couples. When two different metals are immersed, apart from each other, in a solution of an electrolyte, a potential difference is set up between them, and a current will flow if they are connected. This connection can be made either externally or internally.

Consider, for example, zinc and copper. If immersed in dilute sulphuric(VI) acid and connected externally they form the cell

$$-\text{ve pole} \qquad +\text{ve pole}$$
$$\text{Zn} \mid \text{dil. H}_2\text{SO}_4 \mid \text{Cu}$$
$$\text{Zn(s)} \longrightarrow \text{Zn}^{2+} + 2\text{e} \qquad \text{H}^+\text{(aq)} + 1\text{e} \longrightarrow \tfrac{1}{2}\text{H}_2\text{(g)}$$

If the zinc and copper are connected internally, by allowing them to touch within the acid, a similar cell, constantly short-circuited, will be set up, and hydrogen gas will bubble off from the copper. Such an arrangement is known as a galvanic or metallic couple.

A piece of pure zinc will react only very, very slowly with dilute sulphuric(VI) acid, because the conversion of H^+(aq) ions into hydrogen gas is difficult at a zinc surface; hydrogen has a high overvoltage on zinc (p. 440). If the pure zinc is touched, within the dilute sulphuric(VI) acid, by a copper or platinum wire, however, evolution of hydrogen will quickly take place from the copper or platinum surface. Similarly, evolution of hydrogen can be expedited by adding a little copper(II) sulphate(VI) or platinum(IV) chloride solution. This causes a deposit of copper or platinum on the zinc surface, and the establishment of a lot of galvanic or metallic couples. Wiping away the deposited copper or platinum from any part of the zinc surface will stop the evolution of hydrogen at that part of the surface quite noticeably.

2. Galvanised iron. Coating iron with zinc to form galvanised iron is a very common method of rust prevention, but a metallic couple of zinc and iron is immediately set up if the zinc coating becomes imperfect and, if a solution of an electrolyte, or even water, is present, corrosion will take place.

The cell set up will be

$$-\text{ve pole} \quad +\text{ve pole}$$
$$\text{Zn} \mid \text{H}_2\text{O} \mid \text{Fe}$$
$$\text{Zn(s)} \longrightarrow \text{Zn}^{2+}\text{(aq)} + 2\text{e} \quad \text{H}^+\text{(aq)} + 1\text{e} \longrightarrow \tfrac{1}{2}\text{H}_2\text{(g)}$$

and zinc ions will be formed at the zinc surface whilst hydrogen will form at the iron surface. The process can be summarised as in Fig. 154.

The zinc will slowly corrode away, being in fact converted into zinc hydroxide, but the iron will not corrode, i.e. it will not rust.

The zinc is referred to as a *sacrificial coating*, or it is said to provide *cathodic protection*. In the corrosion process, electrons are being taken from the iron,

$$2H^+(aq) + 2e \longrightarrow H_2(g)$$

which is, therefore, acting as the cathode, with the zinc as the anode.

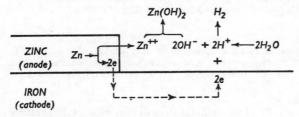

FIG. 154. Cathodic protection of iron by zinc.

Similar cathodic protection of underground iron pipes is provided by driving magnesium rods into the ground and connecting them to the iron pipes by insulated cables. Moisture in the soil provides the electrolyte, but the magnesium corrodes, and not the iron.

3. Tin plate. Coating steel with tin is another very common method of protecting iron, but it is very different in its effect from the use of a zinc coating.

Once a tin coating is pierced, and a solution of an electrolyte, or even water is present, the cell set up will be

$$-\text{ve pole} \quad +\text{ve pole}$$
$$\text{Fe} \mid H_2O \mid \text{Sn}$$
$$\text{Fe(s)} \longrightarrow Fe^{2+}(aq) + 2e \quad H^+(aq) + 1e \longrightarrow \tfrac{1}{2}H_2(g)$$

Fe^{2+} ions will be formed at the iron surface, whilst hydrogen will form at the tin surface (Fig. 155).

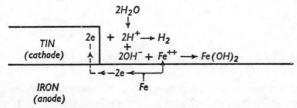

FIG. 155. Corrosion of iron at an imperfection in a tin coating.

The tin is not, in general, a sacrificial coating, and the iron, once corrosion begins, acts as the anode. The Fe^{2+} ions and iron(II) hydroxide formed are ultimately converted into rust (p. 462) so that, although tin plating prevents rusting so long as the plating is perfect, it encourages rusting once any imperfections arise.

Zinc is, therefore, in many ways, a better protective material for iron, but it cannot be used where foodstuffs are concerned because zinc salts are poisonous. That is why tin cans are used so widely for the preservation of food, and why zinc plating is limited to such articles as buckets and dust-bins.

Under certain circumstances, tin does become a sacrificial coating for iron. In the presence of fruit juices, for example, the formation of complex anions of tin with acids such as citric acid makes the tin into the anodic region in relation to iron.

4. Rusting of iron. It is well known that rusting of iron requires the presence of both water and oxygen, but the formation of rust is not fully understood in all its aspects. No two examples of rusting are probably exactly alike, and rust, itself, is of variable composition; it consists of a hydrated form of iron(III) oxide, $Fe_2O_3 . xH_2O$.

Two factors which largely contribute to the origin of rusting are imperfections in the iron surface and the different availability of dissolved oxygen in the solution in contact with the iron.

a. *Imperfections in iron.* Even very small imperfections in an iron surface, caused by chemical impurities or unequal strains during mechanical

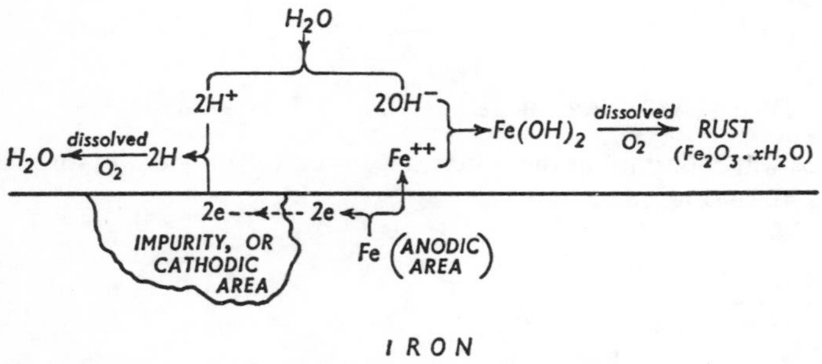

FIG. 156. The rusting of iron initiated at a surface impurity.

treatment, will give rise to a heterogeneous surface in which some points will act as anodes and others as cathodes if the iron is in contact with a solution of an electrolyte.

At the points acting as anodes, iron will form Fe^{2+} ions, and at the points acting as cathodes, hydrogen will be produced. Iron in contact with tin provides a comparison (Fig. 155), only exaggerated by the excessive nature of the impurity (tin).

The hydrogen formed at the cathodic points will hinder the ionisation processes taking place unless it is removed. This is where the oxygen comes in, for dissolved oxygen in the water or solution present, will react with the hydrogen to form water.

The Fe^{2+} ions liberated at the anodic points pass into solution, where, together with the OH^- ions present, they form iron(II) hydroxide, which is oxidised by dissolved oxygen to form rust.

The rust is, in fact, formed by a secondary process. Generally this process takes place within the solution, the Fe^{2+} ions diffusing away from the iron surface. If this is so, the rust does not form a protective layer on the iron. If the solution contains a lot of dissolved oxygen, however, the Fe^{2+} ions will be converted into rust more quickly, and a protective layer may be formed on the iron surface. Excess oxygen can, in this way, retard continued rusting.

The formation of rust may be summarised as in Fig. 156.

b. *Differential aeration*. Even a piece of iron without any surface imperfections will rust if the dissolved oxygen in contact with it is at different concentrations over the surface. This will be so when a piece of iron is half-immersed in water or a solution of an electrolyte. More dissolved oxygen will be available at the surface of the water than in the interior. Conversion of H^+ ions into water will, therefore, be easier at the surface of the water than in the interior, and the iron in the region of the water surface will be cathodic. Consequently, iron below the surface will be anodic, and it is here that rusting takes place.

This accounts for the fact that rust is known to form at points where the oxygen supply is limited. On partially painted iron, for example, rust formation is greatest *underneath* the paint layer. Rust also forms at the centre of a drop of water placed on an iron surface, rather than at the edges of the drop. Rusting, moreover, often causes deep pitting in iron. This is because a pit gets deeper and deeper, once it has started, as oxygen is less available at the bottom of the pit so that the main rusting occurs there.

c. *Summary*. The mechanism of rusting may be summarised as follows:

In regions of lower oxygen concentration

$$Fe(s) \longrightarrow Fe^{2+}(aq) + 2e \qquad \text{(in anodic region)}$$

$$Fe^{2+}(aq) + \begin{matrix} \text{Dissolved} \\ \text{oxygen} \end{matrix} + OH^- \longrightarrow Rust \qquad \begin{matrix} \text{(in solution, in vicinity of} \\ \text{anodic region)} \end{matrix}$$

In regions of higher oxygen concentration

$$H^+(aq) + 1e \longrightarrow H \qquad \text{(in cathodic region)}$$

$$2H + \begin{matrix} \text{Dissolved} \\ \text{oxygen} \end{matrix} \longrightarrow H_2O(l) \qquad \text{(in cathodic region)}$$

5. Experimental demonstrations of rusting. The essential correctness of the mechanism of rust formation outlined in the preceding sections can be demonstrated by the following experiments. Solutions of electrolytes are used instead of water as they speed up the ionisation processes involved.

a. Two similar strips of iron immersed in sodium chloride solution, as in Fig. 157, will exhibit no potential difference, if the oxygen supply is turned

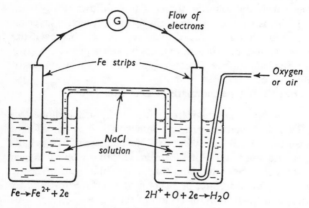

$$Fe \rightarrow Fe^{2+} + 2e \qquad\qquad 2H^+ + O + 2e \rightarrow H_2O$$

FIG. 157. The effect of differential aeration on the rusting of iron.

off. When one of the strips is provided with more oxygen, however, by bubbling oxygen over it, a potential difference can be recorded.

The oxygenated-strip will be the positive pole of the cell, i.e. electrons will be flowing into the oxygenated strip, showing that the discharge of H^+ ions,

$$2H^+(aq) + \frac{\text{Dissolved}}{\text{oxygen}} + 2e \longrightarrow H_2O(l)$$

is taking place more readily at this strip than at the one without the external supply of oxygen.

After a time, removal of H^+ ions from the right-hand beaker will leave the solution with an excess of OH^- ions, i.e. it will be alkaline, and this can be detected by adding an indicator. In the left-hand beaker, the change

$$Fe \longrightarrow Fe^{2+} + 2e$$

will be taking place, i.e. iron will be rusting in the region away from the more readily available oxygen. After a time, the presence of Fe^{2+} ions in the solution in the left-hand beaker can be detected by adding potassium hexacyanoferrate(III) solution.

If the oxygen be supplied to the left-hand strip of iron instead of to the right-hand strip, everything is reversed.

b. A large drop of sodium chloride solution, containing a little potassium hexacyanoferrate(III) and phenolphthalein, placed on a clean sheet of iron will soon be coloured in fairly distinct zones.

The edges of the drop turn pink. The removal of H^+ ions in the regions where oxygen is most readily available leaves these regions with excess OH^- ions which turn the phenolphthalein pink.

The centre of the drop is blue. The Fe^{2+} ions, formed away from the more readily available oxygen, give a blue precipitate with potassium hexacyanoferrate(III).

Between the blue centre and the pink outer ring, a ring of rust forms. Fe^{2+} ions diffusing away from the central area come into contact with excess OH^- ions, and some oxygen, to form rust.

c. Specimens of iron stored in dry air or air-free water will not rust, even after prolonged storage.

d. A strip of clean iron half-immersed in a solution of sodium chloride will be found to form rust below the surface of the solution.

6. Passivity. Some metals, particularly iron, nickel, cobalt and chromium can be rendered passive, i.e. unreactive, by treatment which covers them with a thin coating of oxide.

An iron nail, for example, dipped into copper(II) sulphate(VI) solution is immediately coated with copper. If the same iron nail is first dipped, by a thread, into concentrated nitric(V) acid, and then washed in water, it will not be coated with copper on dipping into copper(II) sulphate(VI) solution. As soon as the nail is touched, however, it loses its passivity due to breaking of the oxide film.

Similarly, a piece of stainless steel will corrode quite noticeably when placed in hot, concentrated hydrochloric acid because the acid breaks down the surface layer. Addition of concentrated nitric(V) acid to the mixture results, at first, in an increased rate of corrosion, but when sufficient nitric(V) acid has been added to give a strongly oxidising mixture, the corrosion suddenly stops as the oxide layer is once again built up on the steel surface.

Other oxidising agents, and electrolytic treatment with the metal as the anode, can also be used for rendering metals passive.

QUESTIONS ON CHAPTER 42

1. What sort of corrosion would you expect at a zinc surface imperfectly coated with a layer of tin?

2. The rusting of iron can be inhibited to some extent by (a) adding sodium sulphate(IV), (b) adding sodium borate(III), (c) adding zinc sulphate(VI), or (d) adding sodium carbonate. Suggest reasons for the inhibiting effect of these substances.

3. A piece of iron immersed in some water is found to rust quite rapidly, but when another sample of the same iron is constantly whirled round in the same sort of water it does not rust. Explain why this is so.

4. Discuss the various methods which have been used to prevent rusting of iron.

5. Write short notes on (*a*) cladding, (*b*) Calorising, (*c*) Sherardising, (*d*) cold galvanising, (*e*) anodising.

6. What factors determine the nature of the reaction (if any) that takes place when a metal is immersed in an aqueous solution of an acid? Suggest explanations of the following observations: (*a*) impure commercial zinc dissolves more rapidly in dilute sulphuric(vi) acid than high-purity zinc, (*b*) iron is not attacked by concentrated nitric(v) acid at room temperature, (*c*) amalgamated aluminium is rapidly attacked by cold water, whereas the unamalgamated metal is almost unattacked. (C. Schol.)

7. White lead, which is important in the paint industry, is manufactured by encouraging the corrosion of lead. Describe how this is done.

8. Collect together some examples of the use of metallic couples in organic chemistry.

9. Explain carefully what will happen if a strip of magnesium ribbon is cleaned with emery paper, dipped into silver nitrate(v) solution, washed in cold water, and then dipped into warm water.

10. Anodic processes are electron-producing whilst cathodic processes consume electrons. Illustrate this statement.

11. How would you determine the composition of a sample of rust?

12. What experiments would you carry out to try to decide whether or not atmospheric carbon dioxide played any part in the rusting of iron?

13. Explain why it is that copper will slowly dissolve in dilute sulphuric(vi) acid, to form copper(ii) sulphate(vi), if air is blown through the mixture.

Chapter 43
Electronic theory of oxidation and reduction

1. Transfer of electrons. The term oxidation was first used to mean the addition of oxygen to an element or compound or the removal of hydrogen from a compound. Reduction meant the addition of hydrogen to an element or compound, or the removal of oxygen from a compound. Such definitions have, however, been extended and, nowadays, many oxidation-reduction, or redox, reactions are best interpreted in terms of transfer of electrons.

The conversion of iron(II) into iron(III) oxide,

$$4FeO(s) + O_2(g) \longrightarrow 2Fe_2O_3(s)$$

is clearly an oxidation of the iron(II) oxide, but the conversion of iron(II) chloride into iron(III) chloride is an analogous change,

$$2FeCl_2(aq) + Cl_2(g) \longrightarrow 2FeCl_3(aq)$$

and is, therefore, regarded as an oxidation of the iron(II) chloride, even though oxygen is not involved. Chlorine is the oxidising agent.

This change can be simplified, and regarded as a conversion of iron(II) into iron(III) ions,

$$Fe^{2+}(aq) \xrightarrow{\text{Oxidation}} Fe^{3+}(aq) + 1e$$

The reverse change will be a reduction,

$$Fe^{3+}(aq) + 1e \xrightarrow{\text{Reduction}} Fe^{2+}(aq)$$

Oxidation can, therefore, *be regarded as the loss of electrons*, and *reduction as the gain of electrons*.

An oxidising agent must be a substance which *will take electrons*, whilst *a reducing agent will give electrons*.

$$\underset{\text{(R.A.)}}{B} \underset{\text{Reduction}}{\overset{\text{Oxidation}}{\rightleftharpoons}} \underset{\text{(O.A.)}}{A} + \text{Electrons}$$

In a redox reaction there is a transfer of electrons, the oxidising agent taking electrons from the reducing agent. Such a transfer can be readily demonstrated using the apparatus shown in Fig. 158. A current flows through the galvanometer because the following changes take place:

a. In right-hand beaker,

$$\underset{\text{(O.A.)}}{Fe^{3+}(aq)} + 1e \xrightarrow{\text{Reduction}} Fe^{2+}(aq)$$

The formation of iron(II) ions in the beaker can be shown by adding potassium hexacyanoferrate(III). The iron(III) ions are gaining electrons and therefore being reduced. In other words, they are taking electrons and acting as oxidising agents.

b. In left-hand beaker,

$$I^-(aq) \xrightarrow{\text{Oxidation}} \tfrac{1}{2}I_2(s) + 1e$$
(R.A.)

The formation of iodine in the beaker can be seen, and demonstrated by adding starch solution. The iodide ions are losing electrons and being oxidised; they are acting as reducing agents.

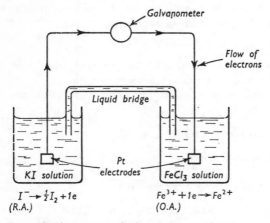

$$I^- \rightarrow \tfrac{1}{2}I_2 + 1e \qquad\qquad Fe^{3+} + 1e \rightarrow Fe^{2+}$$
(R.A.)   (O.A.)

FIG. 158. Electron transfer in a redox reaction.

The overall change taking place is

$$Fe^{3+}(aq) + I^-(aq) \longrightarrow Fe^{2+}(aq) + \tfrac{1}{2}I_2(s)$$
(O.A.)   (R.A.)

the K^+ and Cl^- ions playing no significant part.

The current produced by a Daniell cell (p. 424) also originates from the electron transfer in a redox reaction. The overall change is

$$Cu^{2+}(aq) + Zn(s) \longrightarrow Zn^{2+}(aq) + Cu(s)$$
(O.A.)   (R.A.)

made up of the two stages

$$Zn(s) \xrightarrow{\text{Oxidation}} Zn^{2+}(aq) + 2e \qquad Cu^{2+}(aq) + 2e \xrightarrow{\text{Reduction}} Cu(s)$$
(R.A.)   (O.A.)

2. Potassium manganate(VII) (permanganate) and potassium dichromate(VI). Redox reactions always involve a transfer of electrons, and, in more complicated cases, this may also be accompanied by a rearrangement of atoms. This is so when potassium manganate(VII) and dichromate(VI) act as oxidising agents.

a. *Potassium manganate*(VII). The oxidising action of potassium manganate(VII) in sulphuric acid solution can be represented, in terms of oxygen, as

$$2KMnO_4 \longrightarrow K_2O + 2MnO + [5O]$$

or

$$2KMnO_4 + 3H_2SO_4 \longrightarrow K_2SO_4 + 2MnSO_4 + 3H_2O + [5O]$$

Written ionically, this becomes,

$$2MnO_4^-(aq) + 6H^+(aq) \longrightarrow 2Mn^{2+}(aq) + 3H_2O(l) + [5O]$$

and if the available oxygen be converted into water

$$[5O] + 10H^+(aq) + 10e \longrightarrow 5H_2O(l)$$

it becomes

$$2MnO_4^-(aq) + 16H^+(aq) + 10e \longrightarrow 2Mn^{2+}(aq) + 8H_2O(l)$$

or

$$MnO_4^-(aq) + 8H^+(aq) + 5e \longrightarrow Mn^{2+}(aq) + 4H_2O(l)$$

b. *Potassium dichromate*(VI). The corresponding equations for potassium dichromate(VI) are

$$K_2Cr_2O_7 \longrightarrow K_2O + Cr_2O_3 + [3O]$$
$$K_2Cr_2O_7 + 4H_2SO_4 \longrightarrow K_2SO_4 + Cr_2(SO_4)_3 + 4H_2O + [3O]$$
$$Cr_2O_7^{2-}(aq) + 8H^+(aq) \longrightarrow 2Cr^{3+}(aq) + 4H_2O(l) + [3O]$$
$$[3O] + 6H^+(aq) + 6e \longrightarrow 3H_2O(l)$$
$$Cr_2O_7^{2-}(aq) + 14H^+(aq) + 6e \longrightarrow 2Cr^{3+}(aq) + 7H_2O(l)$$

3. Redox potentials. The strengths of oxidising and reducing agents vary, and the terms are relative ones. Potassium manganate(VII) will generally oxidise hydrochloric acid to chlorine, but potassium dichromate(VI) will not. So far as hydrochloric acid is concerned, potassium dichromate(VI) is not an oxidising agent. Aluminium will reduce chromium(III) oxide to chromium, but hydrogen will not. In relation to chromium(III) oxide, hydrogen is not a reducing agent.

The ionic interpretation of redox reactions enables a numerical comparison of oxidising and reducing powers to be made by using standard electrode potentials.

The standard electrode potential of zinc, for instance, is -0.76 volt, and a zinc electrode will be the negative pole when combined with a standard hydrogen electrode (p. 426). This is because the change

$$Zn(s) \longrightarrow Zn^{2+}(aq) + 2e$$

takes place more readily than the change

$$\tfrac{1}{2}H_2(g) \longrightarrow H^+(aq) + 1e$$

In other words, zinc will give electrons more readily than hydrogen will. Zinc is, therefore, a stronger reducing agent than hydrogen under standard conditions. Alternatively, zinc ions are weaker oxidising agents than H^+ ions.

Copper, with a positive electrode potential, is a weaker reducing agent than hydrogen; copper(II) ions are stronger oxidising agents than H^+ ions.

It will be seen, then, that the standard electrode potentials of the metals (p. 428) give a numerical comparison of the strengths of metals as reducing agents, and of the strengths of their ions as oxidising agents. The standard electrode potentials of metals are also their standard redox potentials.

Similar redox potentials for a change of one ion into another ion, instead of a metal into an ion, can also be obtained experimentally. Fe^{2+} ions, for example, are reducing agents,

$$Fe^{2+}(aq) \longrightarrow Fe^{3+}(aq) + 1e$$

Are they better or poorer reducing agents than hydrogen?,

$$\tfrac{1}{2}H_2(g) \longrightarrow H^+(aq) + 1e$$

This can be decided by setting up a cell represented as

$$-\text{ve pole} \hspace{4cm} +\text{ve pole}$$

$$H_2/Pt \left| H^+(aq)\ (1\ mol\ dm^{-3}) \right| \begin{array}{l} Fe^{3+}(aq)\ (1\ mol\ dm^{-3}) \\ Fe^{2+}(aq)\ (1\ mol\ dm^{-3}) \end{array} \left| Pt \right.$$

with a platinum electrode dipping into a solution containing Fe^{2+} and Fe^{3+} ions, both at a concentration of $1\ mol\ dm^{-3}$, connected to a standard hydrogen electrode. The platinum electrode serves simply as a conductor and a catalyst, and two changes might take place in the right-hand half of the cell,

$$Fe^{3+}(aq) + 1e \longrightarrow Fe^{2+}(aq)$$

or

$$Fe^{2+}(aq) \longrightarrow Fe^{3+}(aq) + 1e$$

The right-hand electrode is found to be the positive pole of the cell, which has an e.m.f. of 0.76 volt, and the change

$$Fe^{3+}(aq) + 1e \longrightarrow Fe^{2+}(aq)$$

must be taking place more readily than the change

$$H^+(aq) + 1e \longrightarrow \tfrac{1}{2}H_2(g)$$

Fe^{3+} ions are, therefore, stronger oxidising agents than H^+ ions, whilst Fe^{2+} ions are weaker reducing agents than hydrogen. $+0.76$ is the redox potential for the change,

$$Fe^{3+} + 1e \longrightarrow Fe^{2+}$$

or for the Fe^{3+}/Fe^{2+} couple.

4. Some uses of redox potentials. Some typical values of redox potentials are given below. Other standard electrode potentials from p. 428 can also be fitted into this list.

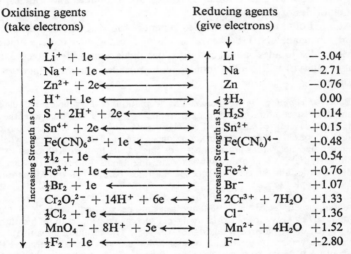

Oxidising agents (take electrons)   Reducing agents (give electrons)

Oxidising agents	Reducing agents	
$Li^+ + 1e$	Li	-3.04
$Na^+ + 1e$	Na	-2.71
$Zn^{2+} + 2e$	Zn	-0.76
$H^+ + 1e$	$\frac{1}{2}H_2$	0.00
$S + 2H^+ + 2e$	H_2S	$+0.14$
$Sn^{4+} + 2e$	Sn^{2+}	$+0.15$
$Fe(CN)_6{}^{3-} + 1e$	$Fe(CN_6)^{4-}$	$+0.48$
$\frac{1}{2}I_2 + 1e$	I^-	$+0.54$
$Fe^{3+} + 1e$	Fe^{2+}	$+0.76$
$\frac{1}{2}Br_2 + 1e$	Br^-	$+1.07$
$Cr_2O_7{}^{2-} + 14H^+ + 6e$	$2Cr^{3+} + 7H_2O$	$+1.33$
$\frac{1}{2}Cl_2 + 1e$	Cl^-	$+1.36$
$MnO_4{}^- + 8H^+ + 5e$	$Mn^{2+} + 4H_2O$	$+1.52$
$\frac{1}{2}F_2 + 1e$	F^-	$+2.80$

Increasing Strength as O.A. (left)   Increasing Strength as R.A. (right)

The figures given refer to standard conditions in aqueous solution at 25°C.

a. *Possibility of reaction.* An oxidising agent will oxidise any reducing agent higher on the list, and a reducing agent will reduce any oxidising agent lower on the list. But these criteria give no indication of the rate at which any particular reaction will take place; that is a matter of kinetics (p. 306).

It will be seen that lithium is the strongest reducing agent given, and it will reduce any of the oxidising agents below it. The well-known metallic replacement reactions are, in this way, to be regarded as redox reactions, e.g.

$$Li + NaCl \longrightarrow LiCl + Na$$
$$\text{or} \quad Li(s) + Na^+(aq) \longrightarrow Li^+(aq) + Na(s)$$
$$\text{(R.A.)} \quad \text{(O.A.)} \quad \text{(Weaker} \quad \text{(Weaker}$$
$$\text{O.A.)} \quad \text{R.A.)}$$

Fluorine is the strongest oxidising agent and it will oxidise any of the reducing agents above it, e.g.

$$\tfrac{1}{2}F_2(g) + Cl^-(aq) \longrightarrow \tfrac{1}{2}Cl_2(g) + F^-(aq)$$

$$\begin{array}{cccc} \text{(O.A.)} & \text{(R.A.)} & \text{(Weaker} & \text{(Weaker} \\ & & \text{O.A.)} & \text{R.A.)} \end{array}$$

b. *Potassium manganate*(VII) *and dichromate*(VI). The oxidising action of these substances involves H^+ ions, and, in measuring the redox potentials given, a H^+ ion concentration of 1 mol dm^{-3} is specified. Under these conditions, manganate(VII) will oxidise chlorides, bromides and iodides, but not fluorides. Dichromate(VI) will oxidise bromides and iodides, but not chlorides or fluorides.

At a pH of 6, however, the redox potential for MnO_4^-/Mn^{2+} is $+0.93$, i.e. the manganate(VII) is a weaker oxidising agent. Under these conditions it will oxidise iodides but not chlorides, bromides or fluorides. At a pH of 3, manganate(VII) will oxidise iodides and bromides, but not chlorides or fluorides.

c. *Effect of complex ion formation.* The formation of a complex ion usually changes the strength of an oxidising or reducing agent. The redox potential for Fe^{3+}/Fe^{2+} is $+0.76$; that for $Fe(CN)_6^{3-}/Fe(CN)_6^{4-}$ is $+0.48$. Hexacyano-ferrate(III) ions are, therefore, weaker oxidising agents than iron(III) ions, whereas hexacyanoferrate(II) ions are stronger reducing agents than iron(II) ions.

Similarly FeF_6^{3-} ions are very weak oxidising agents, and iron(III) ions will not oxidise iodides in the presence of fluorides because of the formation of FeF_6^{3-} complex ions.

Cobalt(II), Co^{2+}, compounds are very weak reducing agents (the redox potential for Co^{3+}/Co^{2+} is $+1.82$) and can only be oxidised to cobalt(III), Co^{3+}, compounds by very strong oxidising agents such as fluorine, sodium bismuthate and ozone, or by electrolytic oxidation (p. 445). The resulting cobalt(III) compounds are very strong oxidising agents which are very unstable because they react even with water to reform cobalt(II) compounds. In the presence of cyanide ions or ammonia, however, complex cobalt(III) ions, e.g. $Co(CN)_6^{3-}$ and $Co(NH_3)_6^{3+}$, are formed. These complexes are so stable, and such weak oxidising agents, that they are formed by the atmospheric oxidation of cobalt(II) compounds in the presence of CN^- ions or ammonia.

5. Oxidation number or oxidation state. *The oxidation number or state of elements in ionic compounds is equal to the charge on their ions in the compound.*

Oxidation numbers can also be allotted to elements in covalent compounds. This is done, on an arbitrary basis, by taking the oxidation numbers to be equal to the charges which the various atoms in a compound would carry if all the bonds in the compound were regarded as ionic instead of covalent. In doing this, a shared pair of electrons between two atoms is assigned to the atom with the greater electronegativity (p. 188). Or, if the two atoms are alike, the shared pair is split between the two, one electron being assigned to each atom. The resulting charges on the various atoms when the bonding electrons are assigned in this way are the oxidation numbers of the atoms.

Some examples of oxidation numbers of atoms in different compounds will illustrate the usage:

$$\underset{-4\ \ +1}{C\ H_4} \qquad \underset{-3\ \ +1}{N\ H_3} \qquad \underset{+1\ -2}{H_2\ O} \qquad \underset{+1\ -1}{H\ F}$$

$$\underset{+1\ -1}{Cl\ F} \qquad \underset{+3\ -1}{I\ Cl_3} \qquad \underset{+5\ -1}{I\ F_5}$$

$$\underset{-2\ +1\ \ -1}{C\ H_3\ Cl} \qquad \underset{0\ \ +1\ \ -1}{C\ H_2\ Cl_2} \qquad \underset{+4\ -1}{C\ Cl_4}$$

The following points should be noted:

i The algebraic sum of the oxidation numbers of all the atoms in an uncharged compound is zero. In an ion, the algebraic sum is equal to the charge on the ion, e.g.

$$\underset{-3\ +1}{N\ H_4^+} \qquad \underset{-2\ +1}{O\ H^-}$$

ii All elements in the elementary state have oxidation numbers of zero, shared pairs between like atoms being split equally between them.

iii As fluorine is the most electronegative element it always has an oxidation number of -1 in any of its compounds.

iv Oxygen, second only to fluorine in electronegativity, has an oxidation number of -2 in almost all its compounds,

$$\underset{+2\ -2}{Mg\ O} \qquad \underset{+3\ -2}{Fe_2\ O_3} \qquad \underset{+4\ -2}{C\ O_2} \qquad \underset{+7\ -2}{Mn_2\ O_7} \qquad \underset{+6\ -2}{Cr\ O_3}$$

Exceptions are provided by fluorine monoxide and the peroxides,

$$\underset{-1\ +2}{F_2\ O} \qquad \left[\underset{-1\quad -1}{:\overset{\bullet\bullet}{O}\overset{\times\times}{\underset{\bullet\,O}{\times}}\overset{\times}{\underset{O\times}{O}}\overset{\times}{\times}} \right]^{2-}$$

the oxidation number of the atoms in the peroxide ion being calculated by splitting the shared pair between the two oxygen atoms linked together.

v In all compounds, except ionic metallic hydrides, the oxidation number of hydrogen is $+1$, e.g.

$$\underset{+1\ -1}{H\ Cl} \qquad \underset{+1\ -2}{H_2\ O} \qquad \underset{-3\ +1}{N\ H_3} \qquad \underset{+1\ -1}{Li\ H} \qquad \underset{+2\ -1}{Ca\ H_2}$$

vi In compounds containing more than two elements, the oxidation number of any one of them may have to be obtained by first assigning reasonable oxidation numbers to the other elements. In sulphuric(vi) acid, H_2SO_4, for example, the most reasonable oxidation numbers for hydrogen and oxygen are $+1$ and -2, which gives sulphur an oxidation number of $+6$. Other examples are,

$$\underset{+1\ +4\ -2}{H_2\ S\ O_3} \qquad \underset{+1\ +7\ -2}{K\ Mn\ O_4} \qquad \underset{+1\ +6\ -2}{K_2\ Cr_2\ O_7} \qquad \underset{+1\ +7\ -2}{K\ Cl\ O_4}$$

vii   Some elements may have widely different oxidation numbers in different compounds as is shown by the following compounds of manganese, chromium, nitrogen and chlorine:

-3	-2	-1	0	1	2	3	4	5	6	7
			Mn		$MnCl_2$	$MnCl_3$ Mn_2O_3	MnO_2		MnO_4^{2-}	MnO_4^- Mn_2O_7
			Cr		$CrCl_2$	$CrCl_3$ Cr_2O_3			CrO_3 CrO_4^{2-} $Cr_2O_7^{2-}$	
NH_3 NH_4^+	N_2H_4	NH_2OH	N_2	N_2O	NO	N_2O_3 HNO_2	NO_2	N_2O_5 HNO_3		
		HCl	Cl_2	$HClO$		$HClO_2$	ClO_2	$HClO_3$		$HClO_4$

viii   *When an element is oxidised its oxidation number is increased. When an element is reduced its oxidation number is decreased.*

     Change in oxidation number can be used to decide whether an oxidation or a reduction has taken place. In the change from chloromethane to dichloromethane, for example,

$$\underset{-2\ +1\ -1}{C\ H_3\ Cl} \longrightarrow \underset{0\ +1\ -1}{C\ H_2\ Cl_2}$$

the oxidation number of carbon is increased from -2 to 0. The carbon is therefore being oxidised.

6. The Stock system of nomenclature. Oxidation numbers provide the basis for the Stock system of naming chemical compounds, which is rapidly becoming widely used and which is recommended by the IUPAC (p. 428). Some examples of Stock names will illustrate the system.

a. *Binary compounds of metals with non-metals:*

$MnCl_2$	Manganese(II) chloride	Fe_2O_3	Iron(III) oxide
$MnCl_3$	Manganese(III) chloride	Fe_3O_4	Iron(II), (III) oxide
MnO_2	Manganese(IV) oxide	PbO_2	Lead(IV) oxide
Mn_2O_7	Manganese(VII) oxide	Pb_3O_4	Dilead(II), lead(IV) oxide

b. *Binary compounds of non-metals with non-metals:*

N_2O	Nitrogen(I) oxide	Cl_2O	Chlorine(I) oxide
NO	Nitrogen(II) oxide	ClO_2	Chlorine(IV) oxide
NO_2	Nitrogen(IV) oxide		

c. *Acids:*

HNO_3	Nitric(v) acid	H_2SO_4	Sulphuric(vi) acid
HNO_2	Nitric(iii) acid	H_2SO_3	Sulphuric(iv) acid

d. *Anions and cations:*

NO_2^-	Nitrate(iii) ion	SO_3^{2-}	Sulphate(iv) ion
NO_3^-	Nitrate(v) ion	SO_4^{2-}	Sulphate(vi) ion
$Cu(H_2O)_4^{2+}$	Tetraquocopper(ii) ion	$Ag(NH_3)_2^+$	Diamminesilver(i) ion

Other examples of Stock names for complex ions and acids are given on pp. 480–1.

The oxidation numbers must be used in naming such substances as iron(iii) chloride, $FeCl_3$ and iron(ii) chloride, $FeCl_2$, where confusion would otherwise arise. But in naming a substance such as Al_2O_3 which ought, systematically, to be called aluminium(iii) oxide the oxidation number can be omitted for it is not really necessary.

7. Balancing redox reaction equations. In balancing a redox reaction equation the reactants and products must, of course, be known, but there are three possible ways (and variations of them) for setting about balancing the equation.

The oxidation of iron(ii) sulphate by potassium manganate(vii) in the presence of dilute sulphuric(vi) acid will be taken as an illustration.

a. *Using half-equations involving oxygen atoms.* Potassium manganate(vii) acts as an oxidising agent in acid solution according to the equation,

$$2KMnO_4 \longrightarrow K_2O + 2MnO + 5O$$

In the presence of dilute sulphuric(vi) acid, the two basic oxides formed react,

$$K_2O + H_2SO_4 \longrightarrow K_2SO_4 + H_2O$$
$$2MnO + 2H_2SO_4 \longrightarrow 2MnSO_4 + 2H_2O$$

so that the equation for the reaction of potassium manganate(vii) as an oxidising agent in the presence of dilute sulphuric(vi) acid becomes,

$$2KMnO_4 + 3H_2SO_4 \longrightarrow K_2SO_4 + 2MnSO_4 + 3H_2O + 5O \qquad \text{i}$$

Iron(ii) sulphate is oxidised, in acid solution, according to the equation,

$$2FeSO_4 + H_2SO_4 + O \longrightarrow Fe_2(SO_4)_3 + H_2O \qquad \text{ii}$$

Combination of the two half-equations, i and ii, gives the final equation, ii being first multiplied by five.

$$2KMnO_4 + 8H_2SO_4 + 10FeSO_4$$
$$\longrightarrow K_2SO_4 + 2MnSO_4 + 8H_2O + 5Fe_2(SO_4)_3$$

b. *Using half-equations involving ions and electrons.* The manganate(VII) ion, in acid solution, acts as an oxidising agent according to the equation,

$$MnO_4^-(aq) \longrightarrow Mn^{2+}(aq)$$

To balance this, electrically and atomically, it must be written as

$$MnO_4^-(aq) + 8H^+(aq) + 5e \longrightarrow Mn^{2+}(aq) + 4H_2O(l) \qquad \text{i}$$

Iron(II) ions are oxidised, in acid solution, according to the equation,

$$Fe^{2+}(aq) \longrightarrow Fe^{3+}(aq) + 1e \qquad \text{ii}$$

To combine the two half-equations, i and ii, it is necessary to multiply ii by five. When this is done, and the equations are combined, the result is

$$MnO_4^-(aq) + 8H^+(aq) + 5Fe^{2+}(aq)$$
$$\longrightarrow Mn^{2+}(aq) + 5Fe^{3+}(aq) + 4H_2O(l)$$

Conversion of the ions into molecules, if required, gives the result,

$$2KMnO_4 + 8H_2SO_4 + 10FeSO_4$$
$$\longrightarrow K_2SO_4 + 2MnSO_4 + 8H_2O + 5Fe_2(SO_4)_3$$

c. *Use of oxidation numbers.* The reaction between potassium manganate(VII) and iron(II) sulphate, in acid solution, is represented, in an unbalanced equation, by

$$MnO_4^-(aq) + H^+(aq) + Fe^{2+} \longrightarrow Fe^{3+} + Mn^{2+}(aq) + H_2O(l)$$
$$+7 \qquad\qquad\qquad +2 \qquad\quad +3 \qquad +2$$

The change in oxidation number of the iron, which is oxidised, is $+1$ whilst that of manganese, which is reduced, is -5. To balance these two changes the equation must be written as

$$MnO_4^-(aq) + H^+(aq) + 5Fe^{2+}(aq)$$
$$\longrightarrow 5Fe^{3+}(aq) + Mn^{2+}(aq) + H_2O(l)$$

Once this has been done, the ratios of oxidising agent and reducing agent must not be altered.

Next, balance the electrical charges by writing,

$$MnO_4^-(aq) + 8H^+(aq) + 5Fe^{2+}(aq)$$
$$\longrightarrow 5Fe^{3+}(aq) + Mn^{2+}(aq) + H_2O(l)$$

Finally, balance the numbers of atoms by writing,

$$MnO_4^-(aq) + 8H^+(aq) + 5Fe^{2+}(aq)$$
$$\longrightarrow 5Fe^{3+}(aq) + Mn^{2+}(aq) + 4H_2O(l)$$

If required, convert the ions into molecules, as follows,

$$2KMnO_4 + 8H_2SO_4 + 10FeSO_4$$
$$\longrightarrow K_2SO_4 + 2MnSO_4 + 8H_2O + 5Fe_2(SO_4)_3$$

8. Common oxidising agents. The functioning of some common oxidising agents is summarised below:

Reagent	Effective change	Decrease in oxidation number
KMnO$_4$ in acid solution . .	MnO$_4^-$ $\longrightarrow$ Mn^{2+}	5
KMnO$_4$ in alkaline solution .	MnO$_4^-$ $\longrightarrow$ MnO$_2$	3
K$_2$Cr$_2$O$_7$ in acid solution . .	Cr$_2$O$_7^{2-}$ $\longrightarrow$ Cr^{3+}	3
Dilute HNO$_3$	NO$_3^-$ $\longrightarrow$ NO	3
Concentrated HNO$_3$ . . .	NO$_3^-$ $\longrightarrow$ NO$_2$	1
Concentrated H$_2$SO$_4$ . . .	SO$_4^{2-}$ $\longrightarrow$ SO$_2$	2
Manganese(IV) oxide . .	MnO$_2$ $\longrightarrow$ Mn^{2+}	2
Chlorine	Cl $\longrightarrow$ Cl$^-$	1
Chloric(I) acid . . .	ClO$^-$ $\longrightarrow$ Cl$^-$	2
KIO$_3$ in dilute acid . .	IO$_3^-$ $\longrightarrow$ I	5
KIO$_3$ in concentrated acid . .	IO$_3^-$ $\longrightarrow$ I$^-$	4

9. Common Reducing Agents. The functioning of some common reducing agents is summarised below:

Reagent	Effective change	Increase in oxidation number
Iron(II) salts (acid) . . .	Fe^{2+} $\longrightarrow$ Fe^{3+}	1
Tin(II) salts (acid) . . .	Sn^{2+} $\longrightarrow$ Sn^{4+}	2
Ethanedioates (acid) . . .	C$_2$O$_4^{2-}$ $\longrightarrow$ CO$_2$	1
Sulphates(IV) (acid) . . .	SO$_3^{2-}$ $\longrightarrow$ SO$_4^{2-}$	2
Hydrogen sulphide . . .	S^{2-} $\longrightarrow$ S	2
Iodides (dilute acid) . . .	I$^-$ $\longrightarrow$ I	1
Iodides (concentrated. acid) .	I$^-$ $\longrightarrow$ I$^+$	2
Metals, e.g. Zn	Zn $\longrightarrow$ Zn^{2+}	2
Hydrogen	H $\longrightarrow$ H$^+$	1

10. Disproportionation. When an element can exist in more than one oxidation state there is always the possibility of it being oxidised or reduced from one state to another. In some substances, too, an atom or an ion appears to undergo simultaneous oxidation and reduction, its oxidation number being both increased and decreased in the same reaction. A change of this sort is known as disproportionation.

a. *Copper(II) compounds.* Copper(II) compounds do not exist in aqueous solution because hydrated cuprous ions undergo disproportionation,

$$2\underset{1}{Cu^+}(aq) \longrightarrow \underset{2}{Cu^{2+}}(aq) + \underset{0}{Cu}(s)$$

to give both oxidation and reduction products,

$$Cu^+(aq) \xrightarrow{\text{Oxidation}} Cu^{2+}(aq) + 1e \qquad Cu^+(aq) + 1e \xrightarrow{\text{Reduction}} Cu(s)$$

The oxidation numbers show the simultaneous oxidation and reduction which is taking place.

The redox potentials,

i $Cu^+(aq) \longleftrightarrow Cu^{2+}(aq) + 1e$ $+0.15$

ii $Cu(s) \longleftrightarrow Cu^+(aq) + 1e$ $+0.52$

show that the Cu^+ ions, acting as reducing agents in a, ought to reduce the Cu^+ ions, acting as oxidising agents in b, and this is what happens. Copper(I) compounds, as a result, are only stable in the presence of water when they are insoluble, e.g. copper(I) chloride, oxide, cyanide, sulphide and thiocyanate, or when they are stabilised by complex formation (p. 485). Thus copper(I) sulphate(VI), made in the absence of water by the action of dimethyl sulphate(VI) on copper(I) oxide, is decomposed by water into copper(II) sulphate(VI) and copper.

b. *Conversion of manganate*(VI) *into manganate*(VII). Manganate(VI) ions are only stable in solutions with a high hydroxyl ion concentration, i.e. high pH. If the pH of the solution is lowered, the manganate(VI) ions disproportionate into manganate(VII) ions and manganese(IV) oxide. The change is represented by the equation,

$$\underset{6}{3MnO_4{}^{2-}(aq)} + 4H^+(aq) \rightleftharpoons \underset{7}{2MnO_4{}^-(aq)} + \underset{4}{MnO_2(s)} + 2H_2O(l)$$

which also shows the change in the oxidation states of the manganese.

This conversion of manganate(VI) to manganate(VII) is encouraged in the preparation of potassium manganate(VII) by passing carbon dioxide into a solution of potassium manganate(VI) obtained by fusing manganese(IV) oxide with potassium hydroxide and an oxidising agent such as potassium nitrate(V).

c. *Disproportionation of free radicals.* Free radicals (p. 343) may undergo disproportionation as, for example, in the formation of ethane and ethene from ethyl radicals formed by the Kolbe electrolysis of a solution of sodium propanoate,

$$2C_2H_5 \longrightarrow C_2H_6 + C_2H_4$$

QUESTIONS ON CHAPTER 43

1. Write down the oxidation numbers of the atoms in the following compounds and ions: (a) H_2SiO_3, (b) $PO_4{}^{3-}$, (c) PH_3, (d) CrF_3, (e) Al_2O_3.

2. What is the oxidation state of the metal in the following compounds: (a) $FeCl_2$, (b) FeO, (c) Fe_2O_3, (d) $HgCl_2$, (e) Hg_2Cl_2, (f) $Hg_3(PO_4)_2$, (g) $CdCl_2$, (h) Co_2S_3, (i) KF?

3. Give examples of sulphur-containing compounds or ions in which the oxidation number of sulphur is (a) -2, (b) -1, (c) 4, (d) 6.

4. Arsenic, antimony and bismuth can exhibit oxidation numbers of -3, 3 and 5. Illustrate this statement.

5. What is the oxidation number of phosphorus in the following compounds: (a) P_4O_{10}, (b) P_4O_6, (c) H_3PO_2, (d) P_2H_4, (e) PH_3, and of nitrogen in the following compounds; (f) N_2O_5, (g) NO_2, (h) N_2O_4, (i) HNO_2, (j) NO, (k) N_2O, (l) NH_2OH, (m) N_2H_4, and (n) NH_3?

6. What are the oxidation states of chlorine in Cl^-, Cl_2, $HClO_4$, HCl, $HClO_3$ and $HClO_2$?

7. Use oxidation numbers to balance the following equations:

$$MnO_2 + HCl \longrightarrow MnCl_2 + H_2O + Cl_2$$
$$Cl_2 + I^- \longrightarrow I_2 + Cl^-$$
$$I_2 + S_2O_3^{2-} \longrightarrow S_4O_6^{2-} + I^-$$
$$HNO_3 + H_2S \longrightarrow NO + S + H_2O$$
$$K_2Cr_2O_7 + HCl \longrightarrow KCl + CrCl_3 + H_2O + Cl_2$$
$$NH_3 + O_2 \longrightarrow NO + H_2O$$
$$KI + H_2SO_4 \longrightarrow K_2SO_4 + I_2 + H_2S + H_2O$$

8. Complete and balance the following equations:

$$I^- + NO_2^- \longrightarrow I_2 + NO \text{ (in acid solution)}$$
$$MnO_4^- \longrightarrow MnO_4^{2-} + O_2 \text{ (in alkaline solution)}$$
$$Fe^{2+} + Cr_2O_7^{2-} \longrightarrow Fe^{3+} + Cr^{3+} \text{ (in acid solution)}$$
$$ClO_2 + OH^- \longrightarrow ClO_2^- + ClO_3^-$$
$$H_2O_2 + MnO_4^- \longrightarrow Mn^{2+} + O_2 \text{ (in acid solution)}$$
$$P \longrightarrow PH_3 + H_2PO_2^- \text{ (in alkaline solution)}$$

9. Write balanced equations for the following reactions: (a) the reduction of iron(III) chloride to iron(II) chloride by aluminium, (b) the addition of tin(II) ions to acidified sodium dichromate(VI) solution, (c) the addition of sodium chlorate(I) to acidified iron(II) sulphate(VI) solution, (d) the addition of sulphate(IV) ions to acidified potassium manganate(VII), (e) the addition of hydrogen sulphide gas to concentrated nitric(V) acid.

10. What will happen (a) when an M solution containing iron(II) and iron(III) ions is added to an M solution containing tin(II) and tin(IV) ions, (b) when an iron rod is placed in a solution M with respect to H^+, Cu^{2+} and Zn^{2+} ions, and (c) when an M solution of iron(II) sulphate is added to an M solution of mercury(II) sulphate(VI)?

11. Explain the following facts: (a) copper(II) ions will oxidise iodide ions, but they will not do so in the presence of ethane–1,2–diamine, (b) chlorine will oxidise a solution of potassium manganate(VI) to potassium manganate(VII), but iodine will not do this, (c) the equilibrium pressure of hydrogen in the reaction of zinc with acid is higher than that of tin with acid, (d) the reducing power of hexacyanoferrate(II) ions is greater than that of iron(II) ions.

12. Illustrate the statement that compounds containing an element in two different oxidation states, e.g. Prussian blue, are highly coloured.

13. Compare and contrast the conceptions of the valency and the oxidation number of an element by choosing selected elements to illustrate your points.

14. What names would be given to the following substances on the Stock system of nomenclature: Cl_2O, HgO, Hg_2O, Fe_2S_3, FeS, MnO_2, PCl_3, PCl_5, $SnCl_2$ and $SnCl_4$?

15. Explain carefully why it is that copper(I) ions disproportionate in aqueous solution whereas iron(II) ions do not.

16. Collect together some examples of disproportionation other than those given in this chapter.

17. Illustrate the statement that the strength of the oxy-acids of an element, X, increases as the oxidation number of X increases.

18. What part does the liquid bridge play in the apparatus shown in Fig. 158? Would any current flow through the galvanometer if the liquid bridge was removed? What would happen if the liquid bridge was replaced by a platinum wire?

19. Illustrate the statement that oxidation may be regarded as a transfer of oxygen, a transfer of electrons, or a change in oxidation state.

Chapter 44
Complex ions

1. Typical examples. An ion formed from a single atom, e.g. Cl^- or Na^+, is known as a simple ion, whereas ions containing more than one atom are known as complex ions. Such complex ions invariably consist of a simple ion linked to other ions, atoms or molecules. Even the very common ions such as the sulphate(VI), nitrate(V) and hydroxide ions, are, properly, complex ions. They can be regarded as $S^{6+}(O^{2-})_4$, $N^{5+}(O^{2-})_3$ and $O^{2-}H^+$ but they are so stable, i.e. they do not dissociate into their component ions, that they function in much the same way as simple ions do, and they are not typical complex ions.

More typical complex ions are provided by the following examples:

$Fe(CN)_6^{4-}$	$SnCl_6^{2-}$	$Ag(NH_3)_2^+$
$Fe(CN)_6^{3-}$	$Fe(CN)_5NO^{2-}$	$Au(CN)_2^-$
$Cu(NH_3)_4^{2+}$	$Co(NO_2)_6^{3-}$	$Zn(OH)_4^{2-}$
SiF_6^{2-}	AlF_6^{3-}	$PbCl_4^{2-}$

Salts containing these, and other similar, ions are known as complex salts, and the groups attached to the central simple ion are known as *ligands*. Complex salts were first studied by Werner (1866–1919) who called them coordination compounds. He used the term *coordination number* to indicate the number of ligands round the central simple ion.

2. Nomenclature of complex ions. The systematic nomenclature adopted for complex ions is based on the following considerations:

a. *Complex cations.* The name to be used begins by giving the number and names of the groups attached to the central atom or ion, i.e. of the ligands, and this is followed by the name of the central atom with its oxidation number (p. 472) indicated by Roman numerals in parentheses.

The $Cu(NH_3)_4^{2+}$ ion, for example, is called the tetraamminecopper(II) ion, and $Co(NH_3)_4Cl_2^+$ is the dichlorotetraamminecobalt(III) ion. Other examples are:

$Ag(NH_3)_2^+$ Diamminesilver(I)	$CrCl_2(H_2O)_4^+$ Dichlorotetraaquo-
	chromium(III)
$Cu(H_2O)_4^{2+}$ Tetraaquocopper(II)	$Ni(NH_3)_6^{2+}$ Hexaamminenickel(II)

b. *Complex anions.* The name used gives the number and names of the ligands, followed by the name of the central atom with an -ate ending and its oxidation number in parentheses. Typical examples are:

$Au(CN)_2^-$	Dicyanoaurate(I)	$CrCl_4(H_2O)_2^-$ Tetrachlorodiaquo-
$Fe(CN)_6^{4-}$	Hexacyanoferrate(II)	chromate(III)
$Fe(CN)_6^{3-}$	Hexacyanoferrate(III)	$PtCl_6^{2-}$ Hexachlorplatinate(IV)

On this basis the full systematic names for the sulphate(VI) and nitrate(V) ions are tetraoxosulphate(VI) and trioxonitrate(V) respectively.

c. *Complex acids.* Acids are named like the anions they form, the -ate ending of the anion being replaced by an -ic ending for the acid, e.g.

$HAuCl_4$ Tetrachloroauric(III) acid H_2PtCl_6 Hexachloroplatinic(IV) acid

In all cases the oxidation number of the central atom is omitted when it only exhibits one oxidation state, e.g.

AlF_6^{3-} Hexafluoroaluminate BH_4^- Tetrahydroborate
H_2SiF_6 Hexafluorosilicic acid $Zn(OH)_4^{2-}$ Tetrahydroxozincate

3. Chelation. The ligands attached to the central atom in the complexes mentioned so far are all molecules or ions which have lone pairs (p. 162), and it is the lone pairs which link the ligands to the central atom.

Simple ligands are attached to the central atom by one bond, but other groups exist which may be attached by more than one bond. Such groups have more than one atom with lone pairs, and the complex compounds they form are known as chelate compounds, from the Greek, χηλή, meaning a crab's claw.

A group capable of forming two links is called a bidentate group, and a common example is provided by ethane–1,2–diamine (ethylene diamine),

usually abbreviated to en⊂, The chelate compounds

are typical. Other bidentate groups are ethane–1,2–diol (glycol), and the ions of ethanedioic (oxalic) acid and aminoethanoic acid (glycine).

Tripyridyl is a tridentate group,

and the hexadentate ion formed from EDTA (p. 486) is an important polydentate group whose use has been developed by Schwarzenbach.

4. Stability of complex ions. There is a marked difference in the stability of various complex ions, and this stability can be measured in terms of a dissociation constant. The tetramminecopper(II) ion, for example, dissociates to some extent,

$$Cu(NH_3)_4^{2+} \rightleftharpoons Cu^{2+} + 4NH_3$$

and the dissociation constant is given by

$$K = \frac{[Cu^{2+}][NH_3]^4}{[Cu(NH_3)_4^{2+}]} = 5 \times 10^{-15}.$$

The higher the value of K for a complex ion, the greater the extent of dissociation, i.e. the less stable the complex ion. K is sometimes referred to as the *instability constant*; a high value means an unstable complex. More commonly, however, numerical values for $1/K$ are known as *stability constants*; if this is done, a high value indicates a stable complex.

Some stability constants for typical complex ions are summarised below:

$Co(NH_3)_6^{2+}$	1.3×10^5 (5.11)		$Ni(NH_3)_6^{2+}$	9.9×10^7 (7.99)
$Zn(NH_3)_4^{2+}$	1.3×10^7 (9.1)		$Cu(NH_3)_4^{2+}$	4×10^{12} (12.6)
$Fe(CN)_6^{4-}$	2×10^8 (8.3)		$Fe(CN)_6^{3-}$	1×10^{31} (31.0)

The values in brackets are the logarithms to the base 10 of the stability constants and these values are often quoted.

5. Solubility and complex ion formation. Silver chloride is generally regarded as being insoluble in water. This is because it has a low solubility product ($K_s(AgCl) = 1 \times 10^{-10}$) so that a saturated solution of silver chloride only contains Ag^+ ions at a concentration of 10^{-5} mol dm^{-3} which for many purposes, is negligible.

If Ag^+ ions can be removed from a saturated solution of silver chloride in contact with solid silver chloride more solid will dissolve. This can be achieved by adding ammonia, which forms a complex ion with Ag^+ ions, as follows:

$$AgCl(s) \rightleftharpoons Ag^+(aq) + Cl^-(aq)$$
$$Ag^+(aq) + 2NH_3 \rightleftharpoons Ag(NH_3)_2^+(aq)$$

In other words, $Ag(NH_3)_2Cl$ is much more soluble than silver chloride.

Addition of acids, i.e. H^+ ions, to a solution of $Ag(NH_3)_2Cl$ reprecipitates silver chloride because the H^+ ions cause the $Ag(NH_3)_2^+$ ions to dissociate by removing NH_3 molecules as NH_4^+ ions,

$$Ag(NH_3)_2^+(aq) \rightleftharpoons Ag^+(aq) + 2NH_3$$
$$2NH_3 + 2H^+(aq) \rightleftharpoons 2NH_4^+(aq)$$

Silver chloride will also dissolve in solutions of potassium cyanide and sodium thiosulphate(VI) because of the formation of $Ag(CN)_2^-$ and $Ag(S_2O_3)_2^{3-}$ ions. Both potassium cyanide and sodium thiosulphate(VI) are used in photography for dissolving silver salts.

There are numerous other examples of an apparently insoluble substance passing into solution through complex ion formation.

a. Iodine, almost insoluble in water, dissolves in solutions containing I^- ions, e.g. in potassium iodide solution,

$$I_2(aq) + I^-(aq) \longrightarrow I_3^-(aq)$$

b. Addition of potassium cyanide solution to iron(II) sulphate(VI) solution at first produces a precipitate of insoluble iron(II) cyanide, but further addition of potassium cyanide solution 'dissolves' the precipitate because of hexacyanoferrate(II) formation,

$$2CN^-(aq) + Fe^{2+}(aq) \longrightarrow Fe(CN)_2(s)$$
$$\text{(insoluble)}$$
$$Fe(CN)_2(s) + 4CN^-(aq) \longrightarrow Fe(CN)_6^{4-}(aq)$$

c. Ammonia solution added to copper(II) sulphate(VI) solution gives, at first, a precipitate of basic copper(II) sulphate(VI), but this dissolves, on further addition of ammonia, forming a deep-blue solution of tetra-amminecopper(II) sulphate(VI),

$$CuSO_4 + 4NH_3 \longrightarrow Cu(NH_3)_4SO_4$$

This solution will dissolve cellulose and is known as Schweizer's solution. Tetraamminecopper(II) sulphate(VI) is less soluble in ethanol/water mixtures than in water and the solid can be precipitated from aqueous solution by adding ethanol.

d. Potassium iodide solution will, at first, precipitate mercury(II) iodide from a solution of a soluble mercury(II) compound. Further addition of potassium iodide produces potassium tetraiodomercurate(II), K_2HgI_4, and this, in a solution made strongly alkaline by adding potassium hydroxide, gives Nessler's solution, used for detecting and estimating ammonia and ammonium salts.

e. Many analytical schemes involve the separation of sulphides of arsenic(III), antimony(III) and tin(II and IV) from those of copper(II), mercury(II), lead(II), bismuth(III) and cadmium. This can be brought about by making use of the fact that the first group of sulphides form soluble complexes with sulphide ions from yellow ammonium sulphide, which is a mixture of $(NH_4)_2S$ and dissolved sulphur. For example,

$$As_2S_3(s) + 3S^{2-}(aq) \rightleftharpoons 2AsS_3^{3-}(aq)$$
$$SnS_2(s) + S^{2-}(aq) \rightleftharpoons SnS_3^{2-}(aq)$$

f. The solubility of gold or platinum in aqua regia is due to the formation of stable $AuCl_4^-$ and $PtCl_6^{2-}$ complex ions. The concentrated nitric(V) acid oxidises the gold or platinum to Au^{3+} or Pt^{4+}, and the concentrated hydrochloric acid provides the chloride ion to form the soluble complexes.

g. The cyanide process for gold extraction depends on the formation of the stable and soluble $Au(CN)_4^-$ complex. Gold is treated with sodium cyanide solution in the presence of air. The oxygen oxidises the gold to Au^{3+} ions, whilst the cyanide provides CN^- ions for the formation of $Au(CN)_4^-$ ions. Silver, similarly, forms $Ag(CN)_2^-$ ions.

Gold, or silver, can be obtained from solutions of their cyanide complexes by adding zinc or aluminium, i.e. by a replacement reaction, or by electrolysis.

Cyanide solutions of gold, silver, copper, zinc and cadmium are widely used for electroplating with these metals, the small concentration of free metallic ions in the solutions favouring the production of a fine-grained, uniform and adherent deposit of metal. Such cyanide solutions are, however, toxic, and they are tending to be replaced by solutions containing complexes of metallic ions formed with salts of amido-sulphuric acid, $HSO_3 \cdot NH_2$, tetrafluoroboric acid, HBF_4, and hexafluorosilicic acid, H_2SiF_6.

h. Lead chloride is less soluble in dilute hydrochloric acid than it is in water because of the common ion effect (p. 382) but, as more and more concentrated hydrochloric acid is used the solubility of lead chloride increases due to the formation of $PbCl_4^{2-}$ complex ions.

i. Rust spots, which always contain iron(III) compounds, and ink stains, which sometimes do (depending on the kind of ink) can be removed by treatment with ethanedioic (oxalic) acid or potassium ethanedioate (oxalate) solution. The ethanedioate forms soluble complex ions, $Fe(C_2O_4)_3^{3-}$ with iron(III) compounds.

6. Precipitation from solutions of complex ions. A precipitate of a substance in solution will not form until something is done to the solution to exceed the solubility product of the substance concerned, and exceeding a solubility product necessitates certain ionic concentrations.

Hydrogen sulphide passed into a solution of silver nitrate will precipitate silver sulphide ($K_s(Ag_2S) = 10^{-50}$), and addition of dilute hydrochloric acid to silver nitrate solution will precipitate silver chloride ($K_s(AgCl) = 1 \times 10^{-10}$). Even very low concentrations of Ag^+ and S^{2-}, or Ag^+ and Cl^-, will cause precipitation because of the low solubility products of silver sulphide and silver chloride.

If silver chloride is dissolved in potassium cyanide solution, however, the resulting solution will contain a high concentration of $Ag(CN)_2^-$ ions, but only a low concentration of Ag^+ ions,

$$Ag(CN)_2^-(aq) \rightleftharpoons Ag^+(aq) + 2CN^-(aq)$$

Addition of dilute hydrochloric acid will not precipitate silver chloride as $K_s(AgCl)$ will not be exceeded, but silver sulphide may be precipitated, on passing hydrogen sulphide in, as $K_s(Ag_2S)$ is so much lower than $K_s(AgCl)$.

A complex ion may, therefore, dissociate sufficiently to undergo some of the reactions of the simple ions concerned, or it may not. A solution of potassium hexacyanoferrate(II), for example, gives no precipitate of iron(II) hydroxide with sodium hydroxide solution, and no precipitate of silver cyanide with silver nitrate solution. The concentrations of Fe^{2+} and CN^- ions in the solution are not high enough. But a solution of tetraamminecopper(II) sulphate(VI) contains a sufficiently high concentration of Cu^{2+} ions to give a precipitate with hydrogen sulphide or with potassium hexacyanoferrate(II) solution.

A simple ion can, then, be prevented from forming a precipitate, i.e. it can be kept in solution, by converting it into a suitable complex ion. This can be useful, as shown in the following examples.

a. *Detection of* Cd^{2+} *ions in presence of* Cu^{2+} *ions.* If hydrogen sulphide is passed into a solution containing both Cd^{2+} and Cu^{2+} ions, both cadmium sulphide (yellow) and copper(II) sulphide (black) will be precipitated but the black copper(II) sulphide will mask the formation of the yellow cadmium sulphide. If, however, the solution containing both simple ions is treated with excess potassium cyanide solution, the resulting mixture contains $Cd(CN)_4{}^{2-}$ and $Cu(CN)_4{}^{3-}$ complex ions formed by the changes,

$$Cd^{2+}(aq) + 2CN^-(aq) \longrightarrow Cd(CN)_2(s) \text{ (white precipitate)}$$
$$Cd(CN)_2(s) + 2CN^-(aq) \longrightarrow Cd(CN)_4{}^{2-}(aq) \text{ (soluble)}$$
$$Cu^{2+}(aq) + 2CN^-(aq) \longrightarrow Cu(CN)_2(s) \text{ (unstable)}$$
$$2Cu(CN)_2(s) \longrightarrow C_2N_2(g) + 2CuCN(s) \text{ (white precipitate)}$$
$$CuCN(s) + 3CN^-(aq) \longrightarrow Cu(CN)_4{}^{3-}(aq) \text{ (soluble)}$$

The $Cu(CN)_4{}^{3-}$ ion is much more stable than the $Cd(CN)_4{}^{2-}$ ion, so that the solution containing both complex ions will give a yellow precipitate of cadmium sulphide, on passing in hydrogen sulphide, but will not precipitate black copper(II) sulphide. The original presence of Cd^{2+} ions can therefore be detected.

b. *Fehling's solution.* Addition of sodium hydroxide solution to a solution of copper(II) sulphate(VI) gives a precipitate of basic copper(II) sulphate(VI), but the precipitate is not formed if Rochelle salt, sodium potassium 2,3-dihydroxybutane–1,4–dioate (tartrate), is first added to the copper(II) sulphate(VI) solution.

The tartrate forms a complex ion with Cu^{2+} ions, in the presence of OH^- ions,

and the mixture containing this complex is known as Fehling's solution. It is used in making copper(I) oxide and in the detection and volumetric estimation of reducing sugars.

c. *E.D.T.A.* This is an abbreviation for ethylenediaminetetracetic acid, or diaminoethanetetracetic acid, which is also known under the trade names Sequestrol and Versene,

$$CH_2 \cdot N \cdot (CH_2 \cdot COOH)_2$$
$$|$$
$$CH_2 \cdot N \cdot (CH_2 \cdot COOH)_2$$

The sodium salt of the acid, $Na_4(C_{10}H_{12}N_2O_8)$ contains the complex ion $(C_{10}H_{12}N_2O_8)^{4-}$, and this forms very stable complexes with Ca^{2+} or Mg^{2+} ions.

Addition of the sodium salt of E.D.T.A. to hard water, containing Ca^{2+} or Mg^{2+} ions, forms stable complexes, e.g. $(CaC_{10}H_{12}N_2O_8)^{2-}$, which 'lock-up' the Ca^{2+} or Mg^{2+} ions and prevent them from forming a scum with soap or a precipitate with any other reagent. E.D.T.A. cannot be used for softening water for drinking as it is too poisonous, but it can be used for softening water used in industrial processes such as the dyeing of textiles. It can also be used in volumetric methods for estimating Ca^{2+} and Mg^{2+} ions in solutions.

d. *Calgon.* Calgon is a trade name, derived from 'calcium gone', for sodium polyphosphate $Na_6P_6O_{18}$. This contains the complex polyphosphate ion, $P_6O_{18}^{6-}$, which, like the sodium salt of E.D.T.A., forms very stable complexes with Ca^{2+} or Mg^{2+} ions, e.g. $[Ca_2P_6O_{18}]^{2-}$. Calgon can, therefore, be used for softening water, and it is available commercially for domestic and industrial use.

Polyphosphates and E.D.T.A. are often referred to as sequestrating agents or complexones and their locking-up of ions in stable complexes as *sequestration*.

7. Insoluble complexes. The examples of complex ion formation given in the preceding two sections have been concerned with soluble complexes.

Insoluble complexes are useful, too, particularly in analysis. If a simple ion in a solution can be converted into an insoluble complex a precipitate will be formed, and this will denote the presence of the original simple ion, particularly if the complex formed is unique for that ion. If the complex formation takes place quantitatively, weighing the resulting precipitate provides a quantitative method for estimating the simple ion.

Many organic reagents which form specific complexes with, ideally, just one simple ion have been developed for such uses.

A typical example is provided by the use of butanedione dioxime (dimethyl glyoxime) in the qualitative and quantitative estimation of Ni^{2+} ions in a solution made just alkaline with ammonia solution. The complex formed has the structure as shown.

Other fairly common reagents include Curpon (α-benzoin oxime) and rubeanic acid (dithio oxamide) for Fe^{2+} ions, *o*-phenanthroline for Fe^{3+} ions, ammonium mercuri-thiocyanate plus cobalt(II) sulphate(VI) for Zn^{2+} ions, and hydroxy-quinoline (Oxine) for Mg^{2+} ions.

8. Detection of complex ion formation. Unexpected solubility observations of a supposedly insoluble substance in a solution, or the failure to obtain a precipitate of a compound of a simple ion, often point to the formation of a complex ion. Other evidence suggesting complex ion formation may be summarised as follows:

a. *Freezing point evidence.* Complex ion formation may change the number of particles dissolved in a solvent and, if so, the freezing point lowering of the solvent will be affected.

A solution of potassium iodide, for example, will contain K^+ and I^- ions. But if mercury(II) iodide is added, and if there is a complete conversion into HgI_4^{2-} ions,

$$2I^+(aq) + HgI_2(s) \longrightarrow HgI_4^{2-}(aq)$$

two ions will be replaced by one with a consequent lowering of the freezing point depression.

Investigation of freezing point depression, and other colligative properties of a a solution (p. 263), can, therefore, be used to study complex ion formation. It is not always applicable, because complex ion formation which does not involve any change in the number of soluble particles, e.g.

$$I_2 + I^-(aq) \longrightarrow I_3^-(aq)$$

will not affect freezing point depression measurements.

b. *Conductivity evidence.* The conductivity of a solution containing simple ions will change if the simple ions are converted into complex ions, for the molar conductivity of the complex ion will be different.

Λ_0 for copper(II) sulphate(VI), for example, will be given by

$$\Lambda_0(CuSO_4) = \Lambda(Cu^{2+}) + \Lambda(SO_4^{2-})$$

but Λ_0 for tetramminocopper(II) sulphate(VI) will be

$$\Lambda_0(Cu(NH_3)_4SO_4) = \Lambda(Cu(NH_3)_4^{2+} + \Lambda((SO_4)^{2-})$$

Moreover, conductivity measurements can indicate, approximately, the number of ions formed by an electrolyte. The Λ_0 values, for the different number of ions formed, are approximately,

No. of ions formed	2	3	4	5
$\Lambda_0/\Omega^{-1}\,cm^2\,mol^{-1}$	125	260	410	540

c. *Transport number measurement.* The existence of complex ions can, sometimes, be shown by transport number measurement (p. 455).

d. *E.m.f. measurement.* The e.m.f. of, for instance, a concentration cell is dependent on ionic concentrations. If, therefore, a simple ion participating in the functioning of such a cell is converted into a complex ion by the addition of another reagent there will be a change in the e.m.f. of the cell.

e. *Partition coefficient measurement.* The formation of a complex ion, and its composition, can often be determined by partition coefficient measurements. The formation of the I_3^- ion from iodine and I^- ions, for example, can be demonstrated by showing that $[I_3^-]$ divided by $[I_2] \times [I^-]$ has a constant value (p. 317).

QUESTIONS ON CHAPTER 44

1. Account for the following facts:

(*a*) Addition of dilute hydrochloric acid to silver nitrate(V) solution gives a white precipitate, but the precipitate dissolves on the addition of ammonia solution.

(*b*) Addition of ammonia solution to a solution of a zinc salt gives a precipitate at first, but the precipitate dissolves when more ammonia solution is added.

(*c*) Copper(I) chloride is insoluble in water and in dilute hydrochloric acid, but it dissolves in concentrated hydrochloric acid.

(*d*) Mercury(II) iodide is more soluble in organic solvents than in water, and is also soluble in potassium iodide solution.

(*e*) Zinc hydroxide dissolves in sodium hydroxide solution, but calcium hydroxide does not.

2. Describe, and explain, the effect of adding both small and large amounts of (i) sodium hydroxide solution, (ii) ammonia solution to solutions containing (*a*) copper(II) ions, (*b*) iron(III) ions, (*c*) iron(II) ions, (*d*) zinc ions, (*e*) cobalt(III) ions, (*f*) nickel(III) ions, (*g*) aluminium ions.

3. Write down the formulae of the following complex ions: (*a*) tetraammine zinc(II), (*b*) tetrachloromercurate(II), (*c*) tetracyanocuprate(I), (*d*) trihydroxo-plumbate(II), (*e*) trinitrotriamminecobalt(III), (*f*) hexanitritocobaltate(III), (*g*) chloroaquo-bisethylenediamminechromium(III), (*h*) dichloroargentate(I), (*i*) tetrahydroxoaluminate.

4. Give the names and formulae of (*a*) four copper complexes, (*b*) six complexes involving halogens, (*c*) three zinc complexes, and (*d*) two hydroxide complexes.

5. Iron(III) salts may not give a red coloration on adding potassium thiocyanate if ethanedioate (oxalate) ions are present, but the red colour reappears on addition of a strong acid. Explain.

6. Account for the following facts: (*a*) Silver iodide is not appreciably soluble in concentrated ammonia solution, though silver chloride is. (*b*) Addition of potassium bromide solution to a dilute ammoniacal solution of silver chloride precipitates silver bromide.

7. What experimental evidence can be advanced in support of the following statements? (*a*) When carbon dioxide dissolves in water, chemical reaction takes place between the two substances. (*b*) The solubility of silver chloride in aqueous ammonia is due to the formation of a complex ion $[Ag(NH_3)_2]^+$. (O. Schol.)

8. What is a complex ion? What experimental observations suggest that such ions exist in certain solutions? (O. Schol.)

9. Zinc will replace silver from a solution of potassium dicyanoargentate(I) but will not replace copper from an equivalent solution of potassium tetracyano-copper(I). Zinc will also replace copper from a solution of tetraamminecopper(II) sulphate(VI) but not from an equivalent solution of potassium tetracyanocopper(I). Account for these facts.

10. What tests do you know in inorganic qualitative analysis which depend on complex formation? Wherever possible, give the formula of the complex formed.

11. Why is it necessary to remove such non-volatile organic compounds as tartrates and citrates before carrying out a qualitative analysis for metallic ions?

12. Explain why copper(II) hydroxide will dissolve both in dilute sulphuric(VI) acid, which is acidic, and in ammonia solution, which is alkaline. Why will silver chloride dissolve in the latter, but not in the former?

13. Explain what happens when the following reagents are added, one after the other, to a solution of copper(II) sulphate(VI): (*a*) sodium hydroxide solution, (*b*) ammonia solution, (*c*) ethanol. Give ionic equations where possible.

14. If the solubility product of silver chloride is 1.5×10^{-10} and the instability constant of the $Ag(NH_3)_2^+$ ion is 6.8×10^{-8}, calculate the approximate solubility of silver chloride in a 3M ammonia solution.

15. 1 mol of ammonia was added to 1 dm³ of a solution of silver nitrate(V) containing 0.005 mol of silver. If the dissociation constant for the $Ag(NH_3)_2^+$ ion is 6×10^{-8}, calculate the concentration of free Ag^+ ions in the mixture.

16. How many grammes of silver iodide will dissolve in a cubic decimetre of M ammonia solution? The solubility product of silver iodide is 1×10^{-16}, and the instability constant of the $Ag(NH_3)_2^+$ ion is 6×10^{-8}.

17. 25 cm³ (excess) of ammonia solution were added to 25 cm³ of 0.1M copper(II) sulphate(VI) and the resulting deep-blue solution was shaken with trichloromethane (chloroform). After the layers had been allowed to settle, 50 cm³ of the trichloromethane layer required 25.5 cm³ of 0.05M hydrochloric acid for neutralisation. 20 cm³ of the blue aqueous layer were neutralised by 33.3 cm³ of 0.5M hydrochloric acid. If the partition coefficient of ammonia between water and trichloromethane is 25, calculate the probable formula of the tetraamminecopper(II) ion.

18. The freezing point of a solution containing 3 g of potassium iodide in 100 g of water was $-0.619°C$. When 4.11 g of mercury(II) iodide were dissolved in the same solution the freezing point rose to $-0.504°C$. Calculate (a) the degree of dissociation of the potassium iodide before the addition of the mercury(II) iodide, and (b) the formula of the complex ion produced when the mercury(II) iodide is added. (Freezing point constant per 100 g = 18.6°C.)

19. How would you demonstrate that copper was present in a complex anion in Fehling's solution?

Chapter 45
Surface effects

THE situation existing at the surface of a liquid or solid is different from that in the interior, and this has important consequences.

A molecule in the interior of a liquid, for instance, is completely surrounded on all sides by other molecules and inter-molecular attractive forces are, on the average, exerted equally in all directions. A molecule at the surface of a liquid, however, is surrounded on one side by molecules in the liquid and, on the other, by the more widely scattered molecules in the vapour state. Such surface molecules will, therefore, experience inward attractive forces which cause surface tension.

Surface tension in liquids affects the properties of a liquid, and similar forces at a solid surface are probably responsible for the adsorptive power of solid surfaces. The effects at a liquid surface are different, in degree, from those at a solid surface, and the two effects will be considered separately.

LIQUID SURFACES

1. Surface tension and surface energy. One of the immediate effects of the surface tension in a liquid is that a liquid surface tends to become as small as it possibly can. As a sphere has the smallest surface-to-volume ratio, freely suspended liquids form spherical drops, whilst bubbles of gas in a liquid are also spherical.

The state of tension at a liquid surface, which is really due to forces acting inwardly, is conventionally regarded as being caused by forces acting in, and parallel to, the surface. On this basis, the surface tension of a liquid is defined as the force in newtons acting, in the surface, upon a line of 1 metre length; the units are $N\ m^{-1}$.

A clearer picture is provided by the idea of surface energy. To move molecules from the interior of a liquid into the surface, i.e. to increase the surface area, must involve doing work against the internal forces within the liquid. In other words, energy must be supplied to increase a surface area. For a surface tension of $\gamma\ N\ m^{-1}$, a surface area increase of $1\ m^2$ would require the expenditure of γ joule of work (1 J = 1 newton metre). The surface energy corresponding to a surface tension of $\gamma\ N\ m^{-1}$ is, therefore, $\gamma\ J\ m^{-2}$

Some typical values for surface tensions in $N\ m^{-1}$, are given below. Water has a surface tension greater than that of most common liquids, but molten metals (including mercury), have much higher values.

Water at 20°C	72.6×10^{-3}	Mercury at 20°C	485×10^{-3}
Benzene at 20°C	28.9×10^{-3}	Zinc at 600°C	770×10^{-3}
Ethanol at 20°C	22.3×10^{-3}	Copper at 1200°C	1160×10^{-3}

The surface tension of liquids usually decreases with increasing temperature.

2. The Parachor. The mass of a molecule is almost equal to the sum of the masses of the component atoms and, historically, much search has been made for other similar additive factors.

Kopp, in particular, investigated the molar volumes of liquids from this point of view. The molar volume of a liquid is defined as the volume occupied by 1 mol, or

$$\frac{\text{Molar}}{\text{volume}} = \frac{\text{Relative molecular mass}}{\text{Density}}$$

The molar volume of gases, at s.t.p., is 22.4136 dm^3, and it was the constancy of this value for gases which suggested that the molar volumes of liquids might be of interest.

In considering values for the molar volumes of liquids it is necessary to decide on the temperature at which the density of the liquid should be measured. Kopp took density values at the boiling points of the liquids concerned, and at 101.325 kN m^{-2} pressure. He found that the molar volumes of many liquids were nearly equal to the sum of the atomic volumes of the constituent atoms. This, however, is not universally true. If it were, all isomers of the same substance would have the same molecular volume, and this is not so. It is clear, then, that the arrangement of atoms within a molecule affects the issue.

Kopp made his calculations at the boiling point of the liquid concerned, but Sugden (1892–1950) took another basis of comparison. He introduced the idea of the parachor which, basically, compares the molar volumes of liquids under conditions where they have equal surface tensions.

Sugden made use of Macleod's empirical equation first put forward in 1923. This equation states that

$$\frac{\gamma^{\frac{1}{4}}}{D - d} = \text{a constant}$$

where γ is the surface tension of the liquid concerned, D the density of the liquid, and d the density of the vapour formed by the liquid, all values being taken at the same temperature. The equation holds quite well for many liquids over a wide range of temperature, but values for associated liquids do not fit the equation very well.[*]

The constancy of $\frac{\gamma^{\frac{1}{4}}}{D - d}$ at different temperatures is, nevertheless, remarkable, and Sugden defined the parachor in terms of it as follows:

$$\text{Parachor } [P] = \frac{M_r \times \gamma^{\frac{1}{4}}}{D - d}$$

where M_r is the relative molecular mass

The value of d is generally very small as compared with D, so that

$$[P] \simeq \frac{M_r \times \gamma^{\frac{1}{4}}}{D} \quad \text{or} \quad [P] \quad \frac{\text{Molar}}{\text{volume}} \times \gamma^{\frac{1}{4}}$$

It will be seen that the parachor of a liquid can be regarded as the molar volume which the liquid would have if its surface tension was unity.

Sugden had hoped to find in the parachor a completely additive quantity, but its usefulness is restricted because, like Kopp's molar volume values, it is not always strictly additive. Sugden was able, however, to allot numerical values to

* The Eötvos equation is similar. It relates the relative molecular mass (M_r), the surface tension (γ), the density (D), and the critical temperature (T_c) of a liquid,

$$\gamma(M_r/D)^{\frac{2}{3}} = K(T_c - T)$$

T being the temperature at which the surface tension and density are measured. K has a constant value for many unassociated liquids.

different atoms and bonds which could be added together to give the parachor value of some substances containing the particular atoms and bonds chosen.

The idea of finding numerical values for bonds and atoms which can be added together to give an overall numerical value for a compound is an attractive one, and spasmodic attempts have been made to replace the parachor by something more reliable, but, so far, with only limited success.

3. Formation of surface films. A freely suspended liquid forms spherical drops to lower its surface energy, and any other process which results in a lowering of surface energy tends to take place. Such processes lead to the formation of surface films.

a. *Surface tension of solutions.* Addition of a little 3-methylbutan–1–ol (iso-amyl alcohol) to water causes a marked lowering of the surface tension, whilst addition of electrolytes to water results in a slight increase in surface tension.

3–methylbutan–1–ol has a lower surface tension (24×10^{-3} N m^{-1}) than water, and in a mixture of two miscible liquids the one with the lower surface tension tends to concentrate at the surface. This is because the surface energy is lowered when the liquid with the lower surface tension is at the surface. The surface is not made up completely of molecules of the lower surface tension liquid, but there is a higher proportion of them at the surface than in the bulk of the mixture. Substances which cause a lowering of surface tension are said to be *surface active* or to be positively adsorbed.

By comparison, electrolytes cause a slight increase in surface tension. This is because their ions, in solution, increase the attractive forces within the liquid as inter-ionic forces are stronger than inter-molecular ones.

b. *Surface films of insoluble substances.* A liquid which is insoluble in water may, or may not, spread over a water surface to form a film. Liquid paraffin, and other hydrocarbons, for example, will not spread on water, but octanol and octadecenoic (oleic) acid will. The spreading can be observed by first dusting a water surface with fine talc. A spreading liquid will push the talc away and a rough estimate of the area of film forced can be obtained by measuring the area of surface not covered by the talc.

Whether or not a liquid will spread on another is a matter of the relative values of the surface tensions of the two liquids. If a liquid, A, is to spread on a liquid, B, the surfaces of A with B and of A with its vapour must be increased, whilst that of B with its vapour must be decreased. If these changes in surfaces produce a resultant lowering of surface energy then A will spread on B; if not, A will not spread on B.

c. *Surface films of insoluble substances with polar groups.* A substance, insoluble in water, will spread over water if it contains a polar group such as —OH or —COOH. Such groups are *hydrophilic*, i.e. have a strong attraction for water; other groups, e.g. alkyl groups, are *hydrophobic*.

When a substance with a polar group forms a surface film on water it is thought that the polar group is within the water surface, or closely linked to it, whilst the hydrophobic group is out of the surface. Such substances, in fact, form surface films which may be regarded as transitional between those formed by soluble and insoluble substances.

4. Uses of compounds with hydrophilic and hydrophobic groups.

a. *In water conservation.* In areas with a shortage of water it is essential to try to prevent undue loss from reservoirs by evaporation. This can be done, to some extent, by spreading hexadecanol (cetyl alcohol) over the surface of the reservoir. The hydrophilic, $-CH_2OH$, group enters the water surface, whilst the alkyl group, $-C_{15}H_{31}$, doesn't. The rate of evaporation is cut down as the surface of the water becomes much more like an alkane surface.

b. *As emulsifiers.* In an emulsion, one phase is dispersed as small droplets in another. The resulting high surface area between the two phases, and the correspondingly high interfacial tension, would lead to an unstable system unless an emulsifier is present. Substances with both hydrophilic and hydrophobic groups can often act as emulsifiers.

In an oil and water mixture, for example, such an emulsifier acts as a go-between. Adsorption of the hydrophobic group by the oil droplets gives them a hydrophilic surface, whilst adsorption of the hydrophilic groups by the water droplets gives them a hydrophobic surface. The oil and water become, in this way, much more nearly miscible.

c. *As detergents.* Detergents are substances with both hydrophilic and hydrophobic groups. Their detailed cleansing action is complicated but, in essence, they enable water and fatty or oily substances to come together into emulsions.

Soap, which in its simplest form, is sodium octadecanoate (stearate), is the traditional detergent. It is not, however, entirely satisfactory when used in hard or acidic waters. In recent years it has been partially replaced by synthetic detergents. These are generally more soluble in water than soap; they do not form a scum with hard water; they enable water to spread and penetrate more fully over, or through, an article being washed; and they can be used under alkaline or acidic conditions.

The hydrophobic group in a detergent is usually a large alkyl group or an alkyl-substituted benzene ring; the hydrophilic group is generally a $-COOH$, $-SO_3H$, $-OSO_3H$ or $-OH$ group. Teepol, with a formula,

$$H_{21}C_{10}-\overset{\displaystyle H}{\underset{\displaystyle CH_3}{\overset{\displaystyle |}{\underset{\displaystyle |}{C}}}}-O{\cdot}SO_3Na$$

provides a typical example.

5. Investigation of surface films. Langmuir and Adam have investigated surface films by containing the film between definite barriers. Adam used an apparatus of the type shown in Fig. 159. AB and CD are strips of waxed paper or glass or mica, passing across a shallow tank containing water or any other liquid being investigated. AB passes right across the tank, penetrates below the liquid surface, and can be moved and held in any position to vary the area ABCD. CD floats on the water surface and is free to move slightly. Strips of platinum foil occupy the spaces between the ends of CD and the sides of the tank. CD is attached to delicate instruments which can measure the force being exerted on it.

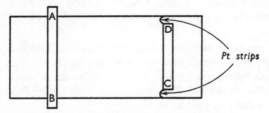

FIG. 159. Apparatus for the study of surface films.

In a typical experiment, the water surface is first cleaned by drawing strips across it to push any surface contamination to one end of the tank, outside the area ABCD. A measured amount of an oil, dissolved in benzene, is then placed on the water surface between AB and CD. As the benzene evaporates, the oil forms a surface film in the area ABCD. This film exerts a force on CD which can be measured, and the variation of the force as the area ABCD is changed can be studied.

If there is simply water on both sides of CD it will not be subjected to any resultant force, but if the surface tension of the liquid in the area ABCD is changed a resultant force will be exerted on CD.

With only a little oil on ABCD there is no force on CD, i.e. no change in the surface tension of the water. As the area ABCD is decreased, however, there comes a point at which a rapidly increasing force is exerted on CD, and further compression of the surface film on ABCD produces even greater forces on CD.

These results are interpreted as meaning that a surface film of oil does not cause much change in surface tension until it is one molecule thick, i.e. a unimolecular layer or a monolayer. Before this stage is reached, the oil cannot completely cover the surface and cannot, therefore, form a complete film. For a given amount of oil, the area of ABCD at which the first marked change in surface tension is observed can be measured. The value obtained is in good agreement with the monolayer theory, because the area covered by a monolayer of a certain amount of oil can be calculated from the molecular dimensions of the oil used. If these molecular dimensions are not known, for a particular substance, they can be measured by recording the area of ABCD when a marked change in surface tension takes place, i.e. when a monolayer is just formed.

A monolayer is probably first formed when the molecules forming the surface film are lying fairly flat on the surface. Compression of the film will cause the molecules to 'stand up' on the surface, and compression beyond this stage will buckle the film.

SOLID SURFACES

6. Surface area. The forces existing at the surface of a solid may enable the solid to absorb a gas or a substance from solution, adsorption being the formation of a surface film on the solid. The greater the surface area of the solid concerned, the more highly adsorptive will it be, and solids with porous structures, and, consequently, high surface areas, such as charcoal or silica gel, are commonly used as adsorbents. Activated charcoal, i.e. charcoal which has been heated in steam, is particularly adsorptive and is used, for example, in gas masks and in decolorising coloured solutions.

Finely divided substances may, too, be very adsorptive because they have large surface areas. A cube of side 1 m, for example, has a surface area of 6 m². If it is subdivided into cubes of side 1×10^{-6} m the surface area will increase to 6×10^6 m².

7. Adsorption of gases. An experimental study of the adsorption of gases on solids shows that there are two main types of adsorption. In the first type, referred to as physical or van der Waals' adsorption, there is no definite chemical binding, of the sort found in ordinary chemical compounds, between the adsorbed surface layer and the solid adsorbent. In the second type, known as chemisorption, there is something in the nature of true chemical binding at the surface of the solid.

a. *Physical adsorption.* Physical adsorption of a gas by a solid is generally reversible. Increase of pressure causes more gas to be adsorbed, but releasing the pressure frees the adsorbed gas. Similarly, decrease of temperature increases adsorption, but gas adsorbed at a low temperature can be freed again on heating.

The extent of physical adsorption is also dependent on the surface area of the solid adsorbent, and on the nature of the gas. In general, the higher the critical temperature of a gas, or the more easily liquefiable or more soluble it is, the more readily will it be adsorbed. Thus 1 g of activated charcoal will adsorb 380 cm³ of sulphur dioxide (critical temperature = 157°C), 16 cm³ of methane (critical temperature = −83°C) and 4.5 cm³ of hydrogen (critical temperature = −240°C).

This relationship is perhaps not surprising as the critical temperature of a gas (p. 74) is related to its intermolecular forces which must play a large part on adsorption.

b. *Chemisorption*. Chemisorption involves the formation of what might be called a surface compound. As it corresponds to something much closer to a chemical change than physical adsorption it is commonly irreversible. In many examples, efforts to free the adsorbed gas produce a definite compound. Oxygen adsorbed on carbon or tungsten, for example, is liberated as carbon monoxide and dioxide or as tungsten oxide. Similarly, carbon monoxide adsorbed on tungsten is liberated as a carbonyl.

Chemisorption is much more specific than physical adsorption and will only occur when there is some possibility of chemical interaction between the gas adsorbed and the solid adsorbent. Moreover, chemisorption, like most chemical changes, often increases with rise of temperature, unlike physical adsorption. On this account, a gas may be physically adsorbed at low temperature and chemisorbed at high temperature. Nitrogen, for example, is physically adsorbed on iron at $-190°C$, but chemisorbed, to form a nitride, at $500°C$.

8. Adsorption isotherms. The amount of gas adsorbed by a given amount of adsorbent depends on both temperature and pressure. At a given temperature, the variation in the amount of gas adsorbed as the pressure is changed can often be summarised in an empirical equation.

a. *Freundlich adsorption isotherm*. This empirical relationship states that

$$\frac{x}{m} = k \times p^n$$

where x is the mass of gas adsorbed by m gramme of adsorbent at a pressure, p; k and n are constants at a particular temperature and for a particular adsorbent and gas.

If the relationship holds, plots of the amount of a gas adsorbed against the pressure will be of the form shown in Fig. 160. These curves also show the decrease in adsorption, at a fixed pressure, as temperature is increased; they are for physical adsorption.

The Freundlich isotherm can also be tested by plotting log x/m against log p. A straight line should result, the value of n being obtainable from the slope of the line, since,

$$\log (x/m) = \log k + n\log p$$

In practice, a reasonably straight line plot is often obtained, particularly if results at low and high pressures are neglected.

b. *Langmuir adsorption isotherm.* Langmuir deduced a theoretical relationship which is in agreement with many experimental measurements over a wider range of pressure than the Freundlich isotherm. It includes the Freundlich isotherm as a special case for intermediate pressures.

Langmuir considered that adsorption could only take place until the adsorbing surface was completely covered with a unimolecular layer of

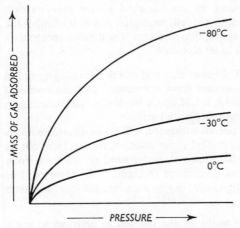

FIG. 160. Freundlich isotherms for the adsorption of nitrogen by charcoal.

adsorbed gas. He considered a kinetic balance between gas molecules striking the surface and being adsorbed and adsorbed molecules being evaporated from the surface.

If the fraction of surface covered by gas at any one moment is θ, the rate at which molecules evaporate from the surface will be proportional to θ, and equal to $k\theta$. At the same time, the rate at which molecules will be adsorbed will be proportional both to the amount of surface not already covered, i.e. $(1 - \theta)$, and to the gas pressure. The rate of adsorption will, therefore, be equal to $k'p(1 - \theta)$.

At equilibrium,
$$k\theta = k'p(1 - \theta)$$
or
$$\theta = \frac{k'p}{k + k'p} = \frac{k''p}{1 + k''p}$$

θ, the fraction of surface covered, will be proportional to the amount of gas, x, adsorbed by a given mass of adsorbent, m, so that
$$\frac{x}{m} = \frac{k'''p}{1 + k''p}$$

This relationship is known as Langmuir's adsorption isotherm. It can be written in the form,
$$\frac{pm}{x} = \alpha + \beta p$$

where α and β are both constants, so that it can be tested by seeing whether or not a plot of $\dfrac{pm}{x}$ against p gives a straight line.

The Langmuir adsorption isotherm gives a satisfactory theoretical interpretation of many experimental results, and holds particularly well for chemisorption where the idea of a limiting unimolecular layer is perhaps more tenable than it is for physical adsorption.

Experimental plots of amount of gas adsorbed against pressure can, however, take a variety of shapes not fully explicable in terms of either the Freundlich or the Langmuir adsorption isotherm. The detailed mechanism of adsorption is, in fact, still to be elucidated.

9. Adsorption from solution. The use of charcoal as a decolorising agent is a well-known example of adsorption from a solution, and charcoal will effectively adsorb many dyestuffs. It will, too, adsorb other substances such as ethanoic(acetic) and ethanedioic (oxalic) acids.

Even glass, which is not a porous substance, will act as an adsorbent. A test tube which has contained methyl violet solution, for instance, can be emptied and repeatedly washed out with water until no trace of violet colour is visible. Addition of propanone (acetone) will, however, free sufficient of the adsorbed methyl violet to give a distinctly coloured acetone solution.

Precipitates, too, often act as adsorbents, and this can be both helpful and a nuisance. It is a nuisance when a pure precipitate is required, or when a precipitate is required for quantitative purposes, for, in these cases, adsorption adds impurities or gives false quantitative results. Alternatively, adsorption on a precipitate is useful in certain confirmatory tests, for a particular precipitate may be formed in the presence of a particular dyestuff to give a distinctive colour. Magnesium hydroxide, for example, will be blue when precipitated in the presence of the dyestuff magneson. Gelatinous precipitates, like magnesium hydroxide, are particularly adsorptive, but granular precipitates will also adsorb. This can be demonstrated by precipitating barium sulphate(VI) in an alkaline solution of potassium manganate(VII). The barium sulphate(VI) precipitate will be pink, even after much washing.

The precise mechanism of adsorption from solution is not known. There is a limit to the adsorption by a given mass of adsorbent, and it is probable that adsorption takes place until a unimolecular layer is built up.

The amount adsorbed is, too, dependent on the concentration of the solution, and, for adsorption from solutions, the concentration of the solute takes the place of pressure in gas adsorption. The Freundlich isotherm (p. 496), using concentration instead of pressure, i.e.

$$\frac{x}{m} = kc^n$$

applies to adsorption from solution more effectively than to adsorption of gases. c is the concentration of the solution when adsorption is completed.

The validity of the Freundlich isotherm in one case can easily be tested by shaking equal masses of activated charcoal with different solutions of ethanedioic (oxalic) acid of known concentration. After standing for a time to allow equilibrium to be established the amount of ethanedioic acid adsorbed in each case can be measured by finding the concentration of ethanedioic acid remaining after adsorption. Plotting the amount adsorbed against this concentration, or the log of the amount adsorbed against the log of this concentration shows the validity of the isotherm.

Alternatively, methylene blue can be adsorbed from solutions of different, but known, initial concentrations, the final concentrations being measured colorimetrically.

EXCHANGE ADSORPTION

10. Ion exchange. An investigation by Spence, in 1845, into the possible causes of the loss of ammonia from manure heaps and the retention of certain substances by soil, led to the discovery that when a solution of ammonium sulphate(VI) percolates through a column of soil the effluent contains no ammonium ions, but, instead, contains calcium ions.

This was the beginning of what has developed into a wide use of ion-exchange substances. In effect, a solid releases one ion and adsorbs another; there is an ion-exchange. When the ions exchanged are cations it is known as *base or cation exchange*; when anions, as *acid or anion exchange*.

The first application of ion-exchange was in water softening. Many naturally occurring substances such as soils and clays act as ion-exchangers and it was discovered, in 1907, that selected substances, such as greensands and certain zeolites, could effectively soften water by exchanging calcium and magnesium ions, which cause hardness, for sodium ions, which don't. Synthetic aluminosilicates, known as permutites, with more effective exchange properties, were soon developed, and have been widely used in water softening. They are made by fusing china clay, sand and sodium carbonate together.

The use of permutites is limited to water softening, and a big step forward was made in 1935, when Adams and Holmes made synthetic resins with much wider ion-exchange properties. At the same time, various methods of improving the ion-exchange capacities of naturally occurring substances were discovered. The sulphonation of certain soft coals, for example, provides useful materials.

On the whole, however, the synthetic resins, and particularly some high-polymers, are better ion-exchangers than natural substances, whether treated or not. Moreover, natural substances are invariably cation exchangers only, whereas synthetic resins can be made which are either cation or anion exchangers.

a. *Cation-exchangers.* These are high-polymers, often derived from phenol-formaldehyde resins, containing an acidic group such as $-HSO_3$, $-COOH$ or $-OH$. They are strong or weak acids, insoluble in water, form insoluble salts, and are generally used in the form of small spherical beads.

A typical example is made from phenylethene (styrene). This is polymerised with diethenylbenzene, and the resulting cross-linked polymer is sulphonated to introduce $-HSO_3$ groups. The product, first made in 1944, is marketed as Dowex 50. Other commercially available cation exchangers are sold under the trade names of Zeo-Karb 225 and Amberlite IR-120.

b. *Anion-exchangers.* Anion exchangers are similar to cation exchangers but contain a basic group such as a secondary amine, a tertiary amine, or a quaternary amine group. They are not so easy to make as cation exchangers. Typical examples are sold as De-Acidite FF IP, Dowex 1−X8, and Amberlite IRA-400.

Ion exchange reactions are reversible so that an ion exchange material can be regenerated when it is exhausted.

11. Applications of ion exchange.

a. *Purification of water.* The first application of ion-exchange materials was in the softening of water. For this purpose a sodium salt of an ion exchange material is used, the process being represented by the equation

$$2NaR \quad + \quad Ca^{2+} \quad \rightleftharpoons \quad CaR_2 \quad + \quad 2Na^+$$

2NaR	Ca²⁺	CaR₂	2Na⁺
(Solid resin)	(In solution)	(Solid resin)	(in
(Insoluble)		(Insoluble)	solution)

where R is the resin used. To regenerate the sodium salt when it has been completely converted into the calcium salt it is only necessary to treat it with a solution of sodium chloride.

Water can, however, be purified much further, and all dissolved salts removed, by using both a cation and an anion exchanger. Such purified water, which is purer than distilled water, is sometimes known as de-ionised or de-mineralised water, the process of purification being referred to as *de-ionisation* or *de-mineralisation.*

In the two-stage process, water is first passed through a column of a cation exchanger in the form of an acid In this stage, any cations in the water are replaced by H^+ ions,

$$HR(s) + X^+(aq) \rightleftharpoons H^+(aq) + XR(s)$$
(Cation
exchanger)

The partially purified water is then passed through an anion exchanger in the form of a base. This replaces any anions by OH^- ions,

$$HOR(s) + Y^-(aq) \rightleftharpoons OH^-(aq) + YR(s)$$
(Anion
exchanger)

When the resins are exhausted they can be regenerated, the cation exchanger by treatment with acid, and the anion exchanger by treatment with alkali.

A one-stage process can be used by passing water down a single column containing a mixture of cation and anion exchangers. For regeneration purposes the two exchangers have to be separated, but this is facilitated by using an anion exchanger with a lower density than the cation exchanger. The two can then be separated, for regeneration, by passing a rapid stream of water up the column. The anion exchanger collects at the top, with the cation exchanger at the bottom. After regeneration, the exchangers are mixed again.

b. *Drinking-water from sea-water.* For most natural waters with a limited salt content, removal of all dissolved salts by ion exchangers is cheaper and easier than distillation, but sea-water contains such a high proportion of salts that it cannot be converted into drinking-water economically by ion-exchangers. Kits are available, however, in life-boats which enable reasonable drinking-water to be obtained from sea-water.

c. *Medical uses.* Removal of excess sodium salts from body fluids can be achieved by giving the patient a suitable ion-exchanger to eat. Weakly basic anion exchangers can also be used to remove excess acid in the stomach and thus relieve indigestion, and blood can be prevented from clotting by exchanging the Ca^{2+} ions which it contains, and which cause it to clot, for Na^+ ions.

d. *Ion-exchange membranes.* Ion-exchange materials supported on paper, fibre or some other material can be used as membranes through which only anions or cations can pass. They act as 'ionic sieves'.

A typical use of such membranes is in an electrical demineralisation of water in a cell as shown in Fig. 161. Both anions and cations can pass out of the central compartment of the cell, but neither anions nor cations can pass into this compartment except by slow diffusion. Electrolytes can, therefore, be removed from the central compartment, and, in this way, salts can be removed from sea-water and other solutions.

Natural membranes such as those in plant cells, blood cells and nerve-fibres probably function in the same way as ion-exchange membranes.

e. *Miscellaneous uses.* Other applications of ion-exchange resins include the removal of harmful substances, e.g. cyanides, from waste water, the refining of sugar, and the removal of potassium hydrogen tartrate from wines. The resins are also extremely useful in the extraction of valuable metals from solutions containing only small concentrations of the metal. Uranium, for example, can be obtained from the dilute sulphate(VI) solutions obtained by leaching uranium ores, silver can be recovered from waste photographic solutions, and valuable metals, such as gold and chromium, can be recovered from waste solutions after electroplating.

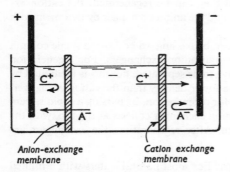

FIG. 161. Use of ion-exchange membrane in an electrolytic cell.

SELECTIVE ADSORPTION. CHROMATOGRAPHY

12. Selective adsorption from solution. Some adsorbents will adsorb the different solutes which may be present in a mixed solution to different extents. If such a solution is passed through a column of the adsorbent, the most highly adsorbed solute will be adsorbed at the top of the column, whilst the weakly adsorbed solutes will be only gradually adsorbed lower down the column. The separation of the different solutes in the column, forming what is known as a *chromatogram*, can be improved by passing some more solvent down the column. This process is known as *developing* the chromatogram.

When separation has been completed, the different solutes can be obtained by removing the column from its container, cutting it up into separate bands, and extracting the various solutes from the different bands. Alternatively, the solutes can be washed right through the column, one at a time, by passing a solvent down it; this is known as *elution*.

Such techniques, known as chromatographic analysis, were first employed by Tswett in 1906 to separate the various pigments present in leaves. The full possibilities of the method have been appreciated only since about 1935, and the technique is now very widely used, and not only for coloured substances, though the term chromatographic still remains.

If coloured substances are used, the separation achieved in the chromatogram can be observed visually. If colourless substances are being separated, ultra-violet illumination may show up the separation, or it may be necessary to use some chemical on the adsorbent to convert the colourless into coloured substances.

The adsorbent materials most widely used include aluminium oxide, calcium carbonate, magnesium oxide, charcoal, starch, calcium phosphate(v) and bentonite. The even packing of a column is an important matter. The solvents used, often mixed together, include propanone (acetone), petroleum ether, benzene, trichloromethane (chloroform), water and hexane. A satisfactory combination of adsorbent and solvent mixture can generally be found for any particular problem.

Porous paper may also be used as the adsorption medium in the technique known as *paper chromatography*. Ion-exchange resins (p. 499) are also used as adsorbents in special cases.

13. Applications of chromatography. Three examples of chromatographic analysis using, respectively, a column of aluminium oxide, paper and an ion-exchange column will illustrate the usage.

a. *Estimation of carotene in grass.* The pigments in a weighed sample of dry grass are extracted by warming with petroleum ether. The solution obtained is green, the yellow colour of the carotene being masked by the green of the chlorophyll.

A vertical glass tube with a pad of cotton wool and a clip at the bottom is packed with a 50–50 mixture of aluminium oxide and sodium sulphate(vi) by pouring in a slurry of the mixed solids with petroleum ether. The column is drained but not allowed to dry.

The cooled solution of pigments is poured on to the top of the column, whereupon a green band of adsorbed chlorophyll forms at the top of the column with a yellow band of adsorbed carotene below it. When all the pigment solution has been poured through the column, the carotene layer is eluted by passing a 2 per cent solution of propanone (acetone) in petroleum ether through. The carotene solution coming from the bottom of the column is collected, and the amount of carotene it contains is estimated colorimetrically.

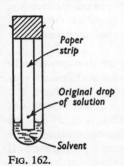

Paper strip

Original drop of solution

Solvent

FIG. 162.

b. *Separation of nickel and cobalt ions.* A small drop of an aqueous solution containing both Ni^{2+} and Co^{2+} ions is applied to a 1 cm wide strip of filter paper arranged as in Fig. 162. The solvent, a mixture of propanone (acetone), water and concentrated hydrochloric acid in the proportions 87:5:8 per cent by volume, rises up the paper. When the solvent front reaches nearly to the top of the strip, the paper is removed and dried. The separation of the Ni^{2+} and Co^{2+} ions cannot be seen at this stage, but it becomes visible when the paper strip is sprayed with, or dipped into, a solution of rubeanic acid. A yellow-brown spot, showing the position of the Co^{2+} ions, is seen well above a blue-violet spot, showing the position of the Ni^{2+} ions.

In other applications the solvent front may be arranged so as to pass down a sheet of paper from a trough, or radially outwards from the centre of a circle of paper.

c. *Separation of transuranium metals.* Intense and protracted neutron irradiation of a sample of Pu^{239} produces a highly complicated mixture of transuranium elements together with many fission products. The problem of separating individual isotopes from such a mixture is enormous, and, without chromatographic techniques, would probably have been insurmountable.

A solution of the transuranium metals in two drops of 0.05M by hydrochloric acid is passed through an ion-exchange column of Dowex-50 kept at 87°C by a vapour jacket. Elution by a solution of 0.4M ammonium 2-hydroxybutanoate gives drops of eluant which are collected, separately, on platinum plates. After evaporation to dryness the nature and quantity of material present on each plate is determined by investigation of the radiations from the material. As with the rare earths, or lanthanons, which can be separated in a similar way, the isotopes of higher relative atomic mass are eluted first. A plot of radioactivity against elution drop number therefore shows a series of peaks corresponding to fermium, einsteinium, californium, berkelium, curium and americium. It was in this way that such elements were first isolated and detected.

14. Gas or vapour chromatography. The type of chromatography described in the preceding sections depends on selective adsorption by a solid from a solution. In gas or vapour chromatography, the same principle is applied to the adsorption of gases or vapours by a liquid

A column of an absorbent powder, such as kieselguhr, is wetted with an involatile liquid such as a hydrocarbon oil, a silicone, or dibutyl benzene–1,2–dicarboxylate (phthalate). A steady flow of a carrier gas, such as nitrogen or hydrogen or carbon dioxide, is then passed through the column, heated if necessary, and the mixture of gases or vapours to be analysed is fed into the stream of carrier gas before it enters the column.

The weakly adsorbed constituents of the mixture are soon carried right through the column, whereas the strongly adsorbed ones pass through only very slowly. Each constituent will, in fact, pass through the column at a different rate, depending on the extent to which it is adsorbed, and will come out from the end of the column at different times. A detector at the end of the column shows the emergence of the various constituents of the original mixture which is thus analysed. The commonest type of detector is a thermal conductivity gauge. The gas emerging from the end of the column is passed round a heated filament. The heat loss from the filament depends on the thermal conductivity of the gas mixture surrounding it, and its electrical resistance, which is measured, changes as the gas mixture changes. The presence of different gases can, in this way, be detected.

Gas or vapour chromatography provides a rapid and cheap method for analysing such complex mixtures as are found in petrols and oils, in gas mixtures, e.g. Calor gas, and in many naturally occurring mixtures of biochemical importance. The method can be used even when only small quantities of mixtures are available for analysis.

QUESTIONS ON CHAPTER 45

1. List ten examples in which quite obvious adsorption can easily be demonstrated.

2. What do you suggest as the likely reason why the surface tension of liquids decreases as the temperature rises?

3. How would you expect the surface tension of a mixture of two miscible liquids to be dependent on the deviation of the two liquids from Raoult's law?

4. Test the validity of the Macleod and Eötvos equations for some common liquids.

5. Find out the details of any one example in which parachor values helped to solve a structural problem.

6. The amount of ethanedioic (oxalic) acid, in moles, adsorbed by 5 g of activated charcoal varies with the equilibrium concentration of acid according to the following figures:

Acid adsorbed/mol	0.29	0.60	0.75	0.90	1.05
Equil. conc. acid/mol dm^{-3}	0.030	0.080	0.115	0.175	0.260

Use these figures graphically to show the validity of the Freundlich isotherm, and to find the value of the constant n. Do the figures also fit the Langmuir isotherm?

7. The volumes of nitrogen, at s.t.p., adsorbed by 1 g of activated charcoal at the same temperature but different pressures are tabulated as follows:

Vol. of N_2	0.987	3.04	5.08	7.04	10.31
Pressure	3.93	12.98	22.94	34.01	26.23

Do these figures agree with the Langmuir isotherm?

8. How would you attempt to establish, experimentally, that the amount of gas adsorbed by charcoal was proportional to the surface area of the charcoal?

9. 500 cm^2 of a water surface is found to be covered by 0.106 mg of octadecanoic (stearic) acid, which has a relative molecular mass of 284 and a density of 0.85 g cm^{-3}. Assuming a close-packed and unimolecular film, estimate the cross-sectional area of a octadecanoic acid molecule, and the thickness of the surface film. What estimate would you make, from this data, of the film thickness to be expected from hexacosanoic (cerotic) acid?

10. Describe the experiments you would carry out to get a numerical measurement of the relative adsorptive powers for ammonia gas of a sample of powdered charcoal and a sample of powdered sulphur.

11. Give an account of the use of adsorption indicators.

12. Distinguish between the meaning of the terms adsorption, absorption, chemisorption and physical adsorption.

13. Explain why (*a*) drops of water are spherical, (*b*) alkalis are good emulsifiers for grease, (*c*) ducks cannot float in water containing too much detergent, (*d*) water will spread on clean glass and wet it, but it will not do so on a waxy or greasy surface.

14. Describe any applications of (i) chromatography, (ii) ion-exchange, which are not mentioned in this chapter.

Chapter 46
Colloids

1. Particle size. A true solution contains small solute particles (molecules or ions) dispersed throughout a solvent; the particles are invisible even under a microscope. A suspension contains very much larger insoluble particles, which may be visible to the naked eye and are certainly visible under a microscope.

Colloidal systems lie between the two extremes of a true solution and a suspension, and contain particles of intermediate size. These may consist of agglomerates of small particles, or of single particles such as large molecules of rubber, plastics, cellulose or proteins.

There is, however, no absolute line of demarcation so far as particle size is concerned. In general, the size of particle encountered in colloidal systems is between about 1 nm and 100 nm in diameter.

As a substance is broken down into smaller and smaller particles the total surface area increases, and surface effects (Chapter 45) become more and more important. Such effects play a predominant role in colloidal systems, and are responsible for the peculiar and distinctive properties of such systems.

The foundations of colloidal science were laid by Thomas Graham (1805–1869), the first President of the Chemical Society. He introduced most of the terms still used in describing colloidal systems.

2. Types of colloidal systems. The terms solute and solvent, commonly used in describing true solutions, are replaced, in colloidal chemistry, by *disperse phase* and *dispersion medium*. The disperse phase is the substance which is distributed, in small particles, throughout the dispersion medium. Just as in true solutions, solutes and solvents can be solids, liquids or gases, so can the disperse phase and the dispersion medium be of different phases, as in the summary of important types of colloidal systems in the table on page 508.

It will be noticed that two gases cannot form a colloidal system, as they are always completely miscible.

The most widely studied type of colloidal system is that in which the dispersion medium is a liquid and the disperse phase a solid, and the discussion in this chapter will be limited, mainly, to such systems. They may be subdivided into *lyophobic* (solvent-hating) colloids and *lyophilic* (solvent-loving) colloids, the latter type giving rise to gels or pastes under some conditions.

a. *Lyophobic sols*, e.g. a sulphur sol formed by reaction between a solution of sodium thiosulphate(VI) and an acid. In lyophobic sols there is little or no interaction between the disperse phase and the dispersion medium. As will be seen (p. 512), the solid particles are kept dispersed throughout the liquid phase because they are electrically charged. Once the solid particles in a lyophobic sol are coagulated and precipitated they cannot be reconverted into sols again simply by remixing with the dispersion medium. For this reason, lyophobic sols are sometimes called *irreversible sols*. They may also be referred to as *suspensoid sols*.

Dispersion medium	Disperse phase	Type of colloid
Gas	Liquid	Aerosol, fog, mist, cloud
Gas	Solid	Aerosol, smoke, dust
Liquid	Gas	Foam, e.g., froth on beer, soap suds, whipped cream
Liquid	Liquid	Emulsion, e.g., milk, rubber latex, salad dressing, haircream
Liquid	Solid	Sols or gels or pastes, e.g., some paints, fruit jellies
Solid	Gas	Solid foam, e.g., pumice, meringues
Solid	Liquid	Solid emulsion, e.g., butter
Solid	Solid	Solid sol or solid gel, e.g., some coloured glasses, wings of butterflies, pearls

b. *Lyophilic sols*, e.g. a sol formed from gelatine and water. In lyophilic sols there is a strong interaction between the disperse phase and the dispersion medium, the latter being strongly adsorbed by the former. It is the adsorbed layer of the dispersion medium on the disperse phase which stabilises the sols, any electrical charge which may exist playing a smaller part. If the disperse phase is separated from the dispersion medium, the sol can be remade simply by remixing. That is why lyophilic sols are called *reversible sols*. They are also called *emulsoid sols* because they are not unlike emulsions.

c. *Gels and pastes*. Lyophilic sols can, under the right conditions, form, or set into, semi-solid masses known as gels or pastes. These may form immediately the disperse phase and dispersion medium are mixed, particularly if the disperse phase is present in large quantities. Alternatively, they may form on cooling a hot sol, or on standing. A fruit jelly or a starch paste provide common examples.

There are a variety of different types of gels but they all have a certain weak rigidity which allows them to be deformed elastically by small forces. Under larger forces, reaching what is known as the yield value, a gel structure breaks up.

A paste is not unlike a gel but generally contains more solid colloidal particles. Examples are provided by Plasticine, putty, wet cements, oil paints, soils, clay and mud.

PREPARATION OF SOLS AND GELS

3. Lyophobic sols by condensation methods. In principle there is no reason why any insoluble solid cannot be obtained in the form of a lyophobic sol, but stable sols are not always particularly easy to prepare or keep. The dispersion medium may be any liquid, but water is most commonly used, the resulting sol sometimes being known as an *aquasol*.

There are two general ways in which lyophobic sols might be made. In condensation methods, atoms or molecules are brought together under conditions in which they can unite and build up into particles of colloidal size. In dispersion methods, coarse powders are broken down into colloidal particles. Clean vessels must always be used.

Typical examples of condensation methods are described below:

a. *By chemical reaction*. Many reactions in which one of the products is an insoluble solid can be carried out in such a way that the solid builds up into particles of colloidal size. Dilute solutions must generally be used and electrolytes must be almost entirely removed from the prepared sol by dialysis (p. 517), if the sol is to remain stable. Complete removal of electrolytes must be avoided as a little electrolyte is generally necessary to stabilise the sol.

The following sols can be made fairly easily by the reactions given:

Sol	Reaction
S sol	(a) $Na_2S_2O_3 + 2HCl \longrightarrow 2NaCl + SO_2 + S + H_2O$
	(b) $2H_2S + SO_2 \longrightarrow 2H_2O + 3S$
Fe(OH)$_3$ sol	$FeCl_3 + 3H_2O \longrightarrow Fe(OH)_3 + 3HCl$.
	Pour a few cm^3 of concentrated iron(III) chloride solution into a large excess of boiling water
As$_2$S$_3$ sol	$As_2O_3 + 3H_2S \longrightarrow As_2S_3 + 3H_2O$.
	Add a 1 per cent solution of As_2O_3 in water to an equal volume of a saturated solution of hydrogen sulphide. Remove excess H$_2$S by passing in hydrogen
Au sol	Reduce a 0.000 5 per cent solution of gold(III) chloride using 0.3 per cent methanal, Rochelle salt, yellow phosphorus and carbon disulphide, or 0.1 per cent tannin solution
Ag sol	Reduce ammoniacal silver nitrate(V) solution by dextrin

b. *By exchange of solvent*. The addition of an ethanolic solution of sulphur to excess water causes the sulphur to separate out as a very fine, and probably colloidal, precipitate, the sulphur being less soluble in the water–ethanol mixture than in ethanol alone. Similarly, addition of a solution of resin in ethanol to excess water gives a colloidal solution.

4. Lyophobic sols by dispersion methods.

a. *Colloid mill*. To break down even the most finely powdered substances into particles of colloidal size is not easy, the difficulty being in transmitting the necessary forces to the particles concerned. Numerous colloid mills have, however, been designed and patented for this purpose. A mixture of the substance to be dispersed and the dispersion medium is fed between two steel surfaces moving in opposite directions with only a small gap between them.

Such a mechanical method produces colloidal particles of widely varied size, and cannot give those of very small size, but it is a convenient industrial method of making colloidal solutions.

b. *Bredig's arc method.* Metallic sols can be made by striking an arc between two electrodes of the metal concerned immersed in the dispersion medium. The metal is vaporised by the arc and the vapour condenses and solidifies into particles of colloidal size. Platinum, gold, silver and other metallic sols can be made in this way.

Currents of 50 to 100 volt and 5 to 10 ampere are satisfactory. The arc should be made and broken to maintain intermittent arcing, and any large particles of metal can be removed by filtration.

A slight trace of electrolyte stabilises the sols formed.

c. *Peptisation.* Particles of a precipitate larger than colloidal size can sometimes be broken down into colloidal particles by adding an electrolyte to provide ions which are adsorbed by the precipitate. Addition of hydrogen sulphide to precipitates of cadmium or nickel(II) sulphides, for example, gives colloidal solutions of the sulphides. Similarly, a precipitate of aluminium hydroxide can be converted into a sol by adding aluminium chloride, and silver chloride precipitates can be broken down into colloidal particles by adding silver nitrate(v) or potassium chloride.

It is probable that the precipitate adsorbs one of the added ions, and that electrically charged particles then split off from the precipitate as colloidal particles.

The process is known as peptisation, the name originating from the fact that egg-white, partially coagulated by heating, can be reformed as a colloidal solution by adding pepsin (extracted from pig's or sheep's stomachs) and a trace of hydrochloric acid.

5. Preparation of lyophilic sols and gels

a. *Sols and gels of gelatine, agar-agar and starch.* Gelatine (extracted from hides, tendons and bones, containing the protein collagen, by boiling with water), agar-agar (extracted from seaweeds found in Japan and the East), and starch form sols simply on mixing with hot water. On cooling, gels are formed which can be reconverted into sols by warming. Cornflour and custard powder are commonly used to give sols and gels of this type.

b. *Pectin gels.* The white rinds of many fruits contain a carbohydrate-like substance called pectin, which may be extracted as a white powder. Pectin, together with weak acids such as 2,3–dihydroxybutane–1,4–dioic (tartaric) and 2–hydroxypropane–1,2,3–tricarboxylic (citric) acids, and sugar, forms gels which are of great importance in jam-making.

c. *Nitrocellulose sols and gels.* Mixtures of nitrocellulose and propanone (acetone) in different proportions form gels (with little propanone) or sols (with excess propanone). The gels are of the type used in such explosives as blasting gelatine.

d. *Rubber sols and gels.* Unvulcanised rubber swells up when placed in benzene or petrol to form a gel. Further addition of benzene or petrol will form a sol of the type generally called rubber solution.

e. *Soap gels.* A block of soap will turn into a gel if left standing in water. Similar gels can be made, too, by neutralising hot ethanolic solutions of fatty acids, such as octadecanoic (stearic) acid, with potassium hydroxide solution.

f. *Barium sulphate*(VI) *gel.* Precipitation of barium sulphate(VI) by mixing saturated solutions of barium thiocyanate and manganese(II) sulphate(VI) in equivalent amounts produces a gel of barium sulphate(VI).

g. *Calcium carbonate gel.* This can be made by mixing saturated solutions of calcium chloride and sodium carbonate.

h. *Sodium chloride gel.* Addition of sulphur diochloride oxide (thionyl chloride), a little at a time, to an equal amount of dry sodium 2–hydroxy-benzene carboxylate (salicylate), with frequent shaking, gives a greenish-yellow gel of sodium chloride. This is a very good example of a substance generally regarded as a crystalline solid being obtained in a colloidal form.

i. *Solid ethanol.* If 10 cm³ of a saturated solution of calcium ethanoate is shaken with 90 cm³ of ethanol, a gel, referred to as solid ethanol, is formed. It will not keep well, but can be stabilised by adding a little octadecanoic (stearic) acid.

j. *Silica gel.* Reaction between dilute hydrochloric acid and a solution of sodium carbonate produces carbonic acid,

$$Na_2CO_3 + 2HCl \longrightarrow 2NaCl + H_2CO_3$$

which, partially, splits up into carbon dioxide and water.

Sodium silicate solution (diluted water glass) reacts similarly, but the silicic acid formed is present in a colloidal state and the mixture readily forms a gel known as silica gel. The gel loses water on heating in a steam oven and the resulting hard, brittle mass can be broken up and used as a dehydrating agent, as a catalyst, and as a catalyst support. The product is very stable to heat and to attack by most chemical reagents.

PROPERTIES OF LYOPHOBIC SOLS

6. Optical Properties. Although the particles in a lyophobic sol cannot be seen by the naked eye, even under a microscope, they are of the right size to cause scattering of light. The visibility of a beam of sunlight in a dusty atmosphere is well known, and is caused by the small dust particles scattering the light.

A beam of light shows up, similarly, when passed through a lyophobic sol, the phenomenon being known as the *Tyndall effect*. The same beam of light passed through a true solution (free of any dust) is not scattered

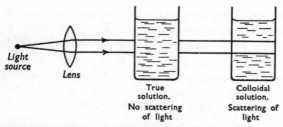

FIG. 163. Formation of Tyndall cone by a colloidal solution.

(Fig. 163). This is one way of distinguishing between a true solution and a lyophobic sol.

If the scattered light is viewed, from the side, through a microscope, points of light originating from the individual colloidal particles can be seen. Such an arrangement (Fig. 164) is known as an *ultra-microscope*, and was first used in studying colloids by Zsigmondy.

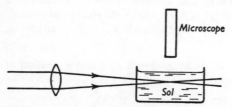

FIG. 164. Principle of the ultra-microscope.

The ultra-microscope shows that the particles in a lyophobic sol undergo Brownian motion (p. 79), and it can be used to obtain some information about the numbers of particles present and about particle size and shape, though any conclusions drawn about size and shape must be tentative as the particles are not being viewed directly.

7. Electrical properties. The stability of a lyophobic sol is due, predominantly, to the fact that the colloidal particles in the sol are electrically charged. The particles therefore repel each other and do not coagulate into larger non-colloidal particles, unless the repulsive forces are overcome or the electrical charges are neutralised.

As a lyophobic sol is much more stable if a trace of electrolyte is present, it is thought that the electrical charge on the particles is caused by the adsorption of ions from the dispersion medium. Either positive or negative ions can be adsorbed to give positive or negative colloidal particles.

The existence of the electrical charge has a marked effect on the properties of lyophobic sols.

a. *Electrophoresis or cataphoresis.* If two platinum electrodes, with a potential difference between them, are immersed in a lyophobic sol, the colloidal particles will move to the positive or negative electrode depending on whether the particles are negatively or positively charged.

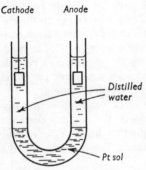

FIG. 165. Electrophoresis.

The movement of the particles in the electrical field can be observed by the ultra-microscope or, alternatively, by observing the boundary between a sol and its dispersion medium. In the arrangement in Fig. 165, for example, the right-hand boundary will move upwards as the negative particles in the platinum sol move towards the positive electrode.

This movement of colloidal particles in an electric field is known as electrophoresis or cataphoresis. The rate with which colloidal particles move depends on their charge, size and shape, but it is of the order of 2×10^{-6} m s^{-1} for a potential gradient of 100 V m^{-1}. The direction of movement can be used to determine whether a colloidal particle is positively or negatively charged.

Using water as the dispersion medium, the charge on the particles of some common sols is as follows:

Negatively charged	Positively charged
Metallic particles	Iron(III) hydroxide
Starch	Aluminium hydroxide
Clay	Haemoglobin
Silicic acid	Basic dyes
Arsenic(III) sulphide	

but particles with opposite charges can be obtained under different conditions (p. 516).

Tiselius, a Swedish scientist, has recently improved the experimental techniques used in studying electrophoresis and modern methods are capable of separating complex mixtures of proteins or amino acids. The separation is dependent on the differing charge of different proteins or amino acids in solution. The separation can be brought about between two electrodes immersed in a buffer solution on both sides of a solution of the mixed proteins or, alternatively, the solution may be subjected to an applied electrical field whilst adsorbed on paper. In the second method electrophoresis is combined with paper chromatography, a combination which has provided a very powerful means for separating amino acids.

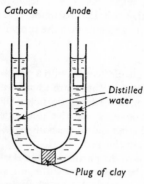

FIG. 166. Electro-osmosis.

b. *Electro-osmosis.* As a colloidal solution is, as a whole, electrically neutral, the charge on the colloidal particles must be balanced by an equal and opposite charge on the dispersion medium. Something in the nature of a Helmholtz double layer (p. 425) is probably set up, a negative charge on a particle being associated with a positive charge in the surrounding dispersion medium.

The movement of particles of the disperse phase towards the positive electrode in electrophoresis will, therefore, be accompanied by the movement of the dispersion medium towards the negative electrode. The movement of the

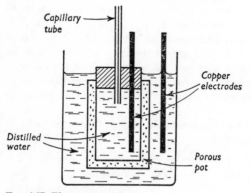

FIG. 167. Electro-osmosis.

dispersion medium is known as electro-osmosis. It can be demonstrated by using stationary colloidal particles, i.e. by preventing electrophoresis.

A plug of clay provides stationary, negatively charged colloidal particles (Fig. 166). Water, in contact with the clay, must carry a positive charge. Under the influence of an electric field, the water will be found to move away from the positive and towards the negative electrode.

A porous pot also provides a stationary medium which is colloidal-like in nature by virtue of its large surface area. If such a pot is assembled as in Fig. 167, application of the electric field in one direction causes a flow of water from inside to outside the pot. If the field is reversed, the flow of water is from outside to inside.

Electro-osmosis and electrophoresis can be shown taking place together by placing two glass tubes into a

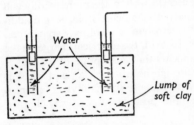

FIG. 168. Electro-osmosis and electrophoresis.

soft lump of moist clay. Water is added to each tube, and electrodes inserted (Fig. 168). On applying a potential difference between the electrodes, clay (negatively charged) will be found to move up one tube towards the positive electrode, whilst water (positively charged) will move up the other tube towards the negative electrode.

c. *Streaming or sedimentation potential.* Motion is produced in electrophoresis and electro-osmosis by the application of an external potential difference. In the reverse effects, a potential difference can be caused by motion.

If, for example, a liquid is made to pass through a capillary tube or through the capillary pores in, say, a porous pot, a potential difference, known as the *streaming potential*, is set up. It can be detected by placing electrodes at the inlet and outlet ends of the capillary.

Similarly, a potential difference, known as the *sedimentation* or *centrifugation potential*, is set up if solid particles are made to pass through a liquid by sedimention or by the application of a centrifugal field. This is sometimes called the Dorn effect.

All these effects in which motion and potential difference are related are known, collectively, as *electro-kinetic phenomena*.

d. *Coagulating effect of electrolytes.* The presence of small traces of electrolyte, i.e. of some ions, is essential for the preparation of a stable lyophobic sol as it is the adsorption of such ions by the colloidal particles which, in most cases, leads to the electrical charge on the particles.

Addition of more ions to a lyophobic sol, however, invariably causes coagulation, the effect depending on the nature of the colloid and on the concentration and nature of the added electrolyte. A negatively charged ion causes the coagulation of a positively charged colloidal particle, and vice versa. For equal concentrations, the coagulating effect depends on the valency of the ion causing the coagulation; this is the *Hardy–Schulze rule*. Al^{3+} ions, for instance, will coagulate negatively charged sols much more effectively than bivalent ions, and bivalent ions are more effective than monovalent ones. Similarly SO_4^{2-} ions coagulate a positive colloid more readily than Cl^- ions. Mixing negative and positive colloids also causes mutual coagulation.

The coagulation of colloidal particles is thought to be due to adsorption of the coagulating ion. Such adsorption will decrease the electrical charge on the colloidal particles until, at the *iso-electric point*, the particles carry no charge and coagulate. That such a mechanism is likely is shown by the fact that the electrophoretic velocity of a colloidal particle decreases as an electrolyte is added until it becomes zero at the iso-electric point.

The greater the concentration of added electrolyte, the more readily will the electrical charge on the colloidal particles be neutralised. Rapid addition of a concentrated solution of an electrolyte to a sol may sometimes, however, not bring about coagulation. Instead, the electrical charge on the colloidal particles is reversed and it appears that adsorption of ions takes place so quickly that the sol passes through its iso-electric point without coagulation. Addition of concentrated aluminium sulphate(VI) solution to a negatively charged gold sol, for example, produces a positively charged gold sol.

Examples in which electrolytes are used to bring about coagulation of sols are provided by (i) the use of alum (a cheap source of Al^{3+} ions) in treating sewage and dirty swimming-bath water, and in styptic pencils which are used to make blood clot, (ii) the formation of deltas when the colloidal particles present in fresh river water are coagulated by the higher concentration of electrolyte in sea-water, (iii) the coagulation of rubber latex by adding methanoic (formic) acid, (iv) the curdling of milk by the 2–hydroxypropanoic (lactic) acid formed as the milk goes sour.

e. *Electrostatic precipitation of smokes or mists.* If a smoke or a mist is passed between highly charged plates, the colloidal particles are deposited. The process is used industrially, under the name of the Cottrell or Cottrell–Möller process for the deposition of smoke, dust and mist-like fumes. Sulphur dioxide is purified in this way, for example, in the Contact process (p. 326). A potential difference of about 50 kV is used.

f. *Adsorption indicators.* Addition of silver nitrate(V) solution to a sodium chloride solution produces a precipitate of silver chloride which is partly in colloidal form. As soon as any excess silver nitrate(V) is added to the mixture, the precipitate preferentially adsorbs Ag^+ ions and becomes positively charged. If a dye, such as fluorescein, is present, the positively charged precipitate will adsorb the fluorescein anion. As a result, the precipitate becomes pink in colour, and, sometimes, coagulates.

Because of this adsorption, fluorescein can be used as an adsorption indicator in the titration of solutions of chlorides, bromides, iodides and thiocyanates against silver nitrate(V). When first added to the solution in the conical flask, the fluorescein imparts a yellowish-green colour. As silver nitrate(V) solution is run in, a white precipitate of silver salt forms and, at the end point, this precipitate turns sharply pink, and may coagulate.

Eosin, dichlorofluorescein and diphenylamine blue are also commonly used as adsorption indicators.

8. Diffusion of lyophobic sols. Dialysis. A solute will diffuse from a true solution into the pure solvent or into a weaker solution. A layer of water, for example, can be placed, with care, on top of a copper(II) sulphate(VI) solution. On standing, the blue colour of the copper(II) sulphate(VI) will diffuse upwards into the water layer (p. 280).

The rate of diffusion of colloidal particles is very, very much slower than that of true solutes, and the boundary between, say, an arsenic(III) sulphide sol and water will remain sharp for a very long time. This is because colloidal particles are very much bigger than the solute particles in a true solution.

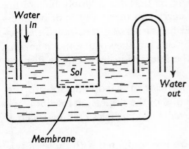

FIG. 169. Dialysis.

The marked difference in the rate of diffusion of colloidal particles and smaller solute molecules or ions can be used for separating sols from true solutions in the process known as dialysis.

An iron(III) hydroxide sol made by the hydrolysis of iron(III) chloride will be mixed with some hydrochloric acid. If the mixture is placed in a container made of cellophane or parchment paper or pig's bladder, and the container is dipped into running water, the acid will diffuse through the membrane and can be detected in the running water. The iron(III) hydroxide particles will diffuse only very, very slowly so that a purified sol will remain in the container. The apparatus is called a dialyser (Fig. 169).

Graham (p. 507) classified substances which would diffuse through parchment paper as crystalloids and those which would not as colloids, but this classification is, nowadays, of little value as the same substance can, in many cases, be obtained in both crystalline and colloidal states.

9. Osmotic pressure of lyophobic sols. The osmotic pressure of a solution depends on the number of solute particles present (p. 263), and a solution containing 1 mol (6.02252×10^{23} molecules) of undissociated and un-associated solute in 22.4 dm³ of solution will have an osmotic pressure of 101.325 kN m⁻² at 0°C.

If the solute molecules are associated two-fold, the number of particles will be halved and the resulting osmotic pressure will be halved, too. In colloidal solutions the particles may consist of very many molecules 'associated' together so that the osmotic pressures are very low.

From measured osmotic pressures of sols it is possible to obtain values of particle mass just as relative molecular mass values can be obtained for true solutions (p. 288). The accurate measurement of the osmotic pressure of a sol is, however, difficult, as a slight trace of any true solute impurity introduces large errors.

10. Viscosity of lyophobic sols. The particles in most lyophobic sols are approximately spherical in shape and there is no interaction between the particles and the dispersion medium. As a result, the viscosity of a lyophobic sol is only very slightly greater than that of the pure dispersion medium.

PROPERTIES OF LYOPHILIC SOLS

11. Comparison with lyophobic sols. The particles in a lyophobic sol are stabilised by the electrical charge they carry and there is little or no interaction between the disperse phase and the dispersion medium. In a lyophilic sol, electrical charges may be present, but they play a much smaller part in rendering the sol stable than the interaction between the disperse phase and the dispersion medium.

It is considered that the colloidal particles in a lyophilic sol are surrounded by an adsorbed layer of molecules of the dispersion medium, and that it is the adsorbed layer which prevents the particles coming together and coagulating.

The general properties of lyophilic sols are compared with those of lyophobic sols below, and a summarising table is given on p. 579.

a. *Optical properties.* In a lyophobic sol the scattering of light by the colloidal particles can only be observed because these particles have a different refractive index from that of the dispersion medium. A lyophilic sol is much more optically homogeneous, the refractive index of the colloidal particles, with their adsorbed layer, being much the same as that of the dispersion medium. Consequently, lyophilic colloids do not exhibit the Tyndall effect at all markedly.

b. *Electrical properties.* If the particles in a lyophilic sol are electrically charged then electrophoresis and electro-osmosis will take place, but many lyophilic sols do not undergo any change in an electric field because they do not carry any electrical charge.

c. *Coagulation of lyophilic sols.* The removal of the electrical charge on a lyophobic colloidal particle, by adsorption of an oppositely charged ion, causes coagulation, and such coagulation can be brought about by low concentrations of added electrolyte. Low concentrations of electrolyte will not, however, coagulate lyophilic sols, and, moreover, the Hardy–Schulze rule does not apply.

To coagulate a lyophilic sol it is necessary to remove or decrease the adsorbed layer of dispersion medium on the disperse phase, by adding something which will compete for the molecules of the dispersion medium. When the dispersion medium is water, this can generally be done by adding a lot of an electrolyte. The ions present become hydrated and, in this way, lower the number of water molecules available for adsorption on the colloidal particles. It is probable that the use of salt in the salting-out of soap (p. 382) helps to coagulate the soap in this way.

Alternatively, addition of alcohol facilitates coagulation by lowering the concentration of water molecules in the dispersion medium.

d. *Diffusion. Dialysis.* Lyophilic sols behave in the same way as lyophobic sols so far as diffusion and dialysis are concerned.

e. *Osmotic pressure.* The osmotic pressure of lyophilic sols is low, as for lyophobic sols.

f. *Viscosity.* The marked interaction between the disperse phase and the dispersion medium in a lyophilic sol causes the sol to have a higher viscosity than the pure dispersion medium, even at low concentration. At higher concentrations the viscosity increases very greatly until a gel is formed.

12. Comparison of lyophobic and lyophilic sols. The main properties of lyophobic and lyophilic sols are compared in the table below.

Lyophobic sols	*Lyophilic sols*
Colloidal particles electrically charged	Particles may or may not be charged
Stability due to electrical charge	Stability due to adsorption of dispersion medium
Not prepared by direct mixing. Irreversible	Often prepared by direct mixing. Reversible
Exhibit Tyndall effect	Do not exhibit Tyndall effect
Undergo electrophoresis and electro-osmosis	Undergo electrophoresis and electro-osmosis if electrically charged
Coagulated by low concentrations of electrolytes	Coagulated by high concentrations of electrolytes
Very low rates of diffusion. Can be separated from true solutes by dialysis	
Viscosity not much greater than that of dispersion medium	Viscosity noticeably higher than that of dispersion medium
Do not generally form gels	Commonly form gels
Low osmotic pressure	Low osmotic pressure

13. Protective action of lyophilic colloids. The coagulating effect of an electrolyte on a lyophobic sol is markedly decreased by the addition of a lyophilic colloid to the lyophobic sol.

It is supposed that the lyophilic material is adsorbed by the colloidal particles of the lyophobic sol, the new particle being, in effect, lyophilic. The lyophobic sol treated in this way is said to be protected, and the lyophilic colloid used is known as a protective colloid.

Various methods have been used to measure the protective action of different materials quantitatively, the commonest being the use of the *gold number*, introduced by Zsigmondy. This is defined as the mass in milligrammes of the protective colloid which must be added to 10 cm³ of a specially prepared red gold sol just to prevent a change to violet (i.e. an increase in particle size) on the addition of 1 cm³ of 10 per cent sodium chloride solution. The gold number of gelatine lies between about 0.005 and 0.01; that of potato starch is about 25.

The use of protective colloids to stabilise colloidal systems is widespread. Typical examples are provided by the addition of egg-white (egg albumen) to olive oil and vinegar in making mayonnaise, and the use of gelatine in ice-cream to prevent the formation of small ice crystals.

14. Sedimentation. The ultra-centrifuge. Solid particles suspended in a liquid will fall or rise depending on the density of the particles. The fall of suspended particles is known as sedimentation; the rise, as creaming.

When the particles concerned are very small, as in colloidal solutions, the Brownian motion will be sufficient to prevent any great sedimentation under the influence of gravity. By using a centrifuge, however, the force exerted on the particles can be greatly increased. In order to study the sedimentation of colloidal solutions, Svedberg developed an ultra-centrifuge with a rotational speed high enough to produce forces over 10^5 times greater than the force of gravity.

Svedberg used the ultra-centrifuge to measure rates of sedimentaion for colloidal particles and, by applying a form of Stokes's law (p. 93), developed one of the most reliable methods for measuring the relative molecular masses of colloids.

Perrin also made use of the sedimentation of colloidal-size particles. By studying the density distribution of such particles after they had been allowed to sediment under the force of gravity, he was able to obtain a value for the Avogadro number.

QUESTIONS ON CHAPTER 46

1. What is the total surface area of a cube of side 1 cm? What is the surface area when the same cube is divided into cubes of side 10^{-6} cm?

2. If the average diameter of gold particles in a sol containing 1 g of gold per dm³ is 4 μm, and the density of gold is 20 g cm⁻³, calculate the approximate number of gold particles per dm³. What osmotic pressure would you expect this sol to have, and what would be its freezing point?

3. If 1 cm³ of gold is completely subdivided into small spheres of radius 2.5 nm, how many spheres will there be and what will be the total surface area?

4. How would you prepare colloidal solutions of (*a*) arsenic(III) sulphide, (*b*) a metal such as gold or platinum? Either describe a method by which the sign of the electrical charge on the particles of a colloid could be determined, or indicate briefly the part played by the electric charge in colloidal phenomena. (O. & C. S.)

5. Describe in detail how you would prepare (*a*) one lyophilic, and (*b*) one lyophobic colloidal solution. What experiments would you use to show three essential differences in character between the two colloidal solutions whose preparation you describe? (O. & C. S.)

6. Write short notes on the following: (*a*) Brownian motion, (*b*) the ultra-microscope, (*c*) protective action, (*d*) gold number.

7. What is the Hardy–Schulze rule? How would you demonstrate its application experimentally?

8. Describe the essential features by which you would recognise a solution as colloidal. (O. & C. S.)

9. Give concisely the meanings of four of the following terms: colloidal solution, dialysis, coagulation, electrophoresis, gel, Tyndall cone. (O. & C.)

10. Explain with examples how you would demonstrate three important properties of colloidal solutions. State whether the properties to which you refer are characteristic of lyophobic colloids, lyophilic colloids, or both. State briefly how you would prepare in a colloidal state one metal and one non-metal or compound. (O. & C.)

11. What do the following terms signify: coagulation, peptisation, endosmosis, Tyndall cone? How would you make a colloidal solution of iron(III) hydroxide and how would you determine whether the charge on the colloidal particles was positive or negative? (O. & C.)

12. What will be the osmotic pressure at 25°C of a solution containing 20 g dm^{-3} of a protein of relative molecular mass 69×10^3?

13. What would you observe if a mixture of a solution of starch and potassium iodide was separated from chlorine water by cellophane in a dialyser?

14. Describe and explain two ways in which the nature of the electrical charge on the particles in an iron(III) hydroxide sol can be determined.

15. Coagulation of colloids is caused by heat, salt solutions, acids or alkalis, electrical precipitation and mixing with an oppositely charged colloid. Illustrate this statement, using specific examples. Is the statement invariably true?

16. Write short notes on the following: peptisation, protective action, dialysis, Brownian motion.

17. What is the Tyndall effect? Why are motor car headlights less effective in a fog than on a clear night? Why are fog lights usually yellow?

18. Compare and contrast, in three columns, the general properties of (i) a solution of sucrose in water, (ii) a suspension of calcium carbonate in water, (iii) an aquasol of sulphur.

19. How would you obtain crystals of pure sodium chloride from a mixture of sodium chloride and starch.

20. Comment on the following: (a) A stable gold sol is obtained by using an organic reducing agent such as tannin whereas the sol obtained using an inorganic reducing agent is less stable. (b) Butter churns more readily from sour than from sweet cream. (c) The electrophoretic velocity of a sol is lowered by adding an electrolyte.

21. If you had to write a chapter on Colloids, what section headings would you choose? Give, in note form, the content of each section.

22. Peptisation is often a nuisance in analysis. Comment.

23. Explain the different principles used in the operation of (i) an optical microscope, (ii) an ultra-microscope, (iii) an electron microscope.

Chapter 47
The phase rule

1. Terms used. The phase rule, put forward by J. Willard Gibbs, between 1875 and 1878, as a result of precise and complicated mathematical and thermodynamic studies, is concerned with the conditions under which equilibria can exist in heterogeneous systems. In particular, it is concerned with the effects which changes of temperature, pressure and concentration will have on the equilibrium of a heterogeneous system. It shows that a simple, unifying principle holds for all systems in equilibrium.

The rule states that the number of phases (P), the number of components (C), and the number of degrees of freedom (F) in a system in equilibrium are always related by the expression

$$F = C + 2 - P$$

To understand the meaning of this it is first necessary to clarify the meanings of the terms phase, component and degree of freedom.

a. *Phases.* A heterogeneous system will consist of various homogeneous parts in contact with each other but with distinct boundaries. The various homogeneous parts, which will also have distinct physical and chemical properties, and which may, if necessary, be mechanically separated from each other, are known as phases.

A phase may, therefore, be defined as *any part of a system which is* (a) *homogeneous and separated from other parts of the system by a distinct boundary,* (b) *physically and chemically different from other parts of the same system, and* (c) *mechanically separable from other parts of the system.*

The following examples will illustrate the use of the term:

System	Phases
Mixtures of gases which do not react	1 gaseous phase (all gases are completely miscible)
Ice and water	2 phases; 1 solid and 1 liquid
Water and water vapour	2 phases; 1 liquid and 1 gaseous
Ice, water and water vapour	3 phases; 1 solid, 1 liquid and 1 gaseous
Two immiscible liquids	2 phases; both liquid
Two miscible liquids	1 liquid phase
Solid naphthalene and naphthalene vapour	2 phases; 1 solid and 1 gaseous
Calcium carbonate, calcium oxide and carbon dioxide	3 phases; 2 solid and 1 gaseous
Monoclinic and rhombic sulphur	2 solid phases

It will be seen that 2-phase systems may consist of solid plus liquid, solid plus gas, liquid plus gas, two liquids or two solids. Miscible gases and miscible liquids constitute a single phase; different solids, except in solid solution (p. 261), are regarded as different phases however intimately they may be mixed.

b. *Components. The components of a system are the least number of separate substances which must be chosen to define fully the composition of* **every** *phase of the system.*

It is important to realise that the components of a system may, or may not, be the same as the actual substances present in the system, i.e. the constituents of the system. Only those constituents of an equilibrium mixture which can undergo *independent* variation are regarded as components.

In the system ice–water–water vapour the only component is water; it is a one-component system. Hydrogen and oxygen are not components of the system because they are not present, as such, in the equilibrium mixture.

In the system calcium carbonate–calcium oxide–carbon dioxide there are three constituents, but only two components. This is because the amount of the third constituent can be expressed in terms of the other two. If calcium oxide and carbon dioxide be chosen, for instance, as the components, the amount of calcium carbonate can be expressed as so much calcium oxide plus so much carbon dioxide, i.e.

$$CaCO_3 = CaO + CO_2$$
(Components)

Alternatively, if calcium carbonate and carbon dioxide be chosen as the components, the amount of calcium oxide can be expressed as so much calcium carbonate minus so much carbon dioxide, i.e.

$$CaO = CaCO_3 - CO_2$$
(Components)

The number of components in any system is fixed and definite, but a choice is allowable in the actual selection of the components, the simplest choice generally being made. For the calcium carbonate–calcium oxide–carbon dioxide system, it is simplest to choose calcium oxide and carbon dioxide as this avoids using negative quantities.

c . *Degrees of freedom. The number of degrees of freedom of a system is the smallest number of variable factors, such as temperature, pressure or concentration of the components, which must be fixed in order that the system can remain permanently in one position of equilibrium.*

The equilibrium between a liquid and its vapour, for example, depends on temperature and on pressure; the amount of liquid present, so long as there is enough to establish a liquid–vapour equilibrium, does not affect the issue. If the temperature is fixed, the vapour pressure is also fixed. Conversely, fixing the pressure fixes the temperature, for a liquid and its vapour can only be in equilibrium under a given pressure at one particular temperature. A liquid–vapour system has, therefore, only one degree of freedom; it may also be described as *univariant*.

For a gas, temperature, pressure and concentration (volume) can all be varied. If only one of these three variable factors is fixed, the other two can still be varied. A gas at a fixed temperature, for example, can still exist under different pressures at different volumes. But if two of the three variables are fixed, the third will also be fixed. A fixed volume of gas at a fixed temperature will have a fixed pressure. A gas, therefore, has two degrees of freedom or is *bivariant*.

ONE-COMPONENT SYSTEMS

2. A gas. The phase rule states that, for every system in equilibrium, the number of degrees of freedom is equal to the number of components plus 2, minus the number of phases,

$$F = C + 2 - P$$

For a single gas, there is one phase and one component. The number of degrees of freedom is, therefore, two, which is in agreement with the conclusion reached at the end of the preceding section.

3. Ice plus water plus water vapour. In a mixture of ice and water, there are two phases and one component. The number of degrees of freedom is, therefore, 1. This means that there is only one temperature at which ice and water can exist in equilibrium at a fixed pressure, or only one pressure at which they will co-exist at a fixed temperature.

In the system ice–water–water vapour there are three phases and one component, giving no degrees of freedom. The system is said to be in-variant. This means that there are no variable factors in this three-phase system. The three phases ice, water and water vapour can, in fact, only exist together in equilibrium at one fixed temperature (0.01°C, p. 36) and one fixed pressure (610.6 N m^{-2}). Even a slight alteration of temperature or pressure will cause one of the three phases to disappear.

The various equilibria which can exist when water is held at various pressures and temperatures can be summarised in a pressure–temperature diagram, as in Fig. 170, which, in order to show the various areas more clearly, is not drawn to scale.

Line OA is the vapour pressure–temperature curve for water; it shows the pressure of water vapour which can exist in equilibrium with liquid water at different temperatures. The upper limit of OA, at A, is the critical point (p. 74) beyond which the liquid phase is no longer distinguishable from the vapour phase.

Line OB is the sublimation pressure–temperature curve for ice, showing the pressure of water vapour which can exist in equilibrium with solid ice. The lower limit of OB will be at 0 K.

Line OC shows the effect of pressure on the melting point of ice or the freezing point of water. So far as is known there is no upper limit to the line, i.e. no point beyond which solid and liquid are indistinguishable.

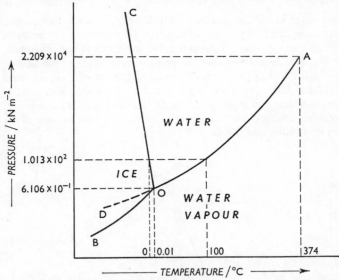

FIG. 170. Pressure–temperature curves for the water system. (Not to scale.)

At O, which is called a *triple point*, solid, liquid and vapour are in equilibrium. The system, at this point, has no degrees of freedom; it is invariant.

Along the lines OA, OB and OC, two phases may coexist in equilibrium; water and water vapour along OA, water and ice along OC, and water vapour and ice along OB. At any point on any of these lines, the system has one degree of freedom.

Within the areas AOC, AOB and BOC only one phase can exist, and at any point within these areas the system has two degrees of freedom. The state of affairs may be summarised as

Position on diagram	Phases in equilibrium	Degrees of freedom
O. Triple point	Solid, liquid and vapour	0
Along OA	Liquid and vapour	1
Along OC	Liquid and solid	1
Along OB	Solid and vapour	1
Within AOC	Liquid phase only	2
Within AOB	Vapour phase only	2
Within BOC	Solid phase only	2

Liquid water can be cooled below its freezing point without solidifying; this is known as *super-cooling*. When water is super-cooled, the line AO is extended along OD, but the system existing at any point on OD is meta-stable (p. 217).

The slope of the line OC shows that the melting point of ice decreases as the pressure increases. This is rather unusual, and is only shown by bismuth and antimony amongst other common substances, but it has important consequences. The decrease of melting point with increase of pressure is related, by Le Chatelier's principle, to an increase in volume, i.e. decrease in density, on solidification. If pressure is applied to ice it will melt, because the decrease in volume on melting will tend to relieve the applied pressure. Such considerations show why ice floats in water, and have an important bearing on ice-skating and the flow of glaciers.

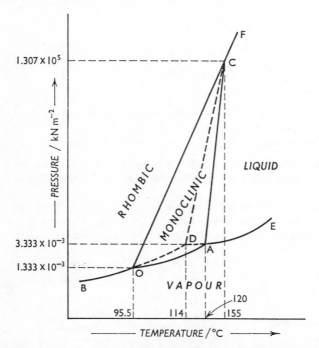

FIG. 171. Pressure–temperature curves for the sulphur system. (Not to scale.)

4. Sulphur. The water–ice–water vapour system is a three-phase system. Sulphur provides an example of a four-phase system, because of the existence of two solid allotropes, rhombic and monoclinic sulphur, representing two solid phases. A full consideration of the sulphur system is complicated by the fact that liquid allotropes of sulphur also exist. In this treatment the conditions under which rhombic, monoclinic or liquid sulphur can or cannot coexist with sulphur vapour are studied.

The pressure–temperature diagram is given in Fig. 171 in a simplified form, and not to scale. What the various lines and points represent is summarised below:

Line or point	What it represents	Phases in equilibrium	Degrees of freedom
AO	Vapour pressure curve of monoclinic S	Monoclinic S and S vapour	1
AE	Vapour pressure curve of liquid S	Liquid S and S vapour	1
BO	Vapour pressure curve of rhombic S	Rhombic S and S vapour	1
CO	Effect of pressure on transition point between rhombic and monoclinic S	Rhombic S and monoclinic S	1
AC	Effect of pressure on m.pt. of monoclinic S	Monoclinic S and liquid S	1
CF	Effect of pressure on m.pt. of rhombic S	Rhombic S and liquid S	1
Point O (95.5°C and 1.333 N m^{-2})	Triple point for rhombic S, monoclinic S and S vapour	Rhombic S, monoclinic S and S vapour	0
Point A (120°C and 3.333 N m^{-2})	Triple point for monoclinic S, liquid S and S vapour	Monoclinic S, liquid S and S vapour	0
Point C (155°C and 130.7 MN m^{-2})	Triple point for monoclinic S, rhombic S and liquid S	Monoclinic S, rhombic S and liquid S	0

Within the areas FCAE, EAOB, FCOB and COA only one phase can exist as shown in the diagram. Monoclinic sulphur can only exist by itself, for instance, within the area OAC.

The dotted lines in the diagram represent metastable systems which, in many cases, can be obtained quite readily because the attainment of equilibrium in the sulphur system is slow. If rhombic sulphur is heated quite rapidly, for example, it will follow the curve BOD rather than BOA, OD being the metastable vapour pressure curve of rhombic sulphur. At D, 114°C, the rhombic sulphur will melt, and then follow the curve DAE, DA being the metastable vapour pressure curve of super-cooled liquid sulphur. DC represents the metastable melting point curve for rhombic sulphur, and point D a metastable triple point for rhombic sulphur, liquid sulphur and sulphur vapour.

It will be noticed that rhombic sulphur, monoclinic S, liquid sulphur and sulphur vapour never exist altogether. If they did, there would be four phases and one component, which, by application of the phase rule

$$F = C + 2 - P$$

would lead to the impossible situation of having a number of degrees of freedom equal to minus 1.

Unlike ice, monoclinic sulphur melts with an increase in volume, i.e. monoclinic sulphur has a higher density than liquid sulphur. The effect of increased pressure on the melting point of monoclinic sulphur, therefore, causes a rise of melting point. Hence the slope of the line AC. Similarly, it is the fact that rhombic sulphur has a higher density than monoclinic sulphur that accounts for the slope of the line OC.

5. Sublimation. The direct change from a solid to a vapour, or vice versa, is known as sublimation. Every solid exerts a vapour pressure, but it is usually very small. It may not be small, however, and it is obvious enough in solids such as naphthalene, which evaporate quite markedly, and in the

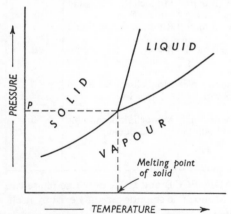

Fig. 172. Sublimition can take place if P is greater than the external pressure on the system.

recognisable smell of many solids. Like the vapour pressure of liquids, that of solids increases with temperature, and if the vapour pressure of a solid reaches atmospheric pressure before the solid melts then the solid will, in fact, vaporise before it melts, i.e. it will sublime at atmospheric pressure.

Substances which will sublime at atmospheric pressure, e.g. naphthalene, iodine, and solid carbon dioxide, can be liquefied at higher pressures. It is only necessary for the melting point of a solid to be reached before its vapour pressure becomes equal to the external pressure for sublimation to be impossible.

Alternatively, solids which do not sublime at atmospheric pressure, may do so at lower pressures.

If the value of P, the triple-point pressure, in Fig. 172, is greater than the external pressure on the system then the vapour pressure of the solid can reach that external pressure before the solid melts, i.e. the solid can sublime.

TWO COMPONENT SYSTEMS

Mixtures of a solid and a liquid, two liquids (completely or partially miscible, or immiscible), or two solids provide examples of two-component systems in which the conditions necessary for equilibrium between various phases raise many interesting questions. Such systems have, however, been considered already by regarding them as solutions.

In this section, two examples of equilibrium between solid and vapour phases of two-component systems will be considered.

6. Thermal decomposition of calcium carbonate (p. 318). On heating calcium carbonate, calcium oxide and carbon dioxide are formed. This is a two-component system, as explained on p. 523, so that

$$F = 4 - P$$

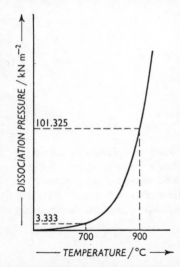

FIG. 173. The dissociation pressure of calcium carbonate.

If the two solid phases, calcium carbonate and calcium oxide, are present, along with gaseous carbon dioxide, there will be three phases, and one degree of freedom. If the temperature is chosen, pressure will be fixed, and the pressure of carbon dioxide in an equilibrium mixture, sometimes known as the *dissociation pressure* of calcium carbonate, rises with temperature in much the same way as the vapour pressure of a liquid, as shown in Fig. 173. The dissociation pressure is 101.325 kN m^{-2} at a temperature of about 900°C. At, or above, this temperature calcium carbonate dissociates much more rapidly than at lower temperatures, on being heated in an open vessel. At 700°C, the dissociation pressure is 3.333 kN m^{-2}. If carbon dioxide is added to a sample of calcium oxide at 700°C, no carbon dioxide will react and no calcium carbonate will be formed until the gas pressure reaches 3.333 kN m^{-2}.

7. Salt hydrates. A saturated solution of copper(II) sulphate(VI) will be in equilibrium with water vapour, and the vapour pressure, at 50°C, is 12 kN m^{-2}. If water vapour is removed, by a vacuum pump, crystals of $CuSO_4 \cdot 5H_2O$ will form and when all the water has been removed from the liquid phase the solid phase, consisting of these crystals, will itself be in equilibrium with water vapour. The vapour pressure, at this stage, at 50°C, will be about 6.27 kN m^{-2}, Further loss of water will form lower hydrates, with correspondingly lower vapour pressures, as summarised in Fig. 174.

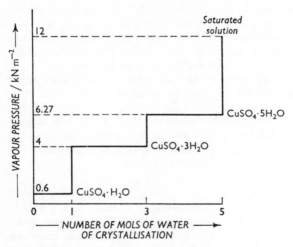

Fig. 174. The vapour pressure of various hydrates of copper(II) sulphate(VI) at 50°C.

Thus, as water is pumped away from a saturated solution of copper(II) sulphate(VI) the vapour pressure will undergo a series of abrupt changes, the various vapour pressures recorded depending on the temperature.

A saturated solution of copper(II) sulphate(VI) will consist of three phases—solid $CuSO_4 \cdot 5H_2O$, liquid and water vapour, and two components copper(II) sulphate(VI) and water. Thus

$$F = 4 - 3$$

giving one degree of freedom. At a fixed temperature, the vapour pressure will also be fixed, but the vapour pressure will increase with rising temperature as shown in Fig. 175, line 4.

$CuSO_4 \cdot 5H_2O$ crystals will also lose water, at certain temperatures, to set up an equilibrium between $CuSO_4 \cdot 5H_2O$ crystals, $CuSO_4 \cdot 3H_2O$ crystals and water vapour. There will still be three phases, and one degree of freedom, and the lowered vapour pressure of this system will vary with temperature as shown in line 3 of Fig. 175. Similarly the vapour pressure curves of the systems $CuSO_4 \cdot 3H_2O-CuSO_4 \cdot H_2O$–water vapour and $CuSO_4 \cdot H_2O-CuSO_4$–water vapour are shown in lines 2 and 1 in Fig. 175.

At points on each of the lines in Fig. 175 three phases coexist in equilibrium. Below curve 1, anhydrous copper(II) sulphate(VI) and water vapour coexist; the water vapour pressure is too low to form any monohydrate. Between curves 1 and 2, the monohydrate and water vapour coexist, and so on.

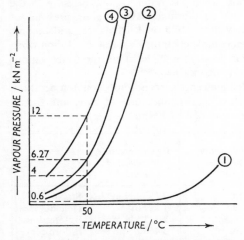

FIG. 175. Pressure–temperature curves for various hydrates of copper(II) sulphate(VI)

8. Deliquescence and efflorescence. Solid $CuSO_4 \cdot 5H_2O$ and solid $CuSO_4 \cdot 3H_2O$ can only exist together, in equilibrium with water vapour, at one pressure of water vapour if the temperature is fixed. If the actual water vapour pressure in the atmosphere is lower than this equilibrium water vapour pressure the pentahydrate will pass over into the trihydrate. In other words the pentahydrate will lose some of its water of crystallisation to the atmosphere, i.e. it will effloresce.

The pentahydrate will, therefore, effloresce, at 50°C, when the water vapour pressure in the atmosphere surrounding it is below 6.27 kN m^{-2}. If the external water vapour pressure is above 12 kN m^{-2} at 50°C, the pentahydrate will absorb water vapour from the atmosphere, i.e. it will deliquesce. At 50°C, the pentahydrate is stable, and will neither effloresce nor deliquesce, in atmospheres with water vapour pressure between 6.27 and 12 kN m^{-2}.

At 20°C, the vapour pressure of the pentahydrate is about 670 N m^{-2} and that of a saturated solution of copper(II) sulphate(VI) is about

2.133 kN m^{-2}. The pentahydrate will, therefore, be stable, at 20°C, in atmospheres with water vapour pressure between 670 and 2133 N m^{-2}. The average water vapour pressure in the atmosphere in this country at 20°C is about 1.87 kN m^{-2}, so that the pentahydrate will neither effloresce nor deliquesce under these normal conditions. In a more humid climate, however, the pentahydrate might deliquesce, and in a very dry climate it might effloresce.

To summarise, a hydrate will effloresce, and become coated with a lower hydrate or the anhydrous salt, when the water vapour pressure in the atmosphere is lower than the equilibrium water vapour pressure of the hydrate. A hydrate, or any other solid, will deliquesce, and become coated with a layer of, or turn completely into, a saturated solution when the atmospheric water vapour pressure is higher than the water vapour pressure of the saturated solution concerned.

Both efflorescence and deliquescence depend, too, on temperature. Some examples are given below, at a temperature of 20°C and for an atmospheric water vapour pressure of 1.87 kN m^{-2}.

Substance	Equilibrium water vapour pressure of hydrate	Vapour pressure of saturated solution	
$CuSO_4 \cdot 5H_2O$	670 N m^{-2}	2.133 kN m^{-2}	Stable
$CaCl_2 \cdot 6H_2O$	335 N m^{-2}	1 kN m^{-2}	Delinquescent
$Na_2SO_4 \cdot 10H_2O$	2.17 kN m^{-2}	2.213 kN m^{-2}	Efflorescent
$Na_2CO_3 \cdot 10H_2O$	3.23 kN m^{-2}	3.466 kN m^{-2}	Efflorescent

QUESTIONS ON CHAPTER 47

1. Explain the meanings of the terms phase and component.

2. Apply the phase rule to a consideration of the solubility of oxygen in water.

3. Apply the phase rule to show that a eutectic mixture of two solids A and B must be a mixture of two solid phases.

4. Consider the distribution of iodine between water and benzene from the point of view of the phase rule. Neglect the vapour phase, but show that the concentration of iodine in the aqueous layer is fixed, at a fixed temperature, if the concentration in the benzene layer is fixed.

5. A pair of metals may separate out from the molten mixture (a) in the pure state, (b) as a definite compound, (c) as a solid solution. Show the form of the freezing point–composition diagrams in each case.

6. The phase equilibria are shown on a pressure–temperature diagram for one-component systems, and on temperature–composition diagrams at constant pressure of pressure–composition diagrams at constant temperature for two-component systems. Give one example of each of these three types of diagram and explain its significance.

7. How many degrees of freedom will there be in the following systems: (a) a single gas in equilibrium with a solution of the gas in a single liquid, (b) two

partially miscible liquids in the absense of vapour, (c) two partially miscible liquids in the presence of vapour, (d) a solution of one solid in one liquid in equilibrium with the solvent vapour?

8. What do you think is the main difference between the pressure–temperature diagram for ice and that for iodine?

9. To what extent is it true to say that 900°C is the boiling point of calcium carbonate?

10. Explain the meaning of dissociation pressure, using an example other than that of calcium carbonate to illustrate your answer.

11. Explain, with examples, the meaning of sublimation.

12. Write notes on (a) solid carbon dioxide, (b) hoar frost.

13. Explain carefully the conditions under which a substance will (a) effloresce, (b) deliquesce.

14. Explain why phosphorus(v) oxide and freshly heated quicklime are more complete drying agents than anhydrous calcium chloride.

15. The vapour pressure of a saturated solution of ammonium nitrate(v) at 10°C, 20°C and 30°C is 860, 1500 and 2533 N m^{-2} of mercury respectively. Explain carefully why powdered ammonium nitrate(v) sets into a mass on storage.

16. The water vapour pressure above a salt hydrate depends only on temperature, so that there is only one degree of freedom. How many components has the system, and how many phases? What would the phases consist of for (a) Glauber's salt, (b) copper(ii) sulphate(vi)–5–water?

17. Magnesium sulphate(vi) forms six hydrates with 1, 2, 4, 5, 6 and 7 molecules of water of crystallisation. What will be the general shape of graph obtained by plotting vapour pressure against the number of molecules of water of crystallisation?

Chapter 48
Free energy and entropy

1. Introduction.* There are, broadly speaking, two points of view from which chemical changes may be considered. The first is generally known as the kinetic–molecular approach, chemical processes being envisaged in terms of atoms and/or molecules and/or ions, which, at normal temperatures, are in constant random motion.

Application of this method requires more and more detailed information about the structures of individual atoms, molecules and ions, and about the laws governing their motion and their interactions. The application of the kinetic–molecular approach to chemistry depends heavily on the detailed development of the atomic, molecular, ionic and kinetic theories, and the limitations of the approach are inherent in the incomplete nature of the theories.

The second approach is that of thermodynamics. This is based on certain broad generalisations derived from observations on matter in bulk and not necessarily associated in any way with atomic, molecular, ionic or kinetic theories. The relationships which can be traced between the thermodynamic behaviour of matter in bulk and the properties of individual atoms and molecules is studied in statistical mechanics or statistical thermodynamics.

Thermodynamics, as its name implies, is concerned with a simultaneous study of the heat and other energy changes which a system can undergo. In ordinary mechanics, problems are usually simplified by neglecting any heat changes. Initially, thermodynamics was applied to steam engines and other heat engines, in an attempt to solve the problem of how to get the maximum useful work out of a given amount of heat. There are still very many engineering applications of thermodynamics, but its application to chemical problems was developed in the second half of the nineteenth century, mainly by J. Willard Gibbs (1983–1903).

In its chemical applications, thermodynamics depends on an understanding of the ideas of free energy and entropy, and on the relationship of these two quantities to the heat content or enthalpy (p. 297).

By making use of these conceptions, thermodynamic arguments, as will be seen, can indicate whether a chemical change will, or will not, tend to take place under stated conditions. Thermodynamics cannot be used, however, to give accurate evidence about the rate at which any change might take place; that is the concern of chemical kinetics.

2. The first law of thermodynamics. It is a generalisation from experience, supported, for instance, by the fact that no perpetual motion machine has ever been designed, that energy can neither be created nor destroyed, and this is the meaning of the first law of thermodynamics. It can be stated in a number of ways, as follows:

---

* A student is advised to re-read Chapter 27 before reading this Chapter.

a. *Energy cannot be created or destroyed.*

b. *The total energy of an isolated system remains constant*, though it may change from one form to another. An isolated system is one which can neither receive energy from nor give energy to anything outside the system.

c. *When energy of one form disappears an equivalent amount of energy appears in some other form.*

The first law of thermodynamics is really a law of conservation of energy. The possibility of converting matter into energy (p. 3) means that the law should really be stated as a law of conservation of mass plus energy. Most applications of thermodynamics, however, are to matter in bulk where any mass changes are neglected.

3. Energy changes. It is sometimes convenient to say, arbitrarily, that a particular system has zero energy, but, more commonly, it is the difference in energy between one system and another, or between the same system under different conditions, that is of most significance.

When a system is changed from a state, A, to a state, B, then

$$\begin{array}{ccc} \text{Change in energy in} & \text{Energy of} & \text{Energy of} \\ \text{passing from } A \text{ to } B & = \text{state } B & - \text{state } A \\ \Delta U & U_B & U_A \end{array}$$

U_A is sometimes referred to as the *internal or intrinsic energy* of the system in state A, and it includes all forms of energy, except kinetic energy, as the system is assumed to be stationary.

The internal energy of a system in a particular state is a constant. It is said to be a *state, or thermodynamic, function* of the system, just as mass, volume, pressure and temperature are. As the internal energy of a system depends on the quantity of material in the system it is also said to be an *extensive* property of the system. Properties, such as density, refractive index and vapour pressure, which do not depend on the quantity of matter considered are called *intensive* properties.

The change in energy of a system in passing from state A to state B depends only on the final and initial states and not on the nature of the paths followed in the change from A to B. If this was not so, it would be possible to change A into B in one way, with a certain change in energy, and then change B back into A in a different way with a different change in energy. Thus, starting and finishing with A, it would be possible either to create or destroy energy and this would be contrary to the first law of thermodynamics.

Energy changes in systems are generally brought about by the evolution or absorption of heat and/or by work being done on or by the system. Work can be done by a system, for example, when it expands against atmospheric pressure or when it produces electrical energy in a cell. For such changes, the first law of thermodynamics leads to the relationship

$$\underset{(\Delta U)}{\underset{\text{energy}}{\text{Increase in internal}}} = \underset{(q)}{\underset{\text{the system}}{\text{Heat absorbed by}}} - \underset{(w)}{\underset{\text{the system}}{\text{Work done by}}}$$

In applying this general relationship it is very important to get the signs right. Heat absorbed is positive, whilst heat evolved is negative. Work done by the system is positive; work done on the system is negative.

If a system absorbs 50 J of heat and does work equivalent to 100 J, the increase in internal energy will be equal to -50 J, i.e. there will be a decrease in internal energy. If a system evolves 50 J of heat, and work of 75 J is done on the system, the increase in internal energy will be 25 J.

4. Changes at constant pressure and constant volume. No work is done by a system in expansion, or on a system in contraction, if the change is brought about at constant volume. The increase in internal energy is, therefore, equal to the heat absorbed at constant volume, i.e. ΔU equals q_v.

If a process is carried out at constant pressure, as, for example, in a vessel open to the atmosphere, the volume may increase, i.e. work may be done on the surroundings, or the volume may decrease, i.e. work may be done on the system. For a volume increase of ΔV, at a constant pressure, the work done by the system is $p\Delta V$, so that

$$\Delta U = q_p - p\Delta V \qquad \text{or} \qquad q_p = \Delta U + p\Delta V$$

If the system changes from a state, 1, to a state, 2, then

$$\Delta U = U_2 - U_1 \qquad \text{and} \qquad \Delta V = V_2 - V_1$$

and
$$q_p = (U_2 - U_1) + p(V_2 - V_1)$$

or
$$q_p = (U_2 + pV_2) - (U_1 + pV_1)$$

The value of $(U + pV)$ for a system is known as the *heat content or enthalpy* of the system; it is represented by the symbol, H, so that

$$H = U + pV$$

The heat content or enthalpy of a system, like the internal energy, is a property or function of state.

The change in the heat content or enthalpy of any system is equal to the heat absorbed at constant pressure, and the change is related to the change in internal energy by the expression

$$\Delta H = H_2 - H_1 = q_p \qquad \Delta H = \Delta U + p\Delta V$$

For reactions involving only solids and/or liquids, volume changes are never very large so that the values of ΔH and ΔU are always almost equal. For reactions involving gases, however, the volume change in the course of the reaction, if carried out at constant pressure, might be large; if so, there would be significant differences between the values of ΔH and ΔU (p. 296).

5. Reversible and irreversible changes. A clear distinction between reversible and irreversible changes must be made in thermodynamic considerations. From a thermodynamic point of view, an irreversible process is one in which a very slight alteration of conditions causes a spontaneous and complete change. Typical examples of such irreversible processes are the flow of water from a higher to a lower level, the flow of heat from a hotter to a colder body, the dissolving of a solute in a solvent, the diffusion of one gas into another or into a vacuum, and the reaction between ammonia and hydrogen chloride. These processes, once started, will continue until completed. They can be reversed, but only by doing work.

By comparison, a reversible change is one which can be made to take place in either direction by a slight change of the conditions. In a reversible process, the forces acting in both directions are nicely balanced; a slight increase in one direction causes movement in that direction, whilst a slight decrease causes movement in the other direction. Systems undergoing reversible changes are never very far away from a position of stable equilibrium. They are never 'out of control', and can, if required, be moved in either direction at will.

In the equilibrium mixture resulting from a chemical reaction

$$A + B \rightleftharpoons C + q \text{ joule}$$

for instance, the system can be moved in either direction by addition of more A, B or C, or by changes in temperature and pressure.

One of the simplest ways to carry out a reaction reversibly, if it is possible, is to carry it out in a cell so that it releases electrical energy and sets up an e.m.f. If this e.m.f. is measured by just balancing it against an external potential difference on a potentiometer wire, the conditions for reversibility are established. Slight increase in the external potential difference would make the cell reaction go one way; slight decrease, and the cell reaction would reverse.

6. Carnot's cycle. The partial conversion of heat energy into mechanical energy is brought about in a heat engine. In such an engine, e.g. a steam engine, heat energy is withdrawn from a source at a high temperature and used to heat a working substance. The hot working substance is then made to do work by, for example, expansion. In doing this work it is cooled, and the cooled working substance is discarded at a lower temperature.

Carnot, in 1824, made a fundamental contribution to thermodynamics by a theoretical consideration of an ideal heat engine in which all the changes could be carried out perfectly reversibly. Such a heat engine cannot, of course, be achieved in practice.

The detailed argument of Carnot is beyond the scope of this book, but the important conclusions he drew were as follows:

a. *The ratio of the heat which can be converted into work to the total initial intake of heat is dependent only on the two temperatures between which the engine works.* It is independent of the working substance used. The ratio is

$$\frac{\text{Heat converted into work}}{\text{Initial intake of heat}} = \frac{q_2 - q_1}{q_2} = \frac{T_2 - T_1}{T_2}$$

where q_2 is the amount of heat absorbed at T_2 and q_1 the amount discarded at T_1. The temperatures are expressed in K.

For an engine operating between, say, 120°C and 20°C, the maximum value for the ratio is 100/393 or 25.4 per cent. Between 220°C and 20°C, it is 40.6 per cent. Such figures, commonly referred to as efficiencies, are theoretical ones for perfectly reversible cycles of changes. Actual engines are never perfectly reversible and, therefore, have lower efficiencies. The lower temperature at which an engine works is, generally, the temperature of the atmosphere. That is why engineers strive for greater efficiency by aiming at higher temperature for the sources from which the heat is taken in.

b. *There is no more efficient cycle of changes than the perfectly reversible Carnot cycle.*

7. The second law of thermodynamics. The second law of thermodynamics is concerned with the conditions under which heat might be converted into work. Like the first law, it can be stated in a variety of ways, and the law is, again, a matter of experience.

a. *Heat cannot be converted into work unless some of the heat is transferred from a higher to a lower temperature.* If heat could be converted into work without the necessity of a temperature difference, a ship could simply withdraw heat from the ocean and use the work it obtained from it for propulsion. No fuel would be necessary. As it is, the fuel is necessary to provide the temperature difference which, alone, enables the conversion, however partial, of heat into work.

The matter can be viewed from a different aspect. The efficiency of an engine depends on the higher and the lower temperatures between which it works, as explained in the preceding section. If T_2 and T_1 are equal, the efficiency would be zero, i.e. no heat would be converted into work.

b. The conclusion of Carnot that his theoretical cycle was the one with maximum efficiency can be generalised into the statement that *a perfectly reversible process is the most efficient of all processes and produces the maximum amount of work for a given amount of heat.*

c. *The maximum amount of work can never be obtained from an irreversible process.* The extra amount of energy which would be converted into work, if the process was reversible, is said to be dissipated, and every irreversible process leads to some dissipation of energy of this sort. No energy is lost (that would be contrary to the first law of thermodynamics), but it appears as heat and not as work.

The energy available from a waterfall, for example, goes, partially, into raising the temperature of the water and cannot, under any circumstances, be converted completely into useful work. Similarly, the energy available from a definite amount of petrol is turned completely into heat if the petrol is burnt, in an irreversible process, in an open container. If the same amount of petrol was burnt in the cylinder of a motor car, some useful work would be obtained. If the process taking place in the cylinder of a motor car could be regarded as reversible (which it is not) then the useful work would be a maximum.

d. As most natural processes are invariably irreversible, the second law of thermodynamics can be summarised in the statement that *there is a tendency in nature for energy to be dissipated in the form of heat.*

8. Free energy. The maximum work which can be obtained from any change will be obtained when the change is brought about reversibly.

The maximum work obtainable from a chemical change brought about reversibly, at a fixed temperature, will, like the internal energy (p. 535), depend only on the initial (1) and final (2) states of the system. Thus

$$\text{Maximum work} = A_2 - A_1 = \Delta A$$

where A_2 and A_1 are extensive properties of state known as the maximum work function, the work function, or the *Helmholtz free energy*. In any natural change, taking place irreversibly, the work done is less than ΔA.

The term maximum work will always include a term due to work done in expansion for any change in which there is an increase in volume. But if the change is carried out at constant pressure and constant temperature, this work done will be constant. The maximum work, less this constant work done in expansion, is known as the net work, i.e.

$$\text{Net work} = \text{Maximum work} - \text{Work done in expansion}$$

Like maximum work, the net work is an extensive property of state, symbolised by G and known as the *Gibbs free energy*. Thus, for a reversible change, at constant temperature and pressure,

$$\text{Net work} = G_2 - G_1 = \Delta G$$

For any irreversible change, the net work is always less than ΔG.

In dealing with solids and liquids, the numerical difference between

Helmholtz and Gibbs free energy is small, but this is not so for gases. Gibbs free energy is the more useful quantity in dealing with most chemical problems for most chemical changes are studied at constant temperatures (in a thermostat) and at constant (atmospheric) pressure.

The free energy per mole, or the molar free energy, of a pure substance is called the *chemical potential* of the substance. It is dependent on the phase in which the substance exists, the temperature and the pressure.

9. Free energy changes in chemical reactions. What is it that enables some chemical reactions to take place, when others will not take place? Thomsen, in 1854, and Berthelot, a little later, noticed that many reactions which took place readily were exothermic, i.e. evolved heat, and they regarded the evolution of heat as the probable driving force behind a chemical reaction. Such ideas are now untenable, for it is well known that many endothermic reactions can take place spontaneously, and reversible reactions can take place with an evolution of heat in one direction and an absorption of heat in the other.

The driving force of a chemical change is now *related to its ability to do work*, not to its ability to evolve heat. If heat could be completely converted into work the two criteria would be the same, but that is not so.

In accordance with the principle, outlined in the second law of thermodynamics, that there is a natural tendency for energy to be dissipated (p. 539), a reaction will only take place of its own accord if it can perform work, for by doing this work some of its available energy becomes less available as it is discarded as heat.

For reactions carried out at constant temperature and constant pressure, i.e. for reactions in open vessels contained in a thermostat, the free energy change in the reaction,

$$\Delta G = G_{Products} - G_{Reactants}$$

provides a measure of the net work associated with the reaction.

Knowledge of ΔG values for a reaction are clearly important, and there are three significant possibilities, for ΔG may be zero, negative or positive.

a. ΔG *is zero*. If the free energy of the products and reactants are equal no net work is obtainable. The system is in a state of equilibrium.

b. ΔG *is negative*. If the free energy of the products is less than that of the reactants, there will be a decrease in free energy as the reaction takes place. The reaction can do net work and it will take place of its own accord, but, whilst a negative value of ΔG indicates favourable thermodynamic conditions for a change to take place, the rate of the change may be so slow that it never really will take place of its own accord (p. 546).

c. ΔG *is positive*. If the free energy of the products is greater than that of the reactants the reaction will not take place, unless something is done, e.g. the temperature raised, to make ΔG negative.

10. Entropy. Entropy is an extensive property of a substance, measured generally in J K^{-1} mol^{-1}, but it is not easy to grasp the full nature of entropy in terms of any pictorial idea. As will be seen, however, the conception of entropy is very important for it is the connecting link between the free energy change for a reaction and the heat of the reaction.

a. *Entropy and disorderliness*. The simplest pictorial idea of entropy is related to order and disorder. The entropy of a substance is, in fact, a measure of the randomness of the substance; what Gibbs once called mixed-up-ness.

The entropy of a pure crystalline substance at absolute zero temperature, for example, is zero. This is because there is no randomness in a pure crystal at absolute zero. The crystal is a perfectly orderly array both geometrically and energetically. As the temperature is increased, the disorderliness increases, eventually through the stage of a liquid to that of a vapour, and the entropy increases accordingly.

Other changes in which increase in entropy are clearly related to an increase in disorderliness include the solution of a crystalline solid in a solvent, a reaction in which a few molecules are converted into a larger number, and the mixing of two gases.

b. *Measure of entropy change*. Clerk Maxwell defined the entropy of a substance as 'a measurable quantity, such that when there is no communication of heat this quantity remains constant, but when heat enters or leaves the substance the quantity increases or decreases respectively'.

For a reversible change, brought about at a fixed temperature, the change in entropy is equal to the heat absorbed or evolved divided by the temperature, i.e. $\Delta S = q/T$.

If heat is absorbed, then ΔS is positive and there is an increase in entropy. If heat is evolved, ΔS is negative and there is a decease in entropy.

The simplest examples of reversible, isothermal changes are those of melting or vaporisation. The specific latent heat of fusion of ice is 334.7 J g^{-1} or 6.025 kJ mol^{-1}. The change from ice to water takes place at 273 K, so that the increase in entropy is $6025/273$, i.e. 22.07 J K^{-1} mol^{-1}.

11. Relationship between free energy and entropy. An absorption of heat, q, at a temperature, T, causes an increase in entropy of ΔS, equal to q/T. Thus

$$q = T \times \Delta S$$

As (p. 536)

$$\Delta H = \Delta U + p\Delta V \quad \text{and} \quad \Delta U = q - \text{Work done}$$

it follows that

$$\Delta H = q - (\text{Work done} - p\Delta V)$$

The work done, less $p\Delta V$, will be equal to the decrease in free energy (p. 540), i.e. to $-\Delta G$, so that

$$\Delta H = q - \Delta G \quad \text{or} \quad \Delta H = T\Delta S + \Delta G \quad \text{or} \quad \Delta G = \Delta H - T\Delta S$$

12. $\Delta G = \Delta H - T\Delta S$. In this important relationship, ΔG can be regarded as the energy which is free to do work, ΔH as the total energy available as heat plus work, and $T\Delta S$ as the energy unavailable as work. $T\Delta S$, in other words, measures the energy which is dissipated as heat.

The expression shows that the free energy change is made up of two factors, one an energy function (ΔH) and the other an entropy function ($T\Delta S$). As it is the value of ΔG which is the driving force behind a chemical change, there must be two directive influences which determine whether a change will take place, one being related to energy and the other to entropy. A system will only change of its own accord if ΔG for the change is negative (p. 540). Such a negative value for ΔG can come about in the following different ways.

a. ΔH *is negative.* $T\Delta S = 0$. In changes where there is no change of entropy, the value of ΔG is equal to that of ΔH.

A smooth marble running to the bottom of a smooth hill, provides an example of this type of change. There is a lowering of potential energy, but no entropy change. The process takes place because the marble wants to achieve a minimum potential energy. The directive influence is solely concerned with the energy term. The marble would not move up the hill of its own accord because the ΔH value for such a change would be positive, giving a positive ΔG value for the change.

At absolute zero, $T\Delta S$ must also be zero, so that ΔG must be equal to ΔH. Under such theoretical conditions ΔG can only be negative when ΔH is negative, i.e. for exothermic reactions. The earlier ideas of Thomsen and Berthelot are, then, correct at absolute zero, or for any change in which the entropy remains constant.

b. $T\Delta S$ *is positive.* $\Delta H = 0$. If there is no overall energy change in any process it can only be an increase in entropy, i.e. a positive value for $T\Delta S$, which dictates that the change shall take place.

An example is provided by the mixing of two gases, without change in volume. The mixing is a spontaneous, irreversible process, but there is no evolution or absorption of heat, i.e. no energy change, in the idealised case. The mixing takes place, in fact, because there is an increase in entropy on mixing. This is because there is a greater state of randomness or disorder in the mixture of the gases than in the two separated gases. The mixed state is the more probable.

In Clausius's words, 'the energy of the universe is constant; the entropy tends always to a maximum'.

c. *Changes in both ΔH and $T\Delta S$.* The examples of the falling marble and the mixing of gases are extreme cases. In the one, energy, but not entropy, changes; in the other, entropy, but not energy, changes. In most changes there is a change in *both* energy and entropy.

If ΔH is negative, i.e. for exothermic reactions, ΔG must also be negative unless $T\Delta S$ has a higher negative value. Exothermic reactions will, then, take place unless there is a high enough lowering of entropy, and it is unusual to find this.

If ΔH is positive, i.e. for endothermic reactions, ΔG can only be negative if $T\Delta S$ is positive, and larger than ΔH. Endothermic reactions are, therefore, most likely to take place when the heat of reaction is low and when the increase in entropy is high. Sufficiently big increases in entropy are most likely in reactions in which solids or liquids are converted into gases, or in reactions giving an increased number of molecules. Large positive values for $T\Delta S$ are also encouraged at high temperatures. That is why many endothermic reactions will only take place at high temperatures.

13. Illustrative examples. The application of these general principles to some typical changes will clarify the ideas.

a. *The water gas reaction.* The heat of reaction at 25°C and 101 325 N m^{-2} is $\Delta H = +131.4$ kJ,

$$C(graphite) + H_2O(g) \longrightarrow CO(g) + H_2(g); \qquad \Delta H = 131.4 \text{ kJ}$$

a figure which can be calculated from standard heats of formation (p. 301). ΔG can only be negative if $T\Delta S$ is positive and greater than 131.4 kJ.

There is an increase in entropy in the reaction, mainly because the products are gaseous and one of the reactants is a solid. At 25°C and 101 325 N m^{-2}, the increase in entropy amounts to 133.9 J per degree per mole of graphite. This gives a value for $T\Delta S$ of 298×133.9 i.e. 39.9 kJ. The corresponding value for ΔG will be $+91.5$ kJ. The reaction is, therefore, not spontaneous at 25°C and 101 325 N m^{-2}.

At 727°C, however, ΔG will be equal to $131.4 - 133.9$ and will have a negative value of -2.5 kJ. The reaction will be spontaneous. This value of ΔG at 727°C is, however, only an estimate, for it is based on ΔH and ΔS values at 25°C whereas the actual values at 727°C will be slightly different. ΔH and ΔS values do not, however, vary very greatly with temperature for most reactions.

b. *Synthesis of methanol.* The water gas reaction, which is endothermic, will take place at high temperatures but not at low ones. The synthesis of methanol from carbon monoxide and hydrogen, which is exothermic, will take place at low temperatures but not at higher ones.

$$CO(g) + 2H_2(g) \longrightarrow CH_3OH(g); \qquad \Delta H = -90.63 \text{ kJ}$$

The ΔH value for the reaction at 25°C and 101 325 N m^{-2} is -90.63 kJ. The ΔS value, under the same conditions, is -221.5 J per degree, giving a $T\Delta S$ value of -66 kJ. The corresponding ΔG value will be -24.63 kJ, so that the reaction will take place at 25°C, though only very slowly.

At 727°C, however, taking the same ΔH and ΔS figures, $T\Delta S$ is equal to -221.5 kJ, so that ΔG is equal to $221.5 - 90.63$, i.e. $+130.87$ kJ. As ΔG is positive the reaction will not take place at this high temperature. It is the reverse reaction which becomes possible.

c. *Rhombic and monoclinic sulphur*. At room temperature, the free energy of monoclinic sulphur is greater than that of rhombic sulphur. Change from monoclinic sulphur to rhombic sulphur will, therefore, give a negative value for ΔG. That is why rhombic sulphur is the stable form at room temperature.

As the temperature is increased, the free energies of both allotropes decrease, but they decrease at different rates. The free energy of monoclinic sulphur, which has the higher entropy, decreases more rapidly than that of rhombic sulphur. At the transition point, 95.5°C (p. 219), the free energies of the two allotropes become equal. That is why the two allotropes can coexist in equilibrium at this temperature.

At temperatures above 95.5°C, the free energy of rhombic sulphur becomes greater than that of monoclinic sulphur, so that the monoclinic sulphur becomes the stabler form.

The entropy difference for rhombic and monoclinic sulphur at the transition point is 0.96 joule per degree per mole. The $T\Delta S$ term has, therefore, a value of 0.96×368.5, i.e. 353.8 J. The value of ΔH, at the transition point, must, therefore, be $+353.8$ J to give ΔG a value of zero.

d. *Energy changes in dissolving ionic solids*. The solution of an ionic solid in a solvent involves the breakdown of the crystal structure and the formation of free ions. This is facilitated by polar solvents, such as water, since they have high dielectric constants so that they lower the interionic forces in the crystal.

The energy necessary to form free ions from a solid crystal is the crystal or lattice energy (p. 173). For sodium chloride this is 777.8 kJ mol^{-1}, i.e.

$$NaCl(s) \longrightarrow Na^+ + Cl^-; \qquad \Delta H = 777.8 \text{ kJ mol}^{-1}$$

If, therefore, 1 mol of sodium chloride was dissolved in 1 dm^3 of water and all the energy to break down the crystal came from the water there would be a fall in temperature of 180°C. This is, of course, very much greater than the measured cooling effect, and the release of energy which accounts for this is the hydration or solvation energy of the ions. The hydration energy for sodium chloride is -774.1 kJ mol^{-1}, i.e.

$$H_2O + Na^+ + Cl^- \longrightarrow Na^+(aq) + Cl^-(aq); \qquad \Delta H = -774.1 \text{ kJ mol}^{-1}$$

which gives a heat of solution for sodium chloride of $(777.8 - 774.1)$ or 3.7 kJ mol^{-1}, i.e.

$$H_2O + NaCl(s) \longrightarrow Na^+(aq) + Cl^-(aq); \qquad \Delta H = 3.7 \text{ kJ mol}^{-1}$$

The entropy change on dissolving sodium chloride at 25°C gives a $T\Delta S$ value of 12.84 kJ mol^{-1} so that the free energy change is given by

$$\Delta G = \Delta H - T\Delta S = 3.7 - 12.84 = -9.14 \text{ kJ mol}^{-1}$$

It is the negative value of the free energy change that accounts for the solubility of sodium chloride, and the value is dependent on the lattice energy, the hydration energy and the entropy change.

For an insoluble solid such as silver chloride, the free energy change is positive. The lattice energy of silver chloride is 857.7 kJ mol^{-1}, and the hydration energy is -790.8 kJ mol^{-1}. This gives a heat of solution of $\Delta H = 66.9$ kJ mol^{-1}. The entropy change at 25°C gives a $T\Delta S$ value of 9.85 giving a free energy change of $+57.05$ kJ mol^{-1}.

14. Electrical measurement of free energy and entropy changes. When a chemical reaction can be made to take place in a cell, the electrical energy liberated can be easily measured. Moreover, if the cell reaction is carried out reversibly, and at a fixed temperature and pressure, the electrical energy is equal to the free energy change in the reaction. If the e.m.f. of the cell is measured using a potentiometer, the e.m.f. is just balanced against an external potential difference enabling the cell reaction to take place reversibly.

The electrical energy produced by a cell is equal to the amount of current multiplied by the e.m.f. of the cell. For 1 equivalent of chemical action, the quantity of electricity will be 96487 coulomb. The electrical energy produced per equivalent will therefore be given by

$$\text{Electrical energy per equivalent} = 96487E \text{ joule}$$

Measurement of E enables values for free energy changes to be obtained, as in the following examples. The method is limited, however, to those changes which can be brought about reversibly in a cell.

a. *Formation of water from its elements.* The reaction between gaseous hydrogen and gaseous oxygen to form liquid water,

$$H_2(g) + \tfrac{1}{2}O_2(g) \longrightarrow H_2O(l); \qquad \Delta H = -285.9 \text{ kJ}$$

can be carried out by burning hydrogen in oxygen, when all the energy liberated will be produced as heat. There will, in fact, be an evolution of 285.9 kJ of heat.

If the reaction is carried out in a cell consisting of a hydrogen electrode and an oxygen electrode, the maximum e.m.f. is found to be 1.227 volt, so that the electrical energy available from the reaction is $96487 \times 1.227 \times 2$, i.e. 236.8 kJ. The 2 is present because two equivalents of water are being formed.

It will be seen that the reaction can produce more heat (285.9 kJ) than work (236.8 kJ), and this is a good numerical illustration of the fact (p. 539) that not all the heat available from a reaction can necessarily be made available as useful work. As the work done by the reaction in the cell is the maximum work available because the reaction is being carried out reversibly, it is equal to the free energy change for the reaction,

$$H_2(g) + \tfrac{1}{2}O_2(g) \longrightarrow H_2O(l); \qquad \Delta G = -236.8 \text{ kJ}$$

The figures given are for a temperature of 25°C, and the large negative value of ΔG suggests that hydrogen and oxygen ought to react readily at this temperature. It is well known, however, that the two gases can be in contact for years at this temperature without reaction. Although the thermodynamic conditions are favourable, the rate of reaction at 25°C, without a catalyst, is unfavourable. Addition of a little platinum catalyst, or the passing of an electric spark, makes the kinetic conditions favourable, so that the reaction takes place, perhaps explosively.

Using the relationship,

$$\Delta G = \Delta H - T\Delta S$$

it follows that the entropy change in the formation of one mole of liquid water from its elements is given by

$$\Delta S = \frac{-285.9 + 236.8}{298} = -164.7 \text{ joule per degree}$$

b. *The Daniell cell*. The chemical reaction taking place in a Daniell cell is represented by the equation,

$$Zn(s) + Cu^{2+}(aq) \longrightarrow Zn^{2+}(aq) + Cu(s); \qquad \Delta H = -209.8 \text{ kJ}$$

If M solutions are used and the reaction is carried out in a calorimeter, the heat evolved is found to be 209.8 kJ. If the reaction is carried out in a cell, at a fixed temperature of 25°C, the maximum e.m.f., given by the difference in the two standard electrode potentials involved (p. 428), is 1.10 volt.

The electrical energy liberated by the reaction is, therefore, $96487 \times 1.1 \times 2$, i.e. 212.25 kJ, and this is equal to the free energy of the reaction,

$$Zn(s) + Cu^{2+}(aq) \longrightarrow Zn^{2+}(aq) + Cu(s); \qquad \Delta G = -212.25 \text{ kJ}$$

The ΔH and ΔG values are very nearly equal, which means that there is very little entropy change in the course of this reaction, but that is simply a coincidence. In other similar cell reactions, ΔG may be greater or smaller than ΔH (p. 548).

15. Standard free energies of formation. It is particularly useful that it is possible to draw up, from experimental measurements, tables of standard values of free energies for different substances which can be treated algebraically for any reaction in the same way as standard heats of formation (p. 300) can.

The standard free energy of formation of a compound is defined as *the free energy change when 1 mol of the compound is formed from its elements under standard conditions.* The choice of these conditions is quite arbitrary, but 25°C and 101.325 kN m^{-2} are generally used. The symbol $\Delta G_{298}^{\ominus}$ is used to indicate standard free energy of formation at 25°C.

It is not possible in a book of this scope to go fully into the methods of measuring such standard free energies of formation, but the use to which they can be put will be illustrated, and one method of measurement has been described on p. 546. Some typical values are quoted below:

$H_2O(g)$	-228.6	$NO(g)$	86.70
$H_2O(l)$	-236.8	$NO_2(g)$	51.84
$CO(g)$	-137.2	$C_2H_2(g)$	209.2
$CO_2(g)$	-394.5	$C_2H_4(g)$	68.11
$NH_3(g)$	-16.63	$C_2H_6(g)$	-32.89

The values are given in kJ mol^{-1}. By convention, the free energies of elements at 25°C and 101.325 kN m^{-2} are taken as zero.

The application of these standard free energy values to find the free energy change in a reaction is illustrated in the hydrogenation of ethyne to ethene,

$$C_2H_2(g) + H_2(g) \longrightarrow C_2H_4(g)$$

The standard free energy of ethyne is 209.2 kJ mol^{-1}, that of hydrogen is 0, and that of ethene is 68.11 kJ mol^{-1}. The free energy change in the reaction will, therefore, be -141.09 kJ at 25°C and 101.325 kN m^{-2}.

Relationships showing how free energy changes with temperture and/or pressure are available so that the free energy change at other temperatures and pressures can be calculated (p. 548).

16. Standard entropies. It is also possible, though the methods are beyond the scope of this book, to obtain values for the standard entropies of elements and compounds, some values, in J k^{-1} mol^{-1} being quoted below:

$O_2(g)$	205.6	Cu	33.3	$CO_2(g)$	213.7
$H_2(g)$	130.4	CuO	43.5	$NH_3(g)$	192.6
$N_2(g)$	191.4	$H_2O(g)$	188.7	$C_2H_2(g)$	209.2
Graphite	5.69	$H_2O(l)$	69.97	$C_2H_4(g)$	219.5
Diamond	2.43	$CO(g)$	197.9	$C_2H_6(g)$	229.8

The symbol $S^{\ominus}$ is used for a standard entropy, and $\Delta S^{\ominus}$ for an entropy change under standard conditions. Like standard free energies of formation, these standard entropy values can be treated algebraically for any reaction.

In the synthesis of liquid water, for example,

$$\underset{130.4}{H_2(g)} + \underset{102.8}{\tfrac{1}{2}O_2(g)} \longrightarrow \underset{69.97}{H_2O(l)}$$

the standard entropies of the reagents and products are as shown, giving an entropy change of $69.97 - (130.4 + 102.8)$, i.e. -163.23 $_0$J k^{-1} mol^{-1} This value for the entropy change is in good agreement with that calculated for the reaction from heats of reaction and free energy values (p. 546).

Combination of standard entropies and standard free energies of formation enable heats of reactions to be calculated

In the synthesis of ammonia from nitrogen and hydrogen, for example,

$$\tfrac{1}{2}N_2(g) + \tfrac{3}{2}H_2(g) \longrightarrow NH_3(g)$$

the sum of the standard entropies of the reactants is $\tfrac{1}{2}(191.4) + \tfrac{3}{2}(130.4)$ i.e. 291.3, whilst the entropy of the product is 192.5. The entropy change for the reaction is, therefore, $192.5 - 291.3$, i.e. -98.8 J K^{-1} mol^{-1} of ammonia at 25°C and 101.325 kN m^{-2}.

The free energy change in this reaction, equal to the standard free energy of formation of ammonia, will be -16.63 kJ. The corresponding $\Delta H°$ value for the reaction, calculated from the relationship

$$\Delta G° = \Delta H° - T\Delta S°$$

will be $-16630 - (298 \times 98.8)$, i.e. -46070 J, which is in good agreement with the experimental value of -46185 J (p. 325).

17. Change of free energy with temperature. The change in free energy in any process is dependent on the temperature, the relationship being given by the *Gibbs–Helmholtz equation*, which must be stated here without derivation:

$$\Delta G = \Delta H + T \times \frac{d(\Delta G)}{dT} \qquad \text{or} \qquad \frac{d(\Delta G)}{dT} = \frac{\Delta G - \Delta H}{T}$$

This equation, which holds at constant pressure enables ΔH values to be calculated if ΔG and $d(\Delta G)/dT$ are known.

Whether or not ΔG is bigger or smaller than ΔH depends on the sign of $d(\Delta G)/dT$, i.e. on whether ΔG increases or decreases with temperature. For a reaction carried out reversibly in a cell, this means that $(\Delta G - \Delta H)$ will be positive if the temperature coefficient of e.m.f., i.e. dE/dT, is positive, and negative if dE/dT is negative. It is because the temperature coefficient of a Daniell cell is almost zero that the ΔG and ΔH values are almost equal (p. 546).

The actual free energy of a single substance always decreases with temperature, and the rate of decrease is equal to the entropy,

$$\frac{dG}{dT} = -S$$

It is not possible, here, to derive the relationship, but it provides a useful alternative way of looking at the meaning of entropy.

The total energy of a single substance increases with temperature, and the rate of increase, in this case, is equal to the heat capacity of the substance.

$$\frac{dH}{dT} = C_p$$

18. Free energy and equilibrium constants. As ΔG for a reaction is zero at equilibrium, it is not surprising that there is a relationship between ΔG and the equilibrium constant for the reaction.

In its simplest form, which has only limited application to ideal gas reactions, the relation is

$$\Delta G_T{}^{\circ} = -RT \ln K_p$$

where $\Delta G_T{}^{\circ}$ is the standard free energy change at $101.325 \text{ kN m}^{-1}$ and temperature, T, and K_p is the equilibrium constant expressed in terms of pressure. This is a simple form of what is known as the *van't Hoff isotherm.*

It enables equilibrium constants to be calculated from ΔG° values, and this is a particularly important possibility for it means that equilibrium constants, which are often difficult to measure directly, can be calculated from purely thermodynamical data. When equilibrium constants can be measured, the relationship also enables free energy changes to be calculated.

For the reaction in which methanol is formed from carbon monoxide and hydrogen (p. 544),

$$CO(g) + 2H_2(g) \longrightarrow CH_3OH(g)$$

the free energy change is -24.63 kJ at 25°C. As this is equal to $-RT \ln K_p$, it follows that $\ln K_p$ is given by

$$\ln K_p = \frac{24\,630}{8.314 \times 298} = 9.94 \quad \text{or} \quad \log_{10} K_p = \frac{9.94}{2.303} = 4.316$$

The value of K_p for this reaction at 25°C is 2.07×10^4. The high value shows how far the reaction can proceed from left to right.

19. Change of equilibrium constant with temperature. The combination of the Gibbs–Helmholtz equation and the van't Hoff isotherm provides an expression, known as the *van't Hoff isochore,* which shows how the equilibrium constant of a reaction is dependent on the temperature (p. 322).

The van't Hoff isotherm,

$$\Delta G_T{}^{\circ} = -RT \ln K_p$$

gives, on differentiation,

$$\frac{d(\Delta G_T{}^{\circ})}{dT} = -R \ln K_p - RT \frac{d(\ln K_p)}{dT}$$

and substitution of this value for $d(\Delta G_T^{\ominus})/dT$ into the Gibbs–Helmholtz equation gives

$$\frac{d(\ln K_p)}{dT} = \frac{\Delta H^{\ominus}}{RT^2}$$

This shows that the K_p values for a reaction measured at different temperatures are related to the standard heat of reaction. Thus, if $\Delta H^{\ominus}$ for a reaction is known, K_p values at 25°C, calculated from free energy changes as in **18**, can be converted into K_p values at other temperatures.

In the simplest use of this relationship it is assumed that $\Delta H^{\ominus}$ does not vary with temperature. If so, then, the values of K_p at two temperatures T_1 and T_2 are related by the expression,

$$\ln K_{p,\ T2} - \ln K_{p,\ T1} = -\frac{\Delta H^{\ominus}}{R}\left\{\frac{1}{T_2} - \frac{1}{T_1}\right\}$$

For the reaction in which methanol is synthesised, K_p at 25°C is 2.07 × 10^4, as in **(18)**. The value of K_p at 300°C will be given by

$$\ln K_{p,\ 573} = \ln K_{p,\ 298} + \frac{90\,630}{8.314}\left\{\frac{1}{573} - \frac{1}{298}\right\}$$
$$= 9.94 - 17.56 = -7.62$$

Therefore, $\log K_{p,\ 573} = \dfrac{-7.62}{2.303} = -3.309 = \bar{4}.691$

giving a value of K_p at 300°C of 4.909 × 10^{-4}. The low value shows that the reaction is not possible.

The assumption that $\Delta H^{\ominus}$ does not change with temperature is only approximate. If the variation of $\Delta H^{\ominus}$ with temperature (p. 549) is taken into account more complicated equations have to be used.

QUESTIONS ON CHAPTER 48

1. Account for the following: (*a*) the entropy of diamond is smaller than that of graphite, (*b*) the entropy of a polyatomic molecule is greater than that of a monatomic molecule, (*c*) the entropy of steam is greater than that of water at 100°C.

2. The ΔH value for the reaction,

$$C(\text{graphite}) + \tfrac{1}{2}O_2 \longrightarrow CO$$

is −110.6 kJ at 25°C and 101.325 kN m^{-2}. The standard entropies of graphite, oxygen and carbon monoxide are given in the table on p. 547. What is the free energy change for the reaction?

3. Using the free energy values given on p. 547, calculate the free energy changes in the following reaction:

> *a* $NO_2(g) \longrightarrow NO(g) + \tfrac{1}{2}O_2(g)$ *b* $CO(g) + \tfrac{1}{2}O_2(g) \longrightarrow CO_2(g)$
>   *c* $C_2H_4(g) + H_2(g) \longrightarrow C_2H_6(g)$

4. Calculate the equilibrium constant at 25°C and 101.325 kN m^{-2} for the change from butane to 2–methylpropane given that the standard free energies

of formation of butane and 2–methylpropane are -15.7 and -17.98 kJ mol^{-1} respectively. What will be the partial pressure of butane in the equilibrium mixture at 101.325 kN m^{-2}?

5. The maximum e.m.f. of a Daniell cell in which the reaction,

$$Zn(s) + Cu^{2+}(aq) \longrightarrow Zn^{2+}(aq) + Cu(s); \Delta H = -210 \text{ kJ}$$

takes place is 1.107 volt. Calculate the entropy change. All values are given at 25°C.

6. Explain what is meant by the terms (a) internal energy, (b) free energy, (c) unavailable energy, (d) maximum work, (e) net work.

7. Taking the specific latent heat of evaporation of water as 2259 J g^{-1}, calculate the entropy change at 100°C for the change from water to steam.

8. The evaporation of water and the solution of ammonium chloride in water both take place spontaneously at room temperature. Both processes absorb heat. Why, then, do they take place? Under what conditions can they be reversed?

9. Explain why it is that most reactions which take place readily at room temperatures are exothermic, whilst endothermic reactions are favoured at temperatures of about 3000°C.

10. Explain why the denser form of an element which exhibits allotropy is expected to have the smaller heat of combustion.

11. 'In the processes of melting, evaporation and dissolution the large gain in entropy on the formation of the more disordered state offsets the positive sign of ΔH.' What does this mean?

12. To what extent are the tables of electrode potentials also tables of free energy values?

13. The chemical reaction taking place in the Clark cell may be represented by the equation,

$$Zn + Hg_2SO_4 = ZnSO_4 + 2Hg$$

The maximum e.m.f. for the cell is 1.4324 volt, and the temperature coefficient of e.m.f. is -119 mV K^{-1}. Calculate the value of ΔH for the reaction at 25°C.

14. Taking the standard electrode potential of zinc as -0.761 volt, calculate the free energy change in the reaction of zinc with hydrochloric acid. Would you expect the heat of the reaction to be bigger or smaller than the free energy change?

15. The standard heats of formation and the standard free energies of formation for some alkanes are given below, in kJ mol^{-1}:

Alkane	CH_4	C_2H_6	C_3H_8	C_4H_{10}	C_5H_{12}	C_6H_{14}
Heat of formation	-74.9	-84.7	-103.9	-124.7	-146.4	-167.2
Free energy of formation	-50.7	-32.9	-23.5	-15.7	-8.20	$+0.21$

Comment on these figures.

16. The standard heats of formation and the standard free energies of formation for the hydrogen halides are given below, in kJ mol^{-1}:

Hydrogen halide	HF	HCl	HBr	HI
Heat of formation	-268.6	-92.3	-362.3	$+25.94$
Free energy of formation	-270.7	-95.5	-53.3	$+1.3$

Comment on these figures.

SI units

1. The Système Internationale d'Unités (SI, for short) is rapidly being adopted internationally. It is based on six arbitrarily defined basic units, from which other units are derived. It is a coherent system because the product or quotient of any two unit quantities on the SI system is the unit of the resultant quantity.

This contrasts with metric systems used in the past, for the older systems though using some basic units also used other additional units, e.g. the calorie and horse-power, which were arbitrarily defined on their own. The older systems were not coherent so that tiresome conversion factors were constantly intruding.

2. The six basic SI units are the metre, the kilogramme, the second, the ampere, the kelvin and the candela. The mole, though not yet fully adopted as a basic SI unit, is commonly used as such and will probably be adopted.

A selected list of basic SI units and some derived units with special names is given below in alphabetical order of the unit name. Some associated older units are also given, for these older units will still be present in older literature and the use of some of them may well persist despite all recommendations to the contrary.

Physical quantity	Name of SI unit	Symbol of unit	Other related units
El. current	ampere	A	
El. charge	coulomb	C	
Volume (see footnote, p. 13)	cubic metre	m^3	1 dm^3 = 1 litre (1 l) 1 cm^3 = 1 millilitre (1 ml)
Energy	joule	J	1 calorie = 4.184 J 1 electron volt = 1.6021×10^{-19} J 1 erg = 10^{-7} J 1 kilowatt hour = 3.6×10^6 J
Temperature	kelvin	K	1°C = 1 K
Mass	kilo-gramme	kg	1 atomic mass unit = 1.66043×10^{-27} kg
Length	metre	m	1 angström (1 Å) = 10^{-10} m
Amount of substance	mole	mol	
Force	newton	N	1 dyne = 10^{-5} N
Pressure	newton per square metre	$N\,m^{-2}$	1 atmosphere = 101 325 N m^{-2} 1 mm of Hg = 133.322 N m^{-2} 1 pascal (Pa) = 1 N m^{-2}
El. resistance	ohm	Ω	
Time	second	s	
El. potential difference	volt	V	
Power	watt	W	

When the unit is named after a person its name is not given a capital initial letter, but the symbol for the unit is.

3. Multiples and sub-multiples of SI units are formed by using the following prefixes:

Multiplication factor	Prefix	Symbol	Multiplication factor	Prefix	Symbol
10^{12}	tera	T	10^{-2}	**centi**	**c**
10^{-9}	giga	G	10^{-3}	**milli**	**m**
10^6	mega	M	10^{-6}	**micro**	**μ**
10^3	**kilo**	**k**	10^{-9}	**nano**	**n**
10^2	hecto	h	10^{-12}	pico	p
10^1	deka	da	10^{-15}	femto	f
10^{-1}	deci	d	10^{-18}	atto	a

Those most likely to be encountered are printed in bold type. Prefixes differing in step by 10^3 are recommended but the use of deci and, particularly, centi will probably remain common.

4. Although it is permissible to quote numerical values in multiples and sub-multiples of SI units, e.g. in g or cm^3, **all data must be 'fed into' equations in strict SI units and not in multiples or sub-multiples of SI units.**

5. The values, in SI units, of some important physical constants are summarised below in alphabetical order:

Constant	Symbol	Value
Avogadro constant	L	$6.02252 \times 10^{23} \text{ mol}^{-1}$
Charge on electron	e	$1.60210 \times 10^{-19} \text{ C}$
Faraday constant	F	$9.64870 \times 10^4 \text{ C mol}^{-1}$
Ice point	T	273.15 K
Mass of electron	m_e	$9.1091 \times 10^{-31} \text{ kg}$
Mass of neutron	m_n	$1.67482 \times 10^{-27} \text{ kg}$
Mass of proton	m_p	$1.67252 \times 10^{-27} \text{ kg}$
Molar gas constant	R	$8.3143 \text{ J K}^{-1} \text{ mol}^{-1}$
Molar volume at s.t.p.		$2.24136 \times 10^{-2} \text{ m}^3 \text{ mol}^{-1}$
Planck constant	h	$6.6256 \times 10^{-34} \text{ J s}$
Triple point of water		273.16 K
Velocity of light	c	$2.997925 \times 10^8 \text{ m s}^{-1}$

Revision questions

1. Comment on the following statements: (a) Volatility is often said to be a criterion of covalency, but ammonium chloride volatilises and diamond does not. (b) Silicates of metals form the greater part of the earth's crust, but metals are hardly ever extracted from silicates on an industrial scale. (c) Lead chloride is less soluble in normal hydrochloric acid than in water, but it is much more soluble in concentrated hydrochloric acid than in either. (d) It is difficult to convert rhombic sulphur directly into monoclinic sulphur, but a sample of monoclinic sulphur changes spontaneously in the cold into rhombic sulphur. (O. & C.)

2. Explain the meaning of the terms italicised and thereby elucidate and illustrate the following statements: (a) Solid *allotropes* can be either *monotropic* or *enantiotropic*. (b) *Osmotic pressure* is the pressure necessary to prevent *osmosis*. (c) A *catalyst* cannot affect the *equilibrium point* of a reaction.

3. Distinguish between the following terms, either giving examples to illustrate them or stating the units in which the quantities concerned could be measured: (a) degree of dissociation and dissociation constant; (b) solubility and solubility product; (c) osmosis and dialysis; (d) conductivity and molar conductivity; (e) drying and dehydration; (f) isomer and isotope. (O. & C.)

4. Explain clearly, with examples, the meaning of the terms diffusion, solubility product, common ion, efflorescence. (Oxf. Prelims.)

5. Explain the meaning of the following statements and describe how you would verify any one of them experimentally: (a) Oxalic acid is dibasic. (b) Formic acid is stronger than acetic. (c) Hydrogen fluoride is associated in the vapour phase. (Oxf. Prelims.)

6. What is meant by the molecular weight of a substance? Discuss the factors which determine it. Explain how you would investigate experimentally the molecular weight of three of the following: (a) phosphorus trioxide; (b) hydrogen chloride; (c) ammonium chloride; (d) mercury; (e) benzoic acid. Mention any interesting conclusions to which your results might lead.

7. Explain briefly the difference between: (a) an atom and its ion; (b) allotropes and isotopes; (c) group and period (in the periodic table); (d) electrolysis and electrolytic dissociation; (e) negative catalyst and catalyst poison. (N.)

8. Suggest explanations for the following facts: (a) Very small solid particles undergo erratic motion when suspended in gases or liquids, but large ones do not. (b) A wet substance dries more rapidly in a vacuum desiccator than in an ordinary one, even though the same drying agent be used. (c) A gas cools on expanding adiabatically. (d) A liquid cools when it evaporates freely.

9. Explain why sodium chloride: (a) dissolves in water; (b) lowers the freezing point of water; (c) is less soluble in solutions of hydrochloric acid than in water; (d) coagulates a colloidal solution of iron(III) hydroxide.

10. Explain concisely the difference between three of the following: (a) a weak and a strong electrolyte; (b) electrovalency and covalency; (c) cooling by adiabatic expansion and the Joule–Thomson effect; (d) dissociation and thermal decomposition; (e) monotropy and enantiotropy. (O. & C. S.)

11. Comment on, illustrate, or explain the following statements: (a) Hydrogen chloride does not obey Henry's law. (b) Metals can displace hydrogen from sodium hydroxide. (c) A strong electrolyte does not obey Ostwald's dilution law. (d) Some allotropes differ in chemical properties, others in physical properties only. (O. & C.)

12. Comment on, illustrate, or explain four of the following statements: (a) A catalyst does not alter the point of equilibrium in a reversible reaction. (b) Combustion is not necessarily accompanied by flame. (c) A colloidal particle carries an electric charge. (d) Amorphous carbon absorbs gases easily, but diamond

will not. (*e*) The end-point of an acid–alkali titration is not always at neutrality. (O. & C.)

13. Define: (*a*) atomic number; (*b*) Avogadro constant; (*c*) the gas constant, *R*; (*d*) Faraday constant. How could the value of one of these be obtained? (Oxf. Schol.)

14. Explain fully what you understand by three of the following: (*a*) a negative catalyst; (*b*) the heat of formation of a metallic oxide; (*c*) the vapour pressure of a liquid; (*d*) the valency of nitrogen in ammonium chloride. (O. & C.)

15. Write down the expression for the equilibrium constant of the reaction

$$N_2 + 3H_2 \rightleftharpoons 2NH_3 + 55.23 \text{ kJ}$$

(*a*) If *a* is the fraction of ammonia present by volume in an equilibrium mixture made from one volume of nitrogen and three volumes of hydrogen, and *P* is the total pressure show that

$$a/(1 - a)^2 = kP$$

where *k* is a constant. (It will be found convenient to express concentrations in partial pressures.)

If 0.25 of the equilibrium mixture at 400°C and 100 atmospheres pressure is ammonia, calculate what fraction of the mixture will be ammonia at 10 atmospheres at that temperature.

(*b*) The equilibrium constant varies with temperature according to the equation

$$\log K_1 - \log K_2 = \frac{55.23}{2.303} \left(\frac{1}{T_1} - \frac{1}{T_2} \right)$$

where K_1 and K_2 are the constants at T_1 and T_2 respectively.

In the light of this equation and your results in (*a*) explain very briefly the conditions used in the manufacture of ammonia. Why is it necessary to use a catalyst? (O. & C. S.)

16. One important industrial method for the production of hydrogen makes use of the reversible reaction

$$H_2O + CO \rightleftharpoons H_2 + CO_2 + 40.58 \text{ kJ}$$

the carbon dioxide then being removed by solution in water.

What factors influence the choice of temperature and pressure at which both parts of the process are carried out? One of the practical difficulties in the first part is the selection of a suitable catalyst; why is a catalyst necessary and what are the criteria of a good industrial catalyst? (S.)

17. A mixture of 1 volume of nitrogen and 3 volumes of hydrogen was heated until equilibrium was attained at pressure *P* and a given temperature *T*. For simplicity consider the mixture resulting from 1 mol of nitrogen and 3 mol of hydrogen. Let the fraction of nitrogen and hydrogen converted to ammonia be *a* and the molar volume at *P* and *T* be *v*.

(*a*) Express the total volume of the gas mixture at *P* and *T* in terms of *a* and *v*.

(*b*) Find the equilibrium constant for the reaction

$$N_2 + 3H_2 \rightleftharpoons 2NH_3$$

at the temperature *T* and show that

$$\frac{a(2 - a)}{(1 - a)^2} = \frac{c}{v}$$

where *c* is a constant.

18. Ammonium sulphate can be made for use as a fertiliser by stirring a solution of ammonium carbonate with finely-divided calcium sulphate. At equilibrium, the mixture is saturated with both calcium sulphate and calcium carbonate, with solubility products of 2.3×10^{-4} and 4.8×10^{-9} respectively. What is the approximate ratio of the concentration of ammonium sulphate to that of ammonium carbonate in the mixture?

19. If a solution of sodium sulphate is stirred with excess barium carbonate until equilibrium is established, what percentage of it will be converted into sodium carbonate? The solubility products of barium carbonate and barium sulphate are 8.0×10^{-9} and 1.1×10^{-10} respectively.

20. Under what circumstances are measurements of the physical properties of solutions suitable for the determination of relative molecular masses?

9.3 gramme of p-cresol in 1000 g of benzene depress the freezing point of benzene by $0.42°C$. 200 g of p-cresol in the same mass of benzene depress the freezing point $5.0°C$. Comment on these observations. (Oxf. Ent.)

21. Discuss the effects of changes of temperature and pressure on the position of equilibrium in a gaseous system. How does the situation change if a heterogeneous system is considered?

22. What do you understand by a 'perfect gas'? Predict the experimental conditions required for a real gas to approach 'perfect' behaviour. (Oxf. Ent.)

23. Explain carefully why it is that heat is evolved when many electrolytes dissolve in water although it would be expected that the work done in separating their ions would be revealed by the absorption of heat.

24. Write short notes on any four of the following: free energy, steam distillation, constant boiling mixtures, critical solution temperatures, fractionating towers, eutectic mixtures.

25. A solution of iodine in carbon tetrachloride gives, on boiling, a mixture of iodine vapour, which is purple, and carbon tetrachloride vapour, which is colourless. The amount of iodine dissolved is shown by the colour of the solution. Design an apparatus using this solution which illustrates the functioning of a fractionating tower.

26. Write short notes on the following: ionic product for water, solubility product, buffer solutions, hydrolysis of salts, radioactivity.

27. The isotopic composition of magnesium is found to be ^{24}Mg (77.4%), ^{25}Mg (11.0%) and ^{26}Mg (11.6%). Assuming the isotopic mass to be equal to the mass number, calculate the relative atomic mass of magnesium.

In most cases the chemical atomic weight of an element is the same from whatever source it comes. What does this imply about the origin of the elements? In which element is there a variation in atomic weight, and why? (D.)

28. In what ways does radioactive disintegration differ from ordinary chemical reactions? Explain briefly what is meant by: (a) a radioactive series; (b) radioactive equilibrium; (c) half-life period.

At each point marked by an asterisk a number has been omitted from the following nuclear equations which represent the net changes occurring in the ultimate disintegration of three radioactive elements P, Q and R. Complete the equations by inserting the appropriate numbers and assign P, Q and R to their correct main groups in the periodic table, briefly explaining how you deduce the necessary information.

$$^{232}_{90}P = \ ^*_*Pb + 6\ ^4_2He + 4\ _{-1}^0e$$
$$^{234}_{92}Q = \ ^{206}_*Pb + 0\ ^4_2He + *\ _{-1}^0e$$
$$^*_*R = \ ^{207}_*Pb + 4\ ^4_2He + 2\ _{-1}^0e$$

Describe one simple application of a radioactive isotope to the solution of a chemical problem. (W.S.)

29. Derive the mathematical expression for Ostwald's dilution law for an electrolyte, one molecule of which ionises to give n cations and m anions.

30. Explain the following: (a) The pH of 0.1M hydrochloric acid is 1.0 but that of 0.1M acetic acid is 2.75. (b) The cations discharged when separate aqueous solutions of copper sulphate and sodium sulphate are electrolysed between platinum electrodes are copper and hydrogen respectively. (c) Pure water is acid to phenolphthalein but alkaline to methyl orange. (d) Strong electrolytes do not obey Ostwald's dilution law.

31. Give and explain a definition of an acid (a) making use of the conception of electrons, and (b) without the use of this concept. Are there any substances covered by one of the definitions you give which would not be covered by the other? Which do you consider to be the better definition, and why? (O. & C. S.)

32. Using the kinetic theory and the concept of reaction velocity, explain the following facts: (a) When nitrogen oxide is heated to 1 000°C it is almost completely decomposed into nitrogen and oxygen. (b) When oxygen and nitrogen are heated to 3000°C the mixture of gases is found to contain about 40 per cent of nitrogen oxide. (c) Hydrogen and oxygen combine to form water with the evolution of considerable heat, but the mixture of gases needs heating before combination takes place. (O. & C. S.)

33. Describe, with a sketch or diagram of the apparatus, how you would measure two of the following: (a) the partition coefficient of iodine between potassium iodide solution and benzene; (b) the molar conductivity of a 0.001M solution of acetic acid; (c) the molecular weight of urea. (O. & C. S.)

34. Describe clearly experiments you would carry out to demonstrate three of the following: (a) that hydrogen diffuses through a porous wall more rapidly than oxygen; (b) that the osmotic pressure of sugar increases with rise of temperature; (c) that the decomposition of hydrogen peroxide in solution is a first-order reaction; (d) that the constant-boiling mixture of hydrogen chloride and water is a mixture and not a compound; (e) that chrome alum and potash alum are isomorphous. (Oxf. Schol.)

35. Suggest methods of investigating one of the following: (a) the rate of decomposition of ammonium nitrite solution at its boiling point; (b) the equilibrium between hydrogen, iodine and hydrogen iodide in the vapour phase; (c) the formula of the copper complex in solution of copper(II) sulphate containing excess ammonia. (Oxf. Schol.)

36. Explain, with examples where possible, what you understand by five of the following: Avogadro constant, active mass, molar conductivity, electrode potential, buffer solution, normal salt. (O. & C.)

37. How would you measure (a) the degree of dissociation of ethanoic acid in water, (b) the degree of association of ethanoic acid in benzene?

38. Many fundamental chemical definitions have had to be modified in the last hundred years. Give some illustrative examples.

39. Explain what is meant by the following statements: (a) The gas constant R is 8.314 J K^{-1} mol^{-1}. (b) The ionic product of water at 25°C is 1.0×10^{-14}. (c) The hydrolysis constant of sodium acetate is 5.5×10^{-10} at 25°C.
    Calculate: (i) the specific heat capacities of argon (atomic weight = 40); (ii) the dissociation constant of acetic acid at 25°C; (iii) the pH value of a deci-normal solution of sodium acetate. (W.)

40. Explain the following: (a) the atomic weight of nickel (atomic number = 28) is less than that of cobalt (atomic number = 27); (b) the rate of simple reactions between gases increases with temperature much more rapidly than does the rate at which the molecules collide; (c) sulphur, when gently heated from room temperature, melts at 119°C, but when strongly heated melts at 113°C; (d) when hydrogen sulphide is passed into a solution of sodium arsenite a yellow precipitate

is obtained, but when passed into dilute aqueous arsenious oxide a faintly opalescent yellow solution results; (e) a Bunsen burner flame sometimes 'lights back' and burns at the bottom of the tube. (W.)

41. Discuss critically the following statements: (a) 'The heat of formation of water is 285 kJ.' (b) 'The properties of the elements are in periodic dependence on their atomic weights.' (c) 'The rate of a reaction and accordingly the yield of product increases with temperature.' (d) 'On electrolysing an aqueous solution containing several cations, the cation present in greatest concentration is preferentially discharged.' (W.)

42. A mole can be defined as the amount of substance containing the same number of atoms as 12 gramme of pure ^{12}C. Do you think this is a good definition, or not? Give your reasons.

43. Give an account of either clathrates or layer lattice structures or graphitic compounds.

44. Stas admitted, in 1887, that there must be something in Prout's hypothesis. What was Prout's hypothesis, and do you think there is anything in it? Give reasons.

45. Compare and contrast the dissociation and association of methane, ammonia, water and hydrogen fluoride.

46. Berzelius wrote, in 1819, that 'it is evident that if analyses are made of all the salts formed by one acid, for instance by sulphuric acid, with all the bases, and of those formed by one base, for example baryta, with all the acids, one would have the necessary data to calculate the composition of all the salts formed by double decomposition, provided that they retained their neutrality.' What, using modern terms, does this mean, and to what extent is it true?

47. Suggest an experimental method for studying the dependence of the rate of the following reaction on the concentration of bromine in the presence of excess formic acid,

$$Br_2 + H \cdot COOH = 2Br^- + 2H^+ + CO_2$$

48. Explain the meaning of the following terms: supercooling, superheating, supersaturation, metastable state.

49. Write an account of the work of any one famous chemist.

50. Give examples in which (a) metals, (b) oxides, (c) enzymes, (d) gases, are used as catalysts.

51. Do you prefer inorganic, organic or physical chemistry? Give your reasons.

52. 'Ein Chemiker, der kein Physiker ist, ist gar nichts.' 'All the interesting scientific developments are now in the field of bio-chemistry.' Comment on one of these sayings.

Answers to questions

Chapter 1 (p. 8)
    7. 63.48×10^{-12} kg
    8. 2.326×10^{-9} kg
    9. 2.293×10^{12} J
  10. 4×10^{10}

Chapter 2 (p. 15)
    4. 12; 50 cm^3
    5. 6.72 dm^3
    6. 8.32
    7. 30.03
    8. 31.78
  12. 31.75
  13. 0.106 g

14. 21.6 per cent
15. 29.74
16. 113.2
17. 127.5
19. 35.35
21. 127.1
22. 107.9; 16.03

Chapter 3 (p. 32)
    3. 89.3 per cent; 27
    4. 40
    5. 23.84; 40.16 per cent
    6. 27.03; M_2Cl_6
    7. 200.6; 199.4

10. Eq. $= 4.489$
11. 193.1
18. 51.80
19. 9.1
21. 137.38
23. 96

Chapter 4 (p. 43)
    1. (a) 123.2 dm^3
      (b) 116 dm^3
    3. 1.168
    4. 903.6 cm^3
    7. 16.22

    8. X_2H_6
    9. 1.555:1
12. 65.86 (N_2); 15.2 (CO_2); 20.26 (O_2)
13. 25.8 per cent

Chapter 5 (p. 52)
    9. C_4H_{10}
  10. 12
  11. 16

12. 12
13. 665.9×10^{12}
14. 104 cm^3
24. 12.0102×10^{-3} kg

Chapter 6 (p. 63)
    2. 200.5
    3. 57.48
    4. 121
    5. MF_6; 238.2
    7. $CHCl_3$
    8. 78.52
    9. 0.27
  10. 24.64 cm^3; 230.9

11. 35.24 per cent
12. 53.94 per cent
13. 70.65 per cent; 47.6 per cent
15. 2 per cent
19. 118.7
20. 62.55 per cent; $K = 2.570$;
    3.08 atm (312×10^5 N m^{-2})

Chapter 7 (p. 82)
    6. 8.2×10^{-3}; 2; 393.5 m s^-
    9. 461 m s^{-1}

Chapter 9 (p. 94)
    8. 200×10^9
    9. 176×10^9

Chapter 10 (p. 102)
    7. 620×10^{21}

Chapter 11 (p. 120)
    11. 2.179×10^{-18} J
    12. 490×10^{-27} nm

Chapter 12 (p. 132)
    1. 13.56×10^{-12}; 36.38×10^9; 0.9831 curie; 1.017 g
    2. (a) 7.66×10^9 s; (b) 169.8×10^9
    3. 870×10^{-6} g; 1.3×10^9 year
    6. 800 s^{-1}
    10. 4.85×10^{-18}
    22. 16.004452

Chapters 13 and 14 (p. 156)
    8. (i) 1.24; (ii) 6.20; 500   15. 1.0041
    10. 470 MeV   16. 11.01127
    11. 93.5 MeV   17. 2.29×10^{12}
    12. 10^{14} g cm^{-3} 18. 728×10^9 J
                          19. 193 MeV

Chapter 20 (p. 224)
    5. 74 per cent

Chapter 21 (p. 235)
    1. 0.2813; 71.88 per cent   15. (a) 8.33 g; (b) 9.6875 g
    2. 25×10^{-3} 17. 3
    4. 20.72   19. 0.19

Chapter 22 (p. 242)
    2. 67.57 per cent
    3. 0.0338
    5. 6
    6. 63.3 per cent N_2; 34.9 per cent O_2; 1.8 per cent A

Chapter 23 (p. 254)
    3. 186.650 kN m^{-2}
    4. (a) 2799 N m^{-2}, 4533 N m^{-2}; (b) 7333 N m^{-2}; (c) 38 per cent CH_3OH
    by volume
    5. Mole fraction of A = 0.167
    6. 164.4 cm^3

Chapter 25 (p. 277)
    1. 181.8   9. 2.066 g
    4. 120.1; 229.9   13. 126
    5. 138.2   14. 390.4°C; 134.6
    6. 385.1   16. −0.0544°C; 891.1 g
    7. S_8; 1.9×10^{-3} 19. (a) 7419 g; (b) 2581 g
    8. 145

Chapter 26 (p. 290)
    2. 345.7   10. 68.1×10^3
    3. (136.2 kN m^{-2}) 1.32 atm; 326 11. 1197
    4. 596   12. 240×10^3
    5. 3.84 g dm^{-3} 13. 30.3×10^3
    9. 34.2×10^3 14. 33.3
                          15. 84.6 per cent

Chapter 27 (p. 303)
1. $\Delta H = -548$ kJ
2. (a) $\Delta H = -597.6$ kJ
   (b) 1255 kJ
3. $\Delta H = 60.2$ kJ
4. $\Delta H = -209.2$ kJ
5. 4720 g
6. $\Delta H = -214.2$ kJ
8. 1: 0.54
9. 2.9°C

10. $\Delta H = -61.52$ kJ
11. 3242 kJ
12. $\Delta H = -192.9$ kJ
13. 10.94 kJ; 10.73 kJ
14. $\Delta H = -460.2$ kJ
15. $\Delta H = -217.6$ kJ
16. (a) 90.4; -40.6 kJ;
    (b) 87.9; -43.1 kJ
17. 23.83 kJ

Chapters 28 and 29 (p. 327)
2. 4; 99.44 g
4. (a) 0.9 mol; (b) 0.845 mol;
   (c) 0.54 mol
7. 2.925:1
9. 109.9 g
16. 303
17. 2.81 per cent
18. 8.044×10^3 N m^{-2}
19. 0.1089; 0.3838 atm

20. (a) 0.037 mol dm^{-3}
    (b) 0.063 mol dm^{-3}
21. (a) 50;
    (b) 5.9 atm;
    (c) 0.65 atm
24. (a) 625, 1156;
    (b) 1:1.848;
    (c) 84.42 mm

Chapter 30 (p. 334)
1. 285.7 kJ
3. 0.34; 0.50
4. 53.590 kJ
7. 57.68 kJ
8. 171.6 kJ

9. 90.16 kJ
10. (a) 32.1; (b) 94.56 kJ
11. 505.4 kJ
12. 103.4 kJ mol^{-1}

Chapter 31 (p. 348)
2. 1
3. 144.5 min
5. 82.4 min
6. 513 s

7. 3
8. 2
13. 1
18. 159.6×10^3

Chapter 33 (p. 371)
2. (a) 1.524;
   (b) 0.02116 Ω^{-1} cm^{-1}
3. (a) 0.5
   (b) 0.06897 Ω^{-1} cm^{-1}, 180.5 Ω

4. 625×10^3 Ω; 16×10^{-6} A
13. (a) 386.0; (b) 246.3
15. 0.495
16. (i) 1:1; (ii) 1:3

Chapter 34 (p. 378)
1. 17.9×10^{-6}
3. (a) 6.03×10^{-3}
   (b) 803×10^{-6}
   (c) 134×10^{-6}
   (d) 828×10^{-9}
6. 443×10^{-6} mol dm^{-3}
7. 95.5 per cent

11. (a) 0.0245; 0.0190;
    (b) 18×10^{-6}, 10.85×10^{-6}
12. -0.187°C
13. 1.342×10^{-3}
15. 2.65×10^{-5}
16. 1.79×10^{-5}
17. 0.036

Chapter 35 (p. 387)
2. 12×10^{-3} g dm^{-3}
3. 1.7×10^{-9} mol dm^{-3}

4. $K = 108x^5/214^5$
5. $\sqrt{x(x + y)}$
6. (a) 10^{-23} mol dm^{-3}, 10^{-21} mol dm^{-3};
   (b) 25×10^{-27}
7. (a) 3.6×10^{-9} mol dm^{-3}, 1.1×10^{-5} mol dm^{-3};
   (b) 1.3×10^{-5} mol dm^{-3}
8. 1.12×10^{-4} mol dm^{-3}; 2.80×10^{-6} mol dm^{-3}
10. (a) 1.332×10^{-8}; (b) 4.038×10^{-4}
11. 0.1586 mg dm^{-3}
12. 1.58×10^{-5} g dm^{-3}
13. 2.04; 2.3×10^{-3}

Chapter 36 (p. 404)
1. 20 per cent
2. (a) 10^{-4} mol dm^{-3}
   (b) 2.5×10^{-4} mol dm^{-3}
3. (a) 2.88; (b) 2; (c) 12; (d) 5.7
8. 0.0132; 2.878

9. 12; 3.162×10^{-6}
10. (a) 0.0365; (b) 3.42;
    (c) 2.29×10^{-3}
11. 2.9×10^{-3}; 2.54
16. 3.89; nil
17. 4.57

Chapter 38 (p. 422)
10. 8.37
11. 0.0075 per cent
12. 8.87

13. 0.73 g
14. 9.43
18. 25.8×10^{-6}

Chapter 39 (p. 432)
2. 1.56 volt
6. 1.1235 volt

7. (a) -0.378 volt; (b) 0.092 volt
12. 0.38 volt

Chapter 40 (p. 448)
1. $4xy/3$
4. 90 per cent
6. 0.383 volt
7. 121.8

8. (a) 50 cm^3; (b) 25 cm^3
9. (a) 14.36 A;
   (b) 14.21 A, 193 coulomb
11. 31.5 μm

Chapter 41 (p. 458)
3. 0.56
4. 0.917
5. Na$^+$, 0.38; Cl$^-$, 0.62

6. 0.527; 0.1517 g
8. 0.419

Chapter 44 (p. 488)
14. 18 g dm^{-3}
15. 3×10^{-10} mol dm^{-3}
16. 9.55×10^{-3} g

Chapter 45 (p. 505)
9. 22×10^{-6} cm^2; 2.5 nm; 3.7 nm

Chapter 46 (p. 520)
1. 6×10^8 m^2
2. 10^{18}; 0.5 mm of water; 4×10^{-6}°C
3. 15×10^{18}; 1.2×10^3 m^2
12. 718.6 N m^{-2}

Chapter 48 (p. 550)
   2. -136.9 kJ
   3. (a) 34.86 kJ; (b) -257.3 kJ; (c) -101 kJ
   4. $K = 2.5$; 0.29 atm
   5. 12.14 J K^{-1}
   7. 109 J K^{-1}
  13. -342.8 kJ
  14. -146.9 kJ

LOGARITHMS

	0	1	2	3	4	5	6	7	8	9	1	2	3	4	5	6	7	8	9
10	0000	0043	0086	0128	0170	0212	0253	0294	0334	0374	4	8	12	17	21	25	29	33	37
11	0414	0453	0492	0531	0569	0607	0645	0682	0719	0755	4	8	11	15	19	23	26	30	34
12	0792	0828	0864	0899	0934	0969	1004	1038	1072	1106	3	7	10	14	17	21	24	28	31
13	1139	1173	1206	1239	1271	1303	1335	1367	1399	1430	3	6	10	13	16	19	23	26	29
14	1461	1492	1523	1553	1584	1614	1644	1673	1703	1732	3	6	9	12	15	18	21	24	27
15	1761	1790	1818	1847	1875	1903	1931	1959	1987	2014	3	6	8	11	14	17	20	22	25
16	2041	2068	2095	2122	2148	2175	2201	2227	2253	2279	3	5	8	11	13	16	18	21	24
17	2304	2330	2355	2380	2405	2430	2455	2480	2504	2529	2	5	7	10	12	15	17	20	22
18	2553	2577	2601	2625	2648	2672	2695	2718	2742	2765	2	5	7	9	12	14	16	19	21
19	2788	2810	2833	2856	2878	2900	2923	2945	2967	2989	2	4	7	9	11	13	16	18	20
20	3010	3032	3054	3075	3096	3118	3139	3160	3181	3201	2	4	6	8	11	13	15	17	19
21	3222	3243	3263	3284	3304	3324	3345	3365	3385	3404	2	4	6	8	10	12	14	16	18
22	3424	3444	3464	3483	3502	3522	3541	3560	3579	3598	2	4	6	8	10	12	14	15	17
23	3617	3636	3655	3674	3692	3711	3729	3747	3766	3784	2	4	6	7	9	11	13	15	17
24	3802	3820	3838	3856	3874	3892	3909	3927	3945	3962	2	4	5	7	9	11	12	14	16
25	3979	3997	4014	4031	4048	4065	4082	4099	4116	4133	2	3	5	7	9	10	12	14	15
26	4150	4166	4183	4200	4216	4232	4249	4265	4281	4298	2	3	5	7	8	10	11	13	15
27	4314	4330	4346	4362	4378	4393	4409	4425	4440	4456	2	3	5	6	8	9	11	13	14
28	4472	4487	4502	4518	4533	4548	4564	4579	4594	4609	2	3	5	6	8	9	11	12	14
29	4624	4639	4654	4669	4683	4698	4713	4728	4742	4757	1	3	4	6	7	9	10	12	13
30	4771	4786	4800	4814	4829	4843	4857	4871	4886	4900	1	3	4	6	7	9	10	11	13
31	4914	4928	4942	4955	4969	4983	4997	5011	5024	5038	1	3	4	6	7	8	10	11	12
32	5051	5065	5079	5092	5105	5119	5132	5145	5159	5172	1	3	4	5	7	8	9	11	12
33	5185	5198	5211	5224	5237	5250	5263	5276	5289	5302	1	3	4	5	6	8	9	10	12
34	5315	5328	5340	5353	5366	5378	5391	5403	5416	5428	1	3	4	5	6	8	9	10	11
35	5441	5453	5465	5478	5490	5502	5514	5527	5539	5551	1	2	4	5	6	7	9	10	11
36	5563	5575	5587	5599	5611	5623	5635	5647	5658	5670	1	2	4	5	6	7	8	10	11
37	5682	5694	5705	5717	5729	5740	5752	5763	5775	5786	1	2	3	5	6	7	8	9	10
38	5798	5809	5821	5832	5843	5855	5866	5877	5888	5899	1	2	3	5	6	7	8	9	10
39	5911	5922	5933	5944	5955	5966	5977	5988	5999	6010	1	2	3	4	5	7	8	9	10
40	6021	6031	6042	6053	6064	6075	6085	6096	6107	6117	1	2	3	4	5	6	8	9	10
41	6128	6138	6149	6160	6170	6180	6191	6201	6212	6222	1	2	3	4	5	6	7	8	9
42	6232	6243	6253	6263	6274	6284	6294	6304	6314	6325	1	2	3	4	5	6	7	8	9
43	6335	6345	6355	6365	6375	6385	6395	6405	6415	6425	1	2	3	4	5	6	7	8	9
44	6435	6444	6454	6464	6474	6484	6493	6503	6513	6522	1	2	3	4	5	6	7	8	9
45	6532	6542	6551	6561	6571	6580	6590	6599	6609	6618	1	2	3	4	5	6	7	8	9
46	6628	6637	6646	6656	6665	6675	6684	6693	6702	6712	1	2	3	4	5	6	7	7	8
47	6721	6730	6739	6749	6758	6767	6776	6785	6794	6803	1	2	3	4	5	5	6	7	8
48	6812	6821	6830	6839	6848	6857	6866	6875	6884	6893	1	2	3	4	4	5	6	7	8
49	6902	6911	6920	6928	6937	6946	6955	6964	6972	6981	1	2	3	4	4	5	6	7	8
50	6990	6998	7007	7016	7024	7033	7042	7050	7059	7067	1	2	3	3	4	5	6	7	8
51	7076	7084	7093	7101	7110	7118	7126	7135	7143	7152	1	2	3	3	4	5	6	7	8
52	7160	7168	7177	7185	7193	7202	7210	7218	7226	7235	1	2	2	3	4	5	6	7	7
53	7243	7251	7259	7267	7275	7284	7292	7300	7308	7316	1	2	2	3	4	5	6	6	7
54	7324	7332	7340	7348	7356	7364	7372	7380	7388	7396	1	2	2	3	4	5	6	6	7

LOGARITHMS

	0	1	2	3	4	5	6	7	8	9	Differences								
											1	2	3	4	5	6	7	8	9
55	7404	7412	7419	7427	7435	7443	7451	7459	7466	7474	1	2	2	3	4	5	5	6	7
56	7482	7490	7497	7505	7513	7520	7528	7536	7543	7551	1	2	2	3	4	5	5	6	7
57	7559	7566	7574	7582	7589	7597	7604	7612	7619	7627	1	2	2	3	4	5	5	6	7
58	7634	7642	7649	7657	7664	7672	7679	7686	7694	7701	1	1	2	3	4	4	5	6	7
59	7709	7716	7723	7731	7738	7745	7752	7760	7767	7774	1	1	2	3	4	4	5	6	7
60	7782	7789	7796	7803	7810	7818	7825	7832	7839	7846	1	1	2	3	4	4	5	6	6
61	7853	7860	7868	7875	7882	7889	7896	7903	7910	7917	1	1	2	3	4	4	5	6	6
62	7924	7931	7938	7945	7952	7959	7966	7973	7980	7987	1	1	2	3	3	4	5	6	6
63	7993	8000	8007	8014	8021	8028	8035	8041	8048	8055	1	1	2	3	3	4	5	5	6
64	8062	8069	8075	8082	8089	8096	8102	8109	8116	8122	1	1	2	3	3	4	5	5	6
65	8129	8136	8142	8149	8156	8162	8169	8176	8182	8189	1	1	2	3	3	4	5	5	6
66	8195	8202	8209	8215	8222	8228	8235	8241	8248	8254	1	1	2	3	3	4	5	5	6
67	8261	8267	8274	8280	8287	8293	8299	8306	8312	8319	1	1	2	3	3	4	5	5	6
68	8325	8331	8338	8344	8351	8357	8363	8370	8376	8382	1	1	2	3	3	4	4	5	6
69	8388	8395	8401	8407	8414	8420	8426	8432	8439	8445	1	1	2	2	3	4	4	5	6
70	8451	8457	8463	8470	8476	8482	8488	8494	8500	8506	1	1	2	2	3	4	4	5	6
71	8513	8519	8525	8531	8537	8543	8549	8555	8561	8567	1	1	2	2	3	4	4	5	5
72	8573	8579	8585	8591	8597	8603	8609	8615	8621	8627	1	1	2	2	3	4	4	5	5
73	8633	8639	8645	8651	8657	8663	8669	8675	8681	8686	1	1	2	2	3	4	4	5	5
74	8692	8698	8704	8710	8716	8722	8727	8733	8739	8745	1	1	2	2	3	4	4	5	5
75	8751	8756	8762	8768	8774	8779	8785	8791	8797	8802	1	1	2	2	3	3	4	5	5
76	8808	8814	8820	8825	8831	8837	8842	8848	8854	8859	1	1	2	2	3	3	4	5	5
77	8865	8871	8876	8882	8887	8893	8899	8904	8910	8915	1	1	2	2	3	3	4	4	5
78	8921	8927	8932	8938	8943	8949	8954	8960	8965	8971	1	1	2	2	3	3	4	4	5
79	8976	8982	8987	8993	8998	9004	9009	9015	9020	9025	1	1	2	2	3	3	4	4	5
80	9031	9036	9042	9047	9053	9058	9063	9069	9074	9079	1	1	2	2	3	3	4	4	5
81	9085	9090	9096	9101	9106	9112	9117	9122	9128	9133	1	1	2	2	3	3	4	4	5
82	9138	9143	9149	9154	9159	9165	9170	9175	9180	9186	1	1	2	2	3	3	4	4	5
83	9191	9196	9201	9206	9212	9217	9222	9227	9232	9238	1	1	2	2	3	3	4	4	5
84	9243	9248	9253	9258	9263	9269	9274	9279	9284	9289	1	1	2	2	3	3	4	4	5
85	9294	9299	9304	9309	9315	9320	9325	9330	9335	9340	1	1	2	2	3	3	4	4	5
86	9345	9350	9355	9360	9365	9370	9375	9380	9385	9390	1	1	2	2	3	3	4	4	4
87	9395	9400	9405	9410	9415	9420	9425	9430	9435	9440	0	1	1	2	2	3	3	4	4
88	9445	9450	9455	9460	9465	9469	9474	9479	9484	9489	0	1	1	2	2	3	3	4	4
89	9494	9499	9504	9509	9513	9518	9523	9528	9533	9538	0	1	1	2	2	3	3	4	4
90	9542	9547	9552	9557	9562	9566	9571	9576	9581	9586	0	1	1	2	2	3	3	4	4
91	9590	9595	9600	9605	9609	9614	9619	9624	9628	9633	0	1	1	2	2	3	3	4	4
92	9638	9643	9647	9652	9657	9661	9666	9671	9675	9680	0	1	1	2	2	3	3	4	4
93	9685	9689	9694	9699	9703	9708	9713	9717	9722	9727	0	1	1	2	2	3	3	4	4
94	9731	9736	9741	9745	9750	9754	9759	9763	9768	9773	0	1	1	2	2	3	3	4	4
95	9777	9782	9786	9791	9795	9800	9805	9809	9814	9818	0	1	1	2	2	3	3	4	4
96	9823	9827	9832	9836	9841	9845	9850	9854	9859	9863	0	1	1	2	2	3	3	4	4
97	9868	9872	9877	9881	9886	9890	9894	9899	9903	9908	0	1	1	2	2	3	3	4	4
98	9912	9917	9921	9926	9930	9934	9939	9943	9948	9952	0	1	1	2	2	3	3	4	4
99	9956	9961	9965	9969	9974	9978	9983	9987	9991	9996	0	1	1	2	2	3	3	3	4

ANTI-LOGARITHMS

	0	1	2	3	4	5	6	7	8	9	Differences								
											1	2	3	4	5	6	7	8	9
·00	1000	1002	1005	1007	1009	1012	1014	1016	1019	1021	0	0	1	1	1	1	2	2	2
·01	1023	1026	1028	1030	1033	1035	1038	1040	1042	1045	0	0	1	1	1	1	2	2	2
·02	1047	1050	1052	1054	1057	1059	1062	1064	1067	1069	0	0	1	1	1	1	2	2	2
·03	1072	1074	1076	1079	1081	1084	1086	1089	1091	1094	0	0	1	1	1	1	2	2	2
·04	1096	1099	1102	1104	1107	1109	1112	1114	1117	1119	0	1	1	1	1	2	2	2	2
·05	1122	1125	1127	1130	1132	1135	1138	1140	1143	1146	0	1	1	1	1	2	2	2	2
·06	1148	1151	1153	1156	1159	1161	1164	1167	1169	1172	0	1	1	1	1	2	2	2	2
·07	1175	1178	1180	1183	1186	1189	1191	1194	1197	1199	0	1	1	1	1	2	2	2	2
·08	1202	1205	1208	1211	1213	1216	1219	1222	1225	1227	0	1	1	1	1	2	2	2	3
·09	1230	1233	1236	1239	1242	1245	1247	1250	1253	1256	0	1	1	1	1	2	2	2	3
·10	1259	1262	1265	1268	1271	1274	1276	1279	1282	1285	0	1	1	1	1	2	2	2	3
·11	1288	1291	1294	1297	1300	1303	1306	1309	1312	1315	0	1	1	1	2	2	2	2	3
·12	1318	1321	1324	1327	1330	1334	1337	1340	1343	1346	0	1	1	1	2	2	2	2	3
·13	1349	1352	1355	1358	1361	1365	1368	1371	1374	1377	0	1	1	1	2	2	2	3	3
·14	1380	1384	1387	1390	1393	1396	1400	1403	1406	1409	0	1	1	1	2	2	2	3	3
·15	1413	1416	1419	1422	1426	1429	1432	1435	1439	1442	0	1	1	1	2	2	2	3	3
·16	1445	1449	1452	1455	1459	1462	1466	1469	1472	1476	0	1	1	1	2	2	2	3	3
·17	1479	1483	1486	1489	1493	1496	1500	1503	1507	1510	0	1	1	1	2	2	2	3	3
·18	1514	1517	1521	1524	1528	1531	1535	1538	1542	1545	0	1	1	1	2	2	2	3	3
·19	1549	1552	1556	1560	1563	1567	1570	1574	1578	1581	0	1	1	1	2	2	3	3	3
·20	1585	1589	1592	1596	1600	1603	1607	1611	1614	1618	0	1	1	1	2	2	3	3	3
·21	1622	1626	1629	1633	1637	1641	1644	1648	1652	1656	0	1	1	2	2	2	3	3	3
·22	1660	1663	1667	1671	1675	1679	1683	1687	1690	1694	0	1	1	2	2	2	3	3	3
·23	1698	1702	1706	1710	1714	1718	1722	1726	1730	1734	0	1	1	2	2	2	3	3	4
·24	1738	1742	1746	1750	1754	1758	1762	1766	1770	1774	0	1	1	2	2	2	3	3	4
·25	1778	1782	1786	1791	1795	1799	1803	1807	1811	1816	0	1	1	2	2	2	3	3	4
·26	1820	1824	1828	1832	1837	1841	1845	1849	1854	1858	0	1	1	2	2	3	3	3	4
·27	1862	1866	1871	1875	1879	1884	1888	1892	1897	1901	0	1	1	2	2	3	3	3	4
·28	1905	1910	1914	1919	1923	1928	1932	1936	1941	1945	0	1	1	2	2	3	3	4	4
·29	1950	1954	1959	1963	1968	1972	1977	1982	1986	1991	0	1	1	2	2	3	3	4	4
·30	1995	2000	2004	2009	2014	2018	2023	2028	2032	2037	0	1	1	2	2	3	3	4	4
·31	2042	2046	2051	2056	2061	2065	2070	2075	2080	2084	0	1	1	2	2	3	3	4	4
·32	2089	2094	2099	2104	2109	2113	2118	2123	2128	2133	0	1	1	2	2	3	3	4	4
·33	2138	2143	2148	2153	2158	2163	2168	2173	2178	2183	0	1	1	2	2	3	3	4	4
·34	2188	2193	2198	2203	2208	2213	2218	2223	2228	2234	1	1	2	2	3	3	4	4	5
·35	2239	2244	2249	2254	2259	2265	2270	2275	2280	2286	1	1	2	2	3	3	4	4	5
·36	2291	2296	2301	2307	2312	2317	2323	2328	2333	2339	1	1	2	2	3	3	4	4	5
·37	2344	2350	2355	2360	2366	2371	2377	2382	2388	2393	1	1	2	2	3	3	4	4	5
·38	2399	2404	2410	2415	2421	2427	2432	2438	2443	2449	1	1	2	2	3	3	4	5	5
·39	2455	2460	2466	2472	2477	2483	2489	2495	2500	2506	1	1	2	2	3	3	4	5	5
·40	2512	2518	2523	2529	2535	2541	2547	2553	2559	2564	1	1	2	2	3	4	4	5	5
·41	2570	2576	2582	2588	2594	2600	2606	2612	2618	2624	1	1	2	2	3	4	4	5	5
·42	2630	2636	2642	2649	2655	2661	2667	2673	2679	2685	1	1	2	2	3	4	4	5	6
·43	2692	2698	2704	2710	2716	2723	2729	2735	2742	2748	1	1	2	3	3	4	4	5	6
·44	2754	2761	2767	2773	2780	2786	2793	2799	2805	2812	1	1	2	3	3	4	4	5	6
·45	2818	2825	2831	2838	2844	2851	2858	2864	2871	2877	1	1	2	3	3	4	5	5	6
·46	2884	2891	2897	2904	2911	2917	2924	2931	2938	2944	1	1	2	3	3	4	5	5	6
·47	2951	2958	2965	2972	2979	2985	2992	2999	3006	3013	1	1	2	3	3	4	5	5	6
·48	3020	3027	3034	3041	3048	3055	3062	3069	3076	3083	1	1	2	3	4	4	5	6	6
·49	3090	3097	3105	3112	3119	3126	3133	3141	3148	3155	1	1	2	3	4	4	5	6	6

ANTI-LOGARITHMS

	0	1	2	3	4	5	6	7	8	9	Differences 1	2	3	4	5	6	7	8	9
·50	3162	3170	3177	3184	3192	3199	3206	3214	3221	3228	1	1	2	3	4	4	5	6	7
·51	3236	3243	3251	3258	3266	3273	3281	3289	3296	3304	1	2	2	3	4	5	5	6	7
·52	3311	3319	3327	3334	3342	3350	3357	3365	3373	3381	1	2	2	3	4	5	5	6	7
·53	3388	3396	3404	3412	3420	3428	3436	3443	3451	3459	1	2	2	3	4	5	6	6	7
·54	3467	3475	3483	3491	3499	3508	3516	3524	3532	3540	1	2	2	3	4	5	6	6	7
·55	3548	3556	3565	3573	3581	3589	3597	3606	3614	3622	1	2	2	3	4	5	6	7	7
·56	3631	3639	3648	3656	3664	3673	3681	3690	3698	3707	1	2	3	3	4	5	6	7	8
·57	3715	3724	3733	3741	3750	3758	3767	3776	3784	3793	1	2	3	3	4	5	6	7	8
·58	3802	3811	3819	3828	3837	3846	3855	3864	3873	3882	1	2	3	4	4	5	6	7	8
·59	3890	3899	3908	3917	3926	3936	3945	3954	3963	3972	1	2	3	4	5	5	6	7	8
·60	3981	3990	3999	4009	4018	4027	4036	4046	4055	4064	1	2	3	4	5	6	6	7	8
·61	4074	4083	4093	4102	4111	4121	4130	4140	4150	4159	1	2	3	4	5	6	7	8	9
·62	4169	4178	4188	4198	4207	4217	4227	4236	4246	4256	1	2	3	4	5	6	7	8	9
·63	4266	4276	4285	4295	4305	4315	4325	4335	4345	4355	1	2	3	4	5	6	7	8	9
·64	4365	4375	4385	4395	4406	4416	4426	4436	4446	4457	1	2	3	4	5	6	7	8	9
·65	4467	4477	4487	4498	4508	4519	4529	4539	4550	4560	1	2	3	4	5	6	7	8	9
·66	4571	4581	4592	4603	4613	4624	4634	4645	4656	4667	1	2	3	4	5	6	7	9	10
·67	4677	4688	4699	4710	4721	4732	4742	4753	4764	4775	1	2	3	4	5	7	8	9	10
·68	4786	4797	4808	4819	4831	4842	4853	4864	4875	4887	1	2	3	4	6	7	8	9	10
·69	4898	4909	4920	4932	4943	4955	4966	4977	4989	5000	1	2	3	5	6	7	8	9	10
·70	5012	5023	5035	5047	5058	5070	5082	5093	5105	5117	1	2	4	5	6	7	8	9	11
·71	5129	5140	5152	5164	5176	5188	5200	5212	5224	5236	1	2	4	5	6	7	8	10	11
·72	5248	5260	5272	5284	5297	5309	5321	5333	5346	5358	1	2	4	5	6	7	9	10	11
·73	5370	5383	5395	5408	5420	5433	5445	5458	5470	5483	1	3	4	5	6	8	9	10	11
·74	5495	5508	5521	5534	5546	5559	5572	5585	5598	5610	1	3	4	5	6	8	9	10	12
·75	5623	5636	5649	5662	5675	5689	5702	5715	5728	5741	1	3	4	5	7	8	9	10	12
·76	5754	5768	5781	5794	5808	5821	5834	5848	5861	5875	1	3	4	5	7	8	9	11	12
·77	5888	5902	5916	5929	5943	5957	5970	5984	5998	6012	1	3	4	5	7	8	10	11	12
·78	6026	6039	6053	6067	6081	6095	6109	6124	6138	6152	1	3	4	6	7	8	10	11	13
·79	6166	6180	6194	6209	6223	6237	6252	6266	6281	6295	1	3	4	6	7	9	10	11	13
·80	6310	6324	6339	6353	6368	6383	6397	6412	6427	6442	1	3	4	6	7	9	10	12	13
·81	6457	6471	6486	6501	6516	6531	6546	6561	6577	6592	2	3	5	6	8	9	11	12	14
·82	6607	6622	6637	6653	6668	6683	6699	6714	6730	6745	2	3	5	6	8	9	11	12	14
·83	6761	6776	6792	6808	6823	6839	6855	6871	6887	6902	2	3	5	6	8	9	11	13	14
·84	6918	6934	6950	6966	6982	6998	7015	7031	7047	7063	2	3	5	6	8	10	11	13	15
·85	7079	7096	7112	7129	7145	7161	7178	7194	7211	7228	2	3	5	7	8	10	12	13	15
·86	7244	7261	7278	7295	7311	7328	7345	7362	7379	7396	2	3	5	7	8	10	12	13	15
·87	7413	7430	7447	7464	7482	7499	7516	7534	7551	7568	2	3	5	7	9	10	12	14	16
·88	7586	7603	7621	7638	7656	7674	7691	7709	7727	7745	2	4	5	7	9	11	12	14	16
·89	7762	7780	7798	7816	7834	7852	7870	7889	7907	7925	2	4	5	7	9	11	13	14	16
·90	7943	7962	7980	7998	8017	8035	8054	8072	8091	8110	2	4	6	7	9	11	13	15	17
·91	8128	8147	8166	8185	8204	8222	8241	8260	8279	8299	2	4	6	8	9	11	13	15	17
·92	8318	8337	8356	8375	8395	8414	8433	8453	8472	8492	2	4	6	8	10	12	14	15	17
·93	8511	8531	8551	8570	8590	8610	8630	8650	8670	8690	2	4	6	8	10	12	14	16	18
·94	8710	8730	8750	8770	8790	8810	8831	8851	8872	8892	2	4	6	8	10	12	14	16	18
·95	8913	8933	8954	8974	8995	9016	9036	9057	9078	9099	2	4	6	8	10	12	15	17	19
·96	9120	9141	9162	9183	9204	9226	9247	9268	9290	9311	2	4	6	8	11	13	15	17	19
·97	9333	9354	9376	9397	9419	9441	9462	9484	9506	9528	2	4	7	9	11	13	15	17	20
·98	9550	9572	9594	9616	9638	9661	9683	9705	9727	9750	2	4	7	9	11	13	16	18	20
·99	9772	9795	9817	9840	9863	9886	9908	9931	9954	9977	2	5	7	9	11	14	16	18	20

Relative atomic masses (1967)

Element			Symbol	Atomic number	Atomic weights	
					Acc.	*Approx.*
Actinium	.	.	Ac	89		227
Aluminium	.	.	Al	13	26.9815	27
Americium	.	.	Am	95		243
Antimony	.	.	Sb	51	121.75	121.5
Argon	.	.	Ar	18	39.948	40
Arsenic	.	.	As	33	74.9216	75
Astatine	.	.	At	85		210
Barium	.	.	Ba	56	137.34	137.5
Berkelium	.	.	Bk	97		249
Beryllium	.	.	Be	4	9.0122	9
Bismuth	.	.	Bi	83	208.980	209
Boron	.	.	B	5	10.811	11
Bromine	.	.	Br	35	79.904	80
Cadmium	.	.	Cd	48	112.40	112.5
Caesium	.	.	Cs	55	132.905	133
Calcium	.	.	Ca	20	40.08	40
Californium	.	.	Cf	98		249
Carbon	.	.	C	6	12.01115	12
Cerium	.	.	Ce	58	140.12	140
Chlorine	.	.	Cl	17	35.453	35.5
Chromium	.	.	Cr	24	51.996	52
Cobalt	.	.	Co	27	58.9332	59
Copper	.	.	Cu	29	63.54	63.5
Curium	.	.	Cm	96		247
Dysprosium	.	.	Dy	66	162.50	162.5
Einsteinium	.	.	Es	99		254
Erbium	.	.	Er	68	167.26	167
Europium	.	.	Eu	63	151.96	152
Fermium	.	.	Fm	100		254
Fluorine	.	.	F	9	18.9984	19
Francium	.	.	Fr	87		223
Gadolinium	.	.	Gd	64	157.25	157
Gallium	.	.	Ga	31	69.72	69.5
Germanium	.	.	Ge	32	72.59	72.5
Gold	.	.	Au	79	196.967	197
Hafnium	.	.	Hf	72	178.49	178.5
Helium	.	.	He	2	4.0026	4
Holmium	.	.	Ho	67	164.930	165
Hydrogen	.	.	H	1	1.00797	1
Indium	.	.	In	49	114.82	115
Iodine	.	.	I	53	126.9044	127
Iridium	.	.	Ir	77	192.2	192
Iron	.	.	Fe	26	55.847	56
Krypton	.	.	Kr	36	83.80	84
Lanthanum	.	.	La	57	138.91	139
Lawrencium	.	.	Lw	103		257
Lead	.	.	Pb	82	207.19	207
Lithium	.	.	Li	3	6.939	7
Lutetium	.	.	Lu	71	174.97	175
Magnesium	.	.	Mg	12	24.305	24.5
Manganese	.	.	Mn	25	54.9380	55
Mendelevium	.	.	Md	101		256

Relative atomic masses (1967)

Element	Symbol	Atomic number	Atomic weight	
			Acc.	Approx.
Mercury . .	Hg	80	200.59	200.5
Molybdenum .	Mo	42	95.94	96
Neodymium . .	Nd	60	144.24	144
Neon . . .	Ne	10	20.179	20
Neptunium . .	Np	93		237
Nickel . . .	Ni	28	58.71	58.5
Niobium . .	Nb	41	92.906	93
Nitrogen . .	N	7	14.0067	14
Nobelium . .	No	102		254
Osmium . .	Os	76	190.2	190
Oxygen . .	O	8	15.9994	16
Palladium . .	Pd	46	106.4	106
Phosphorus .	P	15	30.9738	31
Platinum . .	Pt	78	195.09	195
Plutonium . .	Pu	94		242
Polonium . .	Po	84		210
Potassium . .	K	19	39.102	39
Praseodymium .	Pr	59	140.907	141
Promethium .	Pm	61		147
Protactinium .	Pa	91		231
Radium . .	Ra	88		226
Radon . . .	Rn	86		222
Rhenium . .	Re	75	186.2	186
Rhodium . .	Rh	45	102.905	103
Rubidium . .	Rb	37	85.47	85.5
Ruthenium . .	Ru	44	101.07	101
Samarium . .	Sm	62	150.35	150.5
Scandium . .	Sc	21	44.956	45
Selenium . .	Se	34	78.96	79
Silicon . .	Si	14	28.086	28
Silver . .	Ag	47	107.868	108
Sodium . .	Na	11	22.9898	23
Strontium . .	Sr	38	87.62	87.5
Sulphur . .	S	16	32.064	32
Tantalum . .	Ta	73	180.948	181
Technetium .	Tc	43		99
Tellurium . .	Te	52	127.60	127.5
Terbium . .	Tb	65	158.924	159
Thallium . .	Tl	81	204.37	204.5
Thorium . .	Th	90	232.038	232
Thulium . .	Tm	69	168.934	169
Tin . .	Sn	50	118.69	118.5
Titanium . .	Ti	22	47.90	48
Tungsten . .	W	74	183.85	184
Uranium . .	U	92	238.03	238
Vanadium . .	V	23	50.942	51
Xenon . . .	Xe	54	131.30	131.5
Ytterbium . .	Yb	70	173.04	173
Yttrium . .	Y	39	88.905	89
Zinc . .	Zn	30	65.37	65.5
Zirconium . .	Zr	40	91.22	91

Index

$$CH_3-C\overset{=O}{-}OCH_3 + H_2O \rightleftharpoons CH_3C-OH + CH_3OH$$

$$a \qquad b$$

$$x \qquad b-(a-x)$$

$$(a-x)$$
$$(a-x)$$
$$(a-x)$$